S0-AZX-077

THE HANGING TREE

THE HANGING TREE

Execution and the English People

1770–1868

V. A. C. GATRELL

OXFORD UNIVERSITY PRESS

1994

Oxford University Press, Walton Street, Oxford OX2 6DP

Oxford New York
Athens Auckland Bangkok Bombay
Calcutta Cape Town Dar es Salaam Delhi
Florence Hong Kong Istanbul Karachi
Kuala Lumpur Madras Madrid Melbourne
Mexico City Nairobi Paris Singapore
Taipei Tokyo Toronto

and associated companies in
Berlin Ibadan

Oxford is a trade mark of Oxford University Press

Published in the United States
by Oxford University Press, New York

British Library Cataloguing in Publication Data
Data available

Library of Congress Cataloging in Publication Data
Gatrell, V. A. C., 1941–
The hanging tree... execution and the English people 1770–1868 /
V.A.C. Gatrell.Data available
p. cm.
Includes index.
1. Capital punishment—Great Britain—History.
2. Executions and executioners—Great Britain—History.
3. Hanging—Great Britain—History.
HV8699.G8G38 1994 364.6'6'0941—dc20 94–4108
ISBN 0–19–820413–2

1 3 5 7 9 10 8 6 4 2

Set by Hope Services (Abingdon) Ltd.
Printed in Great Britain
on acid-free paper by
the Bath Press, Avon

PREFACE

It may help to explain some of this book's peculiarities if I say at once that it grew out of a chance discovery. One day I was working in the Public Record Office on something quite different when tedium and a passing curiosity moved me to order from the Home Office catalogues a volume of judges' reports, then unknown to me. I chose the year 1829 at random. The volume fell open at a thick dossier of petitions and papers which I was soon reading with the avidity of one who knew that he really ought to be reading something much less interesting. It was about a rape case set in the iron-making village of Coalbrookdale in Shropshire. A poor man called John Noden had for several years been courting a young woman, Elizabeth Cureton. One June evening he knocked at her cottage window after her parents had gone to bed. She let him in, and soon the couple were having sex on the parlour floor. A few days later she claimed that he had raped her. She prosecuted him at the Shrewsbury assizes, and he was tried, convicted, and sentenced to hang. The papers I read were about the Coalbrookdale community's efforts to persuade the judge, the home secretary, and George IV to give Noden a royal pardon.

The dossier's vividness and complexity surprised me. It contained the judge's transcript of the trial; two huge parchment petitions signed by hundreds of local ironmasters, professionals, and tradesmen; affidavits from neighbours and ex-lovers telling detailed stories about Elizabeth Cureton's earlier sexual adventures; the local doctor's opinion that this was no rape but only a 'sweetheart matter' gone a little wrong; and the correspondence between the judge and the home secretary, Robert Peel, who would decide whether to let Noden die. The whole amounted to some 12,000 words. I found this Hardyesque tragedy poignant and entrancing. Unexpectedly, the normally opaque sexual life of a long-ago community was opened up to me. Also opened was the way the community related to a penal code which hanged people for rape and many offences beside, and the casual way great men made life-and-death decisions. I photostated the documents, visited Coalbrookdale (the locations are astonishingly intact), and, for my own pleasure only, wrote up the microhistory which became Part V of this book.

Then one thing led to another, as usually happens. Reading more deeply into the petition archive, I discovered a mountain of appeals for mercy, untouched since they had been bound in volumes or tied in pink-ribboned bundles nearly two centuries ago. Unknown to historians, there were a couple of thousand for each year of the 1820s alone. Most were submitted by or on behalf of humble people who faced death or transportation for transgressions large or small, real, imagined, or unsatisfactorily proved. The lucky ones got the support of middle-class or gentry patrons: then dossiers were often thick, stories detailed, and opinions about unjust trial or excessive punishment confidently advanced. Even the weakest appeals—scratchily penned, thinly supported, signed by crosses—quivered with emotion. Humble people and their supporters pleaded for the fair hearing denied to them in court, for life itself, or for reprieve from exile across the world. Beneath the formal deference there were fear and desperation, sometimes indignation and disbelief, and occasional anger too. Thus I was propelled into the felt lives of men and women entrammelled by harsh law who had no other historical voices than these. It was a distorted view I gained: I was reading about the law's failures rather than about the many cases satisfactorily dealt with. But the tone of these appeals matched my own mounting incredulity at the fates which befell these obscure people and at the slap-happy justice which sent them to scaffolds or exile for puny as well as casually tried crimes.

At first my reading aimed only to set Cureton's and Noden's story in context, but slowly the context became my main subject and their story just one episode in it. As other stories pressed upon me, it became clear that for many thousands of those past people the law was neither an ideal nor something to which consent was spontaneously given, but an arena of unequal negotiation and despair; moreover, it was knitted into their lives and could shape destinies irretrievably. This realization took me a long way from the detached kinds of criminal justice history which historians have recently debated. It left me with a diminished interest in how far *ancien régime* criminal law expressed class interests, for instance, or contrarily in the tacit exculpations of the law which more conservative historians have advanced. It seemed obvious from my low angle that rich and strong people used the law to rule weaker and poorer people, even if criminal law had other faces and could do nicer things than that. What counted was that these stories fired me with a taste for the lived experiences of past people which criminal justice historians have meanly served. It was an emotional and imaginative engagement with their plights that pushed me forwards, and at last it took me into a history of emotion itself.

The machine which filled my frame as research widened felt dangerous as I drew near it. It was the public scaffold which so many stories tried to evade but upon which so many ended. Few historians have looked closely at this object, and I soon understood why. People did not die on it neatly. Watched by thousands, they urinated, defecated, screamed, kicked, fainted, and choked as they died. I was nauseated to realize this obscenity, and wondered if it should be exposed. Just as polite people learnt not to look in those days, so we shrink away now. Yet I realized that my squeamishness could not be indulged. Not only was what happened upon the scaffold deeply implanted in the contemporary imagination; my disgust also expressed a taboo and feelings which had been organized within the periods and around the very subject I was studying. These feelings could hardly be evaded. They are part of the history itself.

We now know a good deal about the capital code up to the demise of most of its penalties in the 1830s, and then up to the abolition of public executions in 1868. Radzinowicz's wonderful survey established the main chronologies long ago, even if he was too optimistic about 'progress' for modern tastes. Since then, as my footnotes will show, the research of Beattie, Hay, King, Langbein, McGowen, and others has immensely deepened our knowledge of how the law was administered and justified, whom it targeted, and how it related to interests of class. We also know a good deal about what critics *thought* about the capital code in its heyday—or at least what a few vocal critics like Eden and Bentham thought about it, as they recommended that punishment be made more certain: historians of ideas have explored this territory closely. None the less there remain strange gaps in our understanding. What did humble, middling, and middle kinds of people think about hanging? Loose assumptions about 'public opinion' and popular consent in the law often take the place of hard research. And what they *felt* about executions is hardly ever attended to, even though hanging was designed to deliver the most devastating emotional effects upon them.

Parts I and II of the book explore these latter questions by tackling the bottom of the social hierarchy first. They show that the execution crowd received the law's messages far from passively. There was much the crowd consented in (that murderers should die, for example). But people's cries and movements at the scaffold's foot, their feelings about death, the ballads they listened to, and the images they consumed, spoke for something more complex than assent, and often for its opposite, as polite commentators knew. As the last chapter suggests, if there was any single reason why executions were hidden behind prison walls from 1868 onwards it was because the crowd's sardonic commentaries could no longer be borne. Too often that despised crowd denounced justice as murderous in itself.

Complex too were the defences polite people erected against experiencing the scaffold: this is the subject of Part III. The deepening value attached to 'sympathy' in eighteenth-century culture generated tensions which had to be dealt with if consciences were to live easily with hanging. One way of coping was to turn away and not to know—attitudes very different from the ratiocinated and forward-looking kind to which penal reformers appealed. Most middle-class diaries, letters, and newspapers reveal an extraordinary detachment about the spectacle, or else they reveal defences, denials, and rationalizations which spoke for anxiety at the least. Their understanding of the world, or at least of this cruel part of it, was filtered through fantasies, projections, and interests now forgotten. The same went for the higher world addressed in Part VI. In their death-dealing tasks, judges, home secretaries, and the king and his council had a hierarchical order to defend, and they were remote from those they disciplined. But how remote they also seemed in their feelings, sending people to the noose without anxiety, it appeared. Approaching them from below as I did, I found it difficult to like these aloof cold men.

Yet Part IV shows that a few middle kinds of people (their numbers growing by the 1820s, perhaps) were reacting to the scaffold as if themselves experiencing the terror and pain of those killed upon it. They had material interests in making law-enforcement more efficient and in asserting their moral standing and civility. And their recurrent appeal to 'humanity' concealed many agendas, not all benign. But some rationalists, evangelicals, radicals, and mercy petitioners had apparently internalized the prescriptions of novels and advice books over generations past. As 'humanity' became their watchword, the old order began to yield. Yet their triumph should not be overstated. Two cultures opposed each other in these decades: that of the gallows and the transportation ships, and that of the 'humanity-mongers', as the contemptuous term went. There is no doubt which predominated in the contest. The fact that 'humane' opinion was muted and cautious before the 1830s, and that the most audible disgust at execution was heard when the bloody capital code was on the brink of repeal for other (structural) reasons, suggests that the code did not collapse because of a revolution in sensibilities. It suggests that on this subject there was a revolution in sensibilities because the code was collapsing already. In this, as in other historic shifts in collective sensibilities, culturally accredited feelings acquired a good part of their purchase by accommodating justificatory elements retrospectively.

The book ends with a postscript on the abolition of public executions in 1868, upon which moment many themes converged. But the sixty or so years before 1830 are the book's key period, for this was the era of the bloody code's fullest flowering. (It has been a shock to realize that more English people hanged then than had hanged since the seventeenth century.) I take the English case, and apologize for excluding Scotland and Ireland. But much basic research remains to be done on those countries' legal and criminal histories; luckily, Scotland had few hangings anyway. Moreover, what the English case illuminates broadly applies to most other countries which similarly reconstructed their attitudes to others' sufferings a couple of centuries ago. Our subject-matter can hardly be dissociated from the processes through which a brand of civility has been fragilely attained in the West.

The book is not only a history of emotions. It is also (in a measured sense I hope) an emotional book. Or, better, it is a sensational book—in the older, literal sense of that word. It can also only be a book to argue with, since in a history of mentalities one achieves no certainty and runs many risks. Sometimes I release my own incredulity and anger into the writing as I contemplate what the law's agents did to people. Some readers will not share those feelings, for they will know that hangings were not a wanton device of state power: they made sense to many in an age when the feared alternative was the Frenchified and potentially despotic surveillance of policemen. In any case, now as then, many people as a matter of principle care more about the pain suffered by crime's victims than about the pain inflicted on criminals. In my quest for the sources of the sympathetic repertoires which operated contrarily I might not have sufficiently acknowledged the ethical and emotional foundations of that position.

Risks are compounded by the fact that my writing refracts emotions and methods of articulating emotions inherited from the subject and period under study. At times I resort to 'humanitarian narratives' as needfully as those past people did and for the same reasons: to convey the pathos of the subject, or to do justice to those not given it at the time. This is not the way of working conventionally recommended in the academy, and many readers might prefer a more concerted striving for that 'objectivity' upon which historians' reputations are most safely erected. My only defence for working otherwise is the simplest one: had I neutralized my own feelings the book would not have been written. As things are, it has sometimes seemed that I have entered a maze of mirrors in which past feelings reflect into the present and present feelings reflect back into the past; and within this maze I have re-experienced that past as I have not experienced a historical subject before.

For encouragement and advice, I owe many thanks to colleagues, friends, and family. Ulinka Rublack drew my attention to several texts and episodes unknown to me; her research on analogous topics gave her rare insight into the book's aims and purposes, and I clarified a good deal in discussion with her. Boyd Hilton was more helpful than he knew when he offered a generous and thorough critique of early drafts. He made me think more clearly about Peel and evangelicalism than I was inclined to at first, although needless to say what results is no fault of his, any more than it is of others who have helped me. Adam Fox, Nicky Blandford, and Alison Winter gave early drafts close and supportive readings; Steven Sedley kindly lent me his unpublished study of ballads; and Vyvyen Brendon, Tia Cockrell, Eva-Maria Lassen, Naomi Tadmor, and Deborah Thom also cheered me on. Tony Morris has been the most encouraging of editors. And to Pam, Alexander, and Anna I give special thanks for their uncommon tolerance and support while my mind has been on strange and distant things.

V. A. C. G.

Cambridge
1993

CONTENTS

List of Illustrations xvi
Abbreviations xix

Introduction: Taking Broad Views 1
1. Suffering others 1
2. Measuring the subject 6
3. Material bases 11
4. Pressures on structures 18
5. Chronologies of change 21

PART I: THE SCAFFOLD AND THE CROWD

1. Hanging People 29
   1. Dying bravely 29
   2. The last days of Joseph Harwood 40
   3. 'Finish me tidily' 45

2. Death and the Scaffold Crowd 56
   1. 'One continued fair' 56
   2. Ghoulishness? 67
   3. Fearing death 74
   4. Scaffold rituals 80

3. Carnival or Consent? 90
   1. The question of carnival 90
   2. The question of consent 99

PART II: THE PLEBEIAN TEXTS

4. Scaffold Culture and Flash Ballads 109
   1. Making a day of it 109
   2. Collective memory 111
   3. The world of Mr Punch 119
   4. Flash ballads and Francis Place 123
   5. Promiscuous audiences 127

6. The 'criminal' and bawdy ballads 133
7. Tyburn ballads and the game death 138
8. Aftermath 144
Appendix: Francis Place's 'Songs within memory': A selection 149

5. Broadsides and the Gallows Emblem 156
1. The rise of the execution sheet 156
2. Messages 161
3. Purchasers 168
4. The gallows emblem 175

6. The Prerogative of Mercy and the Practices of Deference 197
1. The home secretary's other papers 197
2. The prerogative of mercy 200
3. Deference? 208

PART III: THE LIMITS OF SENSIBILITY

7. Sympathy and the Polite Classes: Arguments and Interpretations 225
1. An argument 225
2. Sympathy and sensibility 228
3. Impeded narratives and resistant mentalities 236

8. Watching from Curiosity 242
1. The unembarrassed witness 242
2. A sanctioned curiosity 250

9. Anxiety and Defence 259
1. Decorum and repression 259
2. Squeamishness and denial 266
3. Darwin's dream and gallows wit 272

10. Executing 'Social Others' 280
1. Social distance, hanging, and 'normality' 280
2. Boswell and the limits of sympathy 284
3. The death of Dr Dodd, 1777 292
4. Thackeray: The modern man of feeling? 294

11. Executing Traitors 298
1. Cato Street monsters 298
2. Punishing treason 315

PART IV: PUBLIC OPINION

12. Opinion and Emotion 325
1. Inventing public opinion 325
2. Acting from feelings 331
3. Chivalry and the female victim 334

13. Wronged Women: The Stories of Sarah Lloyd and Eliza Fenning 339
1. Introduction 339
2. Capel Lofft and the execution of Sarah Lloyd, 1800 340
3. The beatification of Eliza Fenning, 1815 353
4. Residues and conclusions 367

14. Piety and Benevolence 371
1. Belief, pain, and execution 371
2. Washing felons in the blood of the Lamb 376
3. Ordinaries of Newgate 382
4. Piety and sensibility 389

15. Fabricating Opinion 396
1. Diffusing knowledge on the punishment of death 396
2. Propaganda and reality 403
3. Quakers, forgers, and others 408

16. Appealing for Justice 417
1. Appeals and opinion 417
2. Indignation 423
3. Mercy, justice, and the lawyers 429
4. Attorneys ascendant 433
5. Influences and effects? 439

PART V: MERCY, JUSTICE, AND COMMUNITY

17. The Rape of Elizabeth Cureton: A Microhistory 447
1. Microhistories 447

2. The crime 451
3. The community 454
4. Men and women 459
5. Noden's trial 466
6. The law on rape 470
7. The first appeal 480
8. The second appeal 484
9. Outcomes 491

PART VI: THE OLD ORDER RESISTS

18. Furred Homicides, Sable Bigots: The Judges 497
1. Hanging kinds of men 497
2. Mediocrity 502
3. Bad tempers 506

19. Qualities of Justice 515
1. Rough justice? 515
2. Containing illegitimate influence 523
3. Justice in the courtroom 529

20. The King in his Council 543
1. Mercy in London 543
2. Mercy and the king 550
3. The king and Mr Peel 554

21. Mercy and Mr Peel 566
1. The great reformer 566
2. Peel and crime 572
3. An ethical pessimism 576

EPILOGUE

1868: Ending the Spectacle 589
1. A civilizing moment 589
2. Hanging for murder 591
3. Squeamishness and released sensation 595
4. Scapegoating the crowd 601

Appendix 1: The Petition Archive 613
Appendix 2: Execution and Mercy Statistics 616
Index of Persons 621
General Index 627

LIST OF ILLUSTRATIONS

1. *The trial and execution of John Any Bell aged 14*, 1831 — 4
   Broadside, St Bride Printing Library
2. *The Newgate drop in action*, 1809 — 30
   A. Knapp and W. Baldwin, *New Newgate calendar*, 1810
3. *Horsemonger Lane prison, Surrey*, 1809 — 31
   A. Knapp and W. Baldwin, *New Newgate calendar*, 1810
4. *Anne Hurle for forgery and M. Spalding on their way to execution at Newgate*, 1804 — 49
   W. Jackson, *New and complete Newgate calendar*, 1818
5. *The new drop at Newgate* — 53
   Gentleman's Magazine, 1783
6. Robert and George Cruikshank, *In the press yard at Newgate* — 66
   Pierce Egan, *Life in London: or, the day and night scenes of Jerry Hawthorn Esq and his elegant friend, Corinthian Tom, accompanied by Bob Logic and Oxonian in their rambles and sprees through the metropolis*, 1821
7. *Daniel Good's barbarous murder of Jane Jones*, 1842 — 71
   Broadside, British Library, 1881. d. 8
8. *Greenacre's murder and mutilation of Mrs Hannah Brown*, 1837 — 72
   Broadside, British Library, 1881. d. 8
9. *The hanged man's touch* — 82
   A. Knapp and W. Baldwin, *New Newgate calendar*, 1828 edn.
10. *View of the body of John Williams, the supposed Ratcliffe Highway murderer* — 85
    W. Fairburn, *Account of the life, death and interment of John Williams*, 1811
11. *A mother presenting* The malefactor's register *to her son and tenderly intreating him to regard the instructions therein recorded* — 115
    The malefactor's register; or, The Newgate and Tyburn calendar, 1779
12. Joseph Meadows, *Jack Ketch the hangman* — 116
    C. Whitehead, *Autobiography of a notorious legal functionary*, 1836
13. *The death mask of François Courvoisier*, 1840 — 118
    Madame Tussaud's archive

14. George Cruikshank, *Punch outwits Jack Ketch* 122
Cruikshank and Collier, *The tragical comedy or comical tragedy of Punch and Judy*, 1828

15. *The execution of Wild Robert: being a warning to all parents*, 1797 162
Broadside, British Library, 1881. d. 8

16. Thomas Rowlandson, *Last dying speech and confession*, 1799 172
British Museum

17. *Trial, conviction and execution of Samuel Fallows, on 14th April 1823 at Chester* 177

18. Thomas Rowlandson, *A gibbet* (1790s?) 179
Yale Center for British Art, Paul Mellon Collection

19. Thomas Rowlandson, *Crowd by a gibbet* (1790s?) 180
Yale Center for British Art, Paul Mellon Collection

20. Thomas Rowlandson, *Mary Evans hung at York Augst. 10 1799 for poisoning her husband* 181
York City Art Gallery

21. Théodore Géricault, *A public hanging*, 1820 182
Musée des Beaux Arts, Rouen

22. Detail from William Hogarth, *The idle 'prentice* (1747) 184

23. Detail from William Hogarth, *The stages of cruelty* (1751) 185

24. Thomas Bewick, *Gallows vignettes* 186
British birds, 1797–1804, and *The fables of Aesop*, 1818

25. George Cruikshank, *The bank restriction note*, 1819 188
British Museum

26. William Hogarth, *The idle 'prentice executed at Tyburn* (1747) 190

27. *The trial and execution of Martin Clinch and Samuel Mackley for the wicked murder of Mr Fryer, in Islington Fields* 191
Hindley

28. *The sorrowful and weeping lamentation of four lovely orphans left by J. Newton, a wealthy farmer, who was executed at Shrewsbury, on Monday last, for a cruel murder, on the body of his innocent wife and infant child*, 1823 192
Broadside, St Bride Printing Library

29. *A scene in the Old Bailey, immediately before the execution* [of Greenacre] 193
Weekly Chronicle, 7 May 1837

30. William Heath, *Merry England*, 1831 194
British Museum

31. Thomas Rowlandson, *The dissection*, *c.*1775–82 265
Henry E. Huntington Library and Art Gallery

32. Thomas Bewick, *Vignettes* 284
British birds, 1797–1804, and *The fables of Aesop*, 1818

33. *The execution of the Cato Street conspirators, 1 May 1820* 301
G. T. Wilkinson, *An authentic history of the Cato-Street conspiracy*, 1820

34. *Awful execution of the conspirators* 303
The trials of Arthur Thistlewood . . . for high treason, 1820

35. *The Cato Street conspirators' last autographs*, 1820 310–11
Liverpool papers, BL Add. MS 38284

36. *Sarah Lloyd of Bury St Edmunds*, 1800 351
Suffolk Record Office

37. Robert Cruikshank, *Eliza Fenning*, 1815 357
F. W. Hackwood, *William Hone: His life and times*, 1912

38. Joseph Meadows, *Vignette* 500
C. Whitehead, *Autobiography of a notorious legal functionary*, 1836

39. George Cruickshank, *The clerical magistrate* 513
W. Hone, *The political house that Jack built*, 1819

40. *Execution of Henry John Naylor, George Adams, Edward Ward, George Anson, William Bartholomew, John Close, Edward Desmond, and John Davis*, 1822 561

41. John Leech, *The great moral lesson at Horsemonger Lane gaol, November 13* 607
Punch, 1849

Plates are reproduced by permission of: the British Library (2, 3, 4, 7, 8, 11, 15, 29, 35); the British Museum (16, 25, 30); the Syndics of Cambridge University Library (5, 6, 10, 12, 14, 24, 27, 32, 33, 34, 37, 38, 39, 40); the Henry E. Huntington Library and Art Gallery (31); Madame Tussaud's Archive (13); Musée des Beaux Arts, Rouen (21); St Bride Printing Library (1, 28); the Suffolk Record Office (36); the Yale Center for British Art, Paul Mellon Collection (18, 19); the York City Art Gallery (20).

ABBREVIATIONS

Ann. Reg.	*Annual Register*
Beattie	J. Beattie, *Crime and the courts in England, 1660–1800* (Cambridge, 1986)
Borrow	George Borrow, *Celebrated trials and remarkable cases of criminal jurisprudence from the earliest records to the year 1825* (6 vols., 1825)
Cooper	D. D. Cooper, *The lesson of the scaffold: The public execution controversy in Victorian England* (1974)
Griffiths	A. Griffiths, *The chronicles of Newgate* (1883)
Hindley	C. Hindley, *Curiosities of street literature* (1871; facsimile edn., 2 vols., 1966)
HO	Home Office papers (PRO)
Knapp and Baldwin	A. Knapp and W. Baldwin, *The new Newgate calendar* (5 vols., 1809–19)
Mayhew	H. Mayhew, *London labour and the London poor* (4 vols., 1851–62)
OBSP	*Old Bailey Sessions Papers*
PC	Privy Council papers (PRO)
PD1, 2, or 3	Parliamentary Debates (with series)
PH	*Parliamentary History of England*
PP	Parliamentary Papers
PRO	Public Record Office, London
Radzinowicz	L. Radzinowicz, *History of English criminal law and its administration from 1750* (5 vols., 1948–86)
Rayner and Crook	J. L. Rayner and G. T. Crook (eds.), *The complete Newgate Calendar* (5 vols., 1926)
RC	Royal Commission
RC 1866	*Royal Commission on capital punishment*, PP 1866, xxi
RO	Record Office
Romilly	S. Romilly, *Memoirs of the life of Sir Samuel Romilly* (3 vols, 1840)
SC	Select Committee
SC 1819	*Select Committee on the criminal law as relates to capital punishment*, PP 1819, viii
SC 1856	*House of Lords Select Committee to look into the present mode of carrying into effect capital punishments*, PP 1856, vii
SP	State Papers (PRO)
Walpole	*The letters of Horace Walpole, fourth earl of Orford* (ed. P. Cunningham, 9 vols., 1906)

References to documents in HO, PC, and SP are explained in Appendix 1.
Books are published in London except where otherwise indicated.

What a lamentable case it is to see so many Christian men and women strangled on that cursed tree of the gallows?—insomuch, as if in a large field a man might see together all the Christians that, but in one year, throughout England, come to that untimely and ignominious death,—if there were any spark of grace, or charity in him, it would make his heart to bleed for pity and compassion.

Sir Edward Coke, *Institutes* (1641)

Sae rantingly, sae wantonly,
Sae dauntingly gaed he;
He play'd a spring, and danc'd it round,
Below the gallows-tree.

Robert Burns, 'Macpherson's Farewell' (1788)

INTRODUCTION

TAKING BROAD VIEWS

1. *Suffering others*

HOW NOWADAYS MIGHT WE MODERN AND CONSCIENTIOUSLY sympathetic kinds of people (the kinds presumably who read books like this) deal with a small historical tableau like the following? More to the point, how did people deal with it in 1831? At Rochester assizes in August that year an illiterate and pauperized country lad curiously called John Amy Bird Bell was sentenced to death for murder. Early on the Monday following he was taken out of Maidstone prison to a scaffold. Four or five thousand people assembled to watch. The hangman put a noose around his neck, he was prayed at, a platform was pulled away, and he dropped and choked to death. Afterwards his body was dissected by surgeons. It was all common enough for the times, except that John Bell was aged 14.

Two very different narratives attach to the lad's killing. Each seems to see the world and its pains quite differently. They come from different cultures, one might say. One speaks to us closely, the other not. The first gives the account we might prefer to hear, if we do not squeamishly turn away. Its directness hurts us; yet we accept the hurt, perhaps to redeem the ugliness of the killing it describes. The hurt is compounded by shock as we realize that it is the hangman himself who here supposedly writes so kindly, so gently:

> Last year I was called out of town, to hang a little boy for killing with malice aforethought. [He] was the youngest fellow-creature I ever handled in the way of our business; and a beautiful child he was too, as you may have seen by the papers, with a straight nose, large blue eyes and golden hair. . . . [The crowd] saw the stripling lifted fainting on to the gallows, his smooth cheeks of the colour of wood-ashes, his limbs trembling, and his bosom heaving sigh after

> sigh, as if body and soul were parting without my help. It was not a downright murder; for there was scarce any life to take out of him. When I began to pull the cap . . . over his baby face, he pressed his small hands together (his arms, you know, were corded fast to his body) and gave me a beseeching look; just as a calf will lick the butcher's hand. But cattle do not speak: this creature muttered,—'Pray, sir, don't hurt me.' 'My dear,' answered I, 'you should have spoken to my master: I am only the journeyman and must do as I'm bid.' This made him cry, which seemed a relief to him; and I do think I should have cried myself, if I had not heard shouts from the crowd: poor lamb! shame! murder! Quick, said the sheriff; ready, said I; the reverend chaplain gave me the wink: the drop fell: one kick, and he swayed to and fro, dead as the feelings of an English judge.

Alas this was written by no hangman. It was written by Edward Gibbon Wakefield in 1832, in one of his scathing attacks on the capital laws.[1] But our relief lasts. There is controlled irony in this writing, empathy with a boy's terror, and an exploding contempt for the justice that did this thing. We are glad someone protested.

We notice, however, that the horror is accessible to us because Wakefield tells a story, a story, moreover, of a special kind. Its intensely visualized particularity moves the imagination and so taps our pity. We do not yet know how true the story is, but it puts us in touch with the boy's terror; we absorb it within ourselves.

The effect achieved here was rather new. It was not that earlier generations were incapable of pity. Sir Edward Coke is cited at the beginning of this book lest this be lost sight of in the close chronological focus of the work. Coke's was as visionary and compassionate a passage as anything the language of sensibility achieved, and it was written in 1641. Early modern accounts of great men's executions, in state trials for example, similarly attended intimately to their demeanours on scaffolds, the pathos sometimes registered unbearably. Yet there were differences. Coke and his contemporaries would neither have contemplated the biographies of the humblest who hanged nor have thought worse of the law for their hanging. The order of the world depended on these slaughters, and human law must replicate divine will. Wakefield two centuries later was freed from that sacred master plot. Additionally he knew that a vividly visualized narrative engagement, even or especially with an obscure boy's killing, would intensify and communicate emotion, giving it a status it would otherwise lack. He inherited the device from those many empathetic narratives through which eighteenth-

[1] E. G. Wakefield, *The hangman and the judge, or A letter from Jack Ketch to Mr Justice Alderson* (1833), 5–6.

century writers had taken their readers into the closest emotional intimacy with their subjects and which helped establish a modern narrative form.[2] We have also inherited this means of touching the emotions. We are familiar with the projections it mediates and know how to respond to them.

The second narrative invites us into a quite different relationship to the execution. This is no fiction. It is *The Times*'s cool report on John Bell's crime, trial, and hanging.[3] Feelings play no part in this account. Rather it assumes that we know what to *think* about the event, and it stresses the hanging's legitimacy. Withholding affective commentary, it lets us know that justice is being done, the roughest sort perhaps, but justice still, of an age-old kind.

Young John Bell has lived with his father and younger brother in a Rochester poorhouse. They are ignorant, poor, and profligate. One day in the woods, while the younger brother stands watch, John waylays and kills another boy with a knife and robs the corpse of nine shillings, paying his brother 1*s.* 6*d.* for his help. The lads hide the body in undergrowth where weeks later it is found in a state of decay. The knife is found nearby and traced to the Bells. The victim's grave is opened and parish officials force the younger brother (aged 11) to jump into it to search the muddy tangle of flesh and cloth for the missing money. The money is not found; the brothers confess to the crime.

In court John Bell looks like a bumpkin. He is contrasted pointedly with the ladies who attend his trial, 'all well dressed, and some fashionably'. For eight hours he stands in the dock stolidly, as if numbed. Not once does he sit or take refreshment. He drops 'a solitary tear' when the judge announces that his body will be dissected, but he is otherwise impassive. *The Times*'s report makes only one reference to his youth: the jury recommends mercy because Bell is so young, and also because of 'the unnatural manner in which he had been brought up'. Gaselee the judge overrides the recommendation: he had never before tried so atrocious a crime, he declares, and reads the sentence of death. So the boy proceeds to the scaffold, and there too he does not cry. But he knows what he must say. In his last minute he tells the crowd in a firm and loud voice: 'Lord have mercy upon us. All people before me take warning by me.' Then he is launched into eternity, as the phrase went.

[2] P. Brooks, *Reading the plot: Design and intention in narrative* (Oxford, 1984), 30–4, 268–9; T. W. Laqueur, 'Bodies, details, and the humanitarian narrative', in L. Hunt (ed.), *The new cultural history* (Berkeley, Calif., 1989), 177.

[3] *The Times*, 30 July, 2 Aug. 1831; the case is summarized in Rayner and Crook, v. 246–9.

1. *The trial and execution of John Any Bell aged 14*, 1831.
Youngsters of fourteen were liable to the death penalty in this period, but they were hardly ever executed, and after John Bell nobody that age was hanged again. Yet both newspaper and broadside reports on Bell's case hardly mention his youth—only the fact that he had murdered another boy cruelly and for gain. These broadside woodcuts also addressed only the crime's atrocity.

The Times's report was only a few inches long; in other papers the case went unnoticed. A penny broadside was published for the London streets, but it used the story only to advise readers to 'keep from evil ways' (Plate 1).[4] This scant attention was odd, given the rarity of child-hangings after 1800. Although there were famous hangings of youngsters in the later eighteenth century, none other than Bell's has been attested in the early nineteenth.[5] There should have been some fuss. But were it not for Wakefield's fiction, Bell would have disappeared from history as quietly as he did from life.

[4] *The trial and execution of John Any [sic] Bell aged 14* (Broadside Collection, St Bride Printing Library, London).

[5] Of 103 Old Bailey death sentences passed upon children under 14 for *theft* between 1801 and 1836, none was executed; of the 5 child-hangings up to the 1830s which are repeatedly cited from J. Laurence, *A history of capital punishment* (1932), 18, only Bell's can be proved (and he was within the age of criminal responsibility). See B. E. F. Knell, 'Capital punishment: Its administration in relation to juvenile offenders in the nineteenth century and its possible administration in the eighteenth', *British Journal of Criminology*, 5 (1965), 198–207. For juvenile hangings in the eighteenth century, see Radzinowicz, i. 13–14.

These narratives foreshadow our subject well. They hint at the diverse ways in which people released, accommodated, or suppressed emotion as they watched, read about, or reported hangings or contemplated capital justice in general terms. Clearly they had innumerable ways of thinking about justice, just as we do. But the two ways depicted here represent the opposing emotional repertoires which are broadly the subject of this book, and which coexist in uneasy compromises to this day.

Reviewing the debates on the gallows, Sir James Mackintosh once observed that he felt he had 'lived in two different countries, and conversed with people who spoke two different languages'. Tories thought the reformers absurdly optimistic about human nature, creating 'in our imaginations a new man, and then [making] laws to fit him'. Tories preferred to 'consult, not the feelings of individuals, but the general good of society'.[6] These were the 'old' voices, as we may call them, though they still speak. They appealed and appeal to retributive imperatives, ethically sanctioned (they claim). Defending the rightness and deterrent utility of harsh penalties for crime, they held in contempt, as Carlyle did, all those who spoke 'beautiful whitewash and humanity and prison-discipline; and such blubbering and whimpering, and soft Litany to divine and also to quite other sorts of Pity, as we have had for a century now'.[7] They stressed and stress the crime's atrocity and the victim's, not the criminal's, pains. An ethical balance must be restored, God's order replicated, and deterrence assured; and order and security justify harsh means. 'Men are not hanged for stealing horses, but that horses may not be stolen,' as the old axiom said; or as an MP put it in 1868, the question was not 'one of softening the heart or saving the souls of murderers, but of preventing the Queen's subjects from being murdered'. 'Society needs to condemn a little more and understand a little less,' a prime minister deeply advises as I write this; we should change 'from being forgiving of crime to being considerate to the victim'.[8] Or as Wordsworth put it more elegantly in his sonnets supporting the punishment of death in 1839–40:

> But O, restrain compassion, if its course,
> As oft befals, prevent or turn aside
> Judgements and aims and acts whose higher source
> Is sympathy with the unforewarned, who died
> Blameless.

[6] Cited in R. McGowen, 'A powerful sympathy: Terror, the prison, and humanitarian reform in early nineteenth-century Britain', *Journal of British Studies*, 25 (1986), 313–17.

[7] T. Carlyle, 'Model prisons' (1850), in *Latter-day pamphlets* (*Collected works*, n.d.), 70.

[8] PD3 cxci (1868), c. 1041; *Mail on Sunday*, 21 Feb. 1993.

On the other hand there were people like Wakefield, registering the cruelty of punishments of that boy-hanging kind more vividly than that of the crime itself. At worst, as sensibility's critics always pointed out, their appeals to compassion could be promiscuous and unprincipled, demanding submission to sentiments whose value was identified in their spontaneity alone.[9] Yet Wakefield and others were more responsible than that. Without denying the crime's atrocity or the need for punishment, they argued that cold-blooded judicial killing was the greater atrocity, counter-productive in its excess, expressing overweening power, and blind to the mitigating biographical circumstances which determined the criminal character or the needful act.

This opposition in the half-dozen decades before John Bell's hanging sets the questions at the heart of this book. Until 1837 felons were executed in public not only for murder but also for crimes which we nowadays deem unheinous. Did people worry about this, were they anxious, how could they watch at all? What understandings now lost to us contained their emotions? And if feelings changed, among whom and to what extent did they change? In what culture were Wakefield's responses hatched? And what resulted from them? Did their diffusion cause the scaffold system to collapse; or was the scaffold already collapsing for reasons which had nothing to do with opinion at all? In exploring questions like these this book might suggest how some of us have become the self-consciously 'humane' people we like to think we are; it might also hint at the fragile conditions which have permitted humanitarian agendas to come into being.

For the remainder of this introduction, we owe it to the subject (and to specialists) to set scenes, discuss contexts, and sketch interpretations —to count and compare the numbers condemned and hanged in this period, to suggest ways of thinking about major shifts in the history of feeling and of punishment alike, and to offer a brief chronological map.

2. *Measuring the subject*

Late eighteenth- and early nineteenth-century English people were very familiar with the grimy business of hanging. This is so large a social fact separating that era from our own that although it is not the most obvious way of defining modern times, it must be one of them. Admittedly the English noose and axe had been at their most active long before

[9] S. Cox, 'Sensibility as argument', in S. M. Conger (ed.), *Sensibility in transformation: Creative resistance to sentiment from the Augustans to the Romantics* (1990), 63–82.

then: 75,000 people are thought to have been executed in the century 1530–1630, and nothing like this was seen again.[10] Execution rates declined in the second third of the seventeenth century as transportation to the American colonies absorbed many who would once have hanged; and political stability kept hanging rates stable across the next half-century. But then, dramatically, in the later eighteenth century it looked as if the bad old killing days were returning. There had been a mere 281 London hangings between 1701 and 1750; there were nearly *five* times as many between 1751 and 1800. The slaughter rate thereafter stayed high. As many were hanged in London in the 1820s as in the 1790s, and twice as many hanged in London in the thirty years 1801–30 as hanged in the fifty years 1701–50. How easily this extraordinary fact has been forgotten—that the noose was at its most active on the very eve of capital law repeals!

I estimate[11] that some 35,000 people were condemned to death in England and Wales between 1770 and 1830. Most were reprieved by the king's prerogative of mercy and sent to prison hulks or transported to Australia. But about 7,000 were less lucky. Eight times a year at Tyburn or Newgate, once or twice a year in most counties, terrified men and women were hanged before large and excited crowds. Audiences of up to 100,000 were occasionally claimed in London, and of 30,000 or 40,000 quite often. Crowds of 3,000–7,000 were standard. When famous felons hanged, polite people watched as well as vulgar.

What they watched was horrific. There was no nice calculation of body weights and lengths of drop in those days; few died cleanly. Kicking their bound legs, many choked over minutes. Until 1790 women hanged for coining or murdering their husbands had their corpses publicly burnt after hanging. As late as 1820 male traitors had their heads hacked off and held up to the crowd. Even though, notoriously, there were over two hundred capital crimes on the statute-books, most of the hanged were strangled straightforwardly for standard crimes which (except for forgery) had been capital for centuries. In the 1820s a fifth were hanged for murder, a twentieth for attempted murder, another twentieth or so for rape, and somewhat fewer for sodomy. Two-thirds were hanged for property crimes: over a fifth of these for burglary and housebreaking, a sixth for robbery, a tenth for stealing horses, sheep, or cattle, and a twelfth for forgery and uttering false coins. Forgery

10 P. Jenkins, 'From gallows to prison? The execution rate in early modern England', *Criminal Justice History*, 7 (1986), 52.

11 For these estimates and all English data in the following see App. 2.

convictions killed off one in five of those hanged between 1805 and 1818. (All this summarizes data in Appendix 2.)

Who were these people? Fewer women were hanged than men. Of the 1,232 people hanged at Tyburn in 1703–72, only 92 were women; and of the 59 people executed in London in 1827–30, only 4 were women, all murderesses. Many men imprisoned in Newgate were listed as 'labourers'. But most claimed craft and trade status: leather-dressers, weavers, wiredrawers, brush-makers, printers, servants, porters, clerks, tailors, errand-boys, smiths, painters, sawyers, brass-founders, upholsterers, grooms, chair-carvers, drapers, whip-makers, steel-polishers, plasterers, glass-cutters, etc.[12] None the less, apart from the execution of a few wealthy forgers or murderers, most of the hanged were poor and marginalized people—'the very lowest and worst of the people . . . the scum both of the city and the country', as Elizabeth Fry amiably described her Newgate charges in 1818. The more rootless the felon, the more likely the execution. Some 90 per cent of men hanged in London in the 1780s were aged under 21. A high proportion were recent immigrants to the city.[13]

The frequency of English executions was widely noted by foreign observers. It was in vain that English commentators replied that this was the price the English cheerfully paid for liberty and prosperity, that a few hangings were better than the ubiquitous police controls of a despotic state, and that most of the condemned were reprieved anyway. Foreign (and Scottish) eyebrows continued to be raised, for despite population differences, comparisons were startling. Scotland effectively excludes itself from this book if only because a meagre four or so a year hanged there in the 1780s; this rose to 5.4 a year in 1805–14, as against the English average of 67. Scotland's (and Ireland's) relative innocence of the noose continued into the mid-nineteenth century.[14] Other countries could crow too. The Prussian code had restricted capital punishment as early as 1743, and after 1794 only murderers were executed. Catherine's reforms to similar effect followed in Russia in 1767 and Joseph II's in

[12] P. Linebaugh, *The London hanged: Crime and civil society in the eighteenth century* (1991), 143; PP 1830–1, xii. 463–91. Of the 416 prisoners locked up (for all reasons) in Newgate in Oct. 1828, 29 were women: Newgate Prison Register, 1829 (London Corporation RO: PD.28.15).

[13] E. Fry to J. J. Gurney, 14 Nov. 1818 (Gurney papers, i. 203: Library of the Society of Friends, Temp. MS 434); Radzinowicz, i. 14; Linebaugh, *The London hanged*, ch. 3; A. R. Ekirch, *Bound for America: The transportation of British convicts to the colonies, 1718–1775* (Oxford, 1987), 47–8.

[14] Scottish executions dropped to little over 1 a year by 1836–42, eight to ten times less than the English rate; 5 murderers hanged in Scotland in 1857–63, as against 96 in England and Wales. In Ireland 81 were hanged in 1813–14, as against 19 in Scotland; but only 15 murderers hanged in Ireland in 1857–63, compared with the English 96 (PP 1812, x. 217 ff.; 1814–15, ix. 293 ff.; 1844, xxix. 367–73; 1865, xlix).

Austria in 1787. Philadelphia Quakers dispensed with capital punishment after the American Revolution. In Amsterdam in the 1780s less than 1 a year were killed; barely 15 were executed annually in Prussia in the 1770s, and a little over 10 in Sweden in the 1780s. Towards 1770 about 300 people a year were condemned in the whole of France; over twice that number were condemned annually between 1781 and 1785 in London alone. Before the guillotine's invention French punishments were crueller than English. Stretchings, flayings, burnings, and breakings on wheels were common; hands were cut off before execution and hanged bodies routinely burnt. Even so, only 32 people were executed in Paris in 1774–7, against 139 in London, and when London hangings rose from annual averages of 48 in the 1770s to 70 in 1783–7, men and women dangled outside Newgate prison up to 20 at a time, a sight unknown elsewhere. There were on average 23 metropolitan hangings a year throughout the 1820s (more if Surrey hangings south of the Thames are included). Berliners would be unlucky to see a couple in the whole decade. While only 9 were executed in Prussia in 1818, 5 in 1822, and 9 in 1831, in no year of the 1820s did English executions fall below the 50 of 1825, and there were 107 in 1820, 114 in 1821, 74 in 1829. In the 1820s, 672 were executed in England and Wales, more (again) than had been hanged in the whole of either the first or the second quarter of the eighteenth century.[15]

Then suddenly—and I mean suddenly—this ancient killing system collapsed. After nearly forty years' exclusion from office, the whigs came to power with large reformist ambitions. The 1832 Reform Act also opened parliament to some hundred independent MPs, largely middle-class advocates of progress and critics of the *ancien régime*, fervently advocating the bloody code's repeal. Hangings shrank to a tenth of their score a decade before. When most capital statutes were at last repealed in 1837, only eight people were killed that year in the whole country, and six in the year following, all murderers, while the numbers sentenced to death dropped from 438 in 1837 to 56 in 1839. Penitentiaries

[15] French executions caught up with English in the 1820s. The Napoleonic code allowed for 36 capital offences (22 in 1832). In 1826–30 French courts delivered fewer death sentences than English (111 a year on average), but they pardoned fewer too; on average 72 a year were guillotined, 32 after 1832: G. Wright, *Between the guillotine and liberty: Two centuries of the crime problem in France* (Oxford, 1983), 39, 168–70. See also G. A. Kelly, *Mortal politics in eighteenth-century France* (Waterloo, Ontario, 1986), 187; J. R. Ruff, *Crime, justice and public order in old régime France* (1984), 60–2; Radzinowicz, i. 288, 290, 295–7; I. Gilmore, *Riot, risings and revolution: Governance and violence in 18th century England* (1992), 180; R. J. Evans, 'Öffentlichkeit und Autorität: Zur Geschichte der Hinrichtungen in Deutschland vom Allgemeinen Landrecht bis zum Dritten Reich', in H. Reif (ed.), *Räuber, Volk und Obrigkeit* (Frankfurt, 1984), 208–28.

and prisons acquired new importance in penal practice. Prison inspectors announced confidently that 'the law intends that the suffering of the offender shall be proportioned to the enormity of his offence'.[16] Uniformity was more striven for than achieved, just as the reformative idea was more alive in theory than in practice (in transportation it was not alive at all), and it addressed lesser offenders chiefly.[17] None the less it was as if England had become another and gentler country—or a little more like other countries. In most northern states of America executions had been confined to murder since the 1780s, and by the end of the 1830s most north-eastern states had already transferred executions from public spaces to the interiors of prisons.[18] England was only catching up. But at least ideas of just and proportionate punishment were tacitly acknowledged. At least, too, only murderers hanged in England and Wales, even if they continued to hang in large numbers compared to other countries (347 between 1837 and 1868 inclusive), and still in public.[19] Finally, public hanging was itself abolished in 1868. Thereafter until 1964 the state's killing business was done discreetly inside prisons, where nobody could see. The civilized public in whose interest punishments were inflicted could keep its emotional and physical distance from them, as it still does.

There has been no greater nor more sudden revolution in English penal history than this retreat from hanging in the 1830s. It was far more dramatic than the invention of the prison, for example, even though the two processes were inextricably linked. Ensuing generations had no doubt that the end of the bloody code marked a watershed, full of moral meaning. People began to remember the old times as if recalling a distant country where unaccountably awful things happened. 'When I was a lad', Samuel Rogers recalled typically,

> I recollect seeing a whole cartload of young girls, in dresses of various colours, on their way to be executed at Tyburn. They had all been condemned, on one indictment, for having been concerned in (that is, perhaps, for having been spectators of) the burning of some houses during Lord George Gordon's riots.

16 *Prison inspectors' third report (Home District)*, PP 1837–8, xxx. 8.

17 Over a quarter of felons convicted at English and Welsh assizes and quarter sessions were transported in the 1830s with no pretence at their reformation: a fifth died on the journey anyway. About 43,500 men and 7,700 women were transported in the 1830s, as against about 9,300 and 2,500 in 1787–1810, 15,400 and 2,000 in 1811–20, and 28,700 and 4,100 in 1821–30. Some third were Irish. The total from England and Wales was about 41,000 in 1787–1830: A. G. L. Shaw, *Convicts and the colonies: A study of penal transportation* (1966), 147–50.

18 L. P. Masur, *Rites of execution: Capital punishment and the transformation of American culture, 1776–1865* (Oxford, 1989), 4–5.

19 Before 1841, this included 3 who had attempted murder: see App. 2.

It was quite horrible.—Greville was present at one of the trials consequent on those riots, and heard several boys sentenced, to their own amazement, to be hanged. 'Never,' said Greville with great naiveté, 'did I see boys cry so.'[20]

Such vignettes seemed to belong to a world 'so far removed from us that we cannot comprehend it ever having existed', Bagehot wrote in 1858.[21]

With hindsight the capital code's demise looked foreordained. Within living memory 'every page of our statute-book smelt of blood', Charles Phillips wrote in 1857: the cursed tree of the gallows had flourished in the land, and Christian men and women had swung on it 'thick as the leaves in Vallambrosa'. But, luckily, 'a sanguinary system, long continued, is sure to exasperate the popular patience. Outraged humanity rises in its might, and spurning the tortoise pace of legislation, stands between the lawgiver and the victim. This is experience. This is history.'[22] A century later the greatest historian of English criminal law agreed that the long retreat from harsh bodily punishment expressed 'the improvement of morals and manners [and] the growth of humanity'.[23] Most people today think the same way. 'They used to hang you for stealing a sheep, didn't they?' The question is complacent about the present day but meaningful too. It distances us from repudiated parts of our history. Our times bear little relationship to those, it implies; progress in this as in other realms is achieved. The easy tolerance of felons' public strangulations, decapitations, and burnings seems light years distant from the humane feelings recommended today. How could those people have done that to each other? How we have advanced since then!

3. *Material bases*

Needless to say nobody can write a history of penal change or any other kind of change in such terms today. Not much of history marches to the tunes of humanity, after all. As the century of the concentration camps closes with new atrocities, we know what a fragile construct civility is. If western societies over long time-spans have generally contained collective passions within their 'civilizing process', there have always been

20 *Recollections of the table-talk of Samuel Rogers* (2nd end., 1856), 183–4.

21 Cited in Cooper, p. 3.

22 C. Phillips, *Vacation thoughts on capital punishments* (1857), 3–4, 15.

23 Radzinowicz, iv. 352–3.

fractures through which violence recurrently breaks free.[24] Come a collapse in the structures of authority or in the material rewards which sustain our social collaborations, and repudiated instincts are easily unleashed. Even in stable times violence is immanent. There is little pity for underdogs; underdogs themselves may be the ones with least respect for others. Social orders today rest on camouflaged violence which most of us choose to know nothing about. The world is full of societies, including the most powerful, which execute their worst criminals without apparent anxiety. Four years after capital punishment was abolished in Britain, 85 per cent of people wanted it back;[25] most Britons today would restore hanging for several violent crimes. Were public executions reintroduced, there would be crowds to attend them and politicians to support them. In the United States they have barbecues outside prisons on the nights of electrocutions. 'Burn, Bundy, Burn!' their T-shirts advised not long ago, when that criminal was killed.

The humane principles which check these instincts rest on unconscious calculations. It has to be a latent premiss of this book that humane feelings prevail when their costs in terms of security or comfort are bearable; when they can be productively acted upon; and when they bring emotional and status returns to the 'humane'. Culturally dominant groups most deplore brutality when the state's authority or their own is strong enough to obviate the need for its outward display. In these conditions the humane people are usually those who eat, prosper, and are safe. (They include the writer and most of his readers no doubt.) But since Europe's history tells us what people even (or especially) like us are capable of should political and economic centres no longer hold, no history of progress can be offered in these pages. Regression, passion, and instinct have to be as much our theme as the fragile altruism and self-regulations of the socialized personality.

That is the dark view. Is there no relief? Well—while centres *do* hold, people's vague remembering of cruel times past has some kind of point. In stable western societies over the past two hundred years there has been a vast consolidation in socially recommended concern for others' bodily pain and terror, and in ways of affirming that concern too. Those of us who eat, prosper, and are safe can afford to stand in a different relationship to others' miseries from our forebears two hundred years

[24] N. Elias, *The civilizing process*, i. *The history of manners*, ii. *State formation and civilization* (orig. pub. 1939; Oxford, 1978, 1982), ii. 244–6.

[25] H. Potter, *Hanging in judgment: Religion and the death penalty in England* (1993), 255 n. 36.

ago. 'Sensibility' broadly conceived has this to its credit, that from that time to this, some of us politer people have learnt how to empathize with distant sufferers, our imaginations trained to feel their pain within ourselves. We accord this capacity the highest moral value, and deplore those who betray it. So ready are we to suffer vicariously on others' behalf that we exhaust ourselves with narratives and images of remote suffering. Just as Hanoverians released their sympathies through novels of sentiment and Victorians through novels about social problems, so we develop electronic and photographic devices to ensure that we absorb the energy of others' suffering to the point of saturation. We also advise more solemnly against violence than our ancestors did, lest by accident or introjection we ourselves are hurt by it. And it is consistent with this culture that it is not only sex but also pain and death that bear the most outlawed excitements, and so are today's tabooed and pornographized subjects.

One way of explaining these dispositions (developed in Chapter 7) is to stress the powers of social learning in reshaping affective relationships between individuals and groups over long time-spans. Recommended by a culture's prophets and teachers and internalized by dominant groups, shifts in affective patterns develop their own momentum as rules are elaborated. Although some reactions operate automatically, the feelings associated with them are intellectually meditated and highly socialized.[26]

Hanoverians understood something of this sort. 'Pity is not natural to man,' Dr Johnson said: 'Pity is acquired and improved by the cultivation of reason.' 'A tear is an intellectual thing,' William Blake would have added. Dugald Stewart believed that new affective bonds between peoples were diffused through processes of 'sympathetic imitation', and Edmund Burke thought so too: 'It is by imitation, far more than by precept, that we learn everything: this forms our manners, our opinions, our lives.'[27] Cultural imitations ensure that societies and generations differ in their patterns of emotional expression or inhibition. Some societies are aggressive, others not; some cultivate pity, others seem innocent of it; some weep, others do not, and even in those which weep,

[26] N. Elias, 'Human beings and their emotions', in M. Featherstone *et al.* (eds.), *The body: Social process and cultural theory* (1991), 116–18; R. Harré (ed.), *The social construction of emotions* (Oxford, 1986); P. Metcalf and R. Huntington, *Celebrations of death: The anthropology of mortuary ritual* (Cambridge, 1991), introd. and ch. 2; D. B. Morris, *The culture of pain* (1991).

[27] W. Blake, 'The grey monk', D. Stewart, *Elements of the philosophy of the human mind* (1827 edn.), in W. Hamilton (ed.), *Works* (1854), iii. 117; E. Burke, *Essay on the sublime and beautiful*, cited by Stewart, *The philosophy of the human mind*.

weeping may signify differing emotional states.[28] The fact that in 1757 both the high and the low people of Paris could gather in the place de Grève to watch the would-be regicide Damiens tortured, torn apart by horse-teams, hacked by knives, disembowelled, and burnt—and this over hours and with no public vomiting—enforces these points about emotional learning. Within decades such a spectacle was repudiated in France as it was in England. 'All history,' Southey wrote in 1807, 'does not present a spectacle more inexpiably disgraceful to the country in which it occurred.'[29]

Such cultural shifts must be allowed some autonomy in so far as they are determined by reactions or emulations; materialist histories often ignore this autonomy to their cost. Yet it is equally true that cultural histories can err in the other direction. They often evade the material conditions which also determine cultural change. Cultural changes may feed reactively off themselves, but the initiation of new emotional repertoires can seldom plausibly be dissociated from the material and political world. Changes here may best explain why we are disgusted by the idea of public execution when our forebears accepted it easily.

The most familiar of the materialist explanations of changing attitudes to violence and suffering might embrace (for example) the medical anaesthetization of pain, the medicalization of death, and the lowering of mortality rates in modern times. These are all assumed to have made death and pain less familiar and less easily faced than they were in the past.[30] Attitudes to pain and violence have also arguably been affected by redistributions in material and educational resources. Whereas in the nineteenth century those who had political influence and a public voice were richer people who feared theft but were protected from violence (so theft was the crime most acted against then), education and enfranchisement have given a voice to those with less property but who know violence in their own streets only too well. Hence, not surprisingly, today's media-fuelled panics tend to be about violence rather than theft. Most ambitious of all materialist explanations of these vast attitudinal shifts, however, are those associated with the names of Foucault and Elias. In their very different ways they ascribed central importance to the long development of the modern state and its changing powers and expectations.

[28] Metcalf and Huntington, *Celebrations of death*, 44–8.

[29] *Letters from England* (orig. pub. 1807: 1951), 220.

[30] D. de Moulin, 'A historical-phenomenological study of bodily pain in western man', *Bulletin of the History of Medicine*, 48/4 (1974), 540–70.

Foucault (first) utterly discounted a humanitarian history of evolving penal policy. Even as punishment is ostensibly humanized, power and control remain what punishment is about. Each era adopts technologies of control appropriate to its resources; the rest is rhetoric. In early modern Europe violent bodily punishments were never primarily a means of reducing crime, for example. With monarchical rule chronically challenged, they had to express a 'political tactic' through which sovereign power was affirmed. What we now construe as meaningless cruelty was integral to the symbolic display of the sovereign's might:

> The public execution is a ceremonial by which a momentarily injured sovereignty is reconstituted. . . . Its aim is . . . to bring into play, as its extreme point, the dissymmetry between the subject who has dared to violate the law and the all-powerful sovereign who displays his strength. . . . The punishment is carried out in such a way as to give a spectacle not of measure, but of imbalance and excess. . . . By breaking the law, the offender has touched the very person of the prince; and it is the prince—or at least those to whom he has delegated his force—who seizes upon the body of the condemned man and displays it marked, beaten, broken. . . . The public execution did not re-establish justice; it reactivated power.

The violence of the punishment 'is one of the elements of its glory; the fact that the guilty man should moan and cry out under the blows is not a shameful side-effect, it is the very ceremonial of justice being expressed in all its force.'[31]

It must be admitted that early modern English punishments were never as gruesomely aggravated as those of many European states. But this was chiefly because gruesome aggravations were less necessary when the English crown had been in control of its domain for centuries; by the seventeenth century the predominance of sovereign over customary law was secure.[32] If English executions declined after the mid-seventeenth century, it was also thanks to the stability then achieved. The decapitations, disembowellings, and displays of spiked heads on Temple Bar which followed the Jacobite risings in 1745 marked the end of any serious challenge, their symbolizing horror consistent with the claim that highly visible state violence was never so necessary as when sovereignty was questioned. In relatively stable conditions thereafter the state had less need of such displays. The one relic of torture which survived into the eighteenth century, pressing with weights, fell into disuse after 1735 and was statutorily abolished in 1772, as were some of the

[31] M. Foucault, *Discipline and punish: The birth of the prison*, trans. A. Sheridan (1979), 9, 23, 34, 48–9.

[32] J. Sharpe, *Crime in early modern England, 1550–1750* (1984), 175.

more extreme marks of ignominy inscribed on the criminal body—branding in 1779, the burning of traitorous women's hanged bodies in 1790.

Foucault's model also broadly fits what ensued (even though he himself hardly addressed the English case). If harsh bodily punishment was further curtailed in the 1830s (execution rates collapsing then) and if public executions were abolished in 1868, it was arguably because the state's consolidation and bureaucratic competence rendered symbolic displays of might even less necessary than hitherto—and less intelligible as well. The 1832 Reform Act brought into parliament new independent members from a middle class which had little truck with the old languages; while policemen, magistrates' courts, prisons, and other regulatory institutions, along with encouragements to private prosecution, were sufficing to guarantee order in new ways. A quantitative expansion of disciplinary power could adequately offset punishment's mitigations. As an abolitionist MP said in 1868: 'improvements in the efficiency of the police or in the character of legislation would more than compensate for the absence of the terror of the death penalty.'[33] As indeed they did.

If a language of humanity found its rhetorical space in these changing conditions, it was largely a legitimizing language. What mattered more was that the old punishments were simply losing their meaning. As scientific rationalism made it clear that death was the last imaginable punishment, aggravations of the death penalty came to look cruelly superfluous. Penal policy was conceived as serving a more secularized ethical system, aiming to regulate the passions and reshape personality. This put the prison at the centre of penal policy, not the noose.[34] At the same time an increasing social distancing by the polite resulted in deepening embarrassment about any unseemly display. Far from making statements about right order, public executions were making statements only about *dis*order: that was all they encouraged and that was all that began to be seen. In all these ways old penal languages and techniques came to be seen as cruel because they were now dispensable. The functions of the old penal ceremony were no longer understood.[35]

[33] Charles Neate, in PD3 cxci (1868), c. 1047.

[34] M. Wiener, *Reconstructing the criminal: Culture, law, and policy in England, 1830–1914* (Cambridge, 1990), 7–11.

[35] Explaining the continuation of public executions in France until 1939 will have to be another's task, not mine. Likewise in Prussia strangling or burning was abolished only in 1818, and as late as 1836 most executed bodies were broken on wheels or decapitated: Wright, *Between guillotine and liberty*, 167–73; Evans, 'Öffentlichkeit und Autorität', 218, 222.

Foucault's easy way with agency and chronology means that his vision of process may impress more than his *histoire événementielle*. He also downplayed the roles of feeling and culture. So a reference to Elias's vision of the 'civilizing process' is called for finally, even though he never addressed punishment directly.[36] Elias argued that from the subjugation of armed fiefdoms down to its modern bureaucratization it has been the expanding state that has chiefly imposed emotional restraint upon subjects, outlawing the free expression of desires first within courtly circles and then outside them. Economic diversification and bureaucratization ensured that self-control became both a mark of status and a condition of efficiency. Through the long elaboration of conduct rules across the generations, a momentum was set up, until *homo clausus* had emerged by the nineteenth century: blocking aggression, capable of shame and embarrassment when passion was exposed, self-concious about the body and its functions, but also newly respectful of others' bodies, as instinct was subordinated to the socialized superego. From these protracted processes people and groups learnt to inhibit their aggressions, and rules became part of the taken-for-granted.

Elias's grand fusion of Freudian psychoanalytical theory and the history of political processes has been largely ignored by Anglophone historians who prefer to work on smaller canvases than this. And certainly state-formation is too blunt an instrument to serve our present purposes. We might prefer to argue that the shaping of the self-regulating and empathetic personality structures of bourgeois man have had less to do with state-imposed constraints than with the ethical outworkings of market capitalism.[37] We shall certainly want later to accommodate the processes of affective learning achieved in the culture of sensibility, remote from and even subversive of the dictates of state as they were. Even so, Elias can hardly be put out of mind (any more than Freud can) when we meet the behaviour this book uncovers. If we do discover retreats from scaffold horrors which helped mediate the birth of 'modern' repertoires of feeling—squeamishness, for example—they can only be associated with the accumulating restraints upon the free-ranging instincts of socialized individuals to which Elias (and the Freud of *Civilization and its discontents*) drew attention.

[36] Elias, *The civilizing process*.

[37] T. L. Haskell, 'Capitalism and the origins of the humanitarian sensibility': I, *American Historical Review*, 90/2 (Apr. 1985), 339–61; II, ibid. 90/3 (June 1985), 547–66. See Ch. 12, below.

4. *Pressures on structures*

Across our focal period of 1770–1830, the population of England and Wales increased from about 7 to nearly 14 million. London's population grew from about 1 to 1.7 million between 1801 and 1831 alone. Urban populations rose from some fifteen per cent of the whole in 1750 to twenty-five per cent in 1801 and sixty per cent a half-century later. In the first thirty years of the nineteenth century national income increased by over a half. Did the criminal law *have* to accommodate itself to these conditions?

It has been a conventional wisdom that 'crime' inevitably increased in these demographically explosive decades, and that in the interests of security, prosperity, and work discipline, there was a growing demand for a broader and more effective punishment than that applied to the few felons who had been prosecuted and convicted hitherto. In conditions of heightening anxiety about order, from Bentham and Colquhoun in the 1780s on to the businessmen petitioning for penal reform after 1811, and on again via whig, radical, or Quaker reformers in the 1820s, the appeal to rational and efficient punishment always sounded out more loudly than the appeal to 'humanity' did. These constituencies recommended more certain and proportionate punishments not only to deter offenders but also 'to circumscribe the scope of personal spontaneity'—that grand moralizing design of the middle half of the nineteenth century which Wiener has recently elucidated. This was a punishment best measured out in penitentiaries, where prisoners could be intimidated into contemplating their immortal souls at the same time.[38]

Yet these pressures were by no means concerted. For a start, many contemporaries knew better than some historians do that 'crime' was less obviously increasing in the early nineteenth century than anxieties about it and the publicity given to it, along with the facilities for its higher prosecution: up went *prosecution* rates, therefore—not crime

[38] Wiener, *Reconstructing the criminal*, 103; also W. J. Forsythe, *The reform of prisoners, 1830–1900* (1987). These historians' claim that these processes expressed ethical rather than disciplinary or social control imperatives overlooks the fact that these goals need not be mutually exclusive. Those who insist that punishment was and is a form of domination do not usually deny that it was and is other things besides. 'Institutions like the prison, or the fine, or the guillotine, are social artefacts, embodying and regenerating wider cultural categories as well as being means to serve particular penological ends. Punishment is not wholly explicable in terms of its purposes because no social artefact can be explained in this way. . . . Punishment has an instrumental purpose, but also a cultural style and an historical tradition': D. Garland, *Punishment and modern society: A study in social theory* (Oxford, 1990), 19.

rates.[39] Moreover, although commercial voices had their say on law-enforcement reform,[40] most commercial voices were either silent or politically ineffectual outside their own localities. Even bankers' views on the desirability of abolishing death for forgery were contradictory.[41] In any case, since the 1770s the most consistent advocates of legal and penal reform were 'enlightened' lawyers and professionals rather than businessmen. They had less interest in labour discipline than in their profession's advancement and the subversion of the élites who excluded them from influence. Admittedly, a few bankers, businessmen, and Quakers jumped on the reform bandwagon which the barrister Romilly set rolling in 1808. But although these last made most noise after 1815 and are usually assumed to have led 'opinion', we shall suggest in Chapter 16 that a surer index of a deepening opinion critical of the law lay in those less demonstrative middle-class people in the 1820s who pleaded for mercy on other and poorer people's behalf.

In short, it looks as if we must look elsewhere than to 'interests' if we are to argue for a linkage between material conditions and a major shift in penal policy. And sure enough we find it best in a structural problem *within* the criminal law itself—in an *overloading* of the capital code which ensured that its repudiation had to come about sooner or later. In a society facing demobilization, high food prices, and popular unrest after the Napoleonic Wars, the key variable was the criminal law's very success in extending its reach to lawbreakers who had hitherto escaped the law. It worked through the prosecution process.

Between 1805 and 1815 felony prosecutions under the capital code increased by 70 per cent in England and Wales, and by another 75 per cent between 1815 and 1820 alone. From the 4,605 prosecuted in 1805 to the 18,107 in 1830, the increase was nearly 300 per cent.[42] Reformers talked endlessly of prosecutors' squeamishness about bringing people to court under capital law, but nothing of the kind is evident here. Nor was an increase in 'crime' provably behind this, and the impact of policing lay in the future. Population increase played a part. But the prosecution boom chiefly expressed a deepening sense that crime was a

[39] SC 1819: 4–5.

[40] D. Hay, 'Manufacturers and the criminal law in the later eighteenth century: Crime and "police" in south Staffordshire' (typescript), *Past and Present Society Colloquium*, Oxford, 1983.

[41] See Ch. 15, below.

[42] V. A. C. Gatrell and T. B. Hadden, 'Nineteenth-century criminal statistics and their interpretation', in E. A. Wrigley (ed.), *Nineteenth-century society: Essays in the use of quantitative methods for the study of social data* (Cambridge, 1972), 392–3.

problem to which those offended against should respond actively.[43] It also reflected the cumulative impact of statutes between 1752 and 1826 which reimbursed prosecutors and witnesses for their time and labour (prosecutors at assizes and quarter sessions at the beginning of the century had to pay fees and charges of between £10 and £20).[44] A critical sequence of causes and effects was then set in motion. As prosecutions escalated, capital convictions and death sentences did too. Death sentences doubled after 1815, and by the second half of the 1820s, when they averaged 1,336 a year, they were three and a half times more frequent than in 1806–10.[45] The implications could not be evaded for long. A consistent proportion of the condemned could not be hanged if the land were not to be covered in gallows.

At this point culture had its independent say. It was understood that there was a threshold beyond which the number of executions could not safely pass:

> the least excess . . . excites a tenderness in the milder sort of people, which makes them consider government in a harsh and odious light. The sense of justice in men is overloaded and fatigued with a long series of executions, or with such a carnage at once, as rather resembles a massacre than a sober execution of the laws. The laws thus lose their terror in the minds of the wicked, and their reverence in the minds of the virtuous.

That was Edmund Burke cautioning against hanging too many Gordon rioters in 1780.[46] The *Gentleman's Magazine* said the same: 'Convicts under sentence of death in Newgate, and the gaols throughout the Kingdom increase so fast, that, were they all to be executed, England would soon be marked among the nations as the *Bloody Country*' (though it already was). A quarter-century later a French observer thought that 'if these sanguinary sentences were rigorously carried into effect, the scaffolds of England would stream with blood, and the whole nation would rise up in horror against them'. 'Carnage', 'massacre',

[43] V. A. C. Gatrell, 'Crime, authority and the policeman-state', in F. M. L. Thompson (ed.), *The Cambridge Social History of Britain, 1750–1950* (1991), iii. 243–310.

[44] Statutes between 1752 and 1815 reimbursed some prosecutors and witnesses in felony cases for the costs of attending court and of having indictments drawn up, and for loss of time if they were poor. In 1818 costs were extended to cover appearances before the grand jury, and in 1826 they covered expenses incurred during the committal hearings before magistrates, as well as those incurred in prosecuting certain misdemeanours: C. Emsley, *Crime and society in England, 1750–1900* (1987), 145–7; D. Hay and F. Snyder, 'Using the criminal law', in Hay and Snyder (eds.), *Policing and prosecution in Britain, 1750–1850* (Oxford, 1989), 36–41; David Jones, *Crime in nineteenth-century Wales* (Cardiff, 1992), 24.

[45] See App. 2.

[46] E. Burke, 'Letters and reflections on the execution of the rioters' (1780), in *Works and correspondence* (8 vols., 1852), v. 580–1.

'blood', 'bloody', 'bloody code'—in its excess the imagery spoke of inhibitions on excess, even as it hinted at excess's illicit excitement. It was calculated in the 1820s that if you hanged all the condemned, you would have to hang four people every day of the year, excluding Sundays. The 'people' might put up with a lot. But they would not put up with this.[47]

Short of repealing the capital code wholesale, there was only one way out of the dilemma. The numbers and proportions of conditional pardons, already increased between 1793 and 1801,[48] must increase again. Pardons had to be given to well over 90 per cent of those sentenced to death in post-war England and Wales and to some 95 per cent in London. This was a far cry from the steady 50 or 60 per cent pardoned in earlier decades. Tensions deepened as the widening disproportion between death sentences pronounced but respited and death sentences executed highlighted the grim fate of the wretches chosen to hang. Romilly's point that English justice was now only a lottery could hardly be gainsaid.[49] Outsiders like Wakefield could mercilessly expose the fact that capital law had come to look randomly cruel and terminally silly. By no means everyone agreed; but it was this disequilibrium which finally rendered the old system unworkable and unbearable too. The bloody code might fairly be said to have collapsed under pressure of the criminal law's mounting prosecutory effectiveness.

5. *Chronologies of change*

Because these imbalances accumulated slowly and were by no means obvious to all, they cannot account for the exact timing of the bloody code's repudiation. There were many resistances in the way. Governments basking in royal favour for most of four decades, paranoid about public tumults and French precedents, and defensive of 'morality, legality, and respect for constituted authorities' (as Croker defined government purposes in 1819), were not going to make easy concessions to those who thought the criminal code a bit irrational or a bit harsh. Men in high office in the 1820s still believed they could live with the imbalances for a generation yet, and that the prerogative of mercy would continue adequately to mitigate any dangerous harshness in punish-

[47] *Gentleman's Magazine*, 2 (1784), 224; M. Cottu, *On the administration of the criminal code in England, and the spirit of the English government* (*Pamphleteer*, 16/31 (1820)), 37–8.

[48] See App. 2. It was to the 'liberality' of Lord Loughborough's chancellorship that Mackintosh attributed the increased use of pardons in the 1790s after the slaughter years preceding: PD1 xxxix (1819), c. 777 ff.

[49] For doubts about the increasing use of the mercy prerogative, see Radzinowicz, i. 131–6.

ment. For every doubt raised about the capital code, expressions of support were many times more numerous, and these were what tory statesmen heard. Reformers were reduced to commenting sardonically on the reactionary forces they were up against. 'The plunderers of the public, the jobbers, and those who sell themselves to some great man, who sells himself to a greater,' Sydney Smith observed in 1821,

> all scent, from afar, the danger of political change—are sensible that the correction of one abuse may lead to that of another—feel uneasy at any visible operation of public spirit and justice—hate and tremble at a man who exposes and rectifies abuses from a sense of duty—and think, if such things are suffered to be, that their candle-ends and cheese-parings are no longer safe.[50]

But penal reform had low priority on parliamentary agenda. Politics was about war, peace, economic policy, the repression of plebeian discontents, Ireland, freeing Catholics and dissenters, and place and connection. Bentham might sound off for decades about the desired relationship between reason, utility, and the disciplinary prison, but statesmen felt no obligation to listen. Quakers might petition against the forgery statutes as much as they liked; Elizabeth Fry might go around doing no end of good in prisons; appeals for mercy might grow in ambition and numbers—but the bulk of the political nation would still always get more excited about relief to Catholics or dissenters than about a few mavericks worried about punishment. Nor would it do for great men to yield openly to a minority voice, or any public voice for that matter. The concessions parliament did make to Romilly's or Mackintosh's attacks on the capital code from 1808 onwards were grudging, limited, and few; and the concessions Home Secretary Peel made in the way of law reform in the 1820s were mainly consolidatory. Peel not only left the strongest hanging statutes untouched; in his long career at the Home Office he let more people hang than any other secretary of state in the century past.

When the door was unlocked in the 1830s it was unexpectedly. Tory government wounded itself irremediably in disagreements about Catholic relief. The whigs came to office on a wave of fervour for parliamentary and other reforms, unleashing commitments accumulated during frustrating decades in the political wilderness. The 1832 Reform Act gave new energy to independent and abolitionist MPs like William Ewart, Fitzroy Kelly, and Bright—new men from outside the charmed circles who bore sober witness to a half-century's radical, rationalist,

[50] *Works* (2 vols., 1859), i. 331.

and evangelical dissent from an *ancien régime* which had marginalized their class and interests. Henceforth it was their commitment to civility and efficiency rather than that of whig ministers which put pressure on capital law. A whig royal commission mooted the repeal of the killing statutes, and codification of the law too: the age of law reform and the age of Bentham might now indeed seem to be one and the same, as Brougham said. But it was Ewart who put heaviest pressure on Lord John Russell to begin the repeals of capital statutes in 1837; and it was he too who in March 1840 first invited the Commons to abolish capital punishment completely. Ninety MPs voted with him on that occasion. The abolitionist cause was soon undermined, however. Independent voting declined as party controls tightened, and abolitionism was outflanked as debate was deflected to the question of public execution and its deterrent effects rather than of execution outright. The abolition of public hanging in 1868 all but silenced the abolitionist cause for near on a century, just as Ewart, Bright, and their supporters feared it would.

Was it merely a coincidence that just as the 1832 Reform Act presaged the end of gibbeting and anatomizing and the repeal of most capital statutes, so the Reform Act of 1867 presaged the end of public execution? These are beguiling conjunctions, even though they pose some difficulty. It does not quite do to say that they reflected the self-confidence of a new (if qualified) democracy, for example.[51] After 1867 nobody said that the populace might henceforth be spared intimidating displays of legal force simply because most of it was now safely incorporated in the political nation. On the contrary, the abolition of public execution spoke for anxiety, not complacency. The poorest quarter of the urban populace was far from assimilated in 1867; voteless, it remained a target of concern, containment, and stigmatization for decades yet. The old fear still held that the intended intimidations of the scaffold were not working upon them well enough. All that was agreed in 1868 was that executions would work upon plebeian imaginations more terribly by being hidden. As Fielding had written a century before: 'a murder behind the scenes, if the poet knows how to manage it, will affect the audience with greater terror than if it was acted before their eyes.'[52]

Further than this (as we shall see in the book's last chapter), the deepest anxiety of the modernizing state was that the unleashed passions of the scaffold crowd mirrored the state's violence too candidly. The crowd

[51] T. W. Laqueur, 'Crowds, carnivals and the English state in English executions, 1604–1868', in A. L. Beier *et al.* (eds.), *The first modern society: Essays in honour of Lawrence Stone* (Cambridge, 1989), 309, 354–5.

[52] H. Fielding in 1751: *An enquiry into the causes of the late increase of robbers and related writings*, ed. M. R. Zirker (Oxford, 1988), 169.

had come to seem like a repudiated *alter ego* or shadow-self which spoke too truthfully for a progressive nation to tolerate. The crowd gave the lie to the great world's representation of itself as civil, benign, and humane. As a newspaper said of the crowd at the last public execution, public hanging 'made the law and its ministers seem to them the real murderers, and Barrett to be a martyred man'.[53] Politeness must be defended, its hollowness not exposed, the mocking chorus silenced. Some outright abolitionists voted to hide execution because that was better than nothing, and perhaps kinder to the victim (they were wrong in this). But the measure was hatched in the culture of civility, not of humanity. It only hid the cruelty, and so intensified it. Men and women were to be hanged more terribly than ever—without supporters, without opportunities for protests of innocence or displays of anger, defiance, courage—coldly, clinically, and in heart-stopping silence behind high prison walls. There must not be too much feeling; a polite nation's brutality must be camouflaged.

It is only when these political moments are related to the law's structural imbalances, and to a developing will and capacity to evolve new systems of policing, prison inspection, and secondary punishment, but above all to the propertied classes' deepening concern for an urban decorum and order which would not subvert their own by making its bases too apparent, that one approaches explanations for penal change which meet the demands of chronology.

We take it for granted throughout the following, then, that hanging came to be repudiated for reasons greater than the fact that some people began to feel bad about it. We also take it for granted that many came to feel bad about it because they could afford to do so—as new controls bit deeper and made them safe. Hostility to the scaffold on humane grounds was never so vehement as when the perceived need for it was waning. Wakefield wrote his letter from the hangman when change was in the air. Thackeray announced his shame and disgust at watching Courvoisier hang in 1840 when most capital statutes had been dismantled and debate on root-and-branch abolition was swelling. So let it be axiomatic throughout what follows that humane opinion had influence chiefly in so far as it bore a plausible and justificatory relationship to processes which were working to change punishment anyway. The derivative as distinct from causative energies of empathetic feeling

[53] *Daily News*, 27 May 1868.

will not otherwise be laboured in what follows, but this clarifies the ambivalences within the history of feeling to which the book attends.

Yet finally we *should* have it both ways. A change of this large order—the demise of the scaffold, the growth of respect for the felonious body, and the rise of alternative disciplinary techniques—was 'overdetermined', we must say.[54] Causes were multiple, and only rash historians would privilege material, or political, or cultural causes without interrelating all three. Although material variables determine cultural representations, we repeat that culture generated its own momentums as well as its own resistances. In the present instance it determined how material exigencies were assimilated both in policy and in emotional repertoires. Then, in the politically efficient and adaptive society which England was, it ensured that the demands of material reality were accommodated rather than denied or betrayed.

[54] Garland, *Punishment and modern society*, 2.

PART I

THE SCAFFOLD AND THE CROWD

CHAPTER 1

HANGING PEOPLE

1. *Dying bravely*

THE WAYS IN WHICH PEOPLE WERE KILLED ON PUBLIC SCAFfolds have always been shrouded in euphemism. Home secretaries would keep an emotional distance from their own orders as they pencilled on the appeal dossiers of those they left to die a phrase sanitized by repetition: 'The law to take its course.' Newspapers likewise would tersely report that felons were 'launched into eternity'. It was as if when justice was done, thought and feeling had to be muffled. We are still reticent on the subject. Now that sexual taboos have fallen, what was and is entailed in death by execution may be one of the last agreed obscenities. Anxiety and distaste accompany a journey into a subject like this. Writing a book about it feels dangerous and contaminating.

The taboo has not served history well. How deeply and on what levels the public scaffold permeated the English imagination in the century before its abolition has only lately been looked at. Even then the subject has been tackled chiefly in relation to the class interests allegedly embedded in eighteenth-century criminal procedure. In this mode historians have usually referred to the scaffold distantly, as to an abstraction or a legal process merely. Some argue that the scaffold was a necessary deterrent when there was no effective policing, and that thanks to pardons not many hanged anyway: but this itself sanitizes the process.[1] Others see in it a device to deliver terror, but what this actually entailed is again seldom felt through. Historians' detachment becomes absurd when the scaffold is narrowly interpreted as a site of carnival humour and comic ineptitude.[2] It was the site of more horrible things than that. If we are to understand how people felt about scaffold

[1] C. Herrup, 'Law and morality in seventeenth-century England', *Past and Present*, 106 (1985), 106; J. Langbein, 'Albion's fatal flaws', *Past and Present*, 98 (1983), 96–120.

[2] T. Laqueur, 'Crowds, carnivals and the English state in English executions, 1604–1868', in A. L. Beier *et al.* (eds.), *The first modern society* (1989).

punishment in the eighteenth and nineteenth centuries, and how they defended themselves against or celebrated their feelings, we need to engage with what happened there, and not treat the scaffold as if it were only an *idea*. We must move closer to the choking, pissing, and screaming than taboo, custom, or comfort usually allow.

Gallows hangings were more than a symbolic device of justice. Until the collapse of the capital code in the 1830s no ritual was so securely embedded in metropolitan or provincial urban life. Nor was any so frequent. There were hangings once or twice a year in many assize towns. In London after each of the eight annual Old Bailey sessions anything between a couple and a dozen (rising to twenty in the 1780s) were hanged at Tyburn or, after 1783, on the drop outside Newgate prison (Plate 2). Yet more Londoners were hanged south of the Thames in Surrey, first on Kennington Common, then on the roof of Horsemonger Lane prison, constructed in the 1790s to display them (Plate 3). Occasionally seamen, mutineers, or pirates were hanged at Execution Dock, and until 1817 people were sometimes executed near

2. *The Newgate drop in action,* 1809.

This offers a closer view of the Newgate scaffold than most depictions do; but the scaffold's height is exaggerated, the crowd is implausibly thin, and quotations from Hogarth get in the way of fresh observation.

3. *Horsemonger Lane prison, Surrey, 1809.*
Built south of the Thames in the 1790s, with a roof designed to display executions, Surrey's county prison staged hangings as famous as any at Newgate. Despard and his fellow conspirators were hanged and decapitated here in 1803, and the Mannings were hanged here in 1849, watched by Dickens and 30,000 others.

the scenes of their crimes. Visibility and example were all, as the careful orderings by the King in Council of where the Gordon rioters of 1780 were to hang makes clear:

Executions to be in the following order:

1st. Wm Packman on Tuesday the 11th July in Coleman Street as near as may be to the house of Robert Charleton.

2nd. Wm Brown same day in Bishopsgate St., as near as may be to the house of Charles Daking.

3rd. William MacDonald, Mary Roberts, Charlotte Gardner, same day on Tower Hill as near as may be to the end of St Catharine's Lane.

1st. Thomas Taplin, Richard Roberts, Wednesday 12th July at the corner of Bow Street Covent Garden.

2nd. James Henry same day in the open part of Holborn as near as may be to the late dwelling house of Thomas Langdale.

1st. Enoch Fleming—Thursday the 13th July in Oxford street opposite and as near as conveniently may be to the end of Woodstock street.

2nd. Christopher Plumley, otherwise John Williams same day at Tyburn usual place of execution.[3]

In theory, a Londoner growing up in the 1780s could by 1840 have attended some four hundred execution days outside Newgate alone, discounting other locations. If he was unimaginably diligent he could have watched 1,200 people hang (and there were such obsessives). I have already given my estimate of 7,000 executions in England and Wales in the decades 1770–1830, and noted that the penal code was at its bloodiest on the very eve of its collapse. The 672 hanged in the 1820s exceeded those hanged in the whole of either the first or second quarters of the eighteenth century. Even if these numbers had been fewer, the scaffold's significance would hardly be diminished: the law's relationship with the populace can as well be read in a dozen executions as in hundreds. Nor can this device be relegated to the margins of English political culture, as an unfortunate aberration in it. As Hay has put it, the sanction of the gallows and the rhetoric of the death sentence were central to all relations of authority in Georgian England.[4] But we shall go further. The gallows were also embedded in the collective imagination, the subject of anxiety, defence, and denial, of jokes, ballads, images, and satire, and of primal gratifications too.

It expresses the anxieties attached to public hangings that even today we take comfort from an exuberant and cheering fantasy of what they were like, and hence blur the memory of what the noose really did to people. A pleasant myth shields us from the reality of the process. It is not that the myth was without basis. It is what it concealed that is in question.

Central to the fantasy is the memory of the felon's procession to Tyburn before 1783. To surface appearances it all seems rather jolly, and in certain dark senses it was so:

> As clever Tom Clinch, while the rabble was bawling,
> Rode stately through Holborn, to die in his calling;
> He stopped at the George for a bottle of sack,
> And promised to pay for it when he came back.[5]

From Newgate prison the condemned were conveyed in open carts along Holborn, St Giles, and Tyburn Road (later Oxford Street) to the

[3] SP37.21, fos. 154–5. The last scene-of-crime execution was of a sailor in 1817, hanged in Skinner Street, opposite the house he had robbed.

[4] D. Hay, 'Property, authority and the criminal law', in D. Hay *et al.* (eds.), *Albion's fatal tree: Crime and society in eighteenth-century England* (1975), 63.

[5] J. Swift, 'Clever Tom Clinch going to be hanged' (1726/7).

triangular gallows at the foot of the Edgware Road, where Marble Arch is now. The major stations in this parodic progress to Calvary were at inns like the Bowl on the corner of St Giles' High Street, or the George in Holborn, where the condemned would be offered wine; then Tyburn itself; and then again at Surgeon's Hall at the Old Bailey, where murderers' bodies were displayed and dissected. Playing as best they could to the crowd's admiration and engaging in parodic dialogue with it, some felons on their way to their doom constructed the illusion that they were the masters of the ceremonies, and not the City marshal, under-sheriff, priest, constables, and javelin-men who were meant to impart solemnity and security to the procession. Lord Ferrers's composure on his journey to Tyburn in 1760 'shamed heroes', Horace Walpole reported. Hanged for murdering his servant, he bore the procession 'with as much tranquillity as if he was only going to his own burial, not to his own execution'. Plebeians also put on fine displays—or at least those did whom contemporaries noticed:

> The vilest rogues, and most despicable villains, may own a thousand crimes, and often brag of the most abominable actions; but there is scarce one, who will confess that he has no courage. . . . The further a man is removed from repentance, nay, the more void he seems to be of all religion, and the less concern he discovers for futurity, the more he is admired by our sprightly people.[6]

When Lewis Avershaw was hanged on Kennington Common in 1795 he appeared 'entirely unconcerned, had a flower in his mouth, his bosom was thrown open, and he kept up an incessant conversation with the persons who rode beside the cart, laughing and nodding to acquaintances in the crowd'. He was afterwards hanged in chains on Wimbledon Common, and 'for several months, thousands of the London populace passed their Sundays near the spot, as if consecrated by the remains of a hero'.[7] 'Sixteen-string' John Rann in 1774 wore a pea-green coat, a nosegay in his buttonhole, and nankeen small-clothes tied at each knee with sixteen strings. At the gallows he sustained the demeanour of his last dinner-party in Newgate, where the company had included seven of his girls and 'all were remarkably cheerful'.[8] Thanks to the crowds and the convivial exchanges *en route*, a popular daredevil like this might take two hours to travel the couple of miles to his Tyburn death. In 1760 Ferrers's journey from the Tower took nearly three, and a fine display he made, for this was an earl: 'a string of

[6] B. Mandeville, *An enquiry into the causes of the frequent executions at Tyburn* (1725), 28, 32.

[7] Borrow, v. 368.

[8] J. T. Smith, *Nollekens and his times*, ed. W. Whitten (1920), i. 20–1; Griffiths, p. 168.

constables; then one of the sheriffs, in his chariot and six, the horses dressed with ribbons; next Lord Ferrers, in his own landau and six, his coachman crying all the way; guards at each side; the other sheriff's chariot followed empty, with a mourning coach-and-six, a hearse, and the Horse Guards'.[9]

Nor did this festive tradition die when the scaffold was removed to Newgate's exterior in 1783. At Holloway's and Haggerty's hanging in 1807 both men 'conducted themselves with the most decided indifference'. Holloway 'with an affected cheerfulness of countenance . . . jumped upon the scaffold when he had ascended the ladder, his arms being pinioned with a rope behind, . . . got his hat between his two hands, and as well as he was able, bowed to the crowd repeatedly . . . with a view to show that *he died game*, as it is expressed.' He announced his innocence, refused to pray, and told Haggerty to ignore the clergyman.[10] Ascending the Newgate scaffold in 1829, Thomas Birmingham 'was instantly greeted by a vast number of girls of dissolute character in the mob, who called out repeatedly—"Good bye, Tom! God bless you, my trump!" '[11] In the 1830s the ballad of the condemned Sam Hall conveyed the tone of these scaffold exchanges:

> I saw Nellie in the crowd,
> And I hollered,—right out loud—
> 'Say Nellie, ain't you proud—
> Damn your eyes'.[12]

These mocking postures were mainly metropolitan but not exclusively so. Before his execution at York in 1739, Dick Turpin employed five mourners to follow his cart to the scaffold.[13]

Self-parody and the display of courage was one way of dealing with terror. Defiance was another. An agricultural worker executed in Kent for arson during the Swing disturbances in 1830 declared his innocence to the last and 'refused to pull the cap down over his eyes, saying he wished to see the people' as he died. Hanged for coining in Birmingham in 1828, Hanbury Price cried: 'I am a murdered man, I am a victim! Here my lads, here's another murdered man for you! Murder! Murder!' Others spurned God and his priests. When the highway robber Norton died game in 1827 he refused religious consolations.[14] When a schoolmaster in Newgate sought to persuade a condemned man that there was a future life, the

[9] Walpole, iii. 303, 308; Griffiths, p. 167.

[10] Knapp and Baldwin, iv. 370; *Ann. Reg.* 1807: 378–84.

[11] *Ann. Reg.* 1829: 65.

[12] Cited in K. Hollingsworth, *The Newgate novel, 1830–1847: Bulwer, Ainsworth, Dickens and Thackeray* (Detroit, 1963), 4. Ch. 4, below, discusses this ballad more fully.

[13] Griffiths, p. 169.

[14] *Ann. Reg.* 1830: 201; 1828: 64; 1827: 147.

reply got to the truth of it: 'Why you too gammon on as well as the parson! They take your life away, and then they think to make amends by telling you of another and a better world; for my part I am very well satisfied with this, if they will let me stay in it.' Robert Savage, hanged in 1826 for housebreaking, treated the ordinary of Newgate 'with contempt'. Robert Hartley at Maidstone in 1823 declared 'his disbelief of a future state' and dismissed the chaplain's attentions. Awaiting execution, his 'time was chiefly employed in making observations which proved the depravity of his heart'. This meant that he 'spoke of his many heinous offences with exultation' and said that, if discharged, he 'should go on the same way again'. 'Don't fret, mates,' he told two men sentenced with him, 'there's nothing the matter.' He bowed to the crowd on the way to the scaffold, smiled when he saw it, and at the last cried loudly: 'Lord Jesus, into thy hands I commit my spirit—pray let this be a lesson to you all—I wish you all a happy new year.'[15]

Then there was the determined care about posture and dress. Gentlemen like Hatfield the so-called Keswick Impostor in 1803 or Fauntleroy and Hunton the forgers in the 1820s sustained awesome composure to the last. Hatfield behaved 'with the utmost serenity and cheerfulness' on the morning of his execution, reading newspapers, writing letters, and announcing that his death was of little consequence. On the scaffold he was 'pale but calm' as with a 'languid and piteous smile' he tied his own handkerchief round his eyes. He had himself buried in a handsome coffin, rejecting priestly attentions. On the eve of his death Hunton composed a prayer for his wife and ascended the scaffold with 'unshaken firmness and deliberation'.[16] And best clothing was worn by those who could afford it. Few men now dressed as Lord Ferrers had in 1760, in his wedding suit of white and silver, or paraded symbols like the white cockade the burglar Waistcott wore in his hat in 1759 'as an emblem of his whole innocence'.[17] Male dress was becoming sober. Hatfield wore a black jacket with waistcoat, fustian pantaloons, and white cotton stockings; Fauntleroy 'a new suit of black, silk stockings of the same colour, and light pumps'.[18] But women continued to affect sartorial gaiety. Elizabeth Fry found that the 'chief thought' of nearly every condemned woman in Newgate 'relates to her appearance

[15] Anon., *Old Bailey experience* (1833), 161; *Ann. Reg.* 1826: 91; *Execution of Robert Hartley* (BL broadside); *Ann. Reg.* 1823: 2–3.

[16] Anon., *The life of John Hatfield, commonly called the Keswick Imposter* (Carlisle, 1846), 23–8; *The Times*, 9 Dec. 1828.

[17] Griffiths, p. 170; J. H. Jesse, *George Selwyn and his contemporaries* (1843), i. 345.

[18] *Life of Hatfield*, p. 26; P. Egan, *Account of the trial of Mr Fauntleroy* (1824).

on the scaffold, the dress in which she shall be hanged'.[19] When Christian Bowman was hanged and burnt outside Newgate in 1789 she was 'drest in a clean striped gown, a white ribbon, and a black ribbon round her cap'. In 1815 Eliza Fenning wore the dress she was to have worn for her wedding, a 'white muslin gown, a handsome worked cap, and laced boots'. On the scaffold she was said to look pretty.[20]

Striking in all this is the victim's effort to maintain dignity to the last and to die *well*, by drawing on a supportive vein of cynicism which ran deep in popular culture. Also striking is the authorities' tolerance of these efforts. Those with money could spend their last days in Newgate in dissipation, as John Rann did, along with the highwayman Paul Lewis in 1763 when he entertained guests in the condemned cell by singing bawdy songs and vilifying the parson.[21] On the scaffold likewise, custom had long entitled the condemned to address the crowd as they pleased, seditiously if they chose. Although every effort was made to force them to public professions of guilt and penitence, they were not checked if they betrayed that role. Jacobites had betrayed the role spectacularly, some making seditious speeches 'plainly calculated', as Dudley Ryder had observed, 'for nothing else but to incense the people against the government. . . . A rogue cannot be hanged but he must become a saint upon the gibbet.'[22] In 1791 d'Archenholz reported the English belief that 'humanity requires that such an alleviation should be permitted to one who is about to be launched out of the world by a violent death'.[23] But it might be better to talk of the loose grip which pre-bureaucratized authorities wielded over public life, or to acknowledge the force of customs and mentalities which cut across classes, as we shall see in the next chapter. Perhaps the tolerance had a loose ideological underpinning also. If so it was an unexpectedly humanistic one, and it takes us to a central meaning of these celebrations.

From Mandeville to Fielding and beyond, observers deplored felons' loose behaviour, not to mention that of the crowd. When Samuel Richardson's country gentleman attended a Tyburn execution in 1741 he

[19] P. Priestley, *Victorian prison lives: English prison biography, 1830–1914* (1985), 238.

[20] Anon., *The life and death of Christian Bowman, alias Murphy, burnt at the stake on 18th March 1789 for high treason* (coining) ('she was a decent looking woman, about 30 years of age. She behaved with great decency, but was much shocked at the dreadful punishment she was to undergo'); J. Watkins and W. Hone, *The important results of an elaborate investigation into the mysterious case of Elizabeth Fenning* (1815), 35–8.

[21] Griffiths, p. 168; Radzinowicz, i. 166–7; W. J. Sheehan, 'Finding solace in eighteenth-century Newgate', in J. S. Cockburn (ed.), *Crime in England, 1550–1800* (1977), 229–45.

[22] W. Matthews (ed.), *The diary of Dudley Ryder, 1715–1716* (1939), 336.

[23] J. d'Archenholz, *A picture of England: Containing a description of the laws, customs and manners of England* (Dublin, 1791), 148.

was struck by the 'unexpected oddness of the scene'—the victims' 'unconcern and carelessness', their 'thoughtless . . . daring and wanton' behaviour as they drank compulsively whenever onlookers handed them wine. All this he was 'wholly unable to account for'.[24] There were some who could account for it, however. Not least of these was Dr Johnson when he lamented the abolition of the Tyburn procession: 'the old method was most satisfactory to all parties; the publick was gratified by a procession; the criminal was supported by it. Why is all this to be swept away?'[25] This comment is usually taken to indicate bluff Augustan heartlessness. But its key word was 'support', and the generosity of Johnson's observation is clarified in Adam Smith's amplification of it:

> A brave man is not rendered contemptible by being brought to the scaffold. The sympathy of the spectators supports him, and saves him from that shame, that consciousness that his misery is felt by himself only, which is of all sentiments the most insupportable. . . . He has no suspicion that his situation is the object of contempt or derision to any body, and he can, with propriety, assume the air, not only of perfect serenity, but of triumph and exultation.[26]

Johnson's and Smith's insights take us at last beyond the jolly surface of these rituals to the bleaker truth which social memory has censored—that most felons went to their deaths in quaking terror. In this light the abolition of the procession and the long shift towards the privatization of execution, commonly understood as a progressive and humane movement, was the reverse of that. To kill felons without ceremony and in private was to deny them the only worldly support they could hope for in their last hours. As evangelicals had their cool say on the best chances of bringing the felon to penitence, the felon was to be left *alone* with his death, that his spirit might break.

While public executions lasted, many knew that outward bravado did not speak for a felt reality, and that the powdered wig, Holland shirt, gloves, and nosegays which some flaunted on their last journey was the only resort they had to 'meliorate the terrible thoughts of the meagre tyrant Death'.[27] The man who did contrive to conduct himself bravely was often actually drunk out of his mind:

> But valor the stronger grows,
> The stronger liquor we're drinking,
> And how can we feel our woes,
> When we've lost the trouble of thinking?

[24] S. Richardson, *Familiar letters on important occasions* (orig. pub. 1741; 1928), 217–20.
[25] James Boswell, *Life of Johnson*, ed. G. E. Hill and L. F. Powell (6 vols., 1934–50), iv. 188–9.
[26] A. Smith, *Theory of moral sentiments*, ed. D. D. Raphael and A. L. Macfie (Oxford, 1976), 60–1.
[27] Broadside cited in Laqueur, 'Crowds, carnivals and the state', 347.

Like the author of the *Beggar's Opera* here, Mandeville had no doubt that 'the terror of death inwardly excruciates' the condemned man: 'his impudence would soon fail him . . . if he took not refuge in strong liquors. These are his only support, and drunkenness the cause of his intrepidity.' When Lord Ferrers asked for wine on the way to the gallows he was refused it because 'great indecencies had been formerly committed by the lower species of criminal getting drunk'.[28]

If drunkenness gave the game away, so did the demands of the imperilled body. For every one who died boldly, forgotten numbers died expelling urine and faeces, sharing kinship in this with the greatest (Marie Antoinette had to squat on the Conciergerie cobbles when she saw her waiting tumbril). Surgeons anatomizing hanged bodies would complain that their effluvia 'rendered the room quite offensive'.[29] Most wretches had already betrayed their terror in the dark silence of the condemned cell. Here on the eve of execution the attempted suicides would take place, like that of the deranged Bousfield who in 1856 tried to burn himself to death in the condemned cell's fire and had to be taken to the scaffold next morning bandaged like a mummy. The gentleman highwayman McLean, adulated by the crowd in 1750, 'is so little a hero', Walpole wrote, that in private 'he cries and begs'.[30] Paul Lewis became abject when he knew he was to be hanged. Holloway and Haggerty spent the night before their deaths in dejection and prayer, and Haggerty was 'deeply affected' on the scaffold, mounting it with 'an unsteady step and pale countenance'.[31] Women wailed frightfully as they were sentenced to death: 'The agitation and cries of the two women were too shocking for description, particularly of her [the coiner Phoebe Harris] who is to be burnt.'[32] In Newgate's condemned cell in 1804 the forger Anne Hurle was 'several times deprived of sensation and supposed to be dead'.[33] Eliza Fenning in 1815 died bravely enough, but she had broken down at the condemned sermon. 'I found [Elizabeth Fricker] much hurried, distressed, and tormented in mind,' Elizabeth Fry reported on the eve of execution in March 1817:

> her hands cold, and covered with something like the perspiration preceding death, and in an universal tremor. . . . There were also six men to be hanged, one of whom has a wife near confinement, also condemned, and seven young children. Since the awful report came down, he has become quite mad, from

[28] J. Gay, *Beggar's Opera* (1728), III. xiii; Mandeville, *Executions at Tyburn*, 34; Walpole, iii. 309.

[29] Mrs Lachlan, *Narrative of the conversion . . . of James Cook, the murderer of Mr Paas* (1832), 249.

[30] Walpole, ii. 218–19.

[31] Anon., *The very remarkable trial of J. Holloway and O. Haggerty* (5th edn., 1807), 23.

[32] *Gentleman's Magazine* (1786), 437.

[33] Knapp and Baldwin, iv. 234.

horror of mind. A strait waistcoat could not keep him within bounds: he had just bitten the turnkey; I saw the man come out with his hand bleeding, as I passed the cell.[34]

Most people would mount the scaffold 'trembling in a very extraordinary manner', their 'whole frame . . . violently convulsed', their minds 'bordering on stupefaction', having to be supported by officials.[35] Elizabeth Godfrey, hanged for murder in 1807, went to the scaffold in a 'state of frenzy'. Greenacre in 1837 refused the attentions of the ordinary of Newgate bravely enough, but at the scaffold he 'was totally unmanned; all his fortitude had left him, he was unable to speak . . . and the officer was obliged to support him or he would have fallen.'[36] One of the five pirates hanged in 1865 had to be hanged seated from a chair, too faint to stand. Because the defiant display was rarer in the provinces (the rural crowd was smaller and more muted than its London counterpart), provincial scenes were if anything more distressing than the metropolitan: there was less support there. Executed at Morpeth in 1821, the highwaymen Wilkinson and Hetherington refused to betray their associates, but they wept despairingly in their cells and prayed manically as they were led to their deaths.[37] Corder, hanged at Bury St Edmunds in 1828 for the Red Barn murder, was 'so weak as to be unable to stand without support'.[38] Even when there was huge public sympathy, as for the four rioters hanged at Bristol in January 1832, the men were 'so overwhelmed with grief that it was with difficulty that they could be supported to the drop'.[39] 'Goodbye, Curley!' someone in the crowd called to George Hearson, one of three Nottingham rioters hanged in 1832, as he stood on the scaffold awaiting the noose around his neck: whereupon Hearson waved his cap 'and began to dance, as if frantic' until called to his senses by one of his fellow victims.[40] An 18-year-old girl had to be dragged to the scaffold at Bristol by half a dozen men in 1849, the clergyman vainly asking her to walk quietly.[41] By 1868 the images were seared in memory: 'half-fainting wretches, sometimes supported between warders and chaplain—sometimes struggling fiercely with the executioner, to plunge, and shriek, and kick, until the lumbering drop falls—sometimes sinking into pitiable syncope, to be hanged, sprawling over a chair'.[42] When men and women were reprieved their release was

[34] *Memoirs of the life of Elizabeth Fry by her daughters* (2 vols., 1847), i. 263.

[35] Broadside in Hindley, i. 176.

[36] *Weekly Chronicle*, 7 May 1837.

[37] *Execution of Wilkinson and Hetherington* (BL broadside).

[38] Broadside in Hindley, i. 189.

[39] BL broadside.

[40] *Ann. Reg.* 1832: 18.

[41] J. Bright: PD3 civ (1849), c. 1076.

[42] *Daily Telegraph*, 27 May 1868.

explosive. Women leaving Newgate at last for transportation set up 'a sort of saturnalia or riot . . . breaking windows, furniture, or whatever came within their reach'.[43]

Scenes of these kinds remind us that when the condemned felon played to the approbation of his fellows and mocked the hangman, he was calling on the last, only, and narrow resource available to him with which to anaesthetize mortal fear. The parodic splendour of the Tyburn procession should not obliterate the truth that at the end of it the law killed people who were powerless to prevent that outcome and whose bodies were dissolving in terror. We must allow the ceremony (and this book) to take its awful dynamic from this fact.

It could hardly be otherwise. In the long weeks or months between arrest, incarceration, trial, and execution, most of these people had no chance of sustaining the equilibrium which would equip them for the brave death of legend. Every circumstance from the conditions of their imprisonment to the blood-curdling exhortations of the clergyman was calculated to break their spirit. Moreover, most of those hanged were far from the swashbucklers of legend and could not behave like heroes if they tried. They were of such obscurity, their crimes so common, their deaths so humdrum, that their executions failed to earn a broadside, a ballad, or a notice in the newspapers. These were the gallows' staple fodder, not the Jack Sheppards, John Ranns, and Paul Lewises. From their mercy petitions and like sources we can sometimes rescue them from their anonymity and recognize the terrifying banality of their deaths. Joseph Harwood may adequately represent them.

2. *The last days of Joseph Harwood*

Harwood was aged 18 when he was hanged outside Newgate in 1824.[44] He was condemned for participating in what then had to be called a highway robbery but would today be regarded as an unremarkable if violent case of hooliganism. At the annual Review on Hounslow Heath, in daylight, a journeyman tailor named Sheehan had been set upon by a dozen youths, kicked, stamped upon, robbed of clothes and money, and left naked and badly hurt. Sheehan later identified and prosecuted Harwood, the other youths having escaped. Harwood was condemned to death at the Old Bailey by the recorder of London. There was no evidence against him other than the prosecutor's statement that he had

[43] F. Cresswell, *A memoir of Elizabeth Fry* (1856), 146.

[44] The following account is based on Harwood's appeal papers (HO17.81) and OBSP (24 Sept. 1824).

seen Harwood in the gang. After his trial Harwood claimed that he had not been among those who did the violence. Sheehan had been drunk and had himself cruelly beaten up a woman whom he had invited to a drink in a booth: he deserved what he got. But Harwood admitted that he had helped himself to half a crown from Sheehan's pockets after the attack. It was this that cost him his life.

Like most of those hanged, Harwood was not a member of the 'criminal classes', if such a thing ever meaningfully existed outside the law's categorizations. As petitions show, his background was typical of the wounded but not entirely defeated London poor. His father had died in an accident while digging a sewer; a small brother had drowned in the Thames. His mother had five children to care for. Hitherto sent to a charity school, Joseph had to make his living 'by going about the country selling cord for clothes lines'—earning enough, his mother swore, not to need to rob. His mother likewise laboured hard 'to stem the overwhelming powers of poverty, misery, severe distress and all their concomitant evils' (thus her petition to the king, signed with a cross). She testified that although her son fell into bad company through drink, he never associated with thieves; every night he slept at home in Southwark. His arrest left her in despair: this 'latter stroke of her cruel fate has overwhelmed her mind with the most pregnant grief and sorrow and paralysed her every effort'. She got twenty neighbours and friends in Southwark to support her petition. Someone not only wrote the text for her but also taught her the terms in which she had to abase herself before her sovereign:

> That to your illustrious throne she most humbly most respectfully and most submissively appeals (trembling alive for her temerity) for an extension of that Heaven strained quality towards her miserable unhappy and now truly penitent son who incarcerated within the gloomy walls of his cheerless cell is visited with every compunction that the human breast is capable of feeling.

As in many such stories, Harwood's main champions turned out to be the City sheriffs. They would have to preside over his hanging if their pleas for him failed. They elicited the committing magistrate's opinion that 'the evidence did not fix on [Harwood] individually the wanton cruelty of trampling on and stifling the person robbed'. They also got a Hounslow surgeon to testify that he had seen Harwood on the day of the robbery selling ropes innocently enough, that Harwood had not been positively identified by the victim of the robbery even when confronted with him the very next morning, and that Sheehan had been persuaded to prosecute Harwood against his will. This testimony might have resulted in a light sentence had it been heard at the trial. But Old Bailey trials were notoriously rushed, defences ill-prepared, and the recorder of London an impatient and a hanging judge. When the sheriffs urgently took the new testimony to the home secretary on the eve of execution, Peel would not bend. 'I told them I would not advise a respite,' he minuted on Harwood's papers, 'seeing nothing in this letter to warrant it.' So the law was left to take its course next morning. In its curt inch-long report *The Times* thought Harwood 'not unworthy of the sympathy and commiseration of the humane'.[45]

Harwood, meanwhile, was crammed with two or three others into a Newgate cell measuring eight feet by six.[46] It was furnished with a rope mat and stable rug for each prisoner, and at night it was lit by a candle. It was ventilated through a hole in the prison's three-feet-thick front wall, crossed by two frames of close iron bars. The cells were 'beastly', the radical free-thinker Haley wrote from inside knowledge in November 1824: the condemned were 'half devoured by vermin of the most loathesome description'.[47] The food was bread, water, and gruel. By day Harwood had access to two communal rooms and the prison's press-yard, a gloomy space flanked by blank stone walls topped with spikes; here a century before felons had been forced to plead by pressing them with weights. A gate of iron bars permitted prisoners to communicate with friends across a short passage terminated by another barred gate, watched by turnkeys.[48]

As they waited for the King in Council to hear the recorder's report and decide who would die, the tatterdemalion and louse-infested score or so in the condemned cells were in terrible suspense. Since only one, two, or three out of their number would be chosen, they were 'engaged in a lottery, of which the blanks are death', as Gibbon Wakefield put it. Some knew they were too young to die or that their crimes never now earned death. But among the rest there was none of that 'hardness' so often presumed by those who sought to dramatize felons' depravity or the comforts of prison life. Wakefield was imprisoned in Newgate for three years shortly after Harwood (he had foolishly abducted a young

[45] *The Times*, 26 Nov. 1824.

[46] Newgate was 'not a house of correction, or penitentiary, but merely a prison of detention,—a sort of metropolitan watchhouse, for the secure custody of persons about to be tried or executed'; so no distinction in treatment was made between those whom the law presumed still innocent and those it had found guilty: E. Gibbon Wakefield, *Facts relating to the punishment of death in the metropolis* (1831), in M. F. Lloyd Prichard (ed.), *The collected works of Edward Gibbon Wakefield* (1968), 228.

[47] *Newgate Monthly Magazine*, 1/3 (Nov. 1824), 141.

[48] This and the next paragraphs draw on Wakefield, *Punishment of death*, 228–35, 247–58.

heiress), so he was able to monitor the condemneds' behaviour closely. Out of nearly every six-weekly batch of prisoners sentenced to death 'one or more' attempted or achieved suicide, he recorded. He watched their demeanours change as the weeks dragged on, as they crept into the condemned pew in the prison chapel, 'stupified' and 'looking around them vacantly as if unconscious of their state':

> I have seen brown hair turned gray, and gray white, by a month of suspense such as most London capital convicts undergo. . . . The smooth face of a man of twenty-five becomes often marked with decided wrinkles on the forehead, and about the eyes and mouth; and, in certainly three cases out of four, one month of the cells of Newgate causes a great diminution of flesh over the whole body. 'How thin he grows!' is the common remark of the other prisoners, when speaking of one who has passed a month in the condemned pew.

On the eve of the King in Council's decision families and friends gathered outside the prison. Often the recorder delayed his report till late, or the Council was late at dinner: then people would wait all night in 'an extremity of mixed hope and fear', a 'protracted agony' mixed with 'a passion of anger'. Wakefield heard 'more than one of those whose lives were spared by the decision of the Council, afterwards express a wish to murder the recorder for having kept them so long in suspense'. It was Newgate's clerical ordinary who at last conveyed the contents of the report to the condemned. As his list sorted out the doomed from the respited, there were 'scenes of passionate joy, wild despair, jealousy, envy, hatred, malice, and brutal rage'. 'My own strongest sentiment on these occasions was one of anger,' Wakefield wrote, 'of that sort of anger, which is commonly produced by watching gross injustice. One sees twenty-five fellow creatures, who yesterday were all under sentence of death—twenty of them are saved and five are utterly condemned. Are the five the most guilty? By no means.'

How Harwood coped with this awful waiting time is conveyed in the ordinary's diary:[49]

> *24 Sept.* Attended Sessions, heard sentence of death passed upon 23 males and 1 female convict . . .
>
> *25 Sept.* Visited the condemned men—one of them a youth of 18 [Harwood] was dreadfully distressed . . .
>
> *27 Sept.* Visited the condemned men—the youth Harwood continues much affected by his situation . . .
>
> *28 Sept.* . . . Harwood continues in much the same state . . .

[49] Newgate Prison Visitors' Books, 1823–5 (London Corporation RO: 209C).

Subsequently Harwood was 'low and unwell'. Next he was 'ill'. By 10 October he was in the infirmary with a violent fever. He stayed there a further fortnight, 'much depressed in mind'. By the 18th, with a month still to wait before the recorder's report, his spirit was broken. He was 'serious and attentive' to his religious exhortations, the ordinary was pleased to report.

The Deity presided indispensably over the condemneds' last days. In Newgate his agent was the Revd Henry Cotton. Cotton's job as ordinary was to insert the fear of God and damnation in his charges. 'The main business of the ordinary is to break the spirits of capital convicts, so that they may make no physical resistance to the hangman,' Wakefield noted. This was mainly achieved in chapel. The pew for the condemned was a black-painted box below the pulpit. A coffin rested meaningfully on the table in its centre. After each Old Bailey sessions this eloquent space was filled with a score or so of the newly condemned. Day by day it emptied as reprieves came in, leaving the handful of appointed victims in deepening isolation. With women prisoners screened by curtains in one gallery, those due for transportation in another, the non-capital convicts beneath them, and the schoolmaster and children in pews by themselves, this artfully arranged assemblage would be required daily to pray 'for those now awaiting the awful execution of the law'. On the Sunday preceding execution the condemned sermon would be preached. The sheriffs in their gold chains attended. So did curious ladies and gentlemen who applied and paid for tickets, seated where they had the best views. When all had gathered, the condemned would be herded up from their cells in varying states of collapse or defiance. The congregation would sing the praise and glory of God, then the ordinary embarked on the service for the dead. Wakefield describes what ensued:

> He talks for about ten minutes of crimes, punishment, bonds, shame, ignominy, sorrow, sufferings, wretchedness, pangs, childless parents, widows and helpless orphans, broken and contrite hearts, and death tomorrow morning for the benefit of society. . . . [The thief] grasps the back of the pew; his legs give way; he utters a faint groan, and sinks to the floor. The hardened burglar moves not, nor does he speak; but his face is of an ashy paleness. . . . The poor sheep-stealer is in a phrensy. . . . [The forger Fenn moves suddenly] like the affected part of a galvanized corpse. Suddenly he utters a short sharp scream and all is still. . . . The women set up a yell, which is mixed with a rustling noise, occasioned by the removal of those whose hysterics have ended in fainting. . . . This exhibition lasts for some minutes, and then the congregation disperses; the condemned returning to the cells; the forger carried off by

turnkeys; the youth sobbing aloud convulsively, as a passionate child; the burglar muttering curses and savage expressions of defiance; whilst the poor sheep-stealer shakes hands with the turnkeys, whistles merrily, and points upward with madness in his look.[50]

Young Harwood's turn came in late November. Since all his fellow-condemned were reprieved in the recorder's report, he was left quite alone. Cotton was pleased that the fortunate ones showed themselves 'very thankful for the Royal mercy'. He then had to break the bad news to the scapegoat chosen to hang: 'Joseph Harwood (a lad) for highway robbery attended with circumstances of cruelty was ordered for execution on Thursday next. . . . He received the dreadful intelligence with resignation, and appeared to be in a proper frame of mind.' As well he might. At the condemned sermon Cotton preached at him alone (the free-thinker Haley recorded), 'and as he was merely a vulgar highwayman, the sermon was in the usual strain—"warm blood coming through the veins . . ." and all that sort of thing.'[51] Harwood's last night is not recorded. At a quarter to eight on the morning of 25 November, two months after sentencing, he was given the sacrament. He thanked the ordinary, the sheriffs, and the governor pathetically. He asked permission to shake hands with Fauntleroy the condemned forger (awaiting the outcome of appeals mounted on his behalf), and was refused it. He was led to the press-yard past Fauntleroy's cell. Then, pinioned and shackled, he was led out of the Debtor's Door to public view. On the scaffold he 'evinced very great contrition', the ordinary noted, 'and I trust was well prepared for his fate'. When the drop fell, according to *The Times*, 'he was but slightly convulsed'.

3. *'Finish me tidily'*

Harwood's ending brings us to another forgotten meaning of the scaffold. The scaffold was the site of physical pain. If Harwood was only slightly convulsed, he was lucky. Throughout our period nobody doubted that hanging was a slow and painful way of killing people. Neither the introduction of the Newgate drop in 1783 nor the lengthy debates a century later about the ratios between body weight and drop ever succeeded in converting the gallows into an efficient instrument of death.

In 1774 Dr Alexander Monro, professor of anatomy at Edinburgh, told James Boswell that 'the man who is hanged suffers a great deal; that he

[50] Wakefield, *Punishment of death*, 256.

[51] *Newgate Monthly Magazine*, 1/5 (1 Jan. 1825), 226.

is not at once stupified by the shock; . . . a man is suffocated by hanging in a rope just as by having his respiration stopped by having a pillow pressed on the face. . . . For some time after a man is thrown over he is sensible and is conscious that he is *hanging*.'[52] Little changed thereafter, apparently. 'In hanging,' it was stated in 1894, 'death takes place either by asphyxia or apoplexy, or by both.' Should the victim's windpipe be ossified, the pressure of the noose would be 'less perfect' and death slow. Even after the introduction of the long drop in the 1880s, designed to dislocate the cervical vertebrae and rupture the spinal cord, consciousness was thought sometimes to be lost only after two minutes 'or thereabouts'; the heart could beat for several minutes longer, while muscular convulsions could set in after a few minutes' pause. If the long drop was misjudged, decapitation might ensue as well. The dead body never failed to betray the nature of its experience, however scientifically dispatched: whence the

> lividity and swelling of the face, especially of the ears and lips, which appear distorted: the eyelids swollen, and of a blueish colour; the eyes red, projecting forwards, and sometimes partially forced out of their cavities . . . a bloody froth or frothy mucus sometimes escaping from the lips and nostrils . . . the fingers are generally much contracted or firmly clenched . . . the urine and faeces are sometimes involuntarily expelled at the moment of death.[53]

The penis might become erect (and ejaculate by some accounts), and the uterus bleed.[54] The Fenian Michael Barrett, the last man publicly executed in England, died in convulsions; newspapers reported his 'protruding tongue and swollen distorted features discernible under their thin white cotton covering, as if they were part of some hideous masquerading'. He had gone to the scaffold with red hair and beard; when he was lifted off it, his hair, oddly, was said to have turned black.[55]

Hanging, a modern royal commission noted, was 'invented rather for its advertisement value than as a more effective way of taking life than other methods of execution'.[56] The guillotine was infinitely more expe-

[52] W. K. Wimsatt and F. A. Pottle (eds.), *Boswell for the defence, 1769–1774* (1960), 304. For Monro's belief that death by hanging was caused not (as often believed) by apoplexy but by asphyxiation, and for evidence that no hanged man was to be found with a dislocated neck, see A. Monro, *Essays and heads of lectures on anatomy . . . by the late Alexander Monro, secundus, MD* (Edinburgh, 1840), pp. xliv–xlv, 97.

[53] A. S. Taylor, *The principles and practice of medical jurisprudence* (4th edn., 1894), 34–6, 40, 44.

[54] T. Magath, *The medicolegal necropsy* (1934), 124–30.

[55] *Daily Telegraph* and *Daily News*, 27 May 1868.

[56] *RC on capital punishment*, 1949–53: 246. The commission was at pains to point out that death by hanging 'can be regarded as speedy and certain' and that there had been no mishaps at 20th-cent. executions. Given secrecy, the truth cannot be known, but for contrary suggestions, see J. Laurence, *A history of capital punishment* (1932), and A. Koestler, *Hanged by the neck* (1961). Dr Harold Hillman, reader in physiology at the University of Surrey, writes: 'With hanging, the condemned is suspended by the

ditious, as the French never ceased saying; but its contamination by association was impossible to surmount. To Romilly the guillotine he saw in the place de Grève in 1802 was 'an object of horror' thanks to 'the ideas which [it] must awaken in everybody's mind'. He found it a 'hideous instrument of death', even though he quoted Montaigne's view that it inflicted 'the least pain upon the malefactor'.[57]

If these are things which modern readers might wish not to know, the gallows' victims could afford no such detachment. Most had seen others choke their ways to death or possessed a folk knowledge never aired in polite drawing-rooms. This was part of the knowledge which highwaymen like Rann had to accommodate in the condemned cell no less than lads like Harwood did. It was from this knowledge that the Cato Street conspirator Ings could tell the hangman: 'Now, old gentleman, finish me tidily: pull the rope tighter; it may slip.' His fellow conspirator Tidd asked the hangman to put the knot under his right ear rather than his left. John Hatfield in 1803 took minutes to position the rope around his own neck, like Tidd putting touching faith in the myth that death was speedier if the knot lay below the ear.[58] In 1828 Robert Hartley, hanged at Penenden Heath, Maidstone, gave the hangman precise instructions:

> 'Do not be long about it—let me feel what drop you have given me.' He then leant forward to try the length of the rope, and said, 'that will do—the knot is too much under my jaw.' The executioner moved it towards his chin, when he said, 'It is now too much under my chin.' When the rope was adjusted, he said, 'Put on the cap now.' When drawn over his face, he said, 'Let me draw it off my mouth.'

These precautions did little good, however. There was a common pattern in what ensued. As an early nineteenth-century broadside representatively declared, the noose of one man's halter 'having slipped to the back part of his neck, it was full ten minutes before he was dead'. So too at the very last Tyburn hanging in 1783: 'the noose of the halter hav-

neck, and his or her weight dislocates the vertebrae supporting the skull, and deprives the brain of oxygen. These actions take some time because the neck muscles are very powerful, and the vertebral and spinal arteries—which supply blood to the brain—are relatively well protected compared with the carotid artery. . . . The victim is likely to suffer severe pain from stretching the skin, strangulation and dislocation of the neck and is unable to cry out because of the rope around the vocal cords. If the person has strong neck muscles, or is very light, death may take some time. . . . It is not known how long a person feels pain': *Guardian*, 15 Dec. 1990.

[57] Romilly, ii. 84.

[58] There had been protracted debate since the 17th cent. about the best location of the knot. Without anatomical justification, it became customary to place it beneath the ear rather than at the back of neck, in the belief that the pressure on the blood vessels caused unconsciousness or death from apoplexy more speedily than strangulation would: Laurence, *Capital punishment*, 44, 48.

ing slipped to the back part of his neck, it was longer than usual before he was dead.'[59] Hatfield's noose slipped twice, and when he did drop it was only eighteen inches, so his death was prolonged and noisy.[60] Ings struggled on the end of his rope for five minutes before he was still.[61] Hartley 'was much convulsed, and struggled for ten minutes after the drop fell'.[62] The knot having slipped behind his neck, Governor Wall in 1802 took fifteen minutes to die in agony, so the hangman pulled on his legs.[63] In 1804 Anne Hurle was driven to a scaffold erected in the widest part of the Old Bailey, the regular drop being out of commission for a while (see Plate 4). When the cart drew away, 'she gave a faint scream, and, for two or three minutes after she was suspended, appeared to be in great agony, moving her hands up and down frequently.'[64] Hanged in Glasgow in 1820 for treason, James Wilson 'died with difficulty, and after he had hung about twenty minutes blood was seen on his cap, opposite the ears'. (A traitor, his head was then chopped off and held up to view.)[65] In 1829 Thomas Birmingham's rope slipped and 'prolonged his sufferings to a considerable extent. He breathed in agony for nearly five minutes. Shouts and screams from the mob caused the executioner to hang on to his legs till life was extinct.'[66] At a 1797 hanging the scaffold platform gave way, precipitating clergyman and executioner to the ground. There was no time to hood the two prisoners, so they 'swung off with their distorted features exposed to the view of the distressed spectators'.[67] An arsonist Charles White struggled incessantly to escape his bonds. He kicked at the executioner and the ordinary, dislodging the cap hiding his face. The crowd yelled in excitement. Partly suspended, he struggled still, reached the platform with his feet, freed his hands and held on to the rope. The executioner had to force him from the platform and pull at his legs. As he choked, the crowd saw his distended tongue and uncovered and distorted features, and they shrieked.[68] 'Give me rope enough, that I may sooner be out of my misery', a Nottinghamshire rioter told the hangman in 1832: vainly, however, for he took five minutes to die.[69] When the three Manchester Fenians hanged in November 1867, *The Times* reported that 'Larkin's suffering was very great and it was nearly two minutes before he ceased beating

[59] Broadside in Hindley, i. 173; *Gentleman's Magazine*, 2 (1783) 974.

[60] *Life of Hatfield*, 26–8.

[61] Laurence, *Capital punishment*, 207.

[62] *Execution of Robert Hartley* (BL broadside).

[63] Laurence, *Capital punishment*, 105; Griffiths, pp. 424–6.

[64] Knapp and Baldwin, iv. 234.

[65] Anon., *The trial of James Wilson for high treason, with an account of his execution at Glasgow, August, 1820* (Glasgow, 1834).

[66] *Ann. Reg.* 1829: 65.

[67] Broadside in Hindley, i. 181.

[68] Ibid. ii. 188; Griffiths, pp. 439–40.

[69] *Ann. Reg.* 1832: 18.

4. *Anne Hurle for forgery and M. Spalding on their way to execution at Newgate,* 1804.

In a fit of economy in 1804, the sheriffs tried to reintroduce hanging in the old-fashioned way. They erected a three-legged scaffold in the Old Bailey, a hundred yards from Newgate, and carted Hurle and Spalding to it. The Newgate drop had to be restored, however, because newspapers insisted that it was more awesome. Aged 22, Hurle had forged a letter of attorney; her trial and death were sympathetically recorded. But of Methuselah Spalding we know only that he was hanged for an 'unnatural crime'. Sodomy or bestiality cases were always deleted from the records, and even the way he died was not reported.

the air in ineffectual struggle'—and then only because Calcraft the hangman pulled at his legs.[70]

Until the very end of public hanging in 1868, and thereafter in prisons, hangmen were unreliable executioners. Sometimes the rope snapped or the cross-beam fell loose. David Evans at Carmarthen in 1829 was one of many who had to be picked up when this happened. 'I claim my liberty,' he shouted. But he was hanged again, the crowd crying 'Shame! Let him go!'[71] When two burglars were to be hanged at Bury St Edmunds the scaffold collapsed. One hanged well enough; the other had to wait twenty minutes while workmen erected a temporary scaffold for him.[72] The worst story is of Robert Johnston's hanging for robbery in Edinburgh in 1818: 'the perpendicular fall was so short that the unhappy man's toes were still touching the surface, so that he remained half-standing, half-suspended, and struggling in the most dreadful manner.' Below the scaffold the executioner chopped furiously at the drop to make it fall further, but the crowd began to shower the magistrates and police with stones, and when a person 'of genteel exterior' jumped on to the scaffold and cut Johnston down, the crowd bore him off still faintly alive, while others tore his coffin to pieces and attacked the executioner. The constables chased those absconding with Johnston and fought with bludgeons for control of the man. They dragged him to the police office at last, and guarded him until a surgeon restored him to consciousness. Then soldiers carried him back to the scaffold. Hanged again, by now half-naked, with his face uncovered, Johnston broke loose from his bonds and tore at the rope around his neck. 'Dreadful cries were now heard from every quarter.' A napkin was thrown over his face for decency's sake as he choked to death. The execution took an hour to complete.[73]

In nearly every year the grim chronicle of bungled executions and lackadaisical hangmen was extended. Sometimes—as in the business of decapitating traitors—clumsiness was excusable: the procedure was unfamiliar. When Despard was decapitated in 1803, a surgeon hacked at his neck with a dissecting knife. He 'missed the particular joint aimed at, and was haggling at it, till one of the executioners took the head

[70] P. Quinlivan and P. Rose (*The Fenians in England, 1865–1872* (1982), 74) repeat the story that a Catholic priest prevented Calcraft from serving the third Fenian, O'Brien, in the same way; so (the story went), on the end of his rope 'for three-quarters of an hour [O'Brien] breathed, and for three-quarters of an hour the good priest knelt, holding the dying man's hands within his own, reciting the prayers for the dying'. This is not noted in newspapers or broadsides and one hopes that it was a propagandist fabrication.

[71] Laurence, *Capital punishment*, 56–7.

[72] *Ann. Reg.* 1822: 142–3.

[73] Rayner and Crook, v. 172–5; *Scotsman*, 2 Jan. 1819; Laurence, *Capital punishment*, 195–8.

between his hands, and twisted it round several times, and even then it was with difficulty separated from the body'.[74] Mostly, however, Calcraft the hangman simply miscalculated the drops required to effect a speedy death. In office since 1829, Calcraft was 'a mild-mannered man of simple tastes, much given to angling in the New River, and a devoted rabbit fancier'.[75] Nice to rabbits, he had a casual way with people. He hanged them like dogs, it was said. Another dismal apotheosis was reached in the Newgate execution of William Bousfield in 1856. The night before his execution Bousfield tried to kill himself in his condemned cell by throwing himself into the fire; next morning he had to be carried to the scaffold swathed in bandages. Calcraft was nervous; he had received a letter threatening his assassination. He pulled the bolt to let the drop fall and disappeared hastily into the prison. Astonishingly, Bousfield drew himself up and lodged his feet on the side of the drop. Pushed off by a turnkey, he again found the side of the drop; and yet again. He was defeated only when Calcraft was summoned back to drag on his legs and 'the strangulation was completed'. In front of an angry crowd, Bousfield gurgled his way to death as church bells rang to celebrate the end of the Crimean War.[76]

In an era when technological innovation was affecting the design of everything from Bath chairs to water-pumps, it is striking that the technique of hanging was so little touched by mechanical improvements and that so few advances were made in scaffold construction. The abstention of science expressed both a distaste for the contaminations of the hangman's trade and a policy of deliberate neglect: hanging was never meant to be a dignified or peaceful quietus.[77] Mary Blandy was hanged in Oxford in 1752 from a beam placed across the branches of two trees, reached by a ladder which was removed when she was in place. This was the medieval way. By then, ostensibly for humane reasons, felons were more often swung off from carts than pushed off ladders. But the fall was still only a few inches, and if anything death was more prolonged than hitherto, since at least you could jump off a ladder and break your neck. Until 1783 at Tyburn, and in some counties for fifty years beyond that, most felons continued to be hanged much as their forebears were a century before:

[74] *The life of Colonel Despard, with an account of the execution . . . for high treason* (1803), 12.

[75] Griffiths, p. 441.

[76] Laurence, *Capital punishment*, 56–7, 214; Griffiths, pp. 440–1.

[77] On the antique and medieval symbolism of hanging as a demeaning punishment (for women: men were anciently supposed to be put to the sword), see C. Naish, *Death comes to the maiden: Sex and execution, 1431–1933* (1991), 81–2.

They put five or six in a Cart . . . and carry them, riding backwards with the Rope about their Necks, to the fatal Tree. The Executioner stops the Cart under one of the Cross Beams of the Gibbet, and fastens to that ill-favour'd Beam one End of the Rope, while the other is round the Wretches Neck: This done, he gives the Horse a Lash with his Whip, away goes the Cart, and there swing my Gentlemen kicking in the Air: The Hangman does not give himself the Trouble to put them out of their Pain; but some of their Friends or Relations do it for them: They pull the dying Person by the Legs, and beat his Breast, to dispatch him as soon as possible.[78]

Such improvements as there were seldom worked. A trapdoor-drop was built at Tyburn for Lord Ferrers in 1760, but for another quarter-century the device seems to have been thought 'too aristocratic a mode for common vagabonds'.[79] This was no bad thing, since it availed Ferrers little. The mechanism stuck when the raised stage was lowered, so that his toes could still touch the platform. Despite the silk cushions and paraphernalia of status with which the scaffold was bedecked, it took Ferrers four minutes to be strangled, as slowly as the meanest.[80]

In 1783 the City sheriffs unveiled their plans for the new drop outside Newgate. At last science did seem to be having its way; their engine was illustrated diagrammatically and as something wondrous in the *Gentleman's Magazine* (Plate 5), while the sheriffs described it with the pride of ingenious mechanics. The description is as notable for its clinical exactitude as Guillotin's description of his machine was a decade later:[81]

The east part of the stage, or that next to the jail, is enclosed by a temporary roof, under which are placed two seats, for the reception of the sheriffs, one on each side of the stairs leading to the scaffold. Round the north, west, and south, sides are erected galleries for the reception of officers, attendants, etc., and, at a distance of five feet from the same, is fixed strong railings, all round the scaffold, to enclose a place for the constables. In the middle of this machinery is placed a moveable platform, in form of a trap-door, 10 feet long by 8 feet wide, over the middle of which is placed the gibbet, standing from the gaol across the Old Bailey. This moveable platform is raised six inches higher than the rest of the scaffold, and on which the convicts stand; it is supported by two

[78] *M. Misson's memoirs and observations in his travels over England, 1698* (1719), 123–5, cited by H. E. Rollins (ed.), *The Pepys ballads* (8 vols., Cambridge, Mass., 1931), vii, pp. x–xi.

[79] Sir Peter Laurie to J. W. Croker, 8 Nov. 1843: *Correspondence and diaries of J. W. Croker*, ed. L. J. Jennings (3 vols., 1884), iii. 15–16.

[80] Walpole, iii. 304.

[81] *Gentleman's Magazine*, 2 (1783), 991; Sir B. Turner and T. Skinner, *An account of some alterations and amendments attempted in the duty and office of sheriff of the county of Middlesex and sheriffs of the City of London* (1784), 27. For Guillotin's description, see D. Arasse, *The guillotine and the terror*, trans. C. Miller (1989), part I.

5. *The new drop at Newgate.*

In 1783 two evangelical London sheriffs transferred executions from Tyburn to the exterior of Newgate, where the demolition of houses had opened a space in the Old Bailey. In the *Gentleman's Magazine* they unveiled this carefully annotated design of their new scaffold with all the pride of ingenious mechanics.

iron bars, six feet long, secured to the under side of the platform; at the lower end of the said bars are fixed two rollers to run upon a sliding bar. This sliding bar runs also upon two rollers, fixed in a groove made in a strong parallel beam near the bottom of the frame. Through this beam and slides are made two holes, for the two irons which support the platform to drop through. Being thus constructed, the platform is raised to its proper height, and the slider, drawn out a little, is firmly supported thereby. At the head of this slider is fixed a lever, whose handle comes above the platform; and the convicts, standing on the platform, being tied to the gibbet, when the signal is given, the executioner, by a very small force applied to the handle of the lever, slides the bar into its place, and the platform falls from under them; and, by the quickness of the motion, it is observed to put the unhappy objects out of pain in much less time than was usual at Tyburn.

But this last was a pious hope. There is no evidence that the drop was introduced to make executions 'much more effective and . . . humane', as Radzinowicz thought.[82] It was to avoid the need to manœuvre

[82] Radzinowicz, i. 203. For many years, expense inhibited the regular use of the drop in the provinces, though one was installed in Northampton in 1818, and Thurtell was executed on an

horses and carts in congested surroundings and to impart greater solemnity to the occasion. Nor was the scaffold moved from Tyburn to Newgate out of compassion for victims; it was moved out of regard for the property developers of Marylebone, as the Epilogue will show.

In 1824 came another small change. The sheriff Peter Laurie felt that many scaffold customs were born of the 'barbarous feeling' of past ages and needlessly 'harassed the minds of those about to suffer'. So he introduced a new scaffold beam, adorned with adjustable chains to avoid delays while the hangman adjusted the length of the rope. He also did away with the prisoner's obligation to carry his own rope wound round the waist to the scaffold.[83] For all this consideration, however, it was still only the victim's height that was measured, not his or her weight. The scaffold had to be a place of shame and casual terror and its processes were not meant to become sophisticated.

So for decades yet men and women dropped inches, to kick their ways to death on the end of short ropes. 'Two or three times when the officials considered the work was done, the powerful frame trembled, and the knees shook convulsively. This was repeated even after the "swinging" had been stopped': thus the death of Barrett, the last man to hang publicly in England, in May 1868.[84] When by the 1880s short ropes were replaced by long ones in an attempt to kill faster, one wretch was decapitated; another caught the rope under his elbow, and had to be drawn up out of the pit and hanged again. As a result, a confidential Home Office committee in 1888 recommended a table of calculated drops and standardized ropes and scaffold structures. But still when one Conway was hanged in 1891 'it was a long drop, and his neck nearly severed'.[85] What horrors ensued over the next three-quarters of a century were largely kept secret. This was exactly what the Act for hiding executions inside prisons was meant to achieve.

From our glasshouse we have no right to throw stones at ancestors. And all these horrors must be kept in proportion. Scaffold cruelty was not wanton. Efforts were made to allow victims some dignity. Within limits their last comforts were cared for. Until 1868 they were allowed their last self-display, not paraded and lampooned as felons still are in China, apparently, before being killed in public stadiums. And any chronicler of

elaborate one in Hertford in 1828. The Newgate platform was not always used. Anne Hurle was hanged for forgery in 1804 on the 'common gallows', being 'put in a cart and drawn to the place of execution in the widest part of the Old Bailey': Knapp and Baldwin, iv. 234.

[83] *Ann. Reg.* 1824: 30.

[84] *Daily News*, 27 May 1868.

[85] H. Potter, *Hanging in judgement: Religion and the death penalty in England* (1993), 102–4.

war or religious conflict encounters larger atrocities than these. They proceed in Europe as I write.

Also it is an unreal shock we deliver upon ourselves by focusing narrowly on these strangulations. Step outside the little patches of misery at Tyburn or Newgate into the bustling daily world and balance returns. Read newspapers, and executions take up a few passing lines in pages given over to healthier doings. We then meet the felt wholeness of social and political life which diminished the concern of those who lived it. Good and reasonable men like Dr Johnson who wrote against the capital statutes could none the less defend a constitution which marshalled cruel resources to protect property and hierarchy and keep barbarity at bay. 'Our great fear is from want of power in government,' Johnson said when Boswell lamented the oppression of the poor: 'Such a storm of vulgar force has broken in. . . . Sir, it has roared, till the Judges in Westminster Hall have been afraid to pronounce sentence in opposition to the popular cry.'[86] Thus horrors can always be condoned in terms of what Johnson called 'the unsettled state of an imperfect constitution'. The state did not rest on these inflictions: at no point in this book do we deal with a nakedly authoritarian power. But in early modern societies a bureaucratically weak sovereign state had to display its power dramatically and visibly. This is how most historians explain harsh punishments in past centuries. So far as it goes, fair enough.

'Law enforcement must be placed within an historically specific social and political context,' we are primly commanded: and who would dissent from such proper advice?[87] The trouble is that highly contextualized explanations can be bent implicitly to validate any politically organized cruelty. What *happened* may then be evaded, judgement upon it made to look naïve. And yet it would not be an anachronistic verdict to say that the English punishments surveyed here delivered an offence against humanity, and that the scaffold and the justice which sent people to it were monstrous devices of power. The reason is that some contemporaries thought this too. They knew that violent punishments survived not because the state was weak—by the early nineteenth century that could not be claimed—but because rulers were inert or indifferent. We must see next that even the despised scaffold crowds knew this, and could pass bitter judgements upon the ceremonies they gathered to watch.

[86] J. Boswell, *The journal of a tour to the Hebrides with Samuel Johnson* (orig. pub. 1785; 1909), 38.
[87] Herrup, 'Law and morality in seventeenth-century England', 104.

CHAPTER 2

DEATH AND THE SCAFFOLD CROWD

1. *'One continued fair'*

WITH THEIR CALENDAR DOTTED WITH FAIRS, PROCESSIONS, assemblies, elections, and the occasional heady excitement of riot, Hanoverian Londoners were used to crowds of themselves. Their rowdy gatherings have always interested historians. One kind of metropolitan gathering has been all but excluded from historical study, however. It is the biggest, most frequent, and most sensational of them all—assembling eight times a year to watch others hang. Until the very end of public execution this insubordinate scaffold crowd touched the deepest anxieties of the polite classes; it takes us to the heart of popular mentalities as well.

Scaffold gatherings could be very substantial. After each Old Bailey sessions the routine strangulation of footpads, burglars, and horse-thieves would attract 'several thousand' to Tyburn or Newgate. But when murderers, traitors, famous thieves, or rich men hanged, the numbers compared with the 100,000 or so who over several days attended Bartholomew Fair.[1] They matched or exceeded those attending famous political meetings—the 20,000 at Spa Fields in 1817, the 100,000 at Copenhagen Fields in 1834, the 50,000–150,000 claimed for Kennington Common on 10 April 1848, the 60,000 at Peterloo in 1819.[2] In 1777 Dr Dodd's hanging was said to have attracted one of the biggest assemblages that London had ever seen. Every housetop, window, and

[1] J. Grant, *Sketches in London* (1838), 290.

[2] J. Stevenson, *Popular disturbances in England, 1700–1870* (1979), 194, 214, 272; D. Goodway, *London Chartism, 1838–1848* (1982), 136–9.

tree was loaded with spectators; 'the whole of London was out on the streets, waiting and expectant'.[3] Crowd estimates are always unreliable, but claims were made for a Tyburn crowd of 30,000 at the Perreau brothers' execution in 1776 and for Newgate crowds of 45,000 to watch the executions of the murderers Holloway and Haggerty in 1807 and of Eliza Fenning in 1815. The hangings of the burkers Bishop and Williams in 1831, of Courvoisier in 1840, and of the Mannings in 1849 were attended by 30,000 or so; and 50,000 went to Muller's in 1864.[4] Near on 100,000 were said to have watched the executions and decapitations of the Cato Street conspirators in 1820, as well as the hangings of the wealthy forgers Fauntleroy in 1824 and Hunton in 1828.

In truth, the space before Newgate's Debtor's Door could not have held such large numbers.[5] But the figures take account of concourses far distant from the scaffold. On great Newgate occasions the crowd would extend in a suffocating mass from Ludgate Hill, along the Old Bailey, north to Cock Lane, Giltspur Street, and Smithfield, and back to the end of Fleet Lane. Surrounding windows and rooftops would be paid for and occupied. More distant streets would be lined with wagons and carts which people paid to stand on to glimpse the distant view.[6] Crowd control was rudimentary and the pressure sometimes fatal. At the hanging of Holloway and Haggerty thirty spectators were crushed to death just before the drop fell. Five years later, placards had to be posted on the avenues leading to Bellingham's execution: 'Beware of entering the Crowd!—Remember Thirty Poor Creatures were pressed to death when Haggerty and Holloway were executed.'[7] Twenty people had to be treated at St Bartholomew's in 1831, when barriers broke at the burkers' execution.[8] We are not dealing with small, infrequent, or insignificant gatherings. These were key moments in the metropolitan year.

Some 70–80 per cent of all English executions in the early nineteenth century took place not in London but in assize towns (see Appendix 2); so execution day was a focal event in the county calendar also, better

[3] Radzinowicz, i. 175 n., 465–6 n.

[4] *Ann. Reg.* 1831 and 1840; Cooper, pp. 7, 11, 14. The 'burkers' were named after Burke and Hare, notorious for supplying surgeons with bodies they had murdered themselves.

[5] The space was opened up by the demolition of houses between the Old Bailey and Little Old Bailey when the prison was rebuilt in 1770: cf. Rocque's (1747) and Horwood's (1799) maps of London.

[6] *The Times*, 9 Dec. 1828 (Hunton's execution); *Ann. Reg.* 1824: 163, and 1828: 173.

[7] Anon., *The trial of J. Bellingham* (1812), 25.

[8] *Ann. Reg.* 1807: 378–84, and 1831: 335. Police control of scaffold crowds outside Newgate continued to be more difficult than it was outside Horsemonger Lane prison, owing to the constriction and hilliness of the site: RC 1866: 111.

attended than most other festivities were. Some assize towns knew few executions: Cambridge, for example, saw barely a dozen between 1780 and 1868. They were so rare in some rural counties that sheriffs pleaded mightily against death sentences lest the county's reputation be blotted, and when assizes had no capital business to do, judges donned white gloves to celebrate the fact. For routine hangings elsewhere only a few hundred might gather if the event did not coincide with market-day (though it usually did). Still, 12,000 watched Daniel Dawson hang at Cambridge in 1812 for maliciously poisoning Newmarket racehorses; 10,000 people arrived in Carmarthen in 1817 to watch the end of a part-time minister who had poisoned his pregnant mistress.[9] Thurtell's hanging at Hertford in 1823 collected 40,000, some travelling from London. The Red Barn murderer Corder's execution at Bury St Edmunds in 1828 was watched by 7,000 onlookers according to one broadside, 20,000 according to another—in any event a very large assembly for a small town. Travelling one day in 1827 Parson Witts could marvel at the

> crowds of people of the lower orders trudging on towards Gloucester, with great eagerness, young and middle aged and many females. We knew not of any Fair or race or merry meeting: at last the truth flashed on my recollection, all these people were hurrying to witness the execution of the wretched brothers Dyer who are to expiate their crimes this morning by their deaths.

When Joseph Misters hanged for murder in Shrewsbury in 1841, 'the town was converted for the day into a fair,' Archdeacon Bickersteth recalled: 'The country people flocked in their holiday dresses . . . walk[ing] very long distances in some cases. . . . They formed parties, and came. . . . The whole town was a scene of drunkenness and debauchery of every kind,' and in the evening the roads leading from Shrewsbury were 'filled with drunken and disorderly persons'. The archdeacon believed execution day was regarded 'rather as a holiday than as a day of humiliation'. 'They come just as they would to a bull-baiting or a cock-fight,' he said. As hangings became rarer after the 1830s, crowds only increased. A dozen women and children died in the crush when a gallows crowd dispersed in Nottingham in 1844. Trains ferried Londoners to Rush's execution in Norwich in 1849, and some 50,000 gathered in Stafford to watch the poisoner Palmer's hanging in 1856.[10]

Commentators like the archdeacon may be forgiven their anxiety. All

[9] Rayner and Crook, v. 145; D. J. V. Jones, *Crime in nineteenth-century Wales* (Cardiff, 1992), 227.

[10] Hindley, ii. 189; D. Verey, *The diary of a Cotswold parson* (1978), 71; SC 1856: 3–11; HO45.681 (Nottingham).

large crowds occupy liminal situations in which rules are suspended and solidarities are enjoyed beyond authorities' sanctions. The central experience may literally be a licentious one as an alternative order is expressed. 'Great mobs are a safeguard to one another, which makes these days jubilees,' Mandeville observed.[11] On execution day the fragility of deference was dangerously exposed. Soldiers, constables, or clergymen supervised them, but time and again the humble cut free. Suffolk yokels ignored their parson's instructions not to attend Corder's execution in 1828. The people of Aylesbury ignored the clergyman's attempt to use a similar occasion 'as a day upon which some moral effects might be produced'. They went to the execution, not to church.[12]

Hostility to scaffold gatherings deepened as time passed. These crowds behaved and spoke in terms which polite observers grew less able to understand. Many crowds acquiesced in what was done by the law and affirmed its rightness. The hanging of murderers was usually approved. But when humbler people hanged for humble crimes, they could act like a Greek chorus, mocking justice's pretensions. They sometimes saw what happened on scaffolds more clearly than the nice people did. We shall see in the book's last chapter that as the Victorian nation reconstructed its representations of itself, it could no longer tolerate so subversive a commentary. But intolerance had a long provenance. The scaffold, Mandeville had written as early as 1725, was 'a summons to all thieves and pickpockets, of both sexes . . . a free mart, where there is an amnesty for outlaws . . . one continued fair, for whores and rogues of the meaner sort. . . . Apprentices and journeymen to the meanest trades are the most honourable part of these floating multitudes. All the rest are worse.' In 1787 Francis Grose, no enemy of plebeian culture, thought the same: 'The indecent behaviour of the common people assembled on these occasions, gives, to one of the most solemn and dreadful scenes imaginable, the appearance of a fair or merry-making, and tends greatly to defeat the end of punishment, which is not so much to torture the unhappy delinquent, as to deter others from committing the like crime.'[13] By the 1840s such comments were standard. Dickens recommended ending public execution not out of pity for the victim but to deny the crowd occasion for its 'odious'

[11] B. Mandeville, *An enquiry into the causes of the frequent executions at Tyburn* (1725), 34.

[12] SC 1856: 3–4, 13.

[13] Mandeville, *Executions at Tyburn*, 20; F. Grose, *A provincial glossary with a collection of local words and popular superstitions* (1787), sig. p. 6.

levity. Attending Courvoisier's hanging in 1840, he saw in the audience no 'emotion suitable to the occasion. . . . No sorrow, no salutary terror, no abhorrence, no seriousness; nothing but ribaldry, debauchery, levity, drunkenness, and flaunting vice in fifty other shapes.' If people cried 'Hats off!' when the condemned appeared, it was 'only as they would at a Play—to see the Stage the better, in the final scene'. In *Nicholas Nickleby* the hostile fantasy is sustained: from the scaffold, 'in the mass of white and upturned faces, the dying wretch, in his all-comprehensive look of agony, has met not one—not one—that bore the impress of pity or compassion.'[14] In 1845, likewise, Coventry Patmore described the scaffold crowd thus:

> Mothers held up their babes to see,
> Who spread their hands, and crow'd for glee; . . .
>
> . . . .
>
> A baby strung its doll to a stick;
> A mother raised the pretty trick;
> Two children caught and hanged a cat;
> Two friends walk'd on, in lively chat;
> And two, who had disputed places,
> Went forth to fight, with murderous faces.[15]

That sober radical G. J. Holyoake described the crowd at a Glasgow execution in 1853 as 'an avalanche of ordure'; at Muller's execution at Newgate he noted chiefly 'mobs of horsey, doggy, thick-necked, bull-headed, turbulent felons in embryo'.[16] Rather than watch an execution, the evangelical Revd Clay said in 1856, 'the generality of persons with properly trained minds would go 100 miles out of their way'.[17] For Lord Cranworth a decade later the trouble with scaffold crowds was that they were too uniformly plebeian:

> There is probably just as much pickpocketing going on at the Derby, and at other large assemblages of pleasure, as in these mobs at executions; but there is this distinction, those great assemblages of persons, though a great deal of criminality may take place, are not mobs in which the very lowest class of persons have it all to themselves; there is always a mixture of all ranks, and that, I think, tends to humanize the assemblage generally. . . . It is very desirable that there should be a mixture of all classes when great masses are assembled, but you can never have that at an execution; you have nothing but the lowest class

14 P. Collins, *Dickens and crime* (1964), 225–7, 249.

15 'The murderer's sacrament: A fact': F. Page (ed.), *The poems of Coventry Patmore* (1941), 56–7.

16 G. J. Holyoake, *Public lessons of the hangman* (1864), 3–4.

17 SC 1856: 24, 28–9, 32.

of persons, and I therefore think that they are assemblages very much to be deprecated.[18]

Over the bridge towards Newgate they came, we are told of the crowd gathering for England's last public execution in 1868: 'a straggling motley procession, whose grotesque figures the etching-needle of Callot could scarcely have caricatured. The "beggars were coming to town"—over the bridge. There was the wretched raggedness, there was the dirt, sloth, scurvy and cretinism of rural vagabondage, trooping over the bridge.'[19]

This contemptuous chorus has had strange historiographical effects, as has been implied. Historians are alert to the potentialities of many other kinds of crowd—cohesive, custom-defensive, celebratory, carnivalesque, parodic, oppositional, insurrectionary, as might be.[20] Scaffold crowds expressed any or all of these characteristics: yet they have been all but ignored.[21] It is as if historians have absorbed the notion that those who attended executions evinced no moral or ideological dispositions of interest. They lacked the dignity of political crowds, for example. Two questions point a way out of this cul-de-sac. Who did attend executions? And how layered and meaningfully diverse were their responses, in fact?

There is no deep record of the Tyburn crowd, but it was certainly more diverse than Mandeville implied. A country gentleman who rode behind the procession and attended the execution was only satisfying 'a curiosity . . . natural to most people', Samuel Richardson wrote in 1741.[22] The fashionable fencing-master Henry Angelo attended many executions from his teen years onwards and was not at all squeamish about joining the throng: 'I followed the crowd as far as Cavendish-square,' he recalled of the traitor La Motte's Tyburn execution in 1782, 'and soon finding the difficulty of proceeding, I hurried round by

[18] RC 1866: 13.

[19] *Daily Telegraph*, 27 May 1868.

[20] Cf. J. Brewer, *Party ideology and popular politics at the accession of George III* (Cambridge, 1976), ch. 9; M. Harrison, *Crowds and history: Mass phenomena in English towns, 1790–1835* (Cambridge, 1988), and the historiography he discusses in chs. 1 and 2; N. Rogers, *Whigs and cities: Popular politics in the age of Walpole and Pitt* (Oxford, 1989), ch. 10; T. Harris, *London crowds in the reign of Charles II* (Cambridge, 1987), ch. 1; M. Baer, *Theatre and disorder in late Georgian London* (Oxford, 1992), ch. 10. The scaffold crowd features in none of these discussions.

[21] Except by P. Linebaugh, 'The Tyburn riot against the surgeons', in D. Hay *et al.* (eds.), *Albion's fatal tree: Crime and society in eighteenth-century England* (1975); T. W. Laqueur, 'Crowds, carnivals and the English state in English executions, 1604–1868', in A. L. Beier *et al.* (eds.), *The first modern society: Essays in honour of Lawrence Stone* (Cambridge, 1989), 305–55. Radzinowicz, i, is a rich quarry of sources and images.

[22] S. Richardson, *Familiar letters on important occasions* (orig. pub. 1741; 1928), 217.

Portman-square, to secure a seat at a window facing the gallows.' Just as Pepys had taken in executions before breakfast as a matter of course, people like Boswell did so a century later. Fox went to watch Dr Dodd die. Indeed,

> parties were formed at the Shakespeare, the Bedford, the Rainbow, or the Rose [taverns], on the eve of a hanging-day, to convene on the morrow, to go and see the sight. Many a one whose fame now makes a figure in biography as an orator, poet, painter, composer, or actor, and some men of rank, philanthropists, and moralists too, are recorded in the list of amateurs of these sad exhibitions.[23]

And although these people sat aloofly in hired seats around the Tyburn scaffold, they did not invariably find the plebeian crowd contemptible. Some saw a distinction between the 'mob' and the 'crowd' which we have lost sight of. After Ferrers's execution, 'the mob tore off the black cloth as relics; but the universal crowd behaved with great decency and admiration': thus Horace Walpole.[24]

By the nineteenth century evidence on crowd composition is richer. After 1837 there were fewer executions to attend (only murderers), and doubtless fewer people, proportionately, did attend—though probably rather more than those 'small knots and specks of people, mere bubbles in the living ocean' of whom Dickens wrote in 1846.[25] Thackeray's closely observed sense of the crowd at Courvoisier's execution in 1840 comes closer to a plausible truth. It included 'all ranks and degrees—mechanics, gentlemen, pickpockets, members of both Houses of Parliament, street-walkers, newspaper-writers. . . . Pickpocket and peer, each is tickled by the sight alike, and has the hidden lust after blood which influences our race.' He saw 'many young dandies . . . with mustachios and cigars, some quiet, fat, family parties of simple honest tradesmen and their wives'.[26] A metropolitan police superintendent similarly told the Lords committee on capital punishment in 1856 that more people of a respectable class of life went to executions 'than your Lordships might imagine'.[27] Another police inspector told the 1866 royal commission that there were always 'thieves, fighting men, costermongers, labourers' at the scaffold; but there were also artisans and 'superior people'. He was one of the few witnesses whose business it was to observe crowds rather than make assumptions about them: 'I have mixed frequently with crowds, at theatres and different places. At

[23] H. Angelo, *Reminiscences* (2 vols., 1828), ii. 366; i. 467.

[24] Walpole, iii. 311.

[25] Collins, *Dickens and crime*, 228.

[26] 'Going to see a man hanged', *Fraser's Magazine*, 22 (Aug. 1840), 150–8.

[27] SC 1856: Q368.

executions I have looked at the upturned faces of the whole crowd, I have been elevated above them, and I have also noticed the faces of a crowd of the lower classes at a theatre.' The people who attended executions were much the kind of people who filled the upper gallery of Drury Lane theatre: 'I cannot find any difference, they always seem to be identical in the two cases.'[28] These were not overwhelmingly the destitute and outcast of London. They included the *menu peuple* and artisanate among whom the young theatre-loving Dickens would once have counted himself.

These impressions are confirmed in reports which stressed that there were 'always at least two or three distinct and several crowds' at the scaffold. First, there was the 'large, loitering and inquisitive crowd' on the eve of execution. These were rough people mainly, lubricating themselves liberally in the drink shops until the scaffold was wheeled from the sessions courtyard to the Debtor's Door early in the morning. But then there were the seven to eight o'clock crowds. These still included the alleged 'scum of the abandoned class', but their main constituent was those many 'workpeople, male and female, who had secured an extra hour for their breakfast in order to take the execution in their way'. By the 1860s 'hats of the chimney-pot or truncated drain-pipe shape, which somehow symbolishes respectability' would be common. The impression of diversity is further confirmed by the kinds of people who appear in the coroners' lists when accidents occurred. Among the thirty killed and fourteen badly injured at the Holloway–Haggerty execution in 1807 were the following: Mr Harrison, 'a respectable gentleman'; Thomas Bradford, aged 16, pupil to Mr Brodrip, pianoforte-maker of Sherrard Street; a wine merchant's son studying at an Islington seminary (his schoolmaster had refused to allow him to attend the execution but he went none the less); young William Platt, apprentice to Mr Robinson, cutler of Drury Lane (his master had given him permission to attend); John Hetherington, broker of Somerstown, along with his son, apprenticed to him; Charlotte Panton of Drury Lane (her husband, identifying her body, said that with female friends she had gone to the execution without his permission). Identifying his son's body, a shoemaker of Golden Lane testified that 'on Monday morning his son asked his leave to go to see the execution; but he refused him, telling him it would not be safe to go'. Other casualties

[28] RC 1866: 13, 112.

included a tavern-keeper, pieman, porter, weaver, brush-maker, carpenter's apprentice, and butcher's son.[29]

We glimpse here the familial negotiations which preceded attendance at executions; the fascination of boys with them (and their vulnerability when the crowd collapsed); the fascination women felt also; the reluctance of fathers and husbands to allow dependants to attend; the impossibility for some of *not* going when the spectacle was so near to work or home. These were workaday people—artisans, tradesmen, and shopmen and their apprentices, clerks, publicans, coach-makers and coachmen, carriers, warehouse-keepers, and broadsheet vendors: the London populace in all its variety, even if it was sprinkled with urchins and prostitutes and pickers of pockets and people in the grubbiest of clothes. The woodcuts of execution broadsides never bear reliable witness, but it helps that the raggedy plebeian witnesses of Hogarth's *Idle 'prentice executed at Tyburn*, a model for many prints, were often altered in early nineteenth-century cuts into a crowd whose foregrounds were occupied by men in top hats and women in bonnets—the artists striving for a rough-hewn realism. Pickpockets would not have been among the crowd if there had not been full pockets to pick. So far was this from being a festival only of the outcast that in 1828 'it was common throughout the whole metropolis, for master-coach-builders, frame-makers, tailors, shoe-makers, and others who had engaged to complete orders within a given time, to bear in mind to observe to their customers "that will be a hanging-day and my men will not be at work".'[30] 'Within my recollection,' Place recorded in the 1820s, 'a hanging day was to all intents and purposes a fair day' enjoyed by apprentices and masters alike.[31]

Nor, *pace* Lord Cranworth, did the common people have it all to themselves. It is true that some among the politer classes absented themselves as a matter of taste and principle—increasingly so as time passed. Johann Christian Bach refused to watch Dr Dodd hang because he could not 'admire the man who shall take a front seat in the Tyburn boxes, to behold a human being die like a dog on a string'.[32] Quakers attending Hunton in his last hours in 1828 would not mount the scaffold or watch what ensued. By the 1860s the abolitionist Charles Gilpin's inability to watch a hanging was doubtless widespread among the

[29] *Daily Telegraph* and *Daily News*, 27 May 1868 (Barrett's execution); Anon., *The very remarkable trial of J. Holloway and O. Haggerty* (5th edn., 1807), 25–8; *Ann. Reg.* 1807: 378–84.

[30] Angelo, *Reminiscences*, i. 472.

[31] Place papers, BL Add. MS 27826, fo. 88.

[32] Angelo, *Reminiscences*, i. 467.

middle classes: 'He had in the grey dawn of the morning witnessed the crowd and the erection of the scaffold, but he turned away with a sick feeling before the criminal was brought out.' 'At present a new sheriff occasionally almost faints,' it was reported in 1866. One sheriff had never seen a man hang before his election to office, and decided never to attend an execution without having sat out a likely victim's trial and hearing him confess the crime: 'If a women had been hanged I should not have gone to the execution.'[33]

But these gentle feelings were not universal. Bucks and swells and their lackies, aristocrats and gentlemen, writers and artists, continued in large numbers to test themselves from the hired windows overlooking Newgate. 'The rich vulgar are but the poor vulgar—without an excuse for their vulgarity,' Mayhew observed.[34] The women with them were often said to be 'fast', but reporters would be disconcerted by the number who were respectable in manner and clothing.[35] Privilege bought closer access. Aristocrats had long sat as observers on the judicial bench at the sentencing of infamous felons, as the duke of Sussex did at the sentencing of Bishop and Williams in 1831. A royal duke attended Courvoisier's sentencing in 1840, in a court 'crowded with ladies dressed up to the eyes, and furnished with lorgnettes, fans, and bouquets'.[36] On execution morning people of rank were admitted to Newgate's press-yard after breakfasting with the governor, there as practised surveyors of form contemplating the demeanours of the victims on their fatal journey to the Debtor's Door (Plate 6). The duke of Cumberland attended the executions of Hepburn and White in 1811, along with Lords Sefton and Yarmouth.[37] Yarmouth was in the courtyard in 1815 as Eliza Fenning passed through to the scaffold too.[38] Holloway the murderer knelt to protest his innocence before these grand people, flattered by their attendance.[39] Even as Victorian decorum began to restrain the sexual and other proclivities of sporting aristocrats, Lords Fitzclarence, Coventry, Paget, and Bruce attended Courvoisier's condemned sermon.[40] The high people were still going to executions in 1866, renting windows for £25. Inspector Kittle thought the numbers and kinds of such people had not changed in forty years.[41] At the last public

33 PD3 cxc (1868), c. 1137; RC 1866: 149, 222.

34 Mayhew, i. 230.

35 Broadside in Hindley, ii. 197.

36 Ibid. 190–1; W. Ballantine, *Some experiences of a barrister's life* (5th edn., 1882), 72.

37 J. Laurence, *A history of capital punishment* (1932), 184.

38 J. Watkins and W. Hone, *The important results of an elaborate investigation into the mysterious case of Eliza Fenning* (1815), 92.

39 *Ann. Reg.* 1807: 378–84; Radzinowicz, i. 168.

40 Laurence, *Capital punishment*, 184.

41 RC 1866: 109–10.

6. Robert and George Cruikshank, *In the press yard at Newgate*, 1821.
'Sympathy' had long licensed fine people's visits to the condemned cells when the inhabitants were not too offensive to contemplate. High-born males were still visiting the condemned in the 1820s, although by then more open voyeurism ruled. Here the Cruikshanks illustrate Newgate's press-yard minutes before two doomed men are led to the scaffold. The visitors have just breakfasted with the governor. On the left, the prison's clerical ordinary, Cotton, consoles one of the condemned, while the other has his irons struck off.

execution, in May 1868, jeering roughs hooted 'a magnificently attired woman, who, accompanied by two gentlemen, swept down the avenue kept open by the police, and occupied a window afterwards right in front of the gallows'. She appeased the roughs by throwing coppers for them to scramble after.[42]

We are not, then, going to be taken in by the images peddled by the crowd's enemies. Just as the crowd's composition was more diverse than was admitted, so crowd reactions to hanging were more discriminating. We shall see in the next chapter how much they depended on the crime and the identity of the criminal. And to be fair not all polite people were taken in by the images. As Thackeray stood amidst the thousands waiting for Courvoisier's hanging in 1840, he felt that despite the 'blackguards' among them, 'the morals of the men are good and hearty'.

[42] *The Times*, 27 May 1868.

They cried shame if a ragamuffin used an indecent expression to a woman. They formed circles round women to protect them from the heaving and pushing. The crowd was 'extraordinarily gentle and good-humoured'. It had opinions too. Contemptuous of the 'wind, hollow humbug, absurd claptraps' of whigs and tories, '*populus* has been growing and growing, till he is every bit as wise as his guardians'. The common man 'has been reading all sorts of books of late years, and gathered together no little information'. 'I never yet . . . have been in an English mob without the same feeling for the persons who composed it, and without wonder at the vigorous, orderly good sense, and intelligence of the people.'[43]

The *Annual Register* was also impressed by the crowd's respect when Courvoisier dropped. It was 'impossible to behold the mob, with their heads all bared, and their eyes all eagerly directed towards the gallows, without the deepest feelings of awe'.[44] In 1865 the *Morning Herald* ritually decried the 'uneducated and vicious mob' at the Newgate hanging of five pirates, but again a *Times* correspondent dissented from 'all which I have heard or read, or which it is the current fashion or folly to express [about such] exhibitions. It was to me the most solemn sight I ever witnessed.'[45] Likewise in parliamentary debates on the abolition of public execution in 1868 some balanced voices protested that 'scandalous or indecorous conduct was the exception and not the rule' in scaffold crowds. There was always that 'awful silence' when the felon mounted the scaffold; then 'the cry of "Hats off!" amid an indescribable mixture of sounds . . . which clearly indicate the horror and terror felt by the multitude'.[46] This might be a mob, but some observers allowed it its empathetic capacities, as we must too, later in the argument.

2. *Ghoulishness?*

Still (at this stage of the argument at least), we may sympathize with most commentators' instinct to keep their distance from scaffold gatherings. What they watched was not immaterial. They assembled for many reasons, but some went to watch others dying in pain and terror and to relish the mishaps which gave the day a passing colour. Many people would do the same today given a chance; but we should still need to work hard to understand such a thing, just as contemporaries had to. Samuel Richardson was not the only one to be puzzled: 'the face of

[43] 'Going to see a man hanged', 152–3.
[44] *Ann. Reg.* 1840: 239.
[45] Cited in Cooper, p. 93.
[46] PD3 cxc (1868), c. 1133.

everyone spoke a kind of mirth, as if the spectacle they had beheld had afforded pleasure instead of pain, which I am wholly unable to account for.'[47] Reading this behaviour like anthropologists contemplating alien rituals, we too may be 'not quite sure where we stand, what position we wish to take up towards what is being said to us, and indeed uncertain about just what has been said'.[48]

When the drop fell, for example, what did the 'shrieks' express, especially those from women? Women were repeatedly said to have attended executions more avidly than men—all the way from Boswell's taking it for granted in 1778 that 'the greatest proportion of spectators is composed of women', down to the *Morning Post*'s lamenting that most of the 12,000 spectators at an execution in 1845 were 'females and boys'.[49] Such reports are encountered so frequently and in so many differing contexts that it cannot have been only that women were simply more noticed than men, though they probably were.

Either way, the ghoulish expectations of both sexes were frustrated at some peril. When in 1801 Basil Montagu achieved a reprieve for two sheep-stealers and travelled to Huntingdon to stop the execution, the high sheriff advised him to leave the town unobtrusively to avoid the disappointed crowd's abuse.[50] Similarly, the crowd's execration of murderers and sexual transgressors was intense. When Esther Hibner was hanged at Newgate in 1829 for starving to death a workhouse child, she was 'assailed with a loud volley of yells from the people, particularly from the females, of which the crowd was in a great measure composed'. Three cheers were given for the hangman Calcraft, whose first hanging it was in his 45-year-long career.[51] The 20,000 attending the execution of Burke (of Burke and Hare fame) in Edinburgh in the same year greeted his 'every convulsive motion' with 'a loud huzza . . . several times repeated, even after the long agonies of humanity were past. . . . Not a single indication of pity was observable among the vast crowd: on the contrary, every countenance wore the lively aspect of a gala-day, while puns and jokes were freely bandied about.'[52] In 1837 when Greenacre was hanged for murdering and dismembering a woman whose parts he then disposed about London, some thousand people

[47] Richardson, *Familiar letters*, 217–20.

[48] C. Geertz, *Local knowledge: Further essays in interpretative anthropology* (New York, 1983), 42.

[49] James Boswell, *The hypochondriack*, ed. M. Bailey (2 vols., Stanford, Calif., 1928), ii. 282; *Morning Post*, cited by M. Wiener, *Reconstructing the criminal: Culture, law, and policy in England, 1830–1914* (Cambridge, 1990), 95.

[50] B. Montagu, *Account of the origin and object of the Society for the Diffusion of Knowledge upon the Punishment of Death* (1812), 1–5.

[51] *Ann. Reg.* 1829: 73.

[52] Ibid. 19–20.

waited outside Newgate all night. By six in the morning numbers had swollen dangerously. In the crush a boy's life was saved by passing him over the heads of the crowd. A fair was set up, piemen sold 'Greenacre tarts', and ballad-singers hawked pictures and the alleged confession of Greenacre and his paramour. The crowd gave three cheers when the scaffold was wheeled out at four in the morning and three cheers more when the executioner appeared at eight to adjust the rope. Greenacre himself was 'greeted by a storm of terrific yells and hisses' and by 'a loud, deep and sullen shout of execration' as he stumbled to the noose. There were more cheers as he dropped:

> As the body hung quivering in mortal agonies, the eyes of the assembled thousands were rivetted upon the swaying corpse with a kind of satisfaction. . . . So loud was the shout which hailed the exit of the poor wretch, that it was distinctly heard at the distance of several streets, and penetrated to the innermost recesses of the prison. . . . The crowd seemed as if they never could satisfy themselves with gazing at the hanging murderer. The women were, if possible, more ruthless than the men.[53]

Then there was the taste for the relics of crime. The hangman's sales of the hanging rope would be to artisans and the like at a shilling or more a piece. Bits of the rope which hanged William Corder at Bury St Edmunds in 1828 sold for a guinea; the barn in which he murdered Maria Marten 'was sold in toothpicks'.[54] After John Holloway's trial in Sussex in 1831 the trees under which he had murdered his wife were stripped of branches for mementoes, their trunks carved with obscene epithets and graffiti of Holloway's body dangling from the gallows. After this hanging 23,000 people filed past his body in the magistrates' room in Lewes town hall, and after dissection the skeleton was preserved in the Sussex county hospital.[55]

These scenes connect easily with other ghastly images of eighteenth- and early nineteenth-century life. Cruelty to animals and people was ubiquitous; in punishment it was state-licensed to boot. It is not easy to ignore such scenes as those enacted at the Charing Cross pillory, with miscreants exposed at the dinner hour when streets were most crowded. Overlooked by a sheriff, the crowd was expected to bring cats, eggs, decayed cabbages, and dung with which to pelt the victim, women being allowed to throw from the front if they tipped the

[53] *Weekly Chronicle*, 7 May 1837.

[54] A. Fonblanque, 'The diseased appetite for horrors', in his *England under seven administrations* (2 vols., 1837), i. 194.

[55] C. Hindley, *The life and times of James Catnach (late of Seven Dials), ballad monger* (orig. pub. 1878; 1970), 235–43; Rayner and Crook, v. 259–62.

constables first. Eyes were lost, blood flowed in mud on these occasions, and some died—usually those pilloried for 'unnatural' crimes. Ann Marrow lost both eyes when pilloried in 1777 for impersonating a man in marriage with three different women (women pelted her more vigorously than men, we are told); a homosexual died there in 1780.[56] How can we understand the excitement with which crowds watched people die like this?

It would be nice to say in the case of executed murderers that some part of this vehemence expressed an ethical dissociation from the crime and approval of the punishment. But we cannot be so sure. The vehemence often had an almost pornographic content to it, especially when it was fuelled by a collective excitement about the murder in question. In the second quarter of the nineteenth century a good quota of murderers (Greenacre, Cook, Good, Holloway) dismembered their female victims, as if there were a fashion that way. Such crimes were lavishly publicized in chap-book and broadside woodcuts which depicted amputated female legs, sawn-off heads, and bared and bloody breasts in lascivious detail. In a broadside on Greenacre's crime, his victim's head dissolves in the flames of the grate in which he burns her dismembered body (Plates 7 and 8). Another broadside illustrates Greenacre sawing off his victim's arm and also shows her head 'preserved in spirits in the workhouse at Paddington'.[57] Some who watched Greenacre die must have consumed *The Paddington tragedy: . . . the lives and trial of James Greenacre and the woman Sale for the murder of Mrs Hannah Brown, his intended wife, which was brought to light by the discovery of her mutilated remains*. This nineteen-page, sixpenny chap-book was published before the execution, and it had a pull-out frontispiece (the British Library copy is luridly watercoloured) showing Mrs Brown's stockinged legs lying on the floor while Greenacre saws energetically at the rest of the body before the parlour fire, blood gushing frightfully. Mrs Brown is naked as Greenacre sets to work on her head in the frontispiece of a *Newgate calendar* published in 1845. Crowds were always most excited at the hangings of those who had committed outrages upon the murdered body, usually a child's or a woman's. But in view of these depictions, who can say that the impulses which witnesses brought to the scaffold were ethical? At some executions the crowd's excitement suggests a perverse celebration as much as a disavowal of the crime, a vicarious participation in the licence which the outrage signified.

[56] Place papers, BL Add. MS 27826, fo. 172; Rayner and Crook, iv. 113.

[57] BL 1881. d. 8.

APPREHENSION OF GOOD
For the BARBAROUS MURDER of JANE JONES.

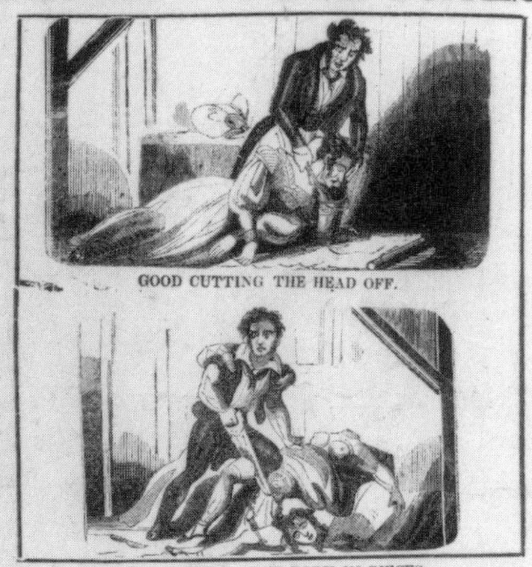

7. *Daniel Good's barbarous murder of Jane Jones*, 1842.
Broadside printers did not shrink from publishing pornographic woodcuts like these when murders were of woman-chopping kinds, as many were in the 1830s and 1840s. Sold before or on execution days, they ensured that many watchers execrated murderers on the scaffold as much because they were excited by the crime as because they disapproved of it ethically.

Other ways of understanding scaffold excitements suggest themselves too. None encourages respect for spectators' delicacy or discrimination. Contemporary commentators were aware of 'primitive' elements in the crowd's witnessings, as we must be also. Some appreciation of crowd psychology had been stirred by Mesmer's discovery of 'animal magnetism' and its implications for what Dugald Stewart termed the 'contagion of sympathetic imitation'. 'In a numerous assembly,' Stewart noted, 'individuals are more subjected, than on other occasions, to their senses and their imagination; and less capable of consulting and obeying the dictates of reason.' The principle of imitation or sympathy might affect large gatherings with such 'electrical rapidity' that 'convulsive and hysterical disorders' were easily propagated, and when the crowd's cause was common and its feelings uniform, 'strong bodily agitations' might be set in motion whose effect will

8. *Greenacre's murder and mutilation of Mrs Hannah Brown*, 1837.

be 'incalculably great'.[58] This was not a bad way of putting it. It comes close to our modern understanding of why individuals in an excited crowd identify with the group against the target of its attentions: namely, that the crowd offers the anonymity which permits individuals to unleash the urgent drives of the needful and angry infant, and the projections which express those drives. 'In the togetherness of mass

[58] D. Stewart, *Elements of the philosophy of the human mind* (in *Works*, ed. W. Hamilton (1854), iii. 151, 155, 157, 169). Stewart's *Principles* were first published in 1792, but the sections on the law of sympathetic imitation, with the crowd as an example, were probably composed as lectures in 1816 and appeared first in the 1827 edition. I am grateful to Alison Winter for these references.

individuals, all individual inhibitions fall away and all the cruel, brutal, destructive instincts which lie dormant in each person as relicts of the primitive era, are awakened for free drive-gratification.'[59]

Contemporary analysis of the scaffold crowd (such as it was) would also often agree that the crowd was hungry for catharsis in or escape from routine-bound, deprived, or resentful lives, and that in such conditions there was release in the very business of collecting together. As J. J. Gurney put it neatly in 1816, the crowd's sympathies were obliterated in 'a feeling of pleasure in the excitement itself'. Bentham thought that when 'the populace runs to an execution . . . this eagerness, which at first seems disgraceful to humanity, is not owing to the pleasure of beholding the unfortunate sufferer in the agonies of death; it is nothing more than the desire to be strongly roused by a tragical exhibition.' Edmund Burke also thought (with Rousseau) that the scaffold offered a free and irresistible street drama without equal in urban life:

> Choose a day on which to represent the most sublime and affecting tragedy we have; appoint the most favourite actors; spare no cost upon the scenes and decorations, unite the greatest efforts of poetry, painting and music; and when you have collected your audience, just at the moment when their minds are erect with expectation, let it be reported that a state criminal of high rank is on the point of being executed in the adjoining square; in a moment the emptiness of the theatre would demonstrate the comparative weakness of the imitative arts, and proclaim the triumph of real sympathy.[60]

Next, there were the multiple gratifications of watching executions which went well beyond a taste for excitement. One was the encounter with repressed wishes which criminals vicariously enacted. The taste for relics surely derived from this, as did some of the minuscule gestures within the crowd. When the scaffold was dismantled after hangings, people would stand on its site for hours, 'rushing to the spot where the execution had taken place', it was noted in 1856.[61] They had stood thus around the dying embers of Phoebe Harris after her hanged body was burnt outside Newgate in 1786; they kicked her ashes about.[62] It was as if people needed to assimilate the aura of the crime and punishment within themselves—through their feet, into their bones, as it were.

[59] Sigmund Freud, 'Group psychology', in *Works*, ed. J. Strachey *et al.* (24 vols., 1953–75), xvii. 79.

[60] Gurney, cited by R. McGowen, 'A powerful sympathy: Terror, the prison, and humanitarian reform in early nineteenth-century Britain', *Journal of British Studies*, 25 (1986), 313; Bentham, cited by B. Montagu, *Thoughts on the punishment of death for forgery* (1830), 68; Burke, 'Essay on sublime and beautiful', in *Works* (1854), i. 81, cited in Radzinowicz, i. 176 n.; Rousseau, in G. A. Kelly, *Mortal politics in eighteenth-century France* (Waterloo, Ontario, 1986), 189.

[61] SC 1856: Q318.

[62] *Universal Daily Register*, 27 June 1786.

Another kind of gratification was evinced by those women who attended scaffolds so avidly. We have seen that their shrieks and excitement mystified polite observers, so offensive were they to conventional views of femininity. But the quasi-erotic fantasies put upon the brave man facing death (impotent in his subjection to death, we note) was well understood by earlier generations:

Beneath the left ear so fit but a cord
(A rope so charming a zone is!),
The youth in his cart hath the air of a lord,
And we cry, 'There dies an Adonis!'[63]

Others attended executions to exorcise a personal guilt projected on to the scapegoat-felon: here was embodied the criminal-god who functioned as the community's redeemer.[64] Execrated women like Esther Hibner might have been the targets of narcissistic rage at the denying mother: a latter-day witch, perhaps.[65] Guilt could be expressed more directly. One Edmund Angelini caused a newspaper sensation in 1824 by volunteering his own life in exchange for that of the forger Fauntleroy: 'his life is useful,' he declared, 'mine a burthen, to the State.'[66]

And so the possibilities multiply. . . . But of all responses to executions, the one of greatest interest to us henceforth relates to the ways in which witnesses evaded the pain threatened in an identification *with* the victim. And it may be at this point that our understanding of the crowd will shift in the direction of generosity, for what is revealed may be more sympathetically intelligible than the manifestations touched on so far. The jocularity, catcalls, and whistles which to polite observers spoke for callousness should not be misconstrued. We suggest now that *strategies of defence* determined many such reactions, and that fear of death was at the centre of them.

3. *Fearing death*

One of the crowd's staunchest enemies was Charles Dickens; and it was he, after watching the unseemly excitement of the crowd at the Mannings' execution in 1849, who wrote that 'the mirth was not hysterical; the shoutings and fightings were not the efforts of a strained excite-

[63] J. Gay, *Beggar's Opera* (1728), I. iv.

[64] Edgar Wind, 'The criminal-god', *Journal of the Warburg Institute*, I (1937–8), 243–5.

[65] Cf. J. Demos, 'Accusers, victims, bystanders: The innerlife dimension', in *Entertaining Satan: Witchcraft and the culture of early new England* (Oxford, 1982).

[66] Griffiths, pp. 457–8.

ment seeking to vent itself in any relief. The whole was unmistakably callous and bad.'[67] But with his own axes to grind, who was Dickens to judge? Our own inclination is to lend greater credit to the MP Charles Newdegate's sharper observation in 1868 that 'much of the disgusting levity exhibited [at executions] was no proof of indifference, but was rather an effort of unregulated minds to efface from their recollection the solemn and impressive scene they had beheld.' Police officers giving evidence to the Lords committee on capital punishment in 1856 were similarly sure that despite outward appearances the vulgar people took home with them the most awful images to ponder.[68]

The truth was that although few scaffold witnesses had access to the repertoires of sensibility sufficient to contain their abrupt responses, they still had to cope with what they watched, and what they watched was fearsome if experienced too closely. Death was on parade, and in plebeian culture death signified more than polite people knew. Hence the strange but revealing fluxes in crowd behaviour at the scaffold. As Southey noted at Governor Wall's execution in 1802, the crowd cheered when Wall appeared on the scaffold, his notoriety such that some feared he might be pulled to pieces. But then a hush descended. Southey was astonished by 'the sudden extinction of that joy, the feeling which at one moment struck so many thousands, stopped their acclamations at once, and awed them into a dead silence when they saw the object of their hatred in the act and agony of death.'[69]

After his three years' incarceration in Newgate, Gibbon Wakefield was well qualified to provide insight on this subject. It was far from the case, he wrote, that the common people were indifferent to death:

> What with funerals all in black, loud grief for the dead and marks of disgust at the sight of a corpse, people are brought up to dread death. After the pains of hell, what so fearful as death? . . . Fearing death so much, we could not live if we expected to die: we should die of the unceasing terror. . . . More or less, all the world tremble at the thought of dying, and, therefore, behave as if they were born to live for ever.

All people must hold the unbearable at bay, he continued; and it was 'just so with the hanging laws':

> When you make a law to punish with death, you fly in the face of nature; and she beats you hollow. You mean to frighten the people, and you frighten them overmuch. You want them to think of the punishment, [but it] is so dreadful

[67] Collins, *Dickens and crime*, 238. [68] PD3 cxc (1868), c. 1138.

[69] R. Southey, *Letters from England* (orig. pub. 1807; 1951), 63.

> that they will not think of it. . . . Fail not to watch the people; the men, women and children, good, bad and indifferent, who have gathered to behold the sacred majesty of the law. You will see such flashing of eyes and grinding of teeth: you will hear sighs and groans, and words of rage and hatred, with fierce curses on yourself [the judge] and me [the hangman]: and then laughter, such as it is, of an unnatural kind, that will make you start; and jests on the dead, to turn you sick.[70]

Wakefield's insight is critical for our argument henceforth. It warns us that the crowd's passion was not always or chiefly celebratory or ghoulish. On the contrary, its passion helped to cancel out terror while camouflaging its submission to the authority that did these things. Defence took many forms. Laughter was one; anger another. Some identified with the sheriffs, the executioner: for to identify with the aggressor against his victim is a standard defence against fellow-feeling when fellow-feeling must be fruitless.[71] 'Hysteria' likewise blocks the acceptance of sympathetic pain. So closer analysis takes us away from the easy distaste for the scaffold crowd with which we began. Although in all these conditions the suffering individuality of the victim might be neither felt nor seen, the reasons had more to do with the crowd's fears and weaknesses than with its innate callousness, ghoulishness, or bullying strength.

Let us follow Wakefield's hint and first dissociate ourselves from a common belief which is at odds with this argument. The belief is that poorer people in past times were more accustomed to cruelty and death than we are, and more callous as a result. In 1851 the Ecclesiological Society's report on burial practices expressed this belief representatively. Amid much lascivious reference to oozing liquids and maggots, it noted how the unfeeling poor often kept family corpses in living-rooms for days before burial. The wealthy classes treated the body with respect and awe, it declared, but 'with the lower classes it is often treated with as little ceremony as the carcass in a butcher's shop. The body is never absent from the sight; eating, drinking, sleeping, it is still by their side.' Then it explained this: 'They have gazed upon death so perpetually, they have grown so intimate with its horrors, that they no longer dread it, even when it attacks themselves.'[72]

This representation of popular callousness is encountered many times thereafter, not least in explaining eighteenth-century punishment:

[70] E. G. Wakefield, *The hangman and the judge: Or a letter from Jack Ketch to Mr Justice Alderson* (1833), 3–5.

[71] See the discussion Ch. 6, below.

[72] Ecclesiological Society, *Funerals and funeral arrangements* (1851), 4.

While . . . the crudity of the criminal law and the brutality of its punishments must shock modern sentiment, it must be remembered that it was largely with brutes that the law had to deal. . . . The little importance attached to human life and suffering must be regarded . . . not as a sign of the brutality of the law but of the general spirit and civilization of the period.[73]

Historians have appropriated the notion too. Trevelyan thought the eighteenth-century English 'a race that had not yet learned to dislike the sight of pain inflicted'. J. H. Plumb likewise believed that 'strong passions could only flourish amidst a callous people, and it is not surprising that the popular sights of London were the lunatics in Bedlam, the whipping of half-naked women at Bridewell, the stoning to death of pilloried men and women, or the hangings at Tyburn.'[74] Ariès similarly discerns an enduringly 'natural' and undramatized 'acceptance' of death on the part of the European commonalty and peasantry across the centuries, allegedly at odds with the more interesting fluxes in attitude supposed to have characterized élite perceptions of the Great Reaper—again as if the common people were too numb and dumb to be fully sentient beings.[75]

These formulas rest mainly on the view that habituation to death put past people in a very different relation to it than our own, thanks mainly to the advanced medicalization of death today and our collective detachment from it, and to high mortality rates in times past, particularly among common people, which supposedly made death a familiar. These arguments are far from watertight, however. It is true that although life expectancy at birth rose from the 1730s onwards, it did not do so as much among the commonalty as among élites. But it was infant mortality that did most damage, a fact which shifted the adult experience of death decisively. Having survived the perils of infancy, a man aged 30 in 1810–11 might still expect to live thirty-two and a half years longer. This quite cheering expectation of adult life would not have induced an attitude to adult death notably more phlegmatic than our own.[76]

[73] P. C. Yorke, *The life and correspondence of Philip Yorke, earl of Hardwicke, lord chancellor of Great Britain* (3 vols., Cambridge, 1913), i. 132–3.

[74] J. H. Plumb, *The first four Georges* (1967), 16; G. M. Trevelyan, *English social history* (1944), 281.

[75] P. Ariès, *The hour of our death* (1981). For criticisms of Ariès's group distinctions and chronological patternings, see L. Stone, 'Death', in his *The past and the present* (1981), 242–59; J. Whaley, Introduction, and J. McManners, 'Death and the French historians', both in J. Whaley (ed.), *Mirrors of mortality: Studies in the social history of death* (1981), 8–9, 117, 122.

[76] E. A. Wrigley and R. Schofield, *The population history of England, 1541–1871: A reconstruction* (1981), 250, 528–9; M. Anderson, 'The emergence of the modern life cycle in Britain', *Social History*, 10/1 (1985), 69–87. How far popular or élite attitudes to death were affected, as Chaunu claimed for France, by the

Today we think of ourselves as peculiarly adept at denying death and hiding it from view. Lamenting our attenuated mourning rituals, we wax nostalgic about death in past times when the divinely ordained process was admitted and was unsqueamishly accommodated in public ritual. The old socialized death with its extended rituals of preparation, burying, and mourning is identified as the healthy and natural one. Death was then honestly faced, we are told, and through this confrontation the adjustments to loss which are essential to health were mediated.[77]

Yet *was* death 'faced' in past times, and was adjustment so easily achieved? What if we face death with an intensity rare in past generations simply because we lack social and religious defences against its terror? Viewed thus, denial might as well be said to have characterized the socialized death of past times as it allegedly characterizes our own forms. The socialized death had ostensible candour to it, and it is true that past generations also refused to draw conceptual, medical, or spatial boundaries between life and death as we draw them today. But this is not to discount the likelihood that socialization and rituals still blocked the pains of death and anaesthetized fear.[78]

Eighteenth-century attitudes to death were not as relaxed as we tend to believe. Go back to the sixteenth century among (say) those who signed wills in Wrightson and Levine's Whickham. There we do apparently find 'communal cultures of death'. The death-bed was a public place, concerned with neighbourhood settlement as much as with the afterlife. The dying were badgered for bequests on their death-beds, and they responded to these requests. They specified how they were to be brought forth and buried, and they got support from religious prescrip-

decline in 18th-cent. mortality rates (allegedly shifting sensibilities from a preoccupation with death to a preoccupation with life and making death more of a 'crisis'), is an open question: Whaley (ed.), *Mirrors of mortality,* 10–12. For agreement that there has never been a time when death was 'natural' or serenely accepted, see M. Vovelle, *Ideologies and mentalities* (1990), 67; Linebaugh, 'Tyburn riot', 116.

[77] G. Gorer, *Death, grief, and mourning in contemporary Britain* (1965), 110; I. Illich, *Medical nemesis: The expropriation of health* (1975), 149 ('the medicalisation of society has brought the epoch of natural death to an end'). See also Ariès, *The hour of our death,* ch. 12; and N. Llewellyn's assertion that in early modern times 'denial was unfamiliar if not impossible', and that past practices were 'not morbid, but therapeutic'. Much of Llewellyn's evidence suggests denial, however; see the idealized image of a dead wife which he cites, which, 'shown as if asleep, undoubtedly helped protect her husband from the shock of her death': *The art of death: Visual culture in the English death ritual,* c. *1500–1800* (1991), 16, 134, 32.

[78] Ariès, *The hour of our death,* 604. Cf. P. Metcalf and R. Huntington, *Celebrations of death: The anthropology of mortuary ritual* (Cambridge, 1992), 62: 'No doubt [elaborate funeral] rites frequently aid adjustment. But we have no reason to believe that they do not obstruct it with equal frequency.' See also J. L. Hockey, *Experiences of death: An anthropological account* (Edinburgh, 1990), 29: the lack of belief in modern funeral rituals may cause survivors' 'grief to deepen into a more pervasive sense of loss [than in earlier times] as "meaning" itself becomes questionable'.

tions on the 'good death'. But all this was in decline by the later seventeenth century, the authors show. Families were turning in on themselves more, while the concern of the dying with the manner and place of burial was suppressed too, as if people wished to think less about their own corpses, and wanted increasingly to evade reality.[79]

Much of the old phlegmatism endured, no doubt. We shall find it in the attitudes of the Hanoverian hanged and their families, for example. Even so, the ways the poor handled their corpses in rituals of washing, watching, and waking to which the Ecclesiological Society was blind, the consoling resurrection and judgement images of hymn-singing,[80] or the festive extravagance of funerals, can all be construed as defences against the unbearable, as sublimations of fear. Callousness is not much in evidence, not to be taken for granted. Doubtless some poor people did sleep with decaying corpses in the house; but the Ecclesiological Society did not pause to think that the very poor had to retain corpses in their living-rooms because they could not afford the reverential displays which living-space, leisure, and money made possible.[81]

Finally, there is the anthropologists' view that most death rituals, in most societies, do not so much confront death as celebrate life, however obliquely.[82] Plebeian funerals had always been disproportionately elaborate in relation to incomes, and might have become more so in the early nineteenth century.[83] But at the jolly funeral of the professional London beggar Jack Stuart in 1815 what was really happening? Was death being faced candidly and 'naturally'? Or was it rather that death was being denied in a compensatory and needful celebration of its opposite:

> his wife and faithful dog, Tippo, as chief mourners, [were] accompanied by three blind beggars in black cloaks. . . . Two blind fiddlers . . . preceded the coffin, playing the 104th Psalm. The whimsical procession moved on, amidst crowds of spectators, from Jack's house in Charlton Gardens, Somers Town, to the churchyard of St Pancras, Middlesex. The mourners afterwards returned to the place from whence the funeral had proceeded, where they remained the whole of the night, dancing, drinking, swearing, and fighting, and occasionally

[79] K. Wrightson and D. Levine, 'Death in Whickham', in J. Walter and R. Schofield (eds.), *Famine, disease and the social order in early modern society* (Cambridge, 1989), 161–5; D. Levine and K. Wrightson, *The making of an industrial society: Whickham, 1560–1765* (Oxford, 1991), 288–93, 341–3.

[80] V. Gammon, 'Singing and popular funeral practices in the eighteenth and early nineteenth centuries', *Folk Music Journal*, 5/4 (1988), 412–47.

[81] After the 1832 Anatomy Act, bodies were often retained to avert the threat of anatomization which now hung over any pauper death: R. Richardson, 'Why was death so big in Victorian England?', in R. Houlbrooke (ed.), *Death, ritual and bereavement* (1989), 117; see also her *Death, dissection, and the destitute* (1987).

[82] Metcalf and Huntington, *Celebrations of death*, 24.

[83] Richardson, 'Why was death so big in Victorian England?', 115–17.

chaunting Tabernacle hymns; for it must be understood, that most of the beggars are staunch Methodists.[84]

4. *Scaffold rituals*

We might debate lengthily whether the meticulous steps danced around the business of execution spoke for a defensive denial of the pain of witnessing others' deaths—or a mystification of death the better to accommodate it—or simply for as candid an encounter with it as surface appearances often suggest. The truth varied with individuals and groups, no doubt. What cannot be disputed is that the symbolic organization of scaffold death expressed impulses which had nothing to do with callousness, and that witnesses participated in that organization in order to draw lessons and support from it. The scaffold's centrality in the collective mind was achieved by the large drama enacted upon it—the struggle of the hanging man or woman. But it was also achieved by the rituals and beliefs which attached to it, and by the fact that to generations highly literate in emblematic meanings the forms and ceremonies of execution mattered crucially. Some of these beliefs and forms addressed the emotions as much as the mind, taming its horrors and blurring the boundaries between death and life. Other forms had agreed and quasi-rational purposes. But, either way, our understanding of the scaffold crowd draws yet further away from our starting-point as its complex languages become manifest.

Take first the processes by which the hanged or about-to-be-hanged were converted into mediators between death and life, and harnessed to good.[85] There were ancient forms of this, apparently lost by the eighteenth century. There are some thin reports of condemned felons winning reprieves if bondless women or harlots offered to marry them, for example. In 1686 'eighteen damsels all in white' allegedly went to James II to plead for the reprieve of one Edward Skelton. Skelton was required to marry one of the damsels, which he wisely did in Newgate's press-yard. This particular episode was remembered in ballad, so it must have been rare if it happened at all; but an intelligible healing and reintegrative rationale informed the story, as marginal people were cemented back into the social fabric.[86] What survived longer in the compensatory

[84] J. T. Smith, *Vagabondiana, or anecdotes of mendicant wanderers through the streets of London, with portraits of the most remarkable drawn from life* (orig. pub. 1817; 1883), 15–16.

[85] C. Gittings, *Death, burial and the individual in early modern England* (1984), 68–9.

[86] H. E. Rollins (ed.), *The Pepys ballads* (Cambridge, Mass., 1931), iii. 249–50, and the ballad 'The mirror of mercy' (1786) that follows.

conversion of evil into good was the notion of late eighteenth- and early nineteenth-century dream-books that to dream of the gallows was an omen of riches and honours to come.[87] What also survived and is much better attested was the belief that the hanged felon could cure tumours and warts. At the Newgate gallows in 1786 a dozen people stroked themselves with a hanged man's hands to cure themselves of wens. Meister in 1799 saw a woman bare her breasts for the dead man's hand to be placed upon them.[88] Three women did this at Newgate in 1814 (Plate 9). After Holloway's execution in Sussex in 1831 a countryman sat trembling for five minutes with the dead man's hands on his forehead. (This might have been among the last such episodes tolerated, for when two women asked for similar treatment the under-sheriff ordered them away).[89] Other superstitions applied too, as they always must to a place so saturated in symbols of sin, redemption, and damnation. The gallows was a place where miracles might happen. Macaulay recounted the story of how the beasts which drew a Monmouth rebel to execution in 1685 'became restive and went back' when the angel of the Lord stood in the way, invisible to the people but not to the animals.[90] In 1763 a man was about to be hanged on Kennington Common when a storm terrified 'the ignorant populace', and the sheriff had to apply for military force to prevent the condemned's rescue. The meaning of this might be revealed in a broadside half a century later. Before Henry Wilhams was hanged he prayed that 'if he was guilty, it might be one of the finest days that could come from heaven; but if he were innocent, that the darkness might overspread the town during the time he was suspended. His supplication reached the throne of grace, for immediately on his being turned off, a dark cloud covered the country for many miles, attended with thunder, lightening, and rain,' whereupon the real murderer was 'stung with guilt and horror' and delivered himself to justice. This might have been fiction, but that its author understood his readers' fantasies is clear.[91]

The real puzzles in scaffold ritual have to do with the (to us) breathtakingly insensitive humiliations inflicted upon the condemned body.

[87] R. Collison, *The story of street literature* (1973), 135–6.

[88] *Universal Daily Register*, 22 June 1786; J. H. Meister, *Letters written during a residence in England* (1799), 62, cited by Gittings, *Death*, 68. In popular belief, all corpses were allowed healing powers: Ariès, *The hour of our death*, 357.

[89] Thomas Hardy recalled the custom in *The withered arm* (1888).

[90] *History of England* (1866 edn.), i. 306.

[91] Radzinowicz, i. 184; Anon., *The trial and execution of Henry Wilhams* (n.d.); Anon., *Execution of Mary White* (Exeter, n. d.) (John Johnston Collection, iv, Bodleian Library, Oxford): I am grateful to Judith Kelman for this reference.

9. *The hanged man's touch*, 1828.

The touch of a hanged man's hand had long been believed to cure cancers and warts. This illustration from 1828 purports to depict an incident in 1814 when some women mounted the Newgate scaffold for the purpose, one of them 'so much affected by the ceremony that she was obliged to be supported'. The picture's comic line (owing much to Rowlandson), and its indifference to topographical accuracy, suggest how few took the old belief seriously by the 1820s.

Although ignominy and dishonouring were officially intended, it is not clear that ignominy was what audiences saw, or that the processes were thought to be cruel. Chiefly expressed was an attitude to death remote from ours. Living *mementos mori*, the about-to-die possessed a totemic stature which exempted them from ordinary treatment. Perhaps to aggravate their deaths was 'somehow to control it and dispose of its mystery' through 'a massive propriation of the divine and the diabolical'.[92] More likely, death was a rite of passage towards which all were moving, not the discrete and demarcated moment which medical sci-

[92] Kelly, *Mortal politics*, 184.

ence makes it now. It was the end of a continuum which began at birth and included the living present:

> As. Soone. As. Wee. To. Bee. Begvnne:
> We. Did. Beginne. To. Bee. Vndone.

Hence the condemned were yoked into ghastly pageants, as if to enact the statement carved on tombstones: 'Behold, I was as you are, and you will be as I am.'[93]

Examples are many. On the way to Tyburn they sat on their coffins wearing their shrouds. Matthew Barker was 'dressed in his shroud' in the cells of Norwich Castle before his last-minute reprieve in 1784.[94] They were exposed also to visitors in the condemned cell. Three thousand people visited the doomed highwayman McLean in his Newgate dungeon in 1750, and Dr Dodd was displayed for two hours on the eve of his execution in 1777.[95] Even in the early nineteenth century people could apply to attend condemned sermons to gawp at the about-to-be-hanged in their segregated pew, a coffin placed meaningly in front of them. It was only in 1845 that the evangelical prison inspector Whitworth Russell managed to stop this practice, old understandings lost at last. Russell thought that the practice merely 'ministers to the indulgence of a morbid taste for spectacles and sensations which have a prejudicial tendency'.[96]

There were, however, cognitive elements in these inflictions too, in that execution rituals bore political meanings. They constructed a drama which defined what law, justice, and sovereignty 'really' were. The state affirmed its potency and ethical basis in symbolic languages and emblematic displays addressed to visually literate audiences. 'Ignominy' had always been the burden of this language. Ignominy carried its charge into the nineteenth century too. For the treasonable offence of murdering her husband in 1807, Martha Alden was drawn on a hurdle to her hanging, her proximity to the earth denoting her status.[97] Eliza Fenning had to walk to the scaffold in 1815 with the halter wound round her waist and its noose in her hand. On his way to hanging and decapitation for treason in Glasgow in 1820, James Wilson, appallingly enough, had to sit on his hurdle alongside the hangman, which latter functionary held the axe which Wilson knew would shortly

[93] Llewellyn, *Art of death*, 10–11.

[94] HO47.1 (Matthew Barker's case), Sept. 1784.

[95] Radzinowicz, i. 168.

[96] Russell to HO, 23 May 1845, proposing rules 'to prevent prisoners under sentence of death being made a shew of': HO45.907.

[97] Borrow, vi. 41.

take his head off.[98] Until 1832 murderers were anatomized to mark their infamy. Nor could they leave the condemned cell until the morning of their deaths, or attend the condemned sermon; the bell of St Sepulchre's would not toll for them.[99]

The corpses of those who escaped the gallows by suicide also got special treatment. Dishonouring was intended here, though again something less consciously desired. It was as if, with the intended execution evaded, compensation must be found in the demonstration that at least the suicide's interment was under control. The burial of ordinary suicides at crossroads with stakes through the heart was all but obsolete by the turn of the century (the last known case was in 1823), so the continued practice of burying criminal suicides thus invites some such explanation. When the Ratcliffe Highway murderer John Williams killed himself in his cell in Coldbath Fields prison in 1811, the magistrate consulted the home secretary on how the 'usual practice' might be elaborated 'in this extraordinary instance of self-murder'. Williams's body, the face 'ghastly in the extreme', was duly arranged on a cart, dressed in white frilled shirt, blue trousers, and blue-and-white-striped waistcoat. A prison iron was attached to the leg, and the body was neatly surrounded by the blood-stained implements used in the murders (Plate 10). Then, preceded by officials, it was paraded through the streets, stopping at each murder scene for fifteen minutes. Near-silent crowds followed the body, but those lining the route greeted it with 'smothered groans' as it passed. Jammed at last into a hole dug at St George's turnpike, the corpse had a stake driven through it, and was then covered with lime and paving-stones to 'loud acclamations' (or as another source put it, 'hideous . . . shouts and execrations') from those watching. Slivers from the stake were sold to spectators as souvenirs.[100]

But if the state and the authorities orchestrated these rituals, they were ensnared by them too. Hearing of Williams's bizarre burial, Romilly, Whitbread, and Montagu protested at its barbarity: 'a spectacle as useless to the public as it was offensive in feeling', Whitbread said.[101] Yet the magistrate who ordered the suicide's burial in this form presumably saw no oddity in thus marking the occasion; he shared the audi-

[98] Anon., *The trial of James Wilson for high treason, with an account of his execution at Glasgow, August, 1820* (Glasgow, 1834), 47–8.

[99] Knapp and Baldwin, iv. 368–70.

[100] *Morning Chronicle*, 1 Jan. 1812; Borrow, vi. 90–102; M. MacDonald and T. R. Murphy, *Sleepless souls: Suicide in early modern England* (Oxford, 1990), 138–9.

[101] B. Montagu, *The debate in the House of Commons, April 5, 1813, upon Sir Samuel Romilly's Bill on the punishment of high treason* (1813), 8–9.

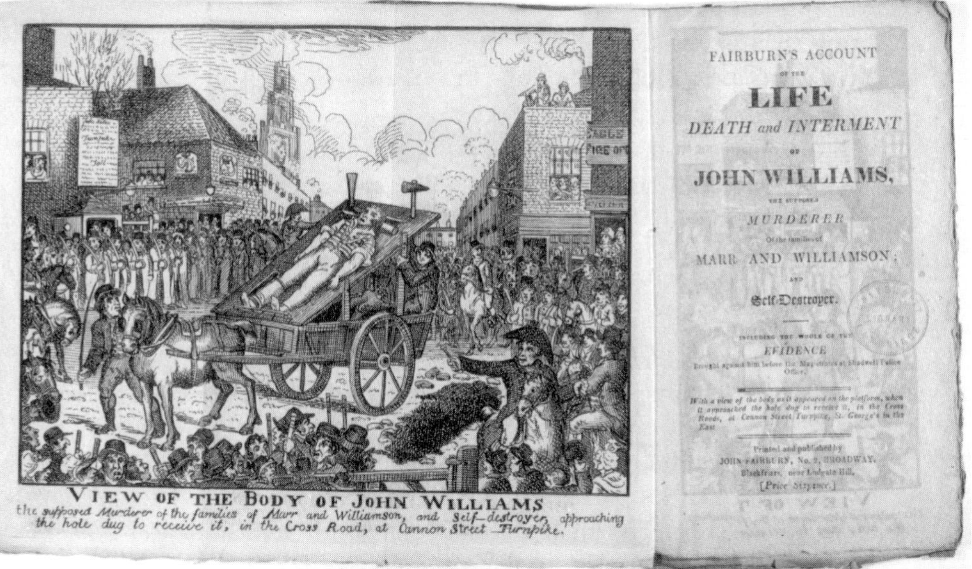

10. *View of the body of John Williams, the supposed Ratcliffe Highway murderer*, 1811. When the chief suspect in the Ratcliffe Highway murders committed suicide in prison, his body was ceremonially carted to ignominious burial at a Wapping crossroads, where a stake was driven through its heart. Fairburn's sixpenny pamphlet was one of three he published on the case, this pull-out frontispiece hastily engraved to catch an excited market.

ence's understandings. Similarly when Colonel Despard and his six co-conspirators were drawn on hurdles around the yard of Horsemonger Lane prison before their execution and decapitation in 1803, the enactment served no public purpose, since nobody outside could see it. But the hurdles still had to be preceded by a *solemn* procession of sheriff, clergyman, and constables; the prison keeper had *solemnly* to bear a white wand, and the executioner had to bear a drawn sword. The procession was speaking to the condemned perhaps; but it was also speaking to itself.[102]

The condemned and their survivors and friends seem to have understood and accepted this symbolic language, not therefore experiencing it as 'cruel'. Some of their collaborative behaviour would be very strange if it were not so. There survives, for example, an account of the expenses that Earl Ferrers's family incurred in 1760 to ensure that he

[102] *Memoirs of the life of Colonel Despard, with his trial at large, and his twelve associates for high treason* (1803).

was executed and buried with the dignity befitting his status. Twenty-four items are costed, amounting to £81. 2*s*. 3*d*.[103] They included:

> To 255¾ yds of Black baize to hang the scaffold @ 20*d*., £21. 6*s*. 3*d*.; To ostrich feathers and velvet for the hearse and six horses £3. 10. 0.; To a strong elm coffin lined with very fine crape quilted with 2 squabs the one under the other over my Lords body, £3. 0. 0.; To a strong leaden coffin, £5. 10. 0.; To a plate with Inscription, £0. 18. 0.; To 6 men in black to carry the coffin to the [Surgeon's dissecting] hall, £0. 15. 0.; To a hearse and pair and 2 coaches and pairs from the hall to Pancras church, £1. 16. 0.

This accounting spoke for rational estate management and the expenditures which family pride demanded. But it also surely reflected a needful illusion that death itself might be controlled. And since it was all planned before the event, it is unlikely that Ferrers was excluded from the discussions. Nor was foresightful practice of this kind class-specific. Poor felons organized their own funerals with similar matter-of-factness:

> But when I am dead, and in my grave,
> A decent funeral let me have,
> Six highwaymen to carry me,
> Given them broad swords and liberty.
>
> Six blooming girls to bear my pall,
> Given them gloves and ribbons all,
> When I am dead, they'll tell the truth
> He was a wild and wicked youth.[104]

Dick Turpin, hanged at York in 1739, hired five men at ten shillings a head to follow his cart dressed as mourners.[105] John Reid, condemned for sheep-stealing in 1774, astonished Boswell by speaking 'very calmly of *the corpse*, by which he meant his own dead body'. Reid told Boswell that he wanted his wife and children to watch him die, and that his wife

> would see him to the last and would *kep* him (i.e., receive his body when cut down); that his son, who was a boy of ten years of age, might forget it (meaning his execution) if he only heard of it, but that he would not readily forget it if he saw it. To hear a man talk of his own execution gave me a strange kind of feeling. . . . He said that his wife was resolved that he should die in white; that

[103] A. Crane, *The Kirkland papers, 1753–1869* (Ashby de la Zouch, 1990), app. iv: 212–13.

[104] 'Wild and wicked youth' was published by Catnach, but, since it includes the line 'Fielding's gang did me pursue', it was almost certainly of 18th-cent. provenance (Macaulay Ballad Collection, Trinity College Library, Cambridge).

[105] Griffiths, p. 169.

it was the custom in his part of the country to dress the dead body in linen, and she thought it would cost no more to do it when he was alive.[106]

Not all of these manifestations need suggest popular belief in the afterlife, but some might have. Since the body's integrity mattered for its resurrection, people worried how it was treated after hanging—anatomized, decapitated, gibbeted, as might be. Surgeons were hated, grave-snatchers execrated. Felons wanted to be buried *decently*, preferably in the home churchyard; 'decently' was always the word used. Here again socially shared understandings operated, for authorities knew how to thwart these expectations cruelly. Anatomization, decapitation, and gibbeting struck at popular anxieties deliberately. Burial policies also struck at them deliberately—particularly the consignment of hanged bodies to unconsecrated and quicklimed graves inside prisons. As James Wilson faced decapitation in 1820 he knew that his body could not rise with its head on at the last trump; but head or no head the sheriffs' promise that he would be buried in his native churchyard 'greatly reconciled him to his fate'. When the sheriffs reneged on their promise and buried him in unhallowed ground, his daughter and niece dug him up by night and took him to his birthplace.[107]

Bodies in gibbets were never safe from nocturnal raids by family or friends. The cages had to be padlocked or riveted and the posts, sometimes thirty feet high, spiked with nails. Anonymous letters would demand that magistrates remove bodies from gibbets: 'if William Whittle that Worthy Man Angs up 10 dayes you may fully Expect to be blown to Damnation'.[108] Again, family or friends would do battle for the hanged body lest the hangman sell it to the anatomists or claim the clothing. There were running fights for Jack Sheppard's corpse in 1724 as his supporters claimed it either for attempts at resuscitation (that never-lapsing hope) or for decent funeral. The cost of neglecting the struggle was shameful: 'One of their bodies was carried to the lodging of his wife, who not being in the way to receive it, they immediately hawked it about to every surgeon they could think of, and when none would buy it, they rubb'd tar all over it, and left it in a field hardly cover'd with earth.'[109]

Those who denied religion were just as determined to control the fate of their own or their relatives' corpses, however. The friends of the

[106] W. K. Wimsatt and F. A. Pottle (eds.), *Boswell for the defence, 1769–1774* (1960), 341, 299, 302.
[107] *The trial of James Wilson*, 48–9.
[108] E. P. Thompson, 'The crime of anonymity', in Hay *et al.* (eds.), *Albion's fatal tree*, 269.
[109] Richardson, *Familiar letters*, 220.

decapitated Cato Street conspirators in 1820 asked the home secretary if they could have the heads and bodies restored to them. They wanted to *display* them, they said—to raise funds for the destitute families. This was quoted as evidence of the radicals' callousness, but it was more like simple economic consideration. The home secretary refused of course. He also refused the wives' request to have the bodies for decent burial. The bodies were covered in quicklime and buried in Newgate instead.[110] The Cato Street friends and families wished only to deny the law its last unqualified triumph over them.

Finally there were moments when hanged bodies were overtly politicized, converted into emblems of injustice and carried off to fine funerals in parodic inversion of the execution's intended significance. Turn-of-the-century radicals did this to triumphantly propagandist effect. After Despard and his co-conspirators were decapitated in 1803 they were buried like heroes. On a hearse drawn by four horses, followed by three mourning coaches for his friends, Despard's nail-studded coffin was borne from Horsemonger Lane gaol across Blackfriars Bridge to St Paul's churchyard, where he was buried in a grave fourteen and a half feet deep to deter body-snatchers. The crowd attending was 'beyond belief'. Followed on foot by long lines of mourners, the bodies of the other conspirators processed no less solemnly to Harper's Chapel in London Road. The doors were locked and they were buried in a locked vault.[111] Even more spectacular was the funeral of young Eliza Fenning in 1815.[112] Popularly believed to have been wrongfully hanged, and the subject of a huge appeal campaign (discussed in Chapter 13), her funeral was a pageant of the very greatest ambition. Cut down from the Newgate scaffold, she was publicly exhibited for three days in the house of a picture-cleaner off Red Lion Square near her home, and people queued to see her as 'she lay in her coffin seemingly as in a sweet sleep, with a smile on her countenance'. Then an 'immense concourse' gathered, singing Methodist hymns as it progressed down Lamb's Conduit Street to the burial ground of St George the Martyr behind the Foundling Hospital. The mourners were led by an undertaker wearing a white hatband, the coffin was carried by six men in black, and a 'rich pall' was suspended over it by six supposed virgins dressed in white.[113]

[110] G. T. Wilkinson, *An authentic history of the Cato Street conspiracy* (1820), 391.

[111] *Memoirs of Despard*, 47.

[112] Anon., *The affecting case of Eliza Fenning, who suffered the sentence of the law, July 26, 1815* (8th edn., 1815), 38–40.

[113] Attendant virgins were standard even at the funerals of the impurest. 12 'marriageable virgins' attended the funeral of Justice Russell of Southwark in 1784, scattering flowers as they walked, notwith-

(This funeral form was standard: Paul Sandby drew just such a cortege when he depicted a spinster's funeral in the 1770s.[114] The magical number six was adhered to here, as in the highwayman ballad already quoted—and at Ferrers's funeral a half-century before: six horses then, and six men to escort Ferrers's body at its several stations before burial.) At Fenning's funeral 'the streets were nearly impassable, every window was thronged, and in many places the tops of the houses were covered with spectators, most of whom seemed to sympathise with the feelings of her afflicted parents.' The parish officers and police had to clear the way for the procession and protect it from the pressing crowd. A riot nearly developed at the burial ground when women turned on a man who shouted insults at the body. They spat on him.

So the scaffold crowd reveals much greater complexity than meets the eye alert only to its ghoulishness. It wore many faces.[115] The regressive behaviour which attended some hangings like Esther Hibner's was exceptional, not the norm, and callousness is the last thing to be assumed. The initial boisterousness and standardized jokes—the cry of 'Hats off!' as the victim appeared and the sudden silence that descended then—the groans for the hangman—the hope that the victim might speak and the straining to hear (*death* spoke, after all)—the shrieks as the drop fell—the breaking of the silence as a thousand heads turned to communicate with neighbours—the uncertain movements (what to do next?)—the reluctant dispersing, or the standing that night on the site of the scaffold—the telling of stories afterwards: all these things reveal themselves as involuntary texts working on many layers, festive, reverential, defensive, defiant, or cowed by turns. Even now, however, we are not done. It requires another chapter to measure the most crucial variable in this behaviour—the crowd's discrimination as it watched different kinds of people hang for differing kinds of crime.

standing the fact that the justice was 'a griping oppressor, detested when living, and now buried amidst riotous confusion': J. W. Frost (ed.), *The records and recollections of James Jenkins* (New York, 1984), 179–90.

[114] Llewellyn, *Art of death*, fig. 41.

[115] Similarly, the mob which enjoyed the pillory, for example, also ensured its demise. When Daniel Isaac Eaton was sent by Ellenborough to the pillory in 1812 for publishing Tom Paine, people applauded him and urged him to stand so that the pillory would shield his head from the sun. Eaton turned his plight to such advantage that when, two years later, the naval hero and critic of naval abuses Lord Cochrane was sentenced to the pillory, the government did not dare execute the sentence. In 1816 the pillory was abolished except for wilful and corrupt perjury and subordination (until 1837); one Dr Bossy was pilloried in 1830: Place papers, BL Add. MS 27826, fo. 172; Griffiths, pp. 148–50.

CHAPTER 3

CARNIVAL OR CONSENT?

1. *The question of carnival*

THE SCAFFOLD CROWD WAS SO COMPLEX IN ITS BEHAVIOUR (AS the last chapter shows) that it could be argued about in quite contrary directions. On the one hand, it could be claimed that the spectacle of execution was awesome enough to induce the watching populace to consent in the law's magnitude, and that they did consent on the whole, as when they applauded murderers' hanging.[1] On the other hand, it has been argued just as plausibly that it was the populace, not the law, that controlled the scaffold arena, converting the ritual to carnival as their parodies turned the world upside down.[2] Each perspective has much to say for itself. None the less, this chapter argues that the truth was more complex than these stark alternatives suggest.

In the first instance, there is no doubt that executions had the capacity to implant the law's presence, power, and moral messages in the collective mind: nor indeed that they achieved this more effectively than the rituals of assizes and quarter sessions did. Poor people would walk great distances to attend county assizes; they were a spectacle to enlighten a dreary and routinized year.[3] But courtroom pomp (such as it was) was constructed as much to sustain judicial dignity as to impress the populace who stank the courtrooms out; not many of the vulgar could fit into courtrooms anyway.

Executions by contrast were more sensationally laden and more numerously and frequently witnessed than trials were. They were mounted *for* the people, and the crowd's function was to bear witness to the might of the law and the wickedness of crime and to internalize

[1] See App. 1.

[2] T. W. Laqueur, 'Crowds, carnivals and the English state in English executions, 1604–1868', in A. L. Beier *et al.* (eds.), *The first modern society: Essays in honour of Lawrence Stone* (Cambridge, 1989), 305–99.

[3] One John Parker in 1822 walked 14 miles from his Norfolk village to watch the Thetford assizes; he took a neighbour's 14-year-old daughter with him for a treat, but raped her on the way back: HO47.63.

those things. It would be silly to doubt that this is what many witnesses did. Of Thackeray's mustachioed tradesmen and their wives it is difficult to imagine otherwise. As much might also be inferred from the huge number of executions on which there is no press report of any crowd response whatever, the reporter not deeming the crowd's response remarkable. Nor need it be denied that the law's stigmatization of the hanged body was echoed in plebeian reactions. If the body carried no significance (as people's friend, as highwayman hero, as unjustly hanged, or as son, husband, father), the populace could usually be counted on to reject it as contaminated. When in 1774 Boswell scoured Edinburgh for a haven in which to attempt the resuscitation of the hanged body of John Reid the sheep-stealer, innkeepers told him that they would rather have their throats cut than allow it in their houses. At the thought of accommodating the body, 'Mrs Bennet screamed, and Andrew said very justly that nobody would come to it [the house] any more if that was done.'[4]

Yet this 'consensual' argument cannot quite suffice. It omits not only the vivid languages of many scaffold crowds which were the obverse of deferential, but also the coercive structures under which consent was given when it was given. What this chapter denies is that the scaffold crowd enjoyed a free autonomy of expression. State power determined crowd reactions at the scaffold; and this will shift the argument radically.

The contrary interpretation of crowd behaviour stresses the scaffold crowd's insubordination. It draws on a sense of the carnivalesque in these gatherings, the 'popular bacchanal' which Ignatieff has referred to and which Peter Burke also discerns in the execution rituals of early modern Europe.[5] It is always pleasant to go to opposite extremes in interpretation, but we should be wary here too. Executions might have been 'carnivalesque'; but the word is an elastic one and it will snap if stretched too far. It snaps deafeningly when Laqueur stretches it towards the claim that it was the crowd rather than the state which controlled the ritual. Let this be the theme we explore first.

In view of the mocking insubordinations of the gallows crowd and the boozing, finery, and rituals which were in popular control, the analogy with carnival is not unappealing. At Tyburn and after 1783 outside

[4] W. K. Wimsatt and F. A. Pottle (eds.), *Boswell for the defence, 1769–1774* (1960), 339.

[5] M. Ignatieff, *A just measure of pain: The penitentiary in the industrial revolution, 1750–1850* (1978), 23; P. Burke, *Popular culture in early modern Europe* (1978), 197.

Newgate we do seem to hear echoes of the travesties and role-reversals of that ancient communal folk comedy which, in Bakhtin's words, had anciently 'made its unofficial but almost legal nest under the shelter of almost every feast', temporarily suspending the prohibitions and inverting the hierarchies of official order. The fact also that carnival in classic guise entailed a ritualized sublimation of fear is not the least reason to associate scaffold behaviour with that tradition. In carnival 'all that was terrifying becomes grotesque. . . . The people play with terror and laugh at it; the awesome becomes a "comic monster".'[6] Yet two things are wrong with this approach: it lacks some chronological credibility; and it begs the key question, who really controlled the event?

Not for nothing has carnival been attended to by early modern historians rather than by historians of the eighteenth and nineteenth centuries.[7] Post-Reformation England was no longer plausibly a land of carnival (if ever it had been). By the eighteenth century we hear only the thinnest echo of that great tradition. Admittedly, Wilkesite crowds in the 1760s still achieved their subversions through self-legitimizing rituals and symbolic appropriations. Admittedly, too, in the cult of the masquerade, polite society played sociable and carnivalesque games around the relationship between the real and the disguised self.[8] Country people retained their mocking customs and skimmington rides, and urban election campaigns were not unlike charivari.[9] London chimney-sweeps paraded the streets dressed as green men on May Day in the 1820s.[10] The three days of London's Bartholomew Fair were described in 1817 as 'the City Carnival—the delight of apprentices, the abomination of their masters—the solace of maid servants, the dread of their mistresses—the encouragement of thieves, the terror of the constables'; or again in 1828 as 'that saturnalia of nondescript noise and nonconformity . . . when the Lord Mayor changes his sword of state

[6] M. Bakhtin, *Rabelais and his world*, trans. Hélène Iswolsky (Cambridge, Mass., 1968), 82, 89, 91.

[7] Cf. E. Le R. Ladurie, *Carnival in Romans* (New York, 1977); N. Z. Davis, 'The reasons of misrule: Youth groups and charivaris in sixteenth-century France', *Past and present*, 50 (1971), 49–75; B. Capp, 'English youth groups and "The Pindar of Wakefield"', *Past and Present*, 76 (1977), 127–33; M. D. Bristol, *Carnival and theatre: Plebeian culture and the structure of authority in Renaissance England* (1985).

[8] J. Brewer, *Party ideology and popular politics at the accession of George III* (Cambridge, 1976), 187–90, 310–11; T. Castle, *Masquerade and civilization: The carnivalesque in eighteenth-century English culture and fiction* (1986).

[9] E. P. Thompson, *Customs in common* (1991), 467–531; F. O'Gorman, 'Campaign rituals and ceremonies: The social meaning of elections in England, 1780–1860', *Past and Present*, 135 (May 1992), 79–115.

[10] R. Southey, *Letters from England* (orig. pub. 1807; 1951), 78–80; C. Phythian-Adams, 'Milk and soot: The changing vocabulary of popular ritual in Stuart and Hanoverian London', in D. Fraser and A. Sutcliffe (eds.), *The pursuit of urban history* (1983), 83–104.

into a sixpenny trumpet and becomes the lord of misrule'.[11] Still, despite all this, carnivalesque language had lost a good deal of the élite support which had once given it meaning, and these exceptions were losing support too. The masquerade was denounced and the carnivalesque fair or wake attacked; Bartholomew Fair itself was soon abolished.[12] Rabelaisian laughter gave way to different forms of humour among élites—irony, sarcasm, and satire, directed at individuals rather than at 'structures'; old grotesqueries were dismissed as 'despicable'.[13] The result of the withdrawal of élite and clerical sanction was that henceforth the old customs were more often witnessed in the village than the town, but even in the village the lord of misrule who had anciently reigned at Christmas was moribund by the 1820s.[14] In London there was a quality of self-conscious antiquarianism and quaintness in the revival of old processions. When the freemasons were mocked in 1742 by rough music (carts drawn by asses, and men riding donkeys with cow-horns in their hands), Angelo recalled the episode as almost 'the last remnant' of mocking public spectacles in the capital.[15] Role inversions which appear to hint at the old laughter turn out to have emulated rather than subverted the rituals of authority. The mock trials conducted by Newgate prisoners, for example, certainly seem to have turned the narrow world of the prison cells upside down:

> When any prisoner committed an offence against the community or against an individual, he was tried by a court in the gaol. A prisoner, generally the oldest and most dexterous thief, was appointed judge, and a towel tied in knots was hung on each side in imitation of a wig. The judge sat in proper form; he was punctiliously styled 'my lord.' . . . A bribe to the judge was certain to secure acquittal, and the neglect of the formality was certainly followed by condemnation. Various punishments were inflicted, the heaviest of which was standing in the pillory. This was carried out by putting the criminal's head through the legs of a chair, and stretching out his arms and tying them to the legs . . .

[11] 'The mirror of months' (1826), in C. Walford, *Fairs, past and present* (1883), 238; *Gentleman's Magazine*, 87/2 (1817), 272, cited by R. Leach, *The Punch and Judy show* (1985), 33, 168.

[12] D. A. Reid, 'Interpreting the festival calendar: Wakes and fairs as carnivals', in R. D. Storch (ed.), *Popular culture and custom in nineteenth-century England* (1982), 125–53.

[13] Bakhtin, *Rabelais*, 117, 120; K. Thomas, 'The place of laughter in Tudor and Stuart England', *Times Literary Supplement*, 21 Jan. 1977: 77–81 (I am grateful to Adam Fox for this reference).

[14] When Strutt collected references to such customs in the 1830s, his sources fell largely silent after the 16th cent.: J. Strutt, *The sport and pastimes of the people of England*, ed. W. Hone (1834).

[15] H. Angelo, *Reminiscences* (2 vols., 1828), i. 408; Strutt, *Sport and pastimes*, 339; W. Hone, *The every-day book* (1825–7), ii. 523–7.

—but this sounds like a persecutory rather than a parodic ritual, less a commentary on the comedy of the great world outside the prison than an appropriation of its cruelties.[16]

The other problem with the interpretation of execution as carnival is that the scaffold was an implausible place at which to represent 'utopia, the image of a future state in which there occurs the "victory of all the people's material abundance, freedom, equality, brotherhood"'.[17] Nor was it convincingly the locus of what Burke discerns as the three major themes in carnival—food, sex, and violence.[18] Most tellingly, the world could hardly turn upside down here, because the harsh realities of worldly power were incontrovertibly affirmed. The state controlled the violence, not the people. Because of this, and for all its mockings, the scaffold crowd could not claim a free celebratory autonomy.

It was not merely that since élites were increasingly critical of the crowd's licence, communal understandings were weakening. The greater fact is that the crowd comprised reactive observers, not initiating agents. Scaffold ritual represented authority, not utopia; execution crowds gathered to witness a statement made by sovereign power. They might subvert that statement, but it was difficult to do this through elaborate parodic enactments, clothings, and symbols. If there was a carnival element at executions, it was witnessed narrowly in the hubbub and movement of people and in the commercialization of the event. Taverns opened early, and 'pye men, and others with gingerbread, and other things bawld about as they do now in the Old Bailey, but to a hundred times the extent, and in much greater variety', as Place recalled in 1824.[19] In country towns likewise, laughter came *after* the hanging, in the pleasures and drunkenness of market or fairground. At execution day in Shrewsbury the performance of a circus would be postponed for two hours so that people could watch both spectacles.[20] The only place where the scaffold could be truly parodied in an old language of mocking inversion was, significantly, outside the real arena of state-inflicted death, and that was in the Punch and Judy show.[21]

Nor can a plebeian control of the event be affirmed in the claim that the 'shabby' and 'risible' rituals of execution were 'unpromising vehicles for the ceremonial display' of sovereign power.[22] It is true that what

[16] T. F. Buxton, *An inquiry, whether crime and misery are produced or prevented, by our present system of prison discipline* (1818), 48–50.

[17] P. Connerton, *How societies remember* (Cambridge, 1989), 50.

[18] Burke, *Popular culture*, 186.

[19] Place papers, BL Add. MS 27826, fo. 97.

[20] C. Phillips, *Vacation thoughts upon capital punishment* (1857), 58.

[21] See Ch. 4, below.

[22] Laqueur, 'Crowds, carnivals and the state', 309.

Walpole called 'the monthly shambles at Tyburn' and Mandeville had termed a 'jubilee . . . one continued fair', were accident-prone affairs. The state's theatre at the scaffold was not 'coherent', as over-polished modern ceremonials strive to be. As Laqueur says, the state was not even conspicuously present, except in the person of the chaplain and sheriffs. But again we must beware of anachronistic judgement in this. None of the shabbiness of execution ritual undermined the state's capacity to make its point, while no state theatre *was* coherent at this time. A great state celebration like that for the Peace of Amiens in 1802 Southey found merely 'preposterous'. 'The English do not understand pageantry,' he declared: 'The poorest brotherhood in Spain makes a better procession on its festival.'[23] Even royal ceremony was inept. As late as 1861 Lord Robert Cecil found royal ceremonials 'ridiculous': 'something always breaks down,' he declared. There was, it has been said of nineteenth-century royal occasions, 'no vocabulary of pageantry, no syntax of spectacle, no ritualistic idiom'. The polished ceremonial 'traditions' to which tele-literate moderns have become accustomed were in truth invented not long ago.[24] This applied to courtrooms too. People chewed and threw nuts at each other in the Old Bailey. Attending the Carlisle assizes in 1803, Coleridge, hungry, had little compunction about 'hallooing to Wordsworth who was on the other side of the Hall—*Dinner*!'[25]

Neither the 'levity' and 'comic foul-ups' which characterized the gallows nor the condemneds' subversion of their roles made for the kinds of theatre satisfying to present-day expectations. Yet they no more obfuscated the real locations of power than the oscillations 'between farce and fiasco' of George IV's or Victoria's coronations, or the mocking of the regent's person when he drove in the park, obfuscated the stature of monarchy itself. However maladroit this ritual of state, at the scaffold the force of de Maistre's dictum was not undermined: 'All grandeur, all power, all subordination rest on the executioner: he is both the object of horror and the bond of all association; remove him, and order at once gives way to chaos, thrones decay, and society disappears.'

In fact, the mayhem at executions can be read as indicating the state's

[23] Southey, *Letters from England*, 54–5.

[24] D. Cannadine, 'The context, performance and meaning of ritual: The British monarchy and the "invention of tradition", *c.*1820–1977', in E. Hobsbawm and T. Ranger (eds.), *The invention of tradition* (Cambridge, 1983), 101–2, 108, 116.

[25] D. V. Erdman (ed.), *The collected works of Samuel Taylor Coleridge: Essays on his times in the* Morning Post *and the* Courier (1978), i. 416.

olympian indifference to the effects achieved, its confidence in itself. It was never the judiciary or the secretary of state who worried about the crowd's disorder. The judiciary remained confident throughout the period that the scaffold delivered its messages well enough. Government was neutral too. When in 1783 the carting of the condemned to Tyburn was abolished and the gallows moved to Newgate, it was not because the state worried about mayhem, but because the City sheriffs deferred to the high-born property developers north and east of Hyde Park.[26] As late as 1866 Lord Cranworth, ex-lord chancellor, was unable to see 'quite all the evils which some people do in public executions'.[27]

Finally, it is insensitive to claim that the insouciance and irreverence of those about to die or of those watching them die expressed 'levity'. We have seen that for every one such act of defiance many more felons died in terror or stupefied by drink. Both the victims of and the audience at the scaffold were subordinated actors, so much so that the striving for a contrary image can only plausibly be read as an ironic rather than as a comic construction, a protest at the roles allotted inescapably to both people and victim.

Illustrations of execution have been made much of in this argument. Noting how in many illustrations the gallows were subordinated to the crowd, Laqueur reaffirms the crowd as the central actor: 'the subject of these pictures is the holiday crowd itself, for which the death of the condemned seems to provide only the occasion. The state is almost wholly unrepresented, but so is death—the gallows and the criminal.'[28] The deliberate understatement of these pictures forbids a reading as innocent as this. What we are actually confronted with is an ironic device common to depictions of many other grand moments. In Rowlandson's *George III and Queen Charlotte driving through Deptford* (1785) the royal coach is relegated obscurely to the picture's margin while the turbulent traffic and rumbustious crowd hold centre stage. But the satirical tension of the picture resides precisely in this imbalance; the royal couple are diminished in visibility and status, but that is the joke: the picture is *about* monarchy.[29] The subordination of centrally causative events to attendant crowds was a device familiar in European art. In pictures of gatherings at shrines, for example, 'it is this very suppression, the reduction of so loaded an act to apparent insignificance, that offers the best proof of the power of the image at the shrine, and of the ultimate bur-

[26] See Epilogue, below. [27] RC 1866: 14. [28] 'Crowds, carnivals and the state', 332, 337.
[29] J. Hayes, *Rowlandson: Watercolours and drawings* (1972), pl. 24, 25.

den it carries, however remotely it may do so.' Bruegel's picture of the Crucifixion (which Laqueur quotes), with Christ almost swamped by the crowd, is unambiguously *about* the Crucifixion, for example.[30] So too the scaffold image was about death, justice, power, and retribution, however subordinated it might be to ambient detail, the pictorial refusal to accept the scaffold's centrality merely endorsing its potency ironically.

The immense ambiguity of execution resided in the crowd's gathering to witness an affirmation of strong men's law over weak. An outward if not always a felt deference was unavoidable, for the audience was powerless to affect the process enacted before it. It could not be *their* festival, like fairs or radical assemblies were. The crowd protested clamorously when Anne Hurle was drawn on a cart to the Newgate gallows in 1804, but 'the sheriff in a loud voice described to them the impropriety of their behaviour, after which they were silent'.[31] The crowd might lampoon authority but could not negate it. They could not stop the process or save a man or woman from the noose.

So little could they control the event that there were few occasions on which they contemplated rescue, even though many knew that 'nothing certainly could have been made more easy'.[32] Very rarely were crowds as openly rebellious as they had been in the Pentez riots against the surgeons in 1749,[33] or as they were in Edinburgh when poor Johnston was so wretchedly hanged in 1818 (see p. 50 above). The usual absence of soldiers at the scaffold upon which Laqueur remarks was an index less of an incapacity to produce state ceremonial than of a simple confidence that they were not needed. There were always protests about the lack of solemnity and pomp at the scaffold, but nothing had to be done about it when a skimpy parade of constables and javelin-men kept order enough and when the hanging of common people was not to be dignified by greater pomp than this. Only the execution of aristocrats was more pompously celebrated. Soldiers ringed the scaffold at

[30] D. Freedberg, *The power of images: Studies in the history and theory of response* (Chicago, 1989), 108–9. Execution pictures with the gallows satisfyingly at the centre of the image or with the 'state' very visibly present are more numerous than Laqueur implies: cf. frontispiece of *A general history of the lives and adventures of the most famous highwaymen [etc.]* (1758). See also Ch. 5, below. Auden's poem 'Musée des Beaux Arts' makes the point economically: *Collected poems*, ed. E. Mendelson (1976), 179.

[31] Knapp and Baldwin, iv. 241.

[32] 'Philonomos', *The right method of maintaining security in person and property* (1751), 51.

[33] P. Linebaugh, 'The Tyburn riot against the surgeons', in D. Hay *et al.* (eds.), *Albion's fatal tree: Crime and society in eighteenth-century England* (1975).

the hanging of Lord Ferrers in 1760, for example.[34] And a little pomp might be allowed in distant counties where hangings were rare and opportunities for any dressing-up few. There was a small military display at John Hatfield's hanging in Keswick in 1803, for example. But this celebrated an event rare in such a place, and was not to protect order. Hatfield's execution was popularly approved.[35]

The exceptions to this rule occurred only in unusually inflammatory circumstances, or when the offence had been against the sovereign. Then the state *did* make its presence felt, quite conclusively. Anticipating huge public sympathy and disturbance at the execution of Dr Dodd in 1777, 'two thousand men were ordered to be reviewed in Hyde Park during the execution, which, however, though attended by an unequalled concourse of people, passed with the utmost tranquillity'.[36] When Colonel Despard and his associates were hanged and decapitated on top of Horsemonger Lane gaol in 1803, Southwark swarmed with constables and mounted patrols of Life Guards.[37] When Bellingham was hanged in 1812 for assassinating Spencer Perceval, 5,000 troops were held ready near Lambeth and several country regiments advanced towards London.[38] When James Wilson was hanged and decapitated in 1820, the crowd of 20,000 was ringed by one of the biggest military displays Glasgow had seen (still the crowd cried 'Murder! Shame!').[39] Foot and Life Guards, the City Light Horse, six light field-pieces of flying artillery, and a civil force of 700 were ready for trouble when the Cato Street traitors were decapitated in 1820.[40] Sir Simon le Blanc tried the Luddites at York in 1813 'with artillerymen with lighted lint-stocks standing by their guns around the court': 'Seventeen men were convicted and ordered for execution the next morning, and as sufficient gallows could not be provided they were hung in relays. The effect was tremendous, for whereas the town had been full of excited mobs of a threatening character, hardly a person was to be seen outside his house on the following day.' As Lieutenant-General Maitland put it with some satisfaction on this bloody occasion (never had so many hanged at once in York

[34] Two prints in the London Guildhall Library illustrate the scene: Print Department, 'Tyburn' folder. One is reproduced in G. R. Scott, *The history of capital punishment* (1950), facing p. 50.

[35] The Carlisle volunteer cavalry ringed the scaffold: Anon., *The life of John Hatfield, commonly called the Keswick Imposter* (Carlisle, 1846), 24–5.

[36] H. Walpole, *Journal of the reign of King George the third* (1859 edn.), ii. 126: cited in Radzinowicz, i. 466.

[37] Rayner and Crook, iv. 266.

[38] Anon., *The trial of J. Bellingham* (1812), 25.

[39] Anon., *The trial of James Wilson for high treason, with an account of his execution at Glasgow, August, 1820* (Glasgow, 1834), 48.

[40] *Morning Chronicle*, 1 May 1820.

since the execution of ten rebels in 1746), the business 'exhibited the appearance of a military execution'. Cavalry regiments waited nearby in case of trouble, and dragoons ringed the scaffolds. Henry Hobhouse, attending these executions on the Home Office's behalf, walked the streets afterwards and noted the defeated and dispirited air of the people.[41] These displays were timeless. Right down to the Fenian executions of 1867 in Manchester (when a thousand police and soldiers kept order and all public buildings were put under such protection that 'the city was in a state of siege'),[42] the presence of the state was evident and substantial enough when required. Carnivals are not commonly celebrated under such circumstances.

2. *The question of consent*

If crowd responses were contained, the carnivalesque terminally muted, what of the crowd's 'consent' in the law they gathered to see executed? If they did consent, we may now appreciate the conditions under which they did so—and the fact that choices were limited. None the less, choices there were, and for all the coerced relationship they bore to the spectacle, witnesses seldom unambiguously affirmed its legitimacy. The rowdier crowds dissociated themselves from what happened much more often and more audibly than they shouted approval at murderers' deaths. Whether or not these hostile passions caught fire depended on three things: the hangman's competence, the identity of the condemned and the presumed justice of the sentence, and the social and political climate of the moment.

The hangman was execrated at most hangings (other than those of murderers, when he might be cheered). He was normally hooted, and sometimes stoned or his life threatened, as Calcraft was threatened before Bousfield's execution in 1856. At politically loaded executions, of traitors or Luddites, the executioner was masked for anonymity and safety. When he travelled on duty he often had difficulty in finding lodgings. He was refused access to the Hertford coach when summoned to perform in that town in 1800, and 'the hint was given all the way along the road' to stop him travelling; the local man hired to perform in his place for a guinea was 'so terrified by the indignant populace' that

[41] A. E. Gathorne Hardy (ed.), *Gathorne Hardy, first earl of Cranbook: A memoir* (1910), i. 22, 267, 252. There were 2 separate hanging days on this occasion, 3 being executed on one day and 14 on the other, the last in batches of 7, the judge thinking they would 'hang more comfortably' that way: R. Reid, *Land of lost content: The Luddite revolt, 1812* (1986), chs. 32–6.

[42] *Ann. Reg.* 1867: 156–71.

he ran away and a convict had to be dragooned into service instead.[43] When in 1805 the hangman petitioned the Court of Aldermen for an increase of salary, it was because he could not obtain other employment, 'however low the calling'.[44] Élites shared this distaste. No judge met the hangman who executed his sentence. Upon the executioner was lodged as upon a scapegoat the evil that justice had to do.

Predictably, not one of the grisly episodes referred to in Chapter 1—when ropes broke, dropping half-asphyxiated victims to the ground, or when toes scrabbled desperately on the sides of the platform to relieve the throttling, obliging the hangman to pull on the victim's legs—failed to elicit cries of outrage directed at hangman and sheriff, or threats to murder the hangman. Calcraft's terror when he bungled Bousfield's hanging was gleefully celebrated in street ballad:

> My name it is Calcraft by every one known
> And a sad life is mine to you I now own,
> For I hang people up and I cut people down,
> Before all the rebel of great London town.
>
> . . . .
>
> For my old friend Cheshire he learned me the trick,
> And I dine in the clouds tonight with Old Nick,
> For the people on earth do use me so bad,
> That with tears I could drown them for I feel now so sad.[45]

When in 1821 the drunken Yorkshire hangman Curry waved the halter at the crowd and jokingly threatened to hang anyone who volunteered a neck, and then bungled the ensuing execution, he was assailed by cries of 'Hang him, hang Jack Ketch!' and later beaten up on his way home.[46] In 1837 the Gloucester quarter sessions had to consider evidence against the executioner of a murderer, Bartlett, who 'had been intoxicated and treated the corpse of the wretched sufferer with brutal levity and coarseness in view of the spectators'.[47]

Secondly, we have seen that the biggest and most approving crowds assembled to watch the hangings of people least like themselves in their presumed psychological construction or in their high social standing. The former were typically murderers or sodomites—people who embodied the repressed other in the self. (Between 1805 and 1832, 295

[43] *Cambridge Chronicle and Journal*, 5 May 1800.

[44] Corporation of London RO, Court of Aldermen papers: petition of John Langley, 17 Dec. 1805.

[45] Sir Frederick Madden Collection of Broadsides, Cambridge University Library, 11/54.

[46] J. Bland, *The common hangman: English and Scottish hangman before the abolition of public executions* (Hornchurch, 1984), 98.

[47] D. Verey (ed.), *The diary of a Cotswold parson* (1978), 137 (26 June 1837).

murderers were hanged in England and Wales, i.e. about 19 per cent of all those hanged, and 50 sodomites were executed.) Or else they were fine people like Governor Wall who had acted murderously against the weak and defenceless; or they were fine villains whose wealth had protected them. People knew things; memories were long, newsmongering energetic. When in 1784 the coffin of the earl of Filney was landed on Custom House Quay on the Thames, 'people ran to look at it'. Filney had escaped the law to live in Italy after committing 'an odious and detestable crime'; and here 'all at once—stifled murmers [*sic*], groans, and hootings, burst out in unison, against the unconscious but detested dead'.[48] It was as if deaths of these kinds were *needed* if the world was to make sense.

That élite criminals seldom escaped execration of this kind suggests a socially constructed discrimination. The crowd was markedly respectful of Courvoisier at his hanging in 1840, for example: no hissing, no hooting as he mounted the scaffold. Since his was a bloody murder for gain and there was no doubt about his guilt, could this respect have been because he was a servant who had murdered an aristocratic master, the uncle of Lord John Russell? So it might seem. For when the criminal came from the high or the quasi-respectable classes (and, oddly, there were few very poor murderers), or when the victims were children, or poor, or women, or defenceless, the crowd showed no respect or pity. When Wall was executed in 1802 (he had had an insubordinate soldier flogged to death while governor of Goree), 'their joy at seeing him appear upon the scaffold was so great, that they set up three huzzas,—an instance of ferocity which had never occurred before'. As his corpse hung, 'the Irish basket-women who sold fruit under the gallows were drinking his damnation in a mixture of gin and brimstone'.[49] The big crowds also came to watch the banker and forger Fauntleroy die in 1824 and the merchant forger Hunton in 1828—social 'others' again, in other words. Despite monumental campaigns for their pardon, the 100,000-strong crowds showed no pity for either of these two when they swung off.

But, conversely, reactions to and the meanings of the scaffold altered radically when humbler people hanged for lesser crimes, or were hanged unjustly. The fact that relatively few people attended the routine hangings of housebreakers, footpads, and stock-thieves suggests that there was no deep psychic return in watching the deaths of men and

[48] J. W. Frost, *The records and recollections of James Jenkins* (New York, 1984), 179.

[49] Southey, *Letters from England*, 63–4.

women rather like themselves, opportunistically thieving or fallen on bad times. And hangings of these humbler kinds were many times more frequent than were the hangings of the great murderers or forgers: the categories mentioned accounted for 46 per cent of all hangings in 1805–32 (see Appendix 2). Sometimes prosecutors paid a heavy price for bringing about such deaths. In 1763 people brought the hanged body of Cornelius Saunders to a Mrs White's house in Lamb Street (she had prosecuted him for stealing £50 from her); they piled up her household furniture in the street and set fire to it. When Eliza Fenning was hanged in 1815 for attempted murder, popular belief in her innocence was intense. Crowds turned their anger against her employer's house in Chancery Lane, and would have pulled it down had they not been stopped by police. When in 1817 the parents of a 15-year-old girl of Ampthill, Bedfordshire, achieved the hanging of a neighbour for raping her, their house was surrounded by a couple of hundred people throwing stones, parading effigies, 'hallowing and charging the Family with having hung the man and that they ought to be hung themselves': they had to appeal to the magistrates for protection.[50] Wakefield referred to the case of John Williams, executed at Newgate in 1827 while bleeding and crippled from an escape attempt: 'respectable shopkeepers in the neighbourhood of the scene of the execution were heard to say that worse than a murder had been committed and that they should like to see the Home Secretary [hang] in the same way.' At every execution during the three years he was in Newgate, Wakefield claimed, 'the assembled crowd sympathised with the criminal and expressed feelings of compassion'. Our best example concerns the south London bricklayer Samuel Wright, condemned in 1863–4 for killing his violent mistress under provocation. He was hanged even though—some said because—working-class opinion was mobilized to save him; for when a procession of workmen went to the Home Office to ask for mercy, Sir George Grey thought this an attempt to 'terrorize' the government. Six hundred Lambeth tradesmen signed one petition for Wright, 2,543 working men and women another, and a public meeting of some 3,000 was held on the eve of the execution, agreeing that if Wright were executed it would be a 'judicial murder', and sending a memorial to the queen at Windsor to that effect. The efforts were in vain, but as the scaffold was erected at Horsemonger Lane prison, a black-bordered handbill was circulated urging people to boycott Wright's execution: 'Let

[50] Radzinowicz, i. 192; *Morning Chronicle*, 28 July 1815, and *The Times*, 1 Aug. 1815; I am grateful to Prof. Clive Emsley for reference to Bedfordshire RO: QSR 23/230, 231.

Calcraft and Co. do their work this time with none but the eye of heaven to look upon their crime. . . . There is one law for the rich and another for the poor.' A thousand policemen were drafted to keep order at the execution, but the crowd was orderly and small: most locals kept away, and shops were shut. There were hoots of protest from the crowd—'Shame!'—'Judicial murder!'—'Where's Townley?' Townley was a well-to-do murderer who had been pardoned on grounds of insanity shortly before. Recalling all this four years later, the abolitionist MP Charles Gilpin 'could not forget the morning of that execution. The people in the neighbourhood, instead of rushing to see the execution, had their blinds drawn down.'[51]

Great men always had to admit that there was a limit to the number of 'ordinary' hangings the people would tolerate. This was why, as prosecutions and hence death sentences mounted in the early nineteenth century, the system unravelled itself and became unworkable. Anticipated reactions also signified in more peaceful parts of the kingdom and in unsung episodes. It is not at all clear that midland miners and yokels went to executions as they did to a bull-baiting, for example. On the contrary their reactions had to be monitored. When Best J. condemned a deformed imbecile for arson at Shrewsbury assizes in 1823, the Shropshire magistrates urged mercy for fear of the 'effect of such a spectacle upon the scaffold, which in the present impression of the people's mind would rather excite sympathy for the sufferer than abhorrence for his crime'. This therefore was one death sentence that home secretary Peel felt it wise to commute.[52]

Finally, broader political contexts always signified. There was never doubt as to where the crowd's sympathies lay when radicals or protesters were executed. The pattern endured from the execution of two silk-weavers for destroying a silk loom in 1769 (bricks and stones were thrown while the scaffold was erected and the sheriff had to tell the crowd that every step had been taken to save the men's lives), on to the cries of 'Murder! Blood!' at the hanging of the Nottingham rioters in 1832. When John Cashman was hanged in Snow Hill for participating in the Spafield riots in 1816, the crowd hooted the hangman and the scaffold had to be guarded by police and soldiers. Sympathy extended even to those whose anti-government postures were insane or quixotic. In 1812 Bellingham hanged for assassinating Spencer Perceval in the Commons to cries of 'God bless you!'[53] The decapitation of traitors was

[51] *The Times*, 11, 12, 13 Jan. 1864; PD3 cxc (1868), c. 1039. [52] HO47.62.

[53] J. Laurence, *A history of capital punishment* (1932), 106; Borrow, vi. 176–9.

always watched by huge crowds, but never without overt or displaced disgust at the law's brutality. When Hardie and Baird were thus executed in Stirling in 1820, 'a sight so appalling as the mangling of the bodies by decapitation, prevented many persons from attending; they might be seen moving along the other streets, under feelings corresponding to the melancholy event.'[54] People applauded Despard in 1803 as he spoke his last defiance on the scaffold; they hissed the executioner, doffed their hats as he hanged, and later burnt one of the witnesses in effigy before, as we have seen, attending 'his body to the grave, as if they had been giving him the honours of a public funeral'.[55]

Granted popular approval of some hangings, and enduring instincts defensively to side with the aggressor against the weak, it is clear that the gallows often and perhaps usually symbolized an illegitimate power, even for those unable to articulate the symbolism. But our analysis still has some way to go before we can fully recognize the resources in plebeian culture which sustained this dangerous perception.

In an age when much of popular culture was as oral as it was literate, when ballads were sung and broadsides read out, and when the pictorial image of the scaffold was ubiquitous, the crowd could never be blind to the ironies of its role or innocent of its own cultural repertoires. Their history had not left them illiterate and vacuous onlookers at the spectacle they gathered to watch. These were knowing people, although much of their knowledge might be beyond words. At the very least they had a collective memory inscribed in a familiar script: the patterns of rumour, movement, and sound which drew them to Newgate; the expectation of holiday; the habits of association in the going and the assembling which custom as well as convenience determined (the drinking in gin shops, the standardized banter), and in the echoes of past such occasions which collective gesture as well as the facts of location might evoke. So the gallows was not only a *memento mori* nor even a symbol of justice (or injustice), though it might be all those things. It was also a place where remembered and expected things happened. We shall see in the next chapters that humble Londoners attended with minds saturated in the sensational, emotional, and quasi-politicized meanings which the gallows had conveyed for a century or more. There is no sense in which their culture could have prepared them to receive the

[54] *An exposition of the spy system, pursued in Glasgow . . . 1816–20 . . by a ten pounder* (Glasgow, 1833), 230–1.

[55] Southey, *Letters from England*, 373.

lessons of the gallows with that uncomplicated 'consent' which conservative historians are fond of projecting upon them. That would be a narrowingly reductionist word, as inadequate to describe their behaviour as it would have been alien to their understanding.

PART II

THE PLEBEIAN TEXTS

CHAPTER 4

SCAFFOLD CULTURE AND FLASH BALLADS

1. *Making a day of it*

IN 1848 THE BARRISTER HEPWORTH DIXON SAW OR HEARD about an episode which perplexed him utterly. A man called Sale was hanged at Newgate for robbery and murder:

When the wretched man came forward on the scaffold, he looked pale and ghastly; but his bearing was insolent, and he died with the apparent insensibility of a dog. 'Bravo!' cried his mother, as the drop fell, and the murderer was launched into eternity: 'I knew he would die game.' A woman who had lived in adulterous intercourse with the malefactor was with her in the crowd; they had made up a party to come and see the last of 'poor Tom'; and when the tragedy was over, sallied off to a public house and made a day of it.[1]

Dixon's readers would have accepted this riveting image at face value, and so might we if we were not otherwise prepared. It is in fact a useful text for measuring middle-class Victorians' inability to read such scenes.

Our difficulties with it begin when *The Times* report of the hanging shows that the chief protagonist was not at all the unfeeling barbarian of Dixon's account. In the condemned cell Sale spent days protesting his innocence before confessing on the eve of execution. He then asked to say farewell to his daughter, aged 6. He wept as he did so. Later he dictated letters to his parents and wife, hoping that his children would be cared for, and sending them locks of his hair: for something of himself must survive. Pinioned before being led to the scaffold, he told the sheriff that he did not feel very well. He took comfort from the sheriff's promise to find funds to send his wife and children back to Yorkshire. On the drop he bowed twice to the 'disorderly' crowd 'with great

[1] W. Hepworth Dixon, *John Howard and the prison world of Europe* (1849), 275–6, and *The London prisons* (1850), 196–7.

firmness'. When he dropped, he 'suffered severely', was left hanging for an hour, and was buried in quicklime in the prison.[2]

All this is so remote from Dixon's report that Dixon becomes as interesting as the scene he described. Dixon's interest was transparent enough, however. He told his version of the story to justify the abolition of public executions, filtering it through this purpose as well as his own prejudices. His generation had travelled a long way from the tacit understanding which urbane gentlemen might once have extended to a scene like this. All Dixon could register were the protagonists' affected insouciance and outward refusal to take the ritual seriously. What he saw was an unbearable subversion, not a way of containing terror.

For what if the mother did cry 'Bravo!' and the party did retire to the pub? The oddity and value of Dixon's report lies in the likelihood that their insouciance is what the women *wanted* to be seen. In which case there would be a deep consistency to their behaviour which made sense in a culture quite alien to Dixon. The inference might be twofold: first, their reaction to the hanging was defensive, in the form of a defiant denial of it; and, secondly, their posture expressed a grimly knowing realism with the deepest purchase on plebeian life. Hiding their true feelings, they refused to play to the intended solemnity of the ritual, knowing that they would be defeated by it if they did.

This latter posture is widely recorded. 'Well, I can die but once. I can't help it,' a condemned man said phlegmatically when he heard the recorder's report against him in 1824; while another 'immediately took up a tobacco pipe . . . with an apparent carelessness'.[3] 'Well, if the worst comes to the worst, I shall but have to dance for an hour,' a condemned woman told Elizabeth Fry, who then commented: 'The terror of the example [at executions] is very generally rendered abortive by the predestinarian notion—vulgarly prevalent among thieves—that "if they are to be hanged, they are to be hanged, and nothing can prevent it".'[4]

Shame and grief could not be admitted when inflicted by systems of authority beyond challenge or comprehension. Rather, a statement had to be made about the life which the felon *had* led and which survivors still had to lead. Death has no universal cultural meaning, anthropologists tell us. In many cultures it is seldom addressed directly: 'it is life that has a certain universal currency, and death appears only as its

[2] *The Times*, 11 Jan. 1848.

[3] Prison visiting diary, cited in M. Cheney, *Chronicles of the damned* (1992), 59.

[4] E. Fry, *Observations on the visiting, superintendence, and government of female prisoners* (1827), 74.

absence.'[5] In this case a refusal to be cowed by death and those who inflicted it was central. *Nil carborundum illegitimis*, as London craftsmen (like my own father) used to say in a different context: don't let the bastards grind you down. The only way to cope with the pain and shame of scaffold death was to display your contempt for it, to applaud the victim's courage, to parade your own courage in fellowship, and to lead life onwards.

These postures in the cultures of the urban *menu peuple* have as yet hardly found space in discussions of the relationship between law and people in this era. How the people felt about the law has usually been addressed in terms of an assumed polarity between 'deference' and 'resistance', or in terms of the 'consent' which some historians so cheerfully infer from the fact that middling kinds of people used the criminal law more than rich people did, in prosecuting humbler people, for example. None of these variables does justice to the multiform mentalities and behaviours of scaffold crowds, as Part I has made clear and as Chapter 6 will make clearer. We need now, in Part II, to reinforce the point, locating scaffold crowds in their more complex cultural contexts by turning to the evidence of *texts*. Urban labourers, artisans, shopkeepers, and the like abundantly heard, read, wrote, and said things about scaffold justice, and they are our subject now. While Chapter 6 explores the language of mercy petitions, this chapter and the next take on collective remembering, orality and literacy, ballads both bawdy and criminal, broadsides moralistic, woodcuts ideographic, and the unspoken understandings encased in these things. We turn to written texts in Chapter 5, execution broadsides particularly. This chapter is about texts which were watched, sung, and remembered.

2. *Collective memory*

Robert Southey wrote this after a walk through London streets in 1807:

> I had a paper thrust into my hand, which proved to be a quack doctor's notice of some never-failing pills. Before I reached home I had a dozen of these. Tradesmen here lose no possible opportunity of forcing their notices upon the public. Wherever there was a dead wall, a vacant house, or a temporary scaffolding erected for repairs, the space was covered with printed bills. Two rival blacking-makers were standing in one of the streets, each carried a boot, completely varnished with black, hanging from a pole, and on the other arm a basket with the balls for sale. On the top of their poles was a sort of standard with

[5] P. Metcalf and R. Huntington, *Celebrations of death: The anthropology of mortuary ritual* (Cambridge, 1992), 6.

a printed paper explaining the virtue of the wares;—the one said that his blacking was the best blacking in the world; the other, that his was so good you might eat it.[6]

It is with effort nowadays that we remind ourselves of the lost noises and images of urban life in past times. We know that streets were awash with signs and placards and filled with cries, songs, spoken words, and jostling interventions. Yet we forget that these things enmeshed communities in shared understandings, teaching people how to see and think about themselves. Even those who could not read or write would be streetwise and visually literate. These experiences were seldom reflected upon, but they shaped town dwellers' negotiations of love, work, and play, and their views of their own and others' chances in life as well. Little was excluded from these evanescent communications, and the prison and scaffold least of all.

The scaffold loomed hugely in the popular imagination before 1830. We meet it at every turn: in ballads, Punch and Judy shows, broadsides, and woodcuts. It appeared in stick-gallows scratched on urban walls and, in smaller communities, in the punitive rituals of the skimmington ride as well, when transgressors against communal norms were hanged in effigy:

> In a cart they dragg'd them through the streets, while the music it did play,
> While C—e and his wife that cursed rogue, was tied to a gallows high,
> While Jack the Executioner, his business did complete,
> They brought them near the place they dwell, and tore him to pieces in
> the street.[7]

The gallows was a key emblem of satire and radical wit too:

> Those lawyers see, with face of brass
> And wigs replete with learning;
> Whose far fetch'd apophthegms surpass
> Republicans' discerning.
> For them to ancient forms be staunch,
> To suit such worthy fellows;
> Oh, spare for them one legal branch,
> I mean, reserve the gallows.[8]

In London even the architecture of punishment was ubiquitous. It stretched from Tyburn in the west, across to the pillory at Charing

[6] R. Southey, *Letters from England* (orig. pub. 1807; 1951), 51.

[7] From an 1827 Isle of Wight ballad about a smuggler turned informer: R. Palmer, *The sound of history: Songs and social comment* (Oxford, 1988), 198.

[8] From the republican song 'Plant, plant the tree, fair freedom's tree': Anon., *The trial of Robert Thomas Crosfield for high treason, 11th and 12th May 1796* (1796), 208–9.

Cross, and thence eastwards to the naval gibbets at Execution Dock on the Thames. It extended from Coldbath Fields and Bedlam in the north, down the axis of prisons stretching with nice symbolism along the open sewer of the Fleet River: Newgate, Ludgate, the Fleet, and Bridewell. Then it crossed the Thames to the Clink and the new Surrey prison built at Horsemonger Lane in the 1790s. This was a city in which satirists and radicals could not but appropriate the imagery of scaffold and prison as part of their language.[9]

What meanings did the image carry for people? What were their consequences for the feelings which poor and middling people brought to the scaffold itself? Pamphleteers endlessly complained that the scaffold failed as a deterrent. If they were right, it was because humbler people had learnt to see things on the scaffold which the authorities neither controlled nor intended. Reactions to executions were moulded by the images and fantasies which the populace brought from their own knowing and experiencing. The scaffold drama was in this sense of the people's own making, with most of the scripts written away from the scaffold itself, inscribed in a collective awareness beyond official control.

We cannot now re-establish the flow of gossip and storytelling within families and neighbourhoods which shaped scaffold expectations. But some of its echoes can be heard at all social levels. The scaffold carried its visceral images to polite as well as humble people, and it may not go amiss to say a word about 'polite' remembering first, for it was intense. When Thackeray went to see Courvoisier hang in 1840, he recalled an apocryphal story of how the executioner twenty years earlier had dropped the last of the decapitated heads of the Cato Street conspirators to the crowd's cries of 'Butterfingers!' In 1882 William Ballantine thought it wonderful that he still remembered from his childhood the murder of a Mrs Donathy, 'a great sensation at the time'. In 1883 Canon Venables recalled the details of the crime for which the murderer Cook was gibbeted in 1832, the last man in England to be so dealt with. 'Who does not remember Liza Fenning?' Charles Phillips asked in 1858, even though she had been hanged in 1815.[10] And if memories faded, there were always the *Newgate calendars* or mementoes in effigy to restore old stories to life.

For over a century after Captain Alexander Smith's *Lives of the most noted highwaymen* made five editions between 1713 and 1719,

[9] P. Rogers, *Grub Street: Studies in a subculture* (1972), 150, 292–5.

[10] W. Ballantine, *Some experiences of a barrister's life* (5th edn., 1882), 3; *Notes and Queries*, 6th ser., 8 (1883), 353; C. Phillips, *Vacation thoughts on capital punishment* (4th edn., 1858), 88.

compilations of criminal biographies were steady earners for publishers. They came into their own in the half-century after the five-volume *Malefactor's register* was published in 1779.[11] Drawing on the ordinary's *Accounts* (which they now eclipsed), Old Bailey sessions papers, chapbooks, and broadsides, and increasingly on each other, the *Calendars* were far from mean productions. The young George Borrow, commissioned for a pittance in 1824 to produce six volumes of *Celebrated trials* for Sir Richard Phillips, paid fair tribute to the 'racy, genuine language' in which they were told, and thought he had learnt his narrative skills from them.[12] Most *Calendars* ran to several hundred pages per volume and six volumes per edition by the end of the eighteenth century (and to eight volumes in the second edition of William Jackson's *New and complete Newgate calendar* of 1818). Often finely engraved with criminal portraits or trial and execution scenes, they were expensive productions beyond reach of the common people, and most professed to be morally didactic (Plate 11). Only in the 1830s and 1840s was this tradition displaced by a kind of bantering facetiousness as the old Tyburn stories began to be seen as quaint; but the stories were still told for all that (in Camden Pelham's *Chronicles of crime* (1841), for example, with its comic illustrations by 'Phiz'). Humorous fictions appeared as well, harnessing Jack Ketch the hangman implausibly to comedy (Plate 12).

If a collection of Newgate lives did not keep memories alive in the middle-class family, Madame Tussaud's waxworks would do just as well. After her move from Paris in 1802 her exhibition grew famously, inspired by her professional familiarity with guillotined heads and her uncle's waxwork *Caverne des grand voleurs* set up in Paris twenty years before. By 1869 what was to become the museum's Chamber of Horrors was already enticing almost as many patrons as it does today. Here collective memory was enshrined indeed, in the form of the plaster-cast or waxwork effigies of the great murderers like Rush, Good, the Mannings, Bousfield, Palmer, Thurtell, Greenacre, Hatto, Barthelemy, Courvoisier, and Burke and Hare—the casts taken from life

[11] For full bibliography up to 1811, see L. B. Faller, *Turned to account: The forms and functions of criminal biography in late seventeenth- and early eighteenth-century England* (Cambridge, 1987), 286–327. Cf. I. A. Bell, *Literature and crime in Augustan England* (1991), ch. 2.

[12] Thus Borrow in his *Lavengro*: 'People are afraid to put down what is common on paper; they seek to embellish their narratives, as they think, by philosophic speculations and abstractions; they are anxious to shine, and people who are anxious to shine can never tell a plain story.—"So I went with them to a music booth, where they made me almost drunk with gin, and began to talk their flash language, which I did not understand," says, or is made to say, Henry Simms, executed at Tyburn some seventy years before the time of which I am speaking. I have always looked upon this sentence as a masterpiece of the narrative style, it is so concise and clear' (cited by E. H. Bierstadt (ed.), [Borrow's] *Celebrated trials* (1928), i, p. xii.)

11. *A mother presenting* The malefactor's register *to her son*, 1779.

The malefactor's register of 1779 was the most ambitious of the *Newgate calendars* published between 1770 and 1830. Its leather-bound volumes, illustrated with fine copper engravings, aimed at the affluent and high-minded readers idealized in this frontispiece—and at their children.

12. Joseph Meadows, *Jack Ketch the hangman*, 1836.

Meadows's spoof portrait of the hangman (the real Ketch died in 1686) illustrated Charles Whitehead's comic novel about that person, *Autobiography of a notorious legal functionary.* The facetiousness of both the image and the novel may hint at the relief felt as the bloody code collapsed in the later 1830s. The portrait so well reflected popular notions of the criminal type that it was copied in broadside woodcuts to depict the supposed features of murderers like Daniel Good.

in the condemned cell or from death after execution (Plate 13). As late as 1917, astonishingly, most of these selfsame heads were still being exhibited many decades after their crimes, with neat accounts of their murders appended.[13] Prisons, medical schools, and museums were also full of the casts of heads of executed criminals, open to viewing by the politest of families. Thurtell in particular remained, Dickens said in 1856, 'one of the murderers best remembered in England', his physiognomy and phrenology representing the criminal type *par excellence* for decades after his hanging in 1824.[14]

But it is the humbler people that are our chief concern. Among them gossip, reminiscence, penny-gaff melodramas, and night-time storytelling about murders, ghosts, and horrors were standard imaginative fodder. These did for them what the *Newgate calendars* did for the affluent, keeping old stories alive for years and so implanting them in memory that Francis Place in the 1820s, for example, could recall tales from his youth a half-century earlier: 'On the 18 March 1741, was hanged at Tyburn the famous and infamous Mary Young, commonly called Jenny Diver, by which name I can remember her being familiarly talked of when I was a boy. . . . [The] merits [of such people] were a constant theme and the subject of continuous conversation.' Thackeray heard his plebeian companions at Courvoisier's hanging in 1840 talk about Lord Ferrers's execution in 1760. Broadside sellers in 1851 remembered Fauntleroy's and Hunton's executions for forgery in the 1820s, and some could still sing verses composed for Thurtell's execution.[15]

To be sure, there was forgetting as well as remembering. Having invoked his memory of Jenny Diver, Place added primly that nowadays 'no one among most people ever thinks of talking about an abandoned prostitute and thief'. Aspiring respectables like him were learning to censor such stories. Then, too, one year's sensation might always be eclipsed by the next year's sensation. The criminal dramas upon which gossip thrived were endlessly re-enacted, so each generation could reinvent its own villains and archetypes.[16] Furthermore, plebeian remembering of crime and punishment was increasingly dependent on print. After 1700 England had become a largely literate society. A regional

[13] *Madame Tussaud's catalogues* (BL).

[14] Mary Cowling, *The artist as anthropologist: The representation of type and character in Victorian art* (Cambridge, 1989), 284–316.

[15] Place papers, BL Add. MS 27826, fo. 26; Mayhew, i. 234, 283.

[16] Walpole noted in 1750 that infamous cases like Sarah Malcolm's in 1733 were 'never mentioned but by elderly folks to their grandchildren, who had never heard of them'. Hogarth had painted her in the condemned cell; Walpole himself bought the picture: Walpole, ii. 104.

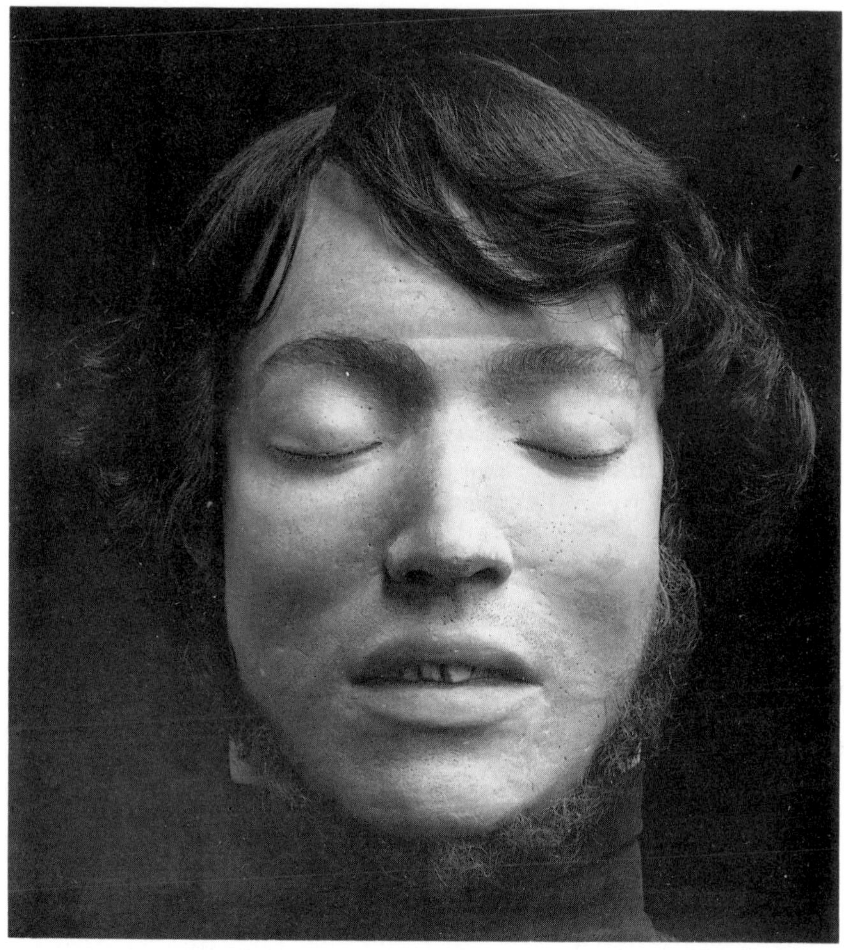

13. *The death mask of François Courvoisier*, 1840.

Courvoisier's death mask was exhibited in Madame Tussaud's well into the twentieth century. Its obscene realism at least reminds us that this book's subjects once lived as we do, and then were killed. Courvoisier was hanged before 30,000 people (including Dickens and Thackeray) for murdering and stealing from his aristocratic master.

average of 30 per cent of men could write their names in 1642, but by the mid-eighteenth century the proportion seems to have risen to about two-thirds. In some industrial regions literacy rates might then have stabilized or declined, but in London and other areas not subject to severe economic dislocation it almost certainly continued to widen.[17] This presaged the great age of the broadside, of the historical and romantic chap-book, the penny dreadful, and gothic romance.[18] As Chapter 5 will show, this is where most popular images of crime were deposited by the second quarter of the nineteenth century.

None the less, oral tradition continued to be symbiotically entwined with literacy rather than defeated by it.[19] In street ballads and street entertainment, remembering and imagining wove a collective idea of the scaffold in the space between print and orality. The tone of these communications was far from uniform. The felon's game death, the ironic approbation of it, the inversion and subversion of the law's rituals, the sentimental tears which sometimes flowed, the execration of but secret wonderment at the murderer, the capacity to assimilate and cope with such horrors and even to approve of them—all these crowd responses had correlatives and sources in street culture. But if there was a dominant tone, it was a sardonic or mocking tone. This was the quality which most mystified observers like Dixon, and we should try now to get as close to it as we can. Street ballads expose it best, but it is encountered in other places too.

3. *The world of Mr Punch*

Take first, for instance, that most transgressive of folk heroes and lords of misrule, the 'most absolute egotist [who] conquers everything by his invincible merriment and laughs at laws, at men, at the devil himself'—Mr Punch.[20] Carnival in England might be moribund, but from lost traditions of the mumming play and the seventeenth-century Italian Pulcinella, Punch carried into the nineteenth century the anarchic, randy, and profane voice of a Rabelaisian mockery which turned the

[17] Schofield estimates that by the 19th cent. up to a half of those signing by marks could read; many children, girls especially, were taught to read and not to write: R. S. Schofield, 'Dimensions of illiteracy, 1750–1850', *Explorations in Economic History*, 10 (1972–3), 437–54; M. Sanderson, *Education, economic change and society in England, 1780–1870* (1983), 10–16.

[18] D. Vincent, *Literacy and popular culture: England 1750–1914* (Cambridge, 1989), 197.

[19] D. Vincent, 'The decline of the oral tradition in popular culture', in R. D. Storch (ed.), *Popular culture and custom in nineteenth-century England* (1982), 20–47.

[20] H. L. H. von Pueckler-Muskau, *Tour in England* (1831), 130–43, cited in R. Leach, *The Punch and Judy show: History, tradition and meaning* (1985), 57, to which study this discussion is much indebted.

world upside down. It was Punch also who embodied the bravado we have observed on the scaffold and shall soon be meeting in ballads:

His money most freely he spends;
  To laugh and grow fat he intends;
With the girls he's a rogue and a rover;
  He lives while he can upon clover;
When he dies—it's only all over;
  And there Punch's comedy ends.[21]

For at least a century before Collier transcribed Piccini's performance in 1828 (but afterwards too), Punch's tale bore the hallmarks of an oral tradition developed on the streets, at Bartholomew Fair, or outside the playhouses whose performances his story sometimes parodied. Punch's was not as yet the children's entertainment that it became. Before 1830 nearly every report and pictorial representation of those who watched him points to the predominance of adult over child spectators (five to one in the pictures, it is calculated). Some were 'decently dressed', but most were those 'least burthened with this world's goods'—'butchers, sweeps, pickpockets, milk-girls . . . Irish labourers, fresh from Munster, roaring with glee'.[22] It was the response of such people that pushed punch-men's improvisations along. Punch's dress, appearance, and squeaky voice seem to have been set by the 1730s, the sexual symbolism of the long hooked nose and the ambiguity of the voice not lost on audiences well versed in puns and hidden meanings.

The story structure which Collier recorded was probably established by the 1780s,[23] but older sources echoed. The association with gallows literature was enduring, the parody and inversion of those texts central to their appeal. One echo was from Gay's *Beggar's Opera*. Like the highwayman Macheath in the *Opera*, Punch in the 1820s sings from behind prison-bars and escapes the gallows through trickery. Again like Macheath's, Punch's mistress is called Polly, and she sings Polly's song from the *Opera*. Moreover, in all versions of the story two features never vary. First, Punch carries his identity as a cold-blooded child- and wife-murderer with determined insouciance:

Who'd be plagued with a wife
  That could set himself free
With a rope or a knife,
  Or a good stick like me?

[21] J. P. Collier's transcription of Piccini's show in 1828, *The tragical comedy or comical tragedy of Punch and Judy* (orig. pub. 1828; 1870).

[22] Leach, *Punch and Judy*, 43–4, 50.

[23] Collier, *Tragical comedy*, 33–9.

—You killed your wife and child.
—They were my own, I suppose; and I had a right to do what I liked with them.

And secondly, as Leach points out, Punch's greatness depended on his battle with the triad of social controls: wife, hangman, and devil—marriage, law, and morality. His triumph in all three battles was indispensable to his popularity. Collier had seen a showman pelted with mud when Punch was not allowed to kill the devil at the end.

The appeal of the story to people whose experience of the law was never benign need not be laboured perhaps. To enjoy Punch's triumph over the law by outrageous trickery was to laugh at a death-dealing force which was never laughable in real life:

LORD CHIEF JUSTICE Hollo! Punch, my boy!
PUNCH Hollo! Who are you with your head like a cauliflower? . . .
LORD CHIEF JUSTICE . . . You're a murderer, and you must come and be hanged.
PUNCH I'll be hanged if I do. [*Knocks down the Chief Justice, and dances and sings*]

And again:

JACK KETCH Why were you so cruel as to commit so many murders?
PUNCH But that's no reason why you should be cruel too, and murder me.

The people who watched this had all watched real hangings. Punch's laughter fed fantasies whose potency in the age of public hanging later generations would never fully realize. Here the despised underdog was ever 'game', refusing to accept 'good manners, good order, good sense, or even goodness itself' (Plate 14):

JACK KETCH Now, Mr Punch, no more delay. Put your head through this loop.
PUNCH Through there! What for?
JACK KETCH Aye, through there . . .
PUNCH What, so? [*poking his head on one side of the noose*]
JACK KETCH Not so, you fool.
PUNCH Mind who you call fool: try if you can do it yourself. Only shew me how, and I do it directly.
JACK KETCH Very well, I will. There, you see my head, and you see this loop: put it in, so [*putting his head through the noose*].
PUNCH And pull it tight, so! [*He pulls the body forcibly down, and hangs Jack Ketch*] Huzza! Huzza! . . .

I've done the trick!
    Jack Ketch is dead—I'm free;
I do not care now, if Old Nick
    Himself should come for me.

14. George Cruikshank, *Punch outwits Jack Ketch*, 1828.
The Punch and Judy show was far from childish as people on the streets understood it. The fact that most had watched real hangings, chiefly of people like themselves, gave an edge to the humour which we have forgotten. Here Punch, convicted child-murderer and wife-beater, persuades the hangman to demonstrate the noose—with fatal consequences for Ketch.

By the mid-nineteenth century Punch was being appropriated to entertain gentlefolk's children in polite drawing-rooms. The devil was replaced by a crocodile or 'bogey', and the bogey carried Punch off to the nether regions to meet the dreadful deserts which conventional morality now required.[24] But there was resistance to this sanitization. Mayhew in 1851 interviewed a punch-man who still had a firm grip on Punch's older meanings. He insisted that the complexities of Punch's tale deserved an audience more worldly than children; he hankered after the kind of audience he had enjoyed before 'opinions had changed'. Punch's real messages, he thought, must be unintelligible to

[24] Leach, *Punch and Judy*, 83–4.

polite people who preferred the 'sentimental' rather than the 'comic' parts of the tale which the street people preferred. The truth was, he told Mayhew cryptically, that Punch was 'like the rest of the world, he has got bad morals, but very few of them'; 'bad morals' were at least free of cant, he implied. They were 'true, just, right, and sound; although he does kill his wife and baby, knock down the Beadle, Jack Ketch, and the Grand Signor, and puts an end to the very devil himself'. Punch had one irresistible message, the punch-man added. It was that, once he had outwitted and killed the devil, 'we can now all do as we like!'[25] For the rest, life was an empty joke.

4. *Flash ballads and Francis Place*

The punch-man's gnomic commentary connects us with the lost mentalities we are looking for; and it is not surprising that we find them sounding most loudly in the arenas in which Punch played. In streets, markets, and fairs the rough ballads commented as sardonically as Punch did on the perils of urban life, its sexual opportunities, and the gallows which could end a life. The ballads' bludgeoning humour was just like Punch's: it lay 'not so much in the point of his replies as of his stick'.[26] Full of bravado and cheerful obscenity, bawdy songs had been an integral part of popular culture at least since the Restoration.[27] 'Flash' ballads—celebrating drink, devil-may-care heroism, and bleak endings at Tyburn—were perhaps even older. It is these that connect us most intimately with the attitudes which common people took to Tyburn or Newgate, explaining what they saw at the scaffold and why they behaved as they did. They remind us unavoidably of the murderer Sale's women; for the songs' most marked characteristic was that they confronted life's horrors without moralizing and sentimentality, and then refused to take them seriously.

Today both forms of ballad, bawdy and flash, are largely forgotten, not least by scholar-collectors who for a century past have earnestly debated the nature of 'traditional' balladry and asserted the superiority of a rural tradition over its gutter offspring. Yet most gutter songs of the later eighteenth century drew vigorously on the old balladic structures.

[25] Mayhew, iii. 43–6, 49, 53, 59.

[26] Anon., 'Horae Catnachianae', *Fraser's Magazine*, 19 (1839), 410.

[27] M. Ingram, 'Ridings, rough music and mocking rhymes in early modern England', in B. Reay (ed.), *Popular culture in seventeenth-century England* (1985), 166–97; R. Thomson, *Unfit for modest ears: A study of pornographic, obscene and bawdy works written or published in England in the second half of the seventeenth century* (1979), ch. 2.

Anonymous and impersonal narratives, they too used 'isolated action shots' to achieve intensity and compression, and stereotypical references to convey images to audiences in whose understandings the narrator had confidence. They too expressed primal attitudes about survival, justice, courage, sex, and death which more self-conscious literature lost sight of.[28]

It is difficult to assess their eighteenth-century currency, because most of the gutter songs were later suppressed.[29] Their ubiquity is clear, however. 'Obscene ballads and songs in praise of thieving were the only ones sung about the streets. . . . No others were circulated in that manner,' one of Francis Place's cronies recalled.[30] Their ubiquity was also witnessed in chronic anxieties about them. In 1750 the bishop of London blamed the late earthquake on 'the infamous and obscene songs and ballads that are openly sung on our public streets, to the great uneasiness of all modest and virtuous persons who are passing by; to the great corruption and depravity of our servants and children and to the total discouragement of virtue among the common people in general'.[31] The *Gentleman's Magazine* called for the licensing of ballad singers to prevent the corruption of young girls among 'our middling gentry'; the *London Magazine* in 1759 wanted the government to employ ballad singers to sing songs of 'a proper tendency'; in 1775 a citizen expected the lord mayor to act against the 'seldom less than five ballad singers at different stands every evening in St Paul's church-yard' who drew 'a crowd of loose and disorderly persons about them', to the peril of respectable people's pockets.[32] When the *Universal Daily Register* in 1785 deplored 'the indecent songs and immoral ballads chanted in almost every street of this metropolis' as 'destructive to the growing youth of both sexes' and 'marked with sentiments diabolical and unnatural', it advised that their singers be whipped and removed from the streets.[33] And a decade later Patrick Colquhoun also insisted that lewd and flash songs be out-

[28] J. S. Bratton, *The Victorian popular ballad* (1975), 7–11; V. Gammon, 'Song, sex and society in England, 1600–1850', *Folk Music Journal*, 4/3 (1982), 208–45; R. Elbourne, *Music and tradition in early industrial Lancashire, 1780–1840* (Woodbridge, 1980).

[29] Most ballads anyway were romances, tales, and songbooks: of the 2,800 chap-books and ballads collected by Boswell and his sons from the mid-18th to early 19th cents. (now at Harvard), only 5% were about crime and criminals: S. Pedersen, 'Hannah More meets Simple Simon: Tracts, chapbooks and popular culture in late eighteenth-century England', *Journal of British Studies*, 25 (1986), 101, 104.

[30] Richard Hayward's notes, in Place papers, BL Add. MS 27825, fos. 165–7. Henceforth references to 'Place' will be to folios or numbered ballads in this volume, unless otherwise indicated.

[31] *Lady's Magazine*, 1750, cited in Place, fo. 143.

[32] R. S. Thomson, 'The development of the broadside ballad trade and its influence upon the transmission of English folk songs', D.Phil. thesis (Cambridge, 1974), 176–9.

[33] *Universal Daily Register*, 8 Apr., 10 June 1785.

lawed. Their singers should give 'a better turn to the lowest classes of people' by being required to sing of 'loyalty to the Sovereign, love to their country and obedience to the laws'.[34]

By the 1790s proposals of these kinds were taken up in the project of counter-revolution. Hannah More pumped out her Cheap Repository Tracts as part of 'a plan for substituting something better for ballad-singing' (Plate 15, p. 162 below),[35] while the old singers were hounded by John Reeves's Association for Preserving Liberty and Property against Republicans and Levellers and then by the Anti-Vice Society, or dragooned into loyalist causes. When William Godwin emerged from Tom Paine's *in absentia* trial in 1792, he 'had not walked three streets, before I was encountered by ballad singers, roaring in cadence rude, a miserable set of scurrilous stanzas upon his [Paine's] private life'; the streets were awash with handbills purporting to answer Paine's reasonings.[36] During the wars with France patriotic and naval songs so swamped the market that the old street chanting declined: 'three-fourths of the demand vanished with the war; and the songs upon home-subjects went but tamely off after the excitement of battle and heroic ordeals'.[37] Vigilantes wielded the threat of summary prosecution convincingly enough to persuade Francis Place that coarse balladry had been 'extinguished'. 'There were probably a hundred ballad singers then [1780s] for one now,' he noted in 1819:

> If any one was found singing any but loyal songs, he or she was carried before the magistrate who admonished and dismissed him or her, they were then told they might have loyal songs for nothing, and that they would not be molested while singing them. Thus the bawdy songs, and those in favour of thieving and getting drunk, were pushed out of existence. . . . The old blackguard songs were in a few years unknown to the youths of the rising generation, thus the taste for them subsided. I have no doubt at all that if the ballad singers were now to be left at liberty by the police to sing these songs . . . the public in the streets would not permit the singing of them. Such songs as even 35 years ago produced applause would now cause the singers to be rolled in the mud.[38]

Place was right that the war years hugely weakened the flash tradition. By the second quarter of the new century Tyburn ballads and

[34] *Treatise on the police of the metropolis* (1795), cited by E. J. Bristow, *Vice and vigilance: Purity movements in Britain since 1700* (1977), 43–4.

[35] Thomson, 'Broadside ballad trade', 127.

[36] W. Godwin, *Uncollected writings, 1785–1822*, ed. J. W. Marken and B. R. Pollin (Gainsville, Fla., 1968), 116.

[37] Charles M. Smith, 'The press of the Seven Dials', in his *The little world of London* (1857), 255.

[38] Place, fos. 144–5.

highwayman romances about Turpin or Sheppard comprised a diminishing part of publishers' output. Most scaffold ballads in the nineteenth century were sanitized songs of repentance pumped out in execution sheets, moralizing as cant songs never did. Songs 'in favour of thieving' were so far forgotten by later generations that W. H. Ainsworth in the 1830s thought it odd that 'criminals' themselves had never produced canting 'poesy', and set about composing some himself.[39]

Place had a special interest in the street ballads. Born in 1771 the son of a sponging-house keeper and publican, and apprenticed to a breeches-maker in 1785–9, the adult Place invested much energy in distancing himself from his dissolute youth. He thought that one measure of how far both he and his class had travelled was the improved character of the people's songs. Affecting incredulity that songs so larded with sexual and thieving terms could ever have been sung as openly as they had been in the 1780s, he and two friends transcribed those which they could remember.[40] This enterprise made its point well. One of his helpers said that it had not occurred to him hitherto that 'the change in the manners of the lower orders had been so great as it really is since my boyhood, say from 1780 to 1792—the period of the breaking out of the French Revolution and the institution of the Sunday Schools'.[41] So we are in Place's debt. The suppression achieved after 1790 was effective enough for the thirty-two ballads or snatches of ballads he transcribed to rank now among the main material upon which analysis can be based. Apart from a few broadsheet reprintings and relics in collections, little would survive from the rough balladry of the later eighteenth century were it not for Place.[42]

[39] Preface to *Rookwood* (1878 edn.), pp. xxix–xxx.

[40] M. Thale (ed.), *The autobiography of Francis Place* (Cambridge, 1972), pp. xxv–xxvi.

[41] Hayward's notes, in Place, fos. 165–7; Place, *Autobiography*, 51. Place headed his transcriptions, 'Specimens of Songs and parts of Songs (from memory) sung about the streets within my recollection and without molestation; Written 1819.' 'The songs that are defective', he added, 'were all known to me when I was an apprentice-boy, but [are] forgotten past recollection. Those made perfect have been so made from the recollection of old fellows whom I have known—mainly by Mr Tijou and Richard Hayward.'

[42] The only other substantial collections of criminal and canting songs mixed literary and broadside material indiscriminately from a period of 350 years and do not reflect what was sung on the streets before 1789. They are J. S. Farmer, *Merry songs and ballads prior to the year AD 1800* (5 vols., 1895–7; facs. reprint, introd. G. Legman, New York, 1964), and the same author's *Musa pedestris: Three centuries of canting songs and slang rhymes* (1896). For survivals of bawdy ballads, see the Madden Collection of Broadsides (Cambridge University Library), particularly 'Bowl away' (21.128) and 'Sandman Joe' (21.229), the latter of which is reproduced from Place in the appendix to this chapter. G. Speaight (ed.), *Bawdy songs of the early music halls* (Newton Abbot, 1975), reprints a selection from William West's bawdy songbooks of the mid-1830s; see also Palmer, *Sound of history*, ch. 6. For pornographic-cum-radical publication in the 1820s (partly street bawdy), see I. McCalman, *Radical underworld: Prophets, revolutionaries and pornographers in London, 1795–1840* (Cambridge, 1988), 205–19.

5. *Promiscuous audiences*

Who heard and knew the rough songs? We should be wrong to assume that audiences were only plebeian, though the assumption is wholly understandable if we go by some accounts of them. Place himself dismissed those who applauded the songs as 'fools, idlers and pickpockets'. They 'were sung in all parts of the town', he reported; 'every one of the songs mentioned might be bought of those who sold ballads in the streets against the walls, each song for a halfpenny.' 'In proportion to the vileness of the songs and the flash manner of singing them was the applause the singers received.' Also, many were sung in the canting tongue, which surely limited their appeal? Allegedly Elizabethan in origin, cant has always been regarded as the peculiar, secret, and protective anti-language of counter-cultures of 'egyptians', rogues, and vagabonds, beyond the understanding of better sorts of people.[43] It was supposed to mystify the uninitiated and sometimes did. In the 1790s the military in charge of convicts in New South Wales needed interpreters to translate convicts' evidence in court, so thick was their slang; and the canting dictionary compiled in 1812 by the twice-transported thief and swindler J. H. Vaux might have been commissioned to meet this need.[44] Always noted, too, is the durability of cant, another alleged measure of its separateness. Many of its terms recur over three or four centuries. This, along with its opacity and its reference to trickery, thievery, imprisonment, and hanging, suggest cant's underworld provenance clearly enough. Here is an example (there are many more to come) for readers who enjoy unknotting puzzles:[45]

[43] 'By cant is meant the specific slang of gipsies and other vagabonds, of thieves and other malefactors': E. Partridge, 'Introduction' to F. Grose, *Classical dictionary of the vulgar tongue*, ed. Partridge (1931), p. ix. A bibliography of canting material is in R. C. Alston, *A bibliography of the English language* (12 vols., 1965–73), x. Readings of cant which assume its opacity, and readings of the allegedly discrete 'criminal class' that is supposed to have used it, shape nearly all discussions of this anti-language, from Mayhew on to M. McIntosh, 'Changes in the organisation of thieving', in S. Cohen (ed.), *Images of deviance* (Harmondsworth, 1971), 103 ff. See A. V. Judges (ed.), *The Elizabethan underworld* (1930); E. Partridge, 'The language of the underworld', in his *Here, there and everywhere: Essays upon language* (1950), 97; G. Salgado (ed.), *Cony-catchers and bawdy baskets* (Harmondsworth, 1977); J. L. McMullan, *The canting crew: London's criminal underworld, 1550–1700* (New Brunswick, NJ, 1984). For a healthily sceptical view of the separateness of the early underworld, see I. W. Archer, *The pursuit of stability: Social relations in Elizabethan London* (Cambridge, 1991), 204–15.

[44] Vaux's *Vocabulary of the flash language*, written in 1812, was published with his *Memoirs* in 1819: see the introd. by N. McLachlan (1964 edn.), p. xvii.

[45] Place no. 11: originally from [Messink], *The choice of harlequin, or, The Indian chief: A pantomimical entertainment . . . as it is acted with the highest applause at the Theatre Royal, Covent Garden* (1782). Place's transcription is corrected and extended from this source. Each verse refers to characters in the performance—an insurance-office keeper, a French macaroni, two genteel harlots, and a hackney coachman— as they are mockingly addressed by the bridewell keeper.

Ye scamps,[a] ye pads,[a] ye divers,[b] and all upon the lay,[c]
In Tothill Fields' gay sheep-walk,[d] like lambs ye sport and play:
Rattling up your darbies,[e] come hither at my call;
I'm jigger dubber[f] here, and you are welcome to mill doll.[g]

The game you've play'd my kiddy, you're always sure to win,
At your insurance-office, the flats[h] you've taken in,
First you touch the shiners[i]—the number up, you break;
With your insuring policies, I'd not insure your neck.

The French with trotters[j] nimble could fly from English blows,
And they've got nimble daddles[k]—monsieur plainly shows.
Be thus the foes of Britain bang'd; ay, thump away, monsieur;
The hemp you're beating now, will make you a solitaire.

My peepers![l] who've we here? Why, this is sure black Moll;
Why, ma'am, you're of the fair sex, and welcome to mill doll;
The cull[m] with you who'd venture into a snoozing-ken,[n]
Like blackamoor Othello, should—'put out the light, and then—'.

I say, my flashy coachman, that you'll better take care,
Nor for a little bub,[o] come the slang[p] upon your fair [fare?];
Your jazy[q] pays the garnish,[r] unless the fees you tip;
Tho' you're a flashy coachman, here the Gagger[s] holds the whip.

Note: Unless indicated otherwise, translations henceforth are from E. Partridge, *A dictionary of the underworld* (3rd edn., 1968).

[a] Highwaymen; [b] pickpockets; [c] engaged in crime; [d] prison; [e] shackles, fetters; [f] warder; [g] pick oakum; [h] false dice, or dupes (both possible); [i] coins; [j] feet; [k] hands; [l] eyes; [m] man, prostitute's dupe; [n] brothel; [o] liquor; [p] play a mean trick; [q] wig; [r] gaoler's fee; [s] turnkey.

The truth is, however, that in Place's youth many people other than plebeians understood these things. Just as those who watched executions were far from uniformly *lumpen*, so the canting songs were accessible to skilled artisans, tradesmen, and the rakish but gentlemanly *demi-monde*. These proximities were to be eroded with time, but in the late eighteenth century 'the two extremes of *very high* and *very low* border close on each other; and the manners, language &c. of the ragged rabble differ in [only] a few instances (and those merely circumstantial) from the vulgar in lace and fringe'.[46] Place himself had learnt the songs at street corners and in cock-and-hen clubs and free-and-easies.[47] He had also learnt them from the 'respectable tradesmen' in his father's public house. Here indecent songs 'were sung with considerable humour . . . every one within hearing was silently listening, and at the conclusion of the song expressed their delight by clapping their hands

[46] G. Parker, *A view of society and manners in high and low life* (2 vols., 1781), i, p. ix.

[47] For Place's description of these venues and the loose promiscuities of apprentice life (but 'the class to which I belonged was by no means the lowest'), see *Autobiography*, 73–8.

and rapping the tables'.[48] He recalled 'when a boy of 10 years of age, being at a party of 20, entertained at a respectable tradesman's, who kept a good house in the Strand, where songs were sung which cannot now be more than generally described from their nastiness, such as no meeting of journeymen in London would [now] allow to be sung in the presence of their families'.[49] His cronies Mr Tijou ('a carver and gilder in a large way of business in Greek Street, Soho') and Richard Hayward (a struggling attorney who later joined the London Corresponding Society along with Place)[50] had learnt the songs in such venues too. Hayward's note in the Place papers insisted that the low ballads were appreciated not only by 'the lower orders but by the middling classes also'. Admittedly they were sung by 'flash and fancy men with the peculiar dress—rollers at the cheeks, striped silk stockings, numerous knee strings, long quartered shoes and all en suite'. But they were also sung by master tradesmen at the Dog and Duck or the Temple of Flora, where they mixed with the flash set promiscuously.[51]

Audiences stretched higher yet. Rakes, half-pay officers, and bucks in the twilit zone beyond Society flirted with the rough trade in unsnobbish conviviality. Place noted that the bawdy 'Sandman Joe' was sung in the Crown and Anchor in the Strand 'by gentlemen'. When George Parker apologized for knowing cant he explained that it resulted from his 'happy knack of conversation . . . as likewise a talent of easily and naturally accommodating myself to the manners of every rank in life'.[52] Captain Francis Grose likewise 'by no means added to his reputation' (William Hone observed) by publishing a *Classical dictionary of the vulgar tongue* in 1785. Despite his military commission and fellowship of the Society of Antiquaries, this 'greatest antiquary, joker, and porter-drinker of his day' flaunted that venturesome taste for slumming it in the boozing kens of St Giles to which we owe his *Dictionary's* conception.[53] So there was a well-heeled market for such publications. John Poulter's *Discoveries* went through twelve editions between 1754 and 1761, and Grose's *Dictionary* went through five between 1785 and 1823. No doubt

[48] Ibid. 77, 57–8. [49] *SC on education*, PP 1835, vii. 69. [50] Place, *Autobiography*, 175, 272.

[51] These were less than genteel venues, according to Place: 'I have seen two or three horses at the door of the Dog and Duck in St George's Fields on a summer evening, and people waiting to see the Highwaymen mount' (BL Add. MS 27826, fo. 189); there too 'I have . . . seen the flashy women come out to take leave of the thieves at dusk, and wish them success' (*SC on education*, PP 1835, vii. 70). The Dog and Duck merited a ballad of its own: 'Come blowen it's past four o'clock, / It's time to repair to the Dog and Duck, / And if with a cull you chance to meet, / Bowl him round to Catherine Street' (notorious for its prostitution): Place no. 3.

[52] G. Parker, *Life's painter of variegated characters in public and private life* (1789), 138.

[53] Grose, *Dictionary of the vulgar tongue*, 385–6.

the songs carried different meanings for these several audiences. The gentleman's foray into the lower culture was in pursuit of a controlled emancipation from which he could return when he chose. Moreover, a gentleman's understanding of cant probably came as much from its long print record as from familiarity with the fast and accented dialogues of the streets. The fact remains that these fascinated visitors to the *risqué* worlds of the taverns and gaming dens were not yet obliged to repudiate plebeian cultures, as Dixon's generation would have to.

Cant had also been opened up by the popular dramatists of Bartholomew Fair and Drury Lane. Middleton, Dekker, and Fletcher had included canting jingles in their plays. A few drinking-songs survive from the seventeenth century. The first canting ballad proper which Farmer records was 'A budg and snudg song', published in 1676.[54] *The new canting dictionary* of 1725 included many canting songs, while Jack Sheppard's execution in 1724 unleashed a spate of pantomimes, biographies, and ballads in which cant was used liberally:[55]

From prigs that snaffle the prancers strong,[a]
 To you of the Peter Lay,[b]
I pray you now listen a while to my song,
 How my boman he kick'd away.[c]

He broke thro' all rubbs in the whitt,[d]
 And chiv'd his darbies in twain;[e]
But filing of a rumbo ken[f]
 My boman is snabbled again.[g]

I Frisky Moll, with my rum coll,[h]
 Wou'd grub in a bowzin ken;[i]
But ere for the scran he had tipt the cole,[j]
 The harman he came in.[k]

A famble, a tattle, and two popps,[l]
 Had my boman when he was ta'en;
But had he not bowz'd in the diddle shops,[m]
 He'd still been in Drury Lane.

Note: The text provided its own key as follows:

[a] Gentlemen of the pad; [b] those that break shop-glasses, or cut portmanteaus behind coaches; [c] her rogue had got away; [d] Newgate, or any other prison; [e] saw'd his chains in two; [f] robbing a pawn broker's shop; [g] taken again; [h] clever thief; [i] would eat in an alehouse; [j] before the reckoning was paid; [k] the constable; [l] a ring, a watch, and a pair of pistols; [m] Geneva shops.

54 *Musa pedestris*, 30–2.

55 Moll Frisky's song in John Thurmond's Drury Lane burlesque, *Harlequin Sheppard*, possible prototype for *The Beggar's Opera*, produced 12 days after Sheppard's execution: H. Bleackley and S. M. Ellis, *Jack Sheppard* (1933), 72–5.

By the later eighteenth century the opacity of songs of this kind was so much diminished that Place provided translations without difficulty. His recollections also make it clear that the songs had moved from the theatres to the streets. Grose implied the open currency of some of the 3,000 or so canting terms he collected by noting how easy it was to hear them among the crowds gathered along 'the Tyburn way': 'Many heroic sentences, expressing and inculcating a contempt of death, have been caught from the applauding populace, attending those triumphant processions up Holborn-hill, with which many an unfortunate hero till lately finished his course; and various choice flowers have been collected at executions.' In fact cant was becoming the humorous argot not only of apprentices but of the smart sets as well. It cohabited easily with 'those burlesque phrases, quaint allusions, and nicknames for persons, things, and places, which, from long uninterrupted usage, are made classical by prescription'.[56] Cambridge swells a generation later were plagiarizing Grose with their own *Dictionary of buckish slang, university wit, and pickpocket eloquence*. Tom Moore published a comic canting poem advising the Congress of Aix-la-Chapelle on the history and skills of pugilism. Pierce Egan's slang-loving gentlemen Corinthians were one of the publishing successes of the 1820s.[57] By the 1830s and 1840s Ainsworth, Lytton, and Dickens were resuscitating cant in their Newgate novels.

Cultural interpenetrations of these kinds, the low colonizing the high, are familiar to all generations. But this one reminds us that demarcations between the 'criminal classes' and the 'respectable' were much hazier in Place's youth than law-enforcers and social investigation and policy made them a century later. We should not exaggerate the eighteenth-century underworld's coherence and impenetrability: indeed, we ought to deny it.[58] If cant had once been the secret language of rogues, such people in the eighteenth century coexisted easily with

[56] Grose, *Dictionary of the vulgar tongue*, 7–8, 9–10.

[57] *Lexicon balatronicum: A dictionary of buckish slang, university wit, and pickpocket eloquence . . . by a member of the Whip Club, assisted by Hell-fire Dick* (1811); *Tom Crib's memorial to Congress . . . by one of the fancy* [Tom Moore] (1819); P. Egan, *Life in London, or The day and night scenes of Jerry Hawthorn, Esq., and his elegant friend Corinthian Tom, accompanied by Bob Logic, the Oxonian, in their rambles and sprees through the metropolis* (1821).

[58] 'Pilfering and Thieving especiall[y] were not then as now almost wholly confined to the very lowest of the people, but were practiced by tradesmens sons, by youths and young men who would now no more commit such act than would the sons of a well bred gentleman, thieving had not as yet [become] a trade to be followed by those who lived by it as it has now become': Place, *Autobiography*, 57. Cf. Beattie, p. 256; P. Linebaugh, *The London hanged: Crime and civil society in the eighteenth century* (1991), ch. 3.

artisans, tradesmen, and apprentices—an oblique hint again at the promiscuous identities of those early scaffold crowds.

Who wrote the songs? Most were doubtless composed by the struggling hacks who had screwed their shillings per ballad from publishers for a century past, people not easily distinguished, except in literacy, from the characters of the *Newgate calendars* which they also helped to compile.[59] Actors were another source. John Harper (d. 1742), who wrote Frisky Moll's song in *Harlequin Sheppard* cited above, played at Bartholomew and Southwark Fairs and at Drury Lane as 'a jolly facetious low comedian'.[60] The soldier-actor George Parker (1732–1800), who likewise composed and published canting ballads, ended his days by selling gingerbread-nuts at fairs and races and died in a Coventry workhouse.[61] To judge from the shaky metres, rhymes, and grammar of some in the Place sample, many were composed by the singers themselves, and later evidence confirms this.[62]

But even if many of the blackguard songs originated in the cultures of literacy, they were rooted deeply in plebeian life. Their anchorage in concrete topographies shows that they were written for people for whom the streets and taverns of London marked the limits of the known universe:

> There was a jolly butcher,
> He liv'd at Norton Falgate,
> He kept a stall at Leadenhall,
> And got drunk at the Dog at Aldgate.[63]

Their bawdiness, their darkly comic or sardonic tone, their fast-moving and concretely grounded narratives, their incessant preoccupation with strategies of coping and surviving, their refusal to moralize, their mistrust of the world as a place of limited good which could not be changed and in which you could prosper only at another's expense—all point to an unforced relationship with singers and listeners who knew

[59] Rogers, *Grub Street*, 290.

[60] R. D. Hume, *Henry Fielding and the London theatre, 1728–1737* (Oxford, 1988), 176–9.

[61] The 3 polished canting ballads in Parker, *Life's painter* (pp. 122–35), were presumably by Parker's hand, although possibly derived from street products.

[62] A song was often composed by the sort of man 'who first chants it about the streets . . . oftener still [by the] man whose chanting and pattering days are over, who has lost his voice and worn out his legs in the trade, and is reduced to his last shifts for a living': Smith, 'The press of the Seven Dials', 254.

[63] Place no. 25. In debt and chronically drunk, the butcher is encouraged by his girl to leave his wild companions—in vain, however: for 'While thus they talked so clever, / The constables came smack in, / To the Bridwell in Clerkenwell / They sent poor Nelly packing. / And as for Dick the butcher / These bailiffs took him napping, / No bail could he get / And he lies for his debt / At a spunging house in Wapping.'

the grimmer realities of life very well, and who knew its sexual and criminal escape routes too.

6. *The 'criminal' and bawdy ballads*

Looked at coolly, the world of the ballads was a world of nightmare. Watchmen were corrupt, magistrates merciless, streets odorous, husbands cuckolded, women treacherous, pox ubiquitous, penises either compulsively questing or disappointingly limp. Significantly, the only affliction not dwelt on in the songs was poverty. Rather, poverty was the taken-for-granted backdrop to all plebeian devices and ingenuities. No political postures sound forth either. 'Jack Chance', for example (Place no. 1, Chapter Appendix), about a highwayman who joined the Gordon riots, was 'made after the executions of the rioters in 1780 and . . . sung about the streets with great applause'. But although twenty-two rioters were hanged then, the song's commentary on the judicial slaughter was only faintly ironic, if irony was meant at all:

> But a victim he fell to his country's laws,
> And died at last in religious cause.
> No popery made the blade to swing,
> And when tuck'd up he was just the thing, just the thing,
> Just the thing, just the thing,
> And when tuck'd up he was just the thing.

Reminding us again of Sale's women, the energy of the songs was put into celebration rather than into protest, and what was celebrated most was triumph over life's adversities, and especially getting something for nothing. 'Drunk the other night' (Place no. 26, Chapter Appendix) celebrates bleak little triumphs over treacherous whores and corrupt watchmen. 'Teddy Blink and Bandy Jack' (Place no. 28, Chapter Appendix) celebrates the excitement of thieving. This had 'a great run', Place's friend Hayward noted; indeed, it was one of the rare thieving songs from that time to be reprinted by Pitts after 1819. The chief hero in the balladeers' pantheon was the highwayman or practised thief; but his reputation also mainly depended on what he got away with. Jack Sheppard survived in song not only because of his famous capture in 1724 by Jonathan Wild but also because in his several escapes from Newgate he achieved the impossible. 'Oh Rare Turpin, Hero' was similarly wondered at because he had ridden from York to London in a day.

Place's era had its highwaymen heroes too. The most notable was 'Sixteen-String' Jack Rann. Place was aged only 3 when Rann met his end at Tyburn in November 1774, but he 'was much talked of since I can remember', he recalled. Rann's crimes were standard—pickpocketing in youth, followed by the increasingly daring thefts of watches and money on the Hounslow Road. It was his panache that they remembered him for: a 'Macheath highly improved and exaggerated', the *Public Advertiser* called him after one lucky acquittal. He rose from employment as an ostler to become the Beau Nash of the common people, changing the fashions of the flash set by wearing sixteen strings at his breeches' knees in place of the buckles favoured hitherto.[64] When he appeared before Sir John Fielding at Bow Street in 1774 he won the turnkeys' admiration by nattily tying blue ribbons to his irons and sporting a bundle of flowers at his breast. Acquitted, he paraded the pleasure-gardens of Bagnigge Wells in a scarlet coat, tamboured waistcoat, white silk stockings, and laced hat. People went to gaze at him and his paramour Miss Roach, and he was mobbed by admirers. At Tyburn executions he would claim a front view out of 'professional interest'. Losing a ring, he declared that it was worth only a hundred guineas which an evening's work would replace. At the trial which condemned him at last he wore a pea-green suit, a ruffled shirt, and a hat bound with silver strings; in the condemned cell of Newgate he dined with seven lady-friends. He went to the scaffold with a nosegay at his breast. The people's bards usually knew better than to strive for elegiac heights, but in honour of such a man's passing an exception was made:

Farewell ye rooks, farewell ye plains,
No more Miss Roach will on you reign.
Your sighs and tears are all in vain.
We part but ne'er shall meet again.

I wish I was a country girl
My cows to milk, my lambs to tell;
And love I'd never took in hand,
I'd never parted with Jack Ran.

'This song was popular notwithstanding Jack had been hanged many years before,' Place noted.[65]

The flash ballads did not normally romanticize the highwayman,

[64] Thereafter 'it was a fashion with costermongers, coal-heavers, drovers and many others to wear breeches very short at the knees—they were always left unbuttoned and the strings with which they should be tied hung down under the knees and stockings were usually fastened with broad red worsted garters. This mode was considered "very knowing" ': Place, fo. 161.

[65] J. Villette, *The annals of Newgate* (1774), iv. 378–88; Anon., *The malefactor's register* (1779), v. 138–46; Rayner and Crook, iv. 99–102; Place papers, 38 (BL Add. MS 27826), fos. 78–86, and 37 (Add. MS 27825), fo. 148 (no. 7).

however. It was the polite classes who did that. Most highwaymen were not at all the genteel swashbucklers of legend. Waifs, strays, and illiterates, their kinship was with footpads and many were brutal.[66] The songs knew this, so what mattered more was the highwayman's full actualization. Curricula vitae must be supplied, even down to addresses (Place no. 8, Chapter Appendix):

> Young Morgan was a lusty blade,
> No lad of better courage;
> Much gold he got on the highway
> Which made him daily flourish.
> In Wentworth Street, his lodgings were
> Among those flashy lasses,
> Until he came a gentleman,
> And left off driving asses.[a]

[a] Stealing handkerchiefs.

Jack Chance's biography was similarly detailed. A foundling, he was brought up by Billingsgate Nan and apprenticed to thievery; he was illiterate, and when he first became a highwayman he had to *hire* a horse. Again, it was not glamour that was celebrated in such people, it was effrontery. You had to defy magistrates if you were to be looked up to, as McFlanagan defied the blind Sir John Fielding at Bow Street when he was offered impressment in the navy in place of a hanging trial:

> 'Says he, McFlanagan, a fine fellow you,
> You're just fit to go to sea.'
> 'Why you lie you old blind bugger,
> I'll be back by Christmas Day.'

And it was consistent with breezy fantasies of this kind that many of the favourite songs would gallop along in jovial doggerel and rollicking rhythm which allowed no room for pathos:

> With my popps[a] in my pocket and a cutlass in my hand
> So I rode up to the diligence and bid the buggers stand . . .
> . . . .
> As we rode o'er Finchley common, the swells were standing there,
> Here comes a bloody scamping blade, only do look there . . .
> . . . .
> It's flash to the cross roads and never make a stand,
> From Finchley up to London, bearing loaded popps in hand.

[a] Pistol.

[66] Faller, *Turned to account*, 175–8; F. McLynn, *Crime and punishment in eighteenth-century England* (1989), ch. 4.

If getting away with something for nothing was one chief pleasure of life, the other chief pleasure was sex. Of all the rough ballads it was the bawdy songs that affirmed life most positively. And what they conceded to women's role in life has some bearing on the unapologetic presence of women at scaffolds, perhaps on Sale's womenfolk's poise too. A few of the bawdy ballads were obscene,[67] but of most of them we can say that in a world which offered few pleasures the celebration of the greatest pleasure was no great shame. Some celebrated it elegantly:

> One night as I came from the play
> I met a fair maid by the way;
> She had rosy cheeks and a dimpled chin,
> And a hole to put poor Robin in.[68]

Others rumbustiously:

> First he niggled her then he tiggled her
> Then with his two balls he began for to batter her
> At every thrust, I thought she'd have burst
> With the terrible size of his Morgan Rattler.[69]

Others aspired to nice conceits, as in the seduction in 'Gee ho Dobbin', when for one of the protagonists the earth really did move:

> Thus down in the waggon this damsel I laid
> But still I kept driving for driving's my trade
> As her bubbies went up her plump buttock went down
> And the wheels seemed to stand and the waggon go round.[70]

Nor was this a culture which denied women their full participation in the sexual comedy.[71] For every song sung by the male in the Place sample, at least one was sung by the female—seldom to male credit. One song Place recalled was 'about tying a mouse to a man's yard which had no erection and letting a cat into the room to run away with both'. In another the wife tired of her husband and looked 'for some lusty fellow / Who's able to give me some reason to laugh'. 'The female volunteers' were so sexually energetic that they defeated the general with his geni-

[67] e.g. 'The phlegm pot' (Place no. 32), a poisonous song about the pox: 'There's many a fine and flashey beau, / That's taking pills for what you know; / They are all beshit from top to toe, / Though they strut St James's Mall; / There's many a lady at Vauxhall, / To go to school must have a call; / Then up in a corner she lets it fall / Where it stinks as strong as Hell.'

[68] Place no. 16. Palmer transcribes a text of 10 verses published in a garland of 1796: *Sound of history*, 217–18.

[69] Hayward notes, in Place papers, BL Add. MS 27825.

[70] Place no. 33; for the music-sheet, see Speaight (ed.), *Bawdy songs*, 22.

[71] It is misleading to dismiss the songs for an 'underlying contempt for women': Thomson, *Unfit for modest ears*, 12.

tal 'cannon and bombs'. And women speculated freely on the qualifications of future husbands:

> Some say that a tailor my husband shall be
> But a tailor good lord why he's no man for me
> For his nose and his arse too near they do meet
> That I think that his heat can hardly be sweet.

Place recalled 'a description of a married man who had a lecherous wife, it described his being a pale fellow reduced by her to a skeleton. I can remember the last two lines in consequence of the shout which was always set up as the song closed with them':

> For which I'm sure she'll go to Hell
> For she makes me fuck her in churchtime.

Among the ballad singers Place remembered best from the 1780s were the pair of women who took up evening pitches on the Strand to sing:

> For my smock's above my knee, she did say, she did say
> You may have a smack at me, bowl away, bowl away.

It was a half-century later that songs began to deny women an equality in sexual enjoyment.[72] A later version of 'Bowl away' (printed in Birmingham probably in the 1830s[73]) ends its explicit and otherwise engaging tale of mutually collaborative seduction and copulation (and indeed of male impotence) with an abrupt intrusion of post-coital male punitiveness:

> O behold my rolling eye, she did say, she did say;
> O behold my rolling eye, she did say:
> Besides my milk white thigh,
> And there's something else that's nigh, bowl away, bowl away,
> And there's something else that's nigh, bowl away.
>
> . . . .
>
> And her cuckoo's nest he spied, [*etc.*]
> As her legs she opened wide, [etc.]
>
> What the devil's in the man [*etc.*]
> It will neither sit nor stand, [*etc.*]
> Tho' I rolled it in my hand, bowl away, [*etc.*]

[72] Even so, Bratton notes in the Victorian sexual ballads a culture which was not 'exclusively the creation of the male mind, in which women . . . played an active part'; the female joke at the male's expense survived: *The Victorian popular ballad*, 160–1.

[73] The printer's imprint is omitted and the dating is inferred from its contiguity in the Madden Collection with other slips of that decade: Madden Collection, 21.128 and 230.740.

When he had spent his store, bowl away, bowl away,
When he had spent his store, bowl away,
What could a man do more?
So he damned her for a whore, bowl away, bowl away.

Place's women are unlikely to have sung the last verse. Nor would it have been favoured by another pair of women street singers whom Place remembered, famous for their chanting of 'Sandman Joe' (Place no. 23, Chapter Appendix). As they concluded the song 'amidst roars of laughter' they bucked their pelvises and shammed orgasm.[74]

7. *Tyburn ballads and the game death*

How can the mentalities witnessed in this material be linked to what happened at the eighteenth-century scaffold? We have noted shared understandings across class, the implied independence of plebeian women, and the celebration of luck, daring, and 'face'. What we note now is that knowing, reality-facing fatalism which Sale's women exhibited—the knowledge that, if life was a joke, its ending was not the worst of outcomes; that among survivors life must continue; and that contrition was the last thing to be displayed.

Our best way into these postures is through the ballads which ended at Tyburn itself. The finest in the Place collection is sung in the canting tongue from inside prison and at Tyburn. Though Place did not know this, its original was *Jack Sheppard's last epistle*, published in the *Daily Journal* after Sheppard's execution in 1724. It survived (presumably in memory) for sixty years before Place learnt it.[75] Place's version differs slightly from the original. Some earlier allusions are expunged, the phrasing is sharper, and an additional verse is supplied. But in both versions the ballad's dark and haunting power derives from the strength of metre and rhyme, from the pathos of understatement, and from the sardonic dying fall of the conclusion. Although written from a developed literacy, and so probably from outside the experience it describes, its dis-

[74] Place no. 23: 'It was usually for a long time on Saturday nights sung in an open space at the back of St Clements in the Strand at the front of an alehouse door call'd the Crooked Billet by two women who used to sham dying away as they concluded the song—amidst roars of laughter'; fo. 149: 'This used to produce great shouts of applause at the end. The women, who sung it, managing the last two lines in a way that may easily be conceived'; also: 'as sung in the [illegible] Society. This was a club, held in the Crown and Anchor Tavern in the Strand, by Gentlemen.' A weakened and bowdlerized version of 'Sandman Joe' was printed in a Birmingham slip of the 1830s with the 4-letter words omitted: Madden Collection, 21.229.

[75] Place no. 24. See *Daily Journal*, 16 Nov. 1724, reproduced in facsimile in C. Hibbert, *The road to Tyburn: The story of Jack Sheppard and the eighteenth-century underworld* (1957), 141.

tinction lies in its confidence in the shared understandings of its audience. It evinces no interest in the antecedent story or the moralizing which characterizes the balladry of the last dying speeches. As it addresses the denizens of Drury Lane and St Giles, it presumes a collective familiarity with the experience it describes and with the attitudes appropriate to it—and moral judgement was not one of them. This was how it was and, familiarly, must be:

To the hundreds[a] of Drury I write,
    And the rest of my flashy companions,
To the buttocks[b] that pad it all night,
    To pimps, whores, bawds and their stallions;
To those that are down in the whit,[c]
    Rattling their darbies[d] with pleasure,
Who laugh at the mum sculls[e] they've bit
    While here they are snacking[f] their treasure.

This time I expect to be nubb'd;[g]
    My duds[h] are grown wondrous seedy;[i]
I pray you now send me some bub,[j]
    A bottle or two, to the needy.
I beg you won't bring it yourself,
    The hangman is at the Old Bailey;
I'd rather you send it by half
    For if they twig[k] you, they'll nail[l] you.

Moll Spriggins came here t' other night,
    She tipp't[m] us a jarum[n] of diddle;[o]
Garnish is the prison'rs delight:
    We footed away to the fiddle.
Her fortune at diving[p] did fail
    For which she has changed habitation,
But now the whore pads in the jail
    And laughs at the fools of the nation.

This time I expect no reprieve;
    The sheriff's come down with the warrants.
An account now behind us we leave
    Of our friends, education and parents.
Our bolts are knocked off in the whit;
    Our friends to die penitent pray us;
The nubbing cull[q] pops from the pit[r]
    And into the tumbril[s] conveys us.

Through the streets our wheels slowly move;
    The toll of the death bell dismays us.
With nosegays and gloves we are deck'd,
    So trim and so gay they array us.

The passage all crowded we see
  With maidens that move us with pity;
Our air all, admiring, agree
  Such lads are not left in the city.

Oh! then to the tree I must go;
  The judge he has ordered that sentence.
And then comes a gownsman you know,
  And tells a dull tale of repentance.
By the gullet we're ty'd very tight;
  We beg all spectators, pray for us.
Our peepers are hid from the light,
  The tumbril shoves off, and we morrice.[t]

Note: Footnotes supplied by Place or by the present author (indicated by square brackets).

[a] The wretched courts lanes and alleys in St Giles; [b] miserable ragged prostitutes; [c] Tothill Fields prison; [d] irons; [e] those whom they robbed; [f] dividing, spending; [g] hanged; [h] clothes; [i] worn out; [j] drink, liquor; [k] see; [l] seize—not for bringing the liquor but because you are known; [m] gave; [n] pot, bowl; [o] punch, mixt liquor &c; [p] picking of pockets; [q] Jack Ketch; [r] gaol; [s] cart; [t] dance on the rope.

The sombre fatalism, the standard motifs and narratives, and the abrupt finale of this song are characteristic of many such, as 'Young Morgan' and 'A new flash song' reprinted in this chapter's appendix show. But perhaps the ballad commemorating the Tyburn hanging of Jack Hall in 1707 should be our last and our key example, for it has enjoyed an astonishing longevity; it illustrates the complex relationship between literacy, orality, and collective memory better than most; and it catches accurately the structures of feeling with which few among the scaffold crowds would be unfamiliar.

The real Jack Hall had not been cast in heroic mould. He was distinguished only by the frequency of his thefts and his several escapes from the gallows. The colourful accounts of his life in the *Account* and *Memoirs* published after his death (including the story that at the age of 7 his parents sold him to a chimney-sweep) were probably largely fictitious.[76] But the fact that he died game guaranteed myth-making. Embroidered in Alexander Smith's *Compleat history of the lives of the most notorious highway-men, footpads, shop-lifts, and cheats* in 1719, Hall's life was even dramatized. Along with a harlequin Sheppard and Punch (significant conjunction!), Hall is drawn as a puppet escaping down a Newgate privy in Hogarth's comment on the way in which the London theatre had been corrupted by the rage for low pantomime, *A just view*

[76] Faller, *Turned to account*, 218–19.

of the British stage (1724). The original Hall was a prototype of the succession of robber heroes from Jack Sheppard onwards, but Hall more than most, more even than Sheppard, was made to carry the wryly defiant commentary on hanging echoed in all these ballads.

The original ballad is no longer extant, but it was almost certainly composed and printed at the time of Hall's execution.[77] Place was the first to record the song, having heard it some eighty years after Hall's death. In 1819 he could remember only two verses, so he was relying on memory to transcribe it, not print. He notes explicitly that it was sung on the streets:

> I furnish'd all my rooms, ev'ry one, ev'ry one,
> I furnish'd all my rooms, ev'ry one.
> I furnish'd all my rooms with mops, brushes, and hair brooms,
> Wash balls and sweet perfumes, them I stole, them I stole.
>
> I sail'd up Holborn Hill in a cart, in a cart,
> I sail'd up Holborn Hill in a cart.
> I sail'd up Holborn Hill, at St Giles's drunk my fill,
> And at Tyburn made my will in a cart, in a cart.[78]

Several decades later our first extant printed version was published as *Jack the chimney sweep* by Pitts.[79] The intervals between Hall's death and Place's hearing and between Place's hearing and Pitts's publication do not mean that the ballad had not been published again between 1707 and the 1820s. But a print record as exiguous as this suggests oral tradition too. 'Any thing written in voice & especially to an Old English tune', as a loyalist lamented in 1792, 'made a more fixed impression on the minds of the younger and lower class of people, than any writen [*sic*] in prose, which was often forgotten as soon as read.'[80] Certainly, the ballad invited the kinds of elaboration common to oral transmission. Pitts's version included these deepeningly sombre verses:

> I sold candles short of weight, that's no joke, that's no joke,
> I sold candles short of weight that's no joke;
> I sold candles short of weight and they nap'd me by the sly,
> All rogues must have their right so must I, so must I,
> All rogues must have their right so must I . . .

[77] The ballad carried the metre and tune of a ballad celebrating the death of Vice-Admiral Benbow in 1702, and there are contemporary references to the tune 'Chimney-sweep', Hall's alleged calling: B. H. Bronson, 'Samuel Hall's family tree', in his *Ballad as song* (Berkeley, Calif., 1969), 18–36. Bronson provides the scores for most variants of the ballad and traces the stanza pattern back to a Leveller song and to ballads of 1618 and 1567.

[78] Place no. 14; Palmer, *Sound of history*, 124.

[79] Pitts's broadsheet is in Macaulay's Ballad Collection (Trinity College Library, Cambridge); reproduced in Palmer, *Sound of history*, 9.

[80] Letter to John Reeves in 1792, cited by Palmer, *Sound of history*, 16–17.

. . . .

O they told me in the jail where I lay, where I lay,
 They told me in the jail where I lay,
They told me in the jail that I should drink no more brown ale,
 But I swore I'd never fail 'till I die, 'till I die . . .

. . . .

Now I must leave the cart toll the bell, toll the bell,
 Now I must leave the cart toll the bell;
Now I must leave the cart sorrowful broken heart,
 And the best of friends must part so farewell, so farewell,
 And the best of friends must part so farewell.

We next hear of the ballad when it was elaborated by the free-and-easy singer W. G. Ross in the 1840s. He renamed it *Sam Hall* and 'sung, or rather recited' it in the 'Cider Cellars' near the Adelphi theatre. Here it met an audience of half-pay officers, seedy lawyers, and university swells not very different from the gentlemen-*roués* who enjoyed such things sixty years before. Ballantine remembered how in Ross's rendering 'the hopelessness of [Hall's] entire life was most dramatically, and I think truly, portrayed', and Edmund Yates also recalled what 'a good bit of character acting' it was—with 'the man made up with a ghastly face, delivered it sitting across a chair, and there was a horrible anathematising *refrain*. . . . For months and months, at the hour when it would be known that Sam Hall would be sung, there was not standing place in the Cider Cellars.'[81] 'The profanity of its expression' prevented Ballantine's quoting the words, but it is probably a Ross version that survives in *Sam Hall, chimney sweep*, a ballad slip published by W. S. Fortey at the Catnach Press in the late 1840s or shortly thereafter. The tone had darkened, and the ballad was now distinguished by swear-words and a more overt contempt for authority and piety.[82] The last three of five verses run as follows:

Then the sheriff he will come,
 He will come,
Then the sheriff he will come,
And he'll look so gallows glum,
And he'll talk of kingdom come,
 Blast his eyes.

[81] Ballantine, *Experiences*, 25–6; E. Yates, *Recollections and experiences* (2 vols., 1884), i. 167–8. See H. Scott, *The early doors: Origins of the music hall* (orig. pub. 1946; 1977), 116 ff. and app. D, for 2 versions of 'Jack/Sam Hall'.

[82] Madden Collection, 11.687.

Then the hangman will come too,
  Will come too,
Then the hangman will come too,
With all his bloody crew,
And he'll tell me what to do,
  Blast his eyes.

And now I goes up stairs
  Goes up stairs,
And now I goes up stairs,
Here's an end to all my cares,
So tip up at your prayers,
  Blast your eyes.

And another version concluded with similar sentiment:[83]

And now I'm going to hell,
  Going to hell,
But what a bloody sell
If you go there as well!
  Damn your eyes!

Thanks to Ross, there was a Sam Hall mania (Yates's term) for a while, which ensured the ballad's wide dissemination. Obscene versions crossed both the Irish Channel and the Atlantic, reaching Harvard University in the 1850s; versions are apparently still current in the USA.[84] More unexpected is the survival of the original strain in areas little touched by music-hall. In the early twentieth century Cecil Sharp collected four versions from folk-singers in Devon and Somerset, and here were verses which neither Pitts nor Fortey had printed:

Up the ladder I did grope, that's no joke, that's no joke,
Up the ladder I did grope, that's no joke.
Up the ladder I did grope, and the hangman spread the rope,
  O but never a word said I coming down, coming down,
  O but never a word said I coming down.[85]

'Sam Hall' owed its longevity to its distinctive metre and tune and to the improvisations it permitted. But even in the early years of Victoria's reign its appeal also rested on the social need to believe that it was possible to face death with such insouciance as this. Here too was a dark variant on Punch's message that life was an empty joke, along with the

[83] Cited in Bratton, *The Victorian popular ballad*, 98.

[84] Bronson, 'Samuel Hall's family tree', 20–1.

[85] Bronson, 'Samuel Hall's family tree', 22; C. Sharp, *Folk songs from Somerset* (5 vols., 1904–9), ii, nos. 31, 239.

insistence that if life had any meaning, it must lie in keeping face in the here-and-now rather than in the anticipation of its aftermath.

Although the highwayman and Tyburn tradition in balladry became attenuated or romanticized in time, reflections of its subversive postures survived in crowd behaviour right to the abolition of public execution in 1868. In blackguard ballads (just as in the felon's game death, the approval of crowds, and the puzzling behaviour of Sale's women), we find a refusal to be defeated, a compulsive cockiness, a vaunting celebration of cleverness. To triumph over affliction, to refuse surrender, to reject mediocrity, and still to mock and laugh was to achieve the main distinction plebeian life could offer. When nemesis did strike in the ballads, it was allowed to strike bitterly, but the final execution was faced without flinching, indeed with extravagant understatement.

In these senses these cultures were coherent, patterned, consistent. Individuals within scaffold crowds knew what there was to say and how to say it; and they knew that the people around them, as well as those on the scaffold, knew this too. So although the murderer Sale's mother in 1848 might have been ignorant of the songs in this chapter's repertoire, she lived in a culture whose attitudes the songs vigorously affirmed. As she watched her son hang, she was not—of course!—without feeling. She had shame, loss, grief, anger, and defiance within her. It was the shame and grief that must not show; face must be kept up. In this she was a true heir to those who had cheered Sixteen-String Jack Rann in 1774, and she would have understood Sam Hall's sardonic defiance very well. The ballad was in her language.

8. *Aftermath*

The memory of canting Tyburn songs like these narrowly survived into the nineteenth century, and a few were reprinted.[86] But most were eclipsed. Oral tradition had its limits; memory depended more and more on print.[87] By mid-century the successors to Catnach's business had half a million ballads in stock, 'and even in these degenerate days, when a ballad makes a real hit, from 20,000 to 30,000 copies of it will go off in a very short time'.[88] But not many of these were about crime, and

[86] Farmer, *Musa pedestris*, reprints several dated Tyburn songs from the early 19th cent., some from *The universal songster, or Museum of mirth* (3 vols., 1825–6), but also 'A leary mot' (from a broadside, *c.*1811); 'The night before Larry was stretched' (*c.*1816); and 'The song of the young prig' (*c.*1819, but with antique references to Tyburn and to transportation to Virginia).

[87] Some 80% of the folk-songs gathered by collectors at the end of the century had an identifiable source in printed broadsides: Thomson, 'Broadside ballad trade', 145–6.

[88] Anon., 'Street ballads', *National Review*, 26 (1861), 400.

fewer still were in cant. Such criminal ballads as there were were more likely to be about transportation or poaching than about highwaymen and hanging: Tyburn was a fading memory.

When, therefore, we do meet criminals in the ballads of the 1830s and 1840s we are usually confronted by fantasies concocted for audiences which had put a good distance between themselves and the streets. As executions tailed off in the 1830s, past criminal cultures and Tyburn rituals came to be encased in nostalgia. 'How poor a thing is a modern execution,' *Tait's Magazine* lamented in 1841: 'A drizzly morning, the Under-Sheriff, and the Patent Drop. What a contrast to the pomp and circumstance that attended on the death of Duval.'[89] There are still glimpses of older pedigrees in the ballads of the 1840s: 'Fielding's gang did me pursue', one of Catnach's few highwayman songs declared.[90] But for the most part the fashion for Sheppard song-sheets and the like in the 1840s says less about the vitality of oral memory than it does about printers' exploitation of the success of Ainsworth's novel *Jack Sheppard* in 1839:

> Years ago, at least a hundred,
> Jack Sheppard lived, the bold and free,
> No braver man e'er cracked a crib, sirs,
> Or swung upon the Tyburn tree.
> Now he lives renowned in story,
> In three volumes is his life;
> Ainsworth shares Jack Sheppard's glory,
> Who murder makes with morals rife.[91]

The most famous of the cant ballads of the early Victorian years was, alas, a pastiche. Sold on the streets in the following version (among others) by the printer Paul in *Jack Sheppard's garland*,[92] the abstruseness of 'Nix my dolly, pals, fake away' at first sounds tantalizingly authentic:

> In a box of the stone jug[a] I was born
> Of a hempen widow[b] the kid forlorn,

[89] *Tait's Edinburgh Magazine*, 8 (1841), 218.

[90] 'Wild and wicked youth': Macaulay's Ballad Collection.

[91] 'The life of Jack Sheppard', in the garland *Jack Sheppard's glory*: Madden Collection, 9.677. Cf. Pitts's *Jack Sheppard's delight* and Catnach's *Jack Sheppard's songster* and *Jack Sheppard's garland* (Madden Collection, 9.678,656,657). Also inspired by Ainsworth's novels were the sentimental laments for Dick Turpin's unfortunate horse which Paul and Co. put out in slip form, *My bonny Black Bess* and *Poor Black Bess*, the latter illustrated with ambitious woodcuts of Turpin and his associate Tom King in Victorian clothes and top hats: 'And in after ages when I'm dead and gone, / This tale will be handed from father to son. / My fate some may pity, but all will confess, / 'Twas in kindness I killed thee, my poor Black Bess' (Madden Collection, 11.625,635.)

[92] Madden Collection 11.615: the sheet contains 4 songs and is illustrated with 2 highwayman woodcuts.

Fake away.[c]
And my noble father, as I've heard say,
Was a famous merchant of capers gay.[d]

Nix my dolly, pals, fake away,
Nix my dolly, pals, fake away.[e]

The knucks[f] in quod[g] did my schoolmen[h] play,
And put me up to the time o' day.
Fake away.
No dummy hunter[f] had forks[h] so sly,
No knuckler[f] so deftly could fake a cly.[i]
But my nuttiest[j] lady one fine day,
To the beaks did her gentleman betray,
Fake away.

And so I was bowled out at last,
And into the jug for a lag was cast.
But I slipped my darbies[k] one fine day,
And gave to the dubsman[l] a holiday,
Fake away.
And here I am, pals, merry and free,
A regular rollicking Romany.

Nix my dolly, pals, fake away,
Nix my dolly, pals, fake away.

[a] Prison cell; [b] her husband hanged; [c] 'go ahead'; [d] Ainsworth's version added a line, 'Who cut his last fling with great applause' (i.e. on the gallows); [e] 'nothing', i.e. go on stealing; [f] pickpocket; [g] prison; [h] fingers; [i] pick a pocket; [j] infatuated; [k] manacles; [l] turnkey.

But the truth is given away by the sanitized outcome of this ballad, so different from the bleaker Tyburn endings of the 1780s. Its original was composed by Ainsworth for his Dick Turpin novel *Rookwood* (1834), the outcome of a painstaking study of Vaux's *Vocabulary of the flash language* on the part of one of those several popular novelists of low life, Dickens among them, of whom it was fairly said that they knew as much about the boozing ken as they did about German metaphysics.[93] For nearly a decade Ainsworth's ballad enjoyed 'extraordinary popularity' in music-hall, but its audience was as different from that of the 1780s as Punch's audience had become: 'Ladies ask familiarly, in china-shops, for chamber utensils by the name of "stone jugs", and accost crossing-sweepers as "*dubs*-men" . . . children, in their nursery sports, are accustomed to

[93] 'Horae Catnachianae', 407.

"nix their dolls"; and the all but universal summons to exertion of every description is "Fake away!" '[94]

The revival of pastiche cant and highwayman stories by Ainsworth and his emulators and their appropriation in the cider cellars entailed the self-conscious construction of a criminal tradition to which a good deal of 'memory' was indebted.[95] The cheap sensation literature of ensuing decades sprang from it. When mid-century parliamentary committees and pamphleteers deplored the fact that the children of the poor knew everything about Turpin and Sheppard and nothing about the identity of the queen, the prime minister, or the Saviour, they knew that this was thanks to the mounting tide of penny dreadfuls, ballads, and performances in penny gaffs or street theatres which Ainsworth's and other Newgate novels had unleashed.

Yet in the last analysis it does not matter that the collective memory and oral tradition were increasingly organized by print and music-hall in these ways. The relationship between popular culture, orality, and literacy had always been symbiotic. *The Beggar's Opera, Moll Flanders*, or *Jonathan Wild* were no more indigenous evocations of criminal culture than Ainsworth's work was. The same might go for songs like 'To the hundreds of Drury I write', or *Jack Sheppard's last epistle*: the word 'write' being the give-away here. These texts from the literary culture, like the criminal biographies in the *Newgate calendars*, had none the less powerfully shaped popular ways of seeing and remembering in the century past. The energy of Pitts and his rival publishers at least offered a bridge for the transmission of some old songs, and the singing venues of London saved the ballad-writer from extinction.[96] Even Ainsworth's fabrications drew on memory. Dick Turpin was 'the hero of my boyhood', he recalled. He had 'listened by the hour to [highwaymen's] exploits, as narrated by my father, and especially to those of "Dauntless Dick", that "chief minion of the moon"'. Nor was he the first to romanticize the Turpin legend. *Richard Turpin, the highwayman* had been acted

[94] 'Flowers of hemp, or The Newgate garland', *Tait's Edinburgh Magazine*, 8 (1841), 218: a compendium of parodic or pastiche cant songs by Sir Theodore Martin. See Bleackley and Ellis, *Jack Sheppard*, 85–7.

[95] Ignorant, presumably, of the suppressed tradition, Ainsworth declared that the best examples of canting songs were produced by such as Pierce Egan, Tom Moore, John Jackson the pugilist (author of 'On the high toby spice flash the muzzle', cited by Byron in *Don Juan*, XI. xvii–xix), and by the author of 'The night before Larry was stretched' (Dean Burrows of Cork, according to Ainsworth, though Farmer attributes it to a Waterford shoemaker): preface to *Rookwood* (1878 edn.), pp. xxix–xxx; Farmer, *Musa pedestris*, 220.

[96] Smith, 'The press of the Seven Dials', 255–6.

at Astley's Amphitheatre in London in 1819.[97] Oral tradition does not become less 'authentic' if its subject-matter flowed into print because it usually flowed out of it too, perhaps to return once more, transformed in the course of each migration.[98]

[97] Preface to *Rookwood*, pp. xxx–xxxi; K. Hollingsworth, *The Newgate novel, 1830–1847: Bulwer, Ainsworth, Dickens, and Thackeray* (Detroit, 1963), 99–100, 106.

[98] Bronson, *Ballad as song*. Many ballads, even in the 19th cent., once printed, endured and evolved in sung repertoires and oral repetition. As late as the 1860s local tradition had to be catered to: ballads like 'The three butchers', published by local printers, had local words and stanza patterns peculiar to the county: 'Street ballads', 400.

APPENDIX

FRANCIS PLACE'S 'SONGS WITHIN MEMORY': A SELECTION

Place papers, BL Add. MS 27825. The numberings are Place's. Cant translations are from E. Partridge, *A dictionary of the underworld* (3rd edn., 1968), unless they are indicated as Place's.

No. 1
Jack Chance

On Newgate steps Jack Chance was found,
And wed up near St Giles's Pound,
My story's true, deny it who can,
By saucy, learing Billingsgate Nan.
Her bosom heav'd with artful joy
When first she beheld the lovely boy;
Then home the prize she straight did bring,
And they all allow'd he was just the thing, just the thing,
 Just the thing, just the thing,
 And they all allow'd he was just the thing.

At twelve years old as we are told,
The boy was sturdy, stout and bold;
He'd learn'd to curse, to swear and fight,
And everything but read and write.
His daddle[a] clean he'd slip between;
In a crowd he'd nap a clout[b] unseen;
And what he got he home would bring,
And they all allow'd he was just the thing, [*etc.*]

But when he grew to man's estate,
His mind did run upon something great
To pad the hoof[c] he seem'd to tramp
So he hired a prad[d] and he went on the scamp.[e]
To shoot in the park it was all his pride,
With a flaming whore stuck by his side;
At clubs he all the flash songs would sing,
And they all allow'd he was just the thing, [*etc.*]

He stood the patter but that's no matter.
He gammon'd the twelve and he work'd [?] the water

Till a pardon he got from his gracious king
Then swaggering Jack he was just the thing, [*etc.*]

With Blue Cockade proclaim'd for war,
With bludgeon stout or iron bar,
To head a mob he never would fail,
At gutting a mug's house or burning a gaol.
But a victim he fell to his country's laws,
And died at last in religious cause.
No popery made the blade to swing,
And when tuck'd up he was just the thing, just the thing,
    Just the thing; just the thing,
    And when tuck'd up he was just the thing.

[a] Hand; [b] steal a handkerchief; [c] wander about on foot; [d] horse; [e] become a highwayman.

No. 8

Young Morgan

(Verses 2–4 are from the Baring Gould Collection of Broadsides, BL I.2, fo. 89 (Catnach printer); the full ballad is also reprinted in 'Horae Catnachianae', *Fraser's Magazine*, 19 (1839).)

Young Morgan was a rattling* blade,
    No lad of better courage.
Much gold he got on the highway
    Which made him daily flourish.
Grand bagnios** was his lodging then
    Among the flashy lasses,
Until he came a gentleman,
    And left of[f] driving asses.[a]

Through Hounslow heath and Putney too,
    Me and my noble poachers,
Me and my prads[b] like lightning flew,
    When we heard the sound of coaches.
Stand and deliver was my word,
    To me make no demand.
Now young Morgan is caught at last,
    At the start to take his trial.

I thought I heard some people say,
    As I rode through the City,
That such a clever youth should die,
    They thought it was a pity.
I thought I heard such human calls,
    That set my tears a flowing,
Oh! now Young Morgan he is tried and cast
    Out of this world is going.

I was the captain of a gang,
    But now in a low condition,
Without the judge or magistrate
    They show on me compassion.
Oh why should I refuse to die,
    For now or ever after,
For now the Captain he is gone,
    His men must follow after!

Note: The asterisks indicate Baring Gould variants on Place's first verse: * 'lusty'; ** 'In Wentworth Street, his lodgings were'.

[a] Stealing handkerchiefs; [b] horses.

No. 26

'Drunk the other night'

Drunk the other night as I reeled home to bed,
I met a young frow[a] just turn out of her kin,[b]
She suddenly seized me, and swore how she'd please me,
If I go would go with her and give her some gin.
Her cheeks look'd so rosy, her eyes look'd so wanton,
Her waist so well shaped, and her bubbies so ripe,
But the gallows young huzzy, while I felt her tuzzy,
Was down with her gropers to unravel my wipe.[c]

I gave her a topper[d] for making so bold,
Then the scouts[e] all came up being flash to the rig,[f]
'Twas the noise of the rattle that made the whore prattle,
So I showed him some cole[g] to bother his wig.[h]
The scouts all came round me while I seem'd amazed,
At last one among them he tipp'd me the wink:
It is one of our party, says he, and he's hearty,
So we all bundl'd into a flash [ken?][i] there to drink.

To do them a kindness it was my intention,
To have a pull[j] on them without more delay.
So without further trouble I tipp'd them the double,[k]
Left the whore and the scouts all the reck[on]ing to pay.

[a] Prostitute; [b] pimp?; [c] handkerchief; [d] blow; [e] watchmen; [f] alert to the trick; [g] money; [h] go away?; [i] underworld dive; [j] advantage; [k] give them the slip.

No. 28

Teddy Blink and Bandy Jack

Place's near-full version is supplemented in the fourth verse and penultimate couplet from Pitts's ballad slip. Pitts's address on the slip shows that it was published after 1819: Madden Collection of Broadsides, Cambridge University Library, 8/927.

On Sunday morning early we went to different chapels,
My pal upon his bended knee the ladies' yacks[a] he grapples;
'Lord grant that we may keep this law', and while she's upward looking,
My pal so ready with his paw, her watch chain is unhooking.

Tol lol etc.

He clings[b] it to his nearest pal, to brush[c] directly after.
The pretty educated lad, first, naps the newest caster;[d]
Then plac'd him in the nearest pew long side of father gray locks;
He brings the yellow bag[e] to view, the tooth-pick case and snuff box.

Now some had lost their pretty rings, and some had lost their lockets,
To rob in Church, Lord what a sin: cries Jane I've lost my pockets.
Poor girl she'd hardly spoke the word when Susan comes out bawling:
Says she I've lost my black silk cloak, and several yards of muslin.

Full five and thirty quid we made among those spooney[f] gorgers;[g]
We did the Welch of all their four[h] as they came from St Georges.
From cly to cly[i] we made a shift ingeniously to enter
And left the Welch without a crook[j] so boldly do then venture.

Now Teddy Blink and Bandy Jack, they laid their heads together,
If they could do the old codger in black, 'twould give them mighty pleasure.
He bless'd the congregation round and through the crowd was passing,
When Teddy drew aside his gown and Bandy spoke to the parson.

But he was seized by the old clerk who thought he had a prigs[k] look,
So he dous'd the glim,[l] tipt him the dark,[m] and boned[n] his wig and prayer book.

. . . .

They work'd the church of what would fence[o] which much alarm'd the people,
For fear they should stone eaters* turn and brush away[c] with the steeple.

* Place noted that 'there was an exhibition at the time of a man who ate stones'.

[a] Watches; [b] throws; [c] run away; [d] cloak; [e] purse; [f] foolish; [g] well-dressed gentlemen; [h] ?; [i] pocket; [j] sixpence; [k] thief's; [l] lantern; [m] gave him the slip?; [n] stole; [o] sell to a receiver.

No. 23
Sandman Joe

Oh, the other day, as Sandman Joe,
  Up Holborn Hill was jogging,
His raw boned steed, it scarce would go,
  But still the dog kept flogging.
His raw boned steed, scarce fit for crows,
  Just starved to death could scarce go,
Whilst Gallows Joe his rump he rubb'd
  And roaring cried, White sand O,
Why here's your lilly lilly lilly lilly white sand O.

Scarce far he'd gone to sell his sand
'Twas near a neighbouring alley,
When turning of his head about,
He spied his flash girl Sally.
His raw boned steed, scarce fit for crows,
Could scarce stand, when he cried wo-o!
But to keep him up his rump he rubb'd
And roaring cried, White sand O, *etc.*

He star'd a while then turn'd his quid,
Why blast you, Sall, I loves you!
And for to prove what I have said,
This night I'll soundly fuck you.
Why then says Sall, my heart at rest
If what you say you'll stand to;
His brawny hands, her bubbies prest,
And roaring cried, White sand O, *etc.*

Said Sall to Joe, where shall we go,
To get some gin to warm us?
Why blast you to St Giles' pond,
For there the gin won't harm us.
His raw boned steed, *etc.*

When to St Giles' they had got,
They made themselves quite merry;
They five times drain'd the quartern pot,
With glorious gin so cherry.
His raw boned steed, *etc.*

O then they kiss'd, and then shook fist:
My dearest Joe I know you;
As sound a dog as ever piss'd
This night I'll doss with Joey.
Then away they went with hearts content
To play the game you all know,
While Gallows Joe he wagg'd his arse
And roaring cried, White sand O
Why here's your lilly, lilly, lilly—lilly—lilly
White sand O.

Addendum

A new flash song

This rare printed survival of a Tyburn ballad, outside Place's recollection, is in the Baring Gould Collection of Broadsides (BL I.2, fo. 61). It is on a ballad slip headed by a woodcut illustrating five figures carrying staves in a landscape.

The illustration and the typography point to an early- to mid-eighteenth-century provenance.

Me and five more we all set up,
    To rob and plunder without doubt,
Away to Hyde Park we did steer
    To light on the culls[a] and rattlers[b] there.

We met with a cull that was just the thing
    With a large pair of wedes[c] and a diamond ring;
We took from him all we could sack,[d]
    With a silver hilted sword, and gold lac'd hat.*

We gammon'd[e] hard our lives to save,
    And to the old beak gossip gave,
But our prosecutor was so hard,
    Unto our youth paid no regard.

. . . .

We were gammon-patter'd,[f] to Newgate sent,
    Which gave our blowings[g] discontent;
I'd rather on the gallows die,
    Than in those dismall cells to lie.

O hark! I hear St Pulchre's toll,
    The Lord have mercy on each soul;
With black hatbands we look'd so neat,
    With weeping eyes, and nosegays sweet.

And as the tumbler[h] mov'd along,
    Some people sung a different song.
Let the young dogs go, they'll leave enough
    To strip us old culls to the buff.[i]

When to the turnpike[j] we had got
    We turn'd our heads and look'd about,
And saw the gibbet so high to look,
    And wish'd our friends advice we took.

When the cap was pull'd over our eyes,
    Unto the Lord then each one cries,
It is a pity, they all did cry,
    Such clever lads as they should die.

Then of these youths don't make your game,[k]
    Although they die in wretched shame;
For oft times you might live to see
    The old grey hairs brought to the tree.

Our blowings[g] in a rattler[b] came,
Our bodies for to carry home,
Then in a hearse we were convey'd
Here's an end of our lives, our debts are paid.

* Several more such robberies are detailed before the ballad reaches the trial and aftermath.
[a] Dupes; [b] coaches; [c] ?; [d] pocket; [e] told lies; [f] out-talked, tried ; [g] mistress-whores; [h] tumbril, cart?; [i] skin?; [j] Tyburn turnpike; [k] deride.

CHAPTER 5

BROADSIDES AND THE GALLOWS EMBLEM

1. *The rise of the execution sheet*

IF, AS TIME PASSED, THE OLD TYBURN AND CANTING BALLADS were heard less frequently on the streets, the same cannot be said of the noisy vending of trial and execution broadsheets. There was a mounting tide of these, swelling first in the mid to later eighteenth century and again in the second quarter of the nineteenth. These make up a second corpus of plebeian texts on the scaffold. They expressed popular understandings about life's chances and death's certainties just as eloquently as the ballads did, but to very different effects.

The Tyburn songs were usually knowing, sardonic, and half-comic celebrations, or else they were lugubriously defiant, as we have seen. Never speaking of contrition or the justice of retribution, they did not so much challenge conventional morality as ignore it. They are the objective correlatives, so to speak, of those phlegmatic or disenchanted or overtly hostile attitudes discernible in many scaffold witnesses when people like themselves hanged—of those attitudes which so disconcerted and mystified observers from Dickens to Hepworth Dixon. But that other (or even the same) members of the crowd could just as well display different repertoires of attitudes we have also seen: acquiescence, approval, identification with the law. And of these repertoires the objective correlatives were the broadsides. As this chapter must show, these almost always paid lip-service to conventional morality; their values were sentimental, not transgressive. The felon's repentance was central to their comment, and readers were warned against error.

No betrayal of right plebeian values need be inferred from the consumption of these things. No false consciousness is in question. The law has always borne many meanings, and doubtless did even to those who sang sardonic Tyburn ballads. Broadside morality aimed only to make sense of the world, after all. It presumed a sequence of criminal cause and punitive effect which was rooted in the commonsensical ethics of the people, giving life an intelligible structure, as much as suspicion of rich men's law might be rooted. The crowd that cheered the highwayman could just as well hiss the sodomist or murderer, stay silent for the footpad, and pityingly accept the infanticide's doom.

Moreover, the broadsides carried as much of a symbolic charge as the ballads did, for they too connected their perusers with the psychic energy of execution. They did this through woodcut illustrations which offered the crowd its only visual mirroring of itself as well as of the scaffold's emblematic power (and recall how relatively hungry people were for illustrations then). Despite much scholarly attention to popular literacy, literature, and images in this period, these illustrations have been looked at only cursorily, just as the purchasers and messages of broadsides have been. This is the justification for the study which now ensues. Like the ballads, the broadsides attest to the fact that the people who attended executions knew what their roles were, and what it was they were watching.

Our view of the broadside trade is shaped by the accidents of survival.[1] With their smudged ink of lampblack and oil and with typefaces and woodcuts mixed at random, most early sheets were printed on tea paper which disintegrated when damp; and before the arrival of the cheap newspaper, they were used as wrappings, bookbindings, and lavatory paper.[2] Some survive from the seventeenth century in the Pepys and other collections, but the first boom in output seems to have occurred around the mid-eighteenth century, helped by the slow demise of the execution *Accounts* of the ordinary of Newgate, hitherto the staple production on the subject.[3] The more expensive *Newgate calendars* for middling and middle-class readers replaced these, enjoying 'a ten

[1] Execution sheets made up only a small proportion of Pepys's broadside collection (V. G. Day (ed.), *The Pepys ballads* (5 vols., 1987)), but scattered references suggest that Tyburn ballads were among the most popular in the 17th cent.: T. Watt, *Cheap print and popular piety, 1550–1640* (Cambridge, 1991), 39; J. Earle, *Microcosmographie, or A piece of the world discovered in essayes and characters*, ed. G. Murphey (Waltham St Lawrence, 1928), 45–6. I am grateful to Adam Fox for these references.

[2] M. Spufford, *Small books and pleasant histories: Popular fiction and its readership in seventeenth-century England* (Cambridge, 1981), 48.

[3] From 1712 the ordinary's *Accounts* were published at 6 folio pages at 2*d*. or 3*d*. a piece, and from 1734 in 16 or 28 quarto pages at 4*d*. or 6*d*. Printings ran into thousands and 'enjoyed one of the widest markets that printed prose narratives could obtain in the eighteenth century': but they were relatively costly purchases for any below the London artisanate: P. Linebaugh, 'The ordinary of Newgate and his *Account*', in J. S. Cockburn (ed.), *Crime in England, 1550–1800* (1977), 247, 250; P. Rawlings, *Drunks, whores, and idle apprentices: Criminal biographies of the eighteenth century* (1992), 24–6.

times greater sale than either the "Spectator", the "Guardian" or the "Rambler"', Angelo recalled.[4] Sixpenny chap-books on major cases also continued to be produced for another three-quarters of a century, flowing from the provinces to London bookshops as well as in the reverse direction,[5] and their price implies that these too were aimed at people with money in pocket. But single-sheet publishers began to expand their cheaper output of broadsheets as well, to capture humbler readers. Noting their proliferation as a novelty, Walpole reported in 1750 that malefactors' lives and deaths were 'set forth with as much parade' as a general's life might be; of the highwayman McLean's hanging 'there are as many prints and pamphlets . . . as about the earthquake'.[6] Place noted that by the 1770s and 1780s the sale of halfpenny last dying speeches aimed at the poor had become 'enormous'.

If this was the period of take-off, self-sustaining growth came with the iron-frame press of 1815. Costing only £30 or so and printing some two hundred sheets an hour, this led to such a proliferation of provincial printers that by 1820 few if any provincial towns were unable to produce their own literature.[7] Printers like Shepherd and the Bonners of Bristol dominated the West Country broadside market for a quarter-century after 1815, while East Anglian towns like Ipswich, Bury, and Norwich each acquired their broadside publishers, seizing such opportunities as capital sentences at local assizes offered and plagiarizing broadsides on cases elsewhere to offset slack trade. Woodcuts too were appropriated without regard to local topography or credibility. The Bonners published sheets on the London Ratcliffe Highway murders in 1811 and on Courvoisier's execution in 1840, on cases tried as far afield as Nottingham or Worcester, and on the Brandreth, Ludlam, and Turner decapitations in Derby in 1817.[8]

Further growth came in and after the 1830s, thanks to the decline in

[4] H. Angelo, *Reminiscences* (2 vols., 1828), i. 467. On the more expensive publications, see L. B. Faller, *Turned to account: The forms and functions of criminal biography in late seventeenth- and early eighteenth-century England* (Cambridge, 1987), app. I, 'Who read the popular literature of crime?' He omits to discuss broadsides.

[5] Typical was *The trial of John and Nathan Nichols father and son for the wilful murder of Sarah Nichols, daughter of the former, and sister of the latter; at the Suffolk Lent assizes, 1794*. Printed and published by Gedge and Rackham in Bury St Edmunds, 12 pages long and priced 6*d*., it was 'sold by Messrs Robinson, Pater Noster Row; Richardson and Axtell, Royal Exchange, London; and all other booksellers in the kingdom'. On Fauntleroy's execution for forgery in 1824, 3 broadsides and a dirge-sheet survive in the BL, but so do half a dozen pamphlets on his trial, execution, and 'dying behaviour', some illustrated and one 35 pages long. For mid-century, see Mayhew, i. 284.

[6] Walpole, ii. 227, 230.

[7] D. Vincent, *Literacy and popular culture: England 1750–1914* (Cambridge, 1989), 201; L. Shepard, *John Pitts, ballad printer of Seven Dials, London, 1765–1844* (1969), 43.

[8] Bristol broadsides: BL 1880. c. 20.

executions, paradoxically. Attention was concentrated on fewer cases now since only murderers were hanged (transportation sheets made up some of the shortfall); but as executions' scarcity value increased, sales could be sensational: 2.5 million sheets were claimed for Rush's and the Mannings' executions in 1849, and just over 1.6 million each for Corder's in 1828, Greenacre's in 1837, Courvoisier's in 1840, and Good's in 1842; Thurtell's execution in 1823 was no less profitable.[9] For a major execution output followed a standard sequence. First came the 'Sorrowful Lamentation', a handbill in quarter-sheet on the fate of the murder victim. (This had become possible only after the 1830s, when the time between death sentence and execution was extended to allow for appeals: 'Before that, sir, there wasn't no time for a Lamentation; sentence o' Friday, and scragging o' Monday,' Mayhew was told.) Then followed a half-sheet detailing particulars of the crime extracted from newspaper reports. On execution day or soon afterwards came the full broadsheet (doubled on great occasions), embellished with accounts of the trial, confession, and execution, verses, and woodcut portraits or gallows scenes. In big cases 'the book' of four, eight, or more pages might follow, replicating the preceding publications. Local competition could be intense. When the poisoner Mary Burdock was executed in Bristol in 1835 the rival Bristol printers Bonner, Davis, Watson, and Taylor each put out single sheets on the trial (Bonner published one daily), followed by sheets on the dying speech and confession and finally on the execution itself. Bonner also published the coroner's examination and the recorder's charge to the grand jury and announced that the whole trial 'may be had in a small pamphlet at this office'.[10]

There was sometimes big money in the trade, and marketing methods were efficient. Sheets were sold to street sellers at *2d.* to *2½d.* per dozen, each vendor pocketing some four-fifths of the receipts in profit. The deal (always in cash) profited the printer since he had no bad debts.[11] All this—and the contemporary expansion of the penny dreadful pioneered by publishers like Edward Lloyd—spoke for the 'march of intellect', Mayhew was told. It certainly spoke for the increasing commercialization of the scaffold drama in response to a demand which had expanded steadily, with literacy itself, since 1750. Only when penny newspapers captured the market in the 1860s did this flood of street

[9] Mayhew, i. 284–5. Hindley accepted Mayhew's figure of 1,650,000 sheets for Greenacre's execution: C. Hindley, *The life and times of James Catnach (late of Seven Dials), ballad monger* (orig. pub. 1878; 1970), 281.

[10] Bristol broadsides (BL).

[11] Charles Smith, 'The press of the Seven Dials', in his *The little world of London* (1857), 264.

literature rapidly decline—though even then ambitious broadsides continued to be produced for major executions.[12]

Jemmy Catnach of Seven Dials dominated the market until his retirement in 1838. It was chiefly he who transformed and diversified the street print. He set up around 1813 as a jobbing master in his back parlour, entering a lasting competition with John Pitts, who had led the trade since 1797 and whom he loathed, but outpacing him in his flair for sensation.[13] Catnach was soon printing a vast range of penny histories and halfpenny songs on subjects romantic, comic, and historical, the verses composed at a shilling a time by one of the 'seven bards of Seven Dials', though he wrote the verses himself when the bards were on the drink. At first he used worn typefaces, cheap ink and tea paper, and battered antique woodblocks. In the 1820s, however, he began to use stronger paper and real printer's ink. Illustrations became more ambitious as boxwood made finer engraving possible.[14] His immediate customers were impoverished street-sellers who survived by selling sheets on a few coppers' outlay. Catnach had to take their pennies to the bank in a hackney coach because no neighbour would change them lest they catch fever from them; so many of the pennies were bad that he paved his back kitchen with them. Catnach was a model of the penny entrepreneur. He began his fortune by capitalizing on the Queen Caroline scandal in 1820, but his next great success came with the publication of his 'Full, true, and particular account' of Thurtell's murder of Weare in 1823. Netting over £500 with this, he worked his men night and day for a week on his four presses to produce about 250,000 copies. Even this was bettered in ensuing productions on Thurtell's trial and execution; 500,000 copies were turned off in eight days, and 'every night and morning large bundles were despatched to the principal towns in the three kingdoms'. Some of his verses on this occasion were ambitious enough to be quoted decades later by one of Mayhew's informants. Others stayed in mind for different reasons. Thomas Hood's parody caught the tone:

[12] Cf. *Trial, sentence and execution of the Fenians for the murder of C. Brett at Manchester, November 23rd, 1867* (St Bride Printing Library, Broadside Collection). The impact of penny newspapers is measured in the fact that only 280,000 broadsides were sold for Muller's execution in 1864, and 60,000 for Jeffery's in 1866; the dying-speech trade also collapsed with the privatization of execution in 1868: Hindley, ii. 159–60.

[13] Shepard, *Pitts*, 37, 44, 60. Shepard's date of 1802 for the start of Pitts's business has been corrected by R. S. Thomson, 'The development of the broadside ballad trade', D.Phil. thesis (Cambridge, 1974), 127.

[14] He used some blocks made by Bewick: Hindley, *Catnach*, 44–7, 391; K. Lindley, *The woodblock engravers* (Newton Abbot, 1970), chs. 1, 2.

They cut his throat from ear to ear,
His brains they battered in;
His name was Mr William Wear,
He dwelt in Lyon's Inn.

By the 1830s Catnach's advertised stocklist of ballads, chap-books, and broadsides was probably the longest in England. When he died in 1842 he was allegedly worth near on £10,000.[15]

2. *Messages*

Verses of Catnach's kind and worse must have been in Mayhew's mind when he deplored broadsheets' 'morbid sympathy and intended apology for the criminal'.[16] Concern about the quality of mass reading material of this kind was chronic, and there had been such wondrous attempts to improve it that we may be forgiven for thinking that it must have been truly subversive. The Cheap Repository for Religious and Moral Tracts in the 1790s, for example, put out rival halfpenny sheets written in simple language which peddled evangelical pieties and sanitized old ballads. For example, the 'Young Morgan' which Place remembered—'Young Morgan was a rattling blade, / No lad of better courage / Much gold he got on the highway / Which made him daily flourish'—was changed by Hannah More in *Execution of Wild Robert: being a warning to all parents* (1787) to:

Wild Robert was a graceless youth,
And bold in every sin;
In early life with petty thefts
His life he did begin.

How well this kind of thing went down may be a moot point, especially since More made it clear that Robert's ensuing execution was basically his mother's fault:

Blame not the law which dooms your son,
Compar'd with you 'tis mild;
'Tis you have sentenc'd me to death,
To hell have doom'd your child.[17]

But these things were helped by decorative borders and better woodcuts (Plate 15), energetic distribution networks, and bulk discounts, as

[15] Hindley, *Catnach*, 51–2, 79, 143–5, 383, 406, 409; Mayhew, i. 283; Madden Collection of Broadsides, Cambridge University Library.

[16] Mayhew, i. 281.

[17] BL broadside, 1872. a. 1.

The Execution of Wild Robert.

Being a Warning to all Parents.

WILD ROBERT was a graceleſs Youth,
  And bold in every ſin;
In early life with petty thefts
  His courſe he did begin.

But thoſe who deal in leſſer ſins]
  In great will ſoon offend;
And petty thefts, not check'd betimes,
  In murder ſoon may end.

And now, like any beaſt of prey,
  Wild Robert ſhrunk from view,
Save when at eve on Bagſhot heath
  He met his harden'd crew.

With this fierce crew Wild Robert there
  On plunder ſet his mind;
And watch'd and prowl'd the live-long night
  To rob and ſlay mankind.

But God, whoſe vengeance never ſleeps,
  Tho' he delays the blow,
Can in a ſingle moment lay
  The proſperous villain low.

One night, a fatal night indeed!
  Within a neighb'ring wood,
A harmleſs paſſenger he robb'd,
  And dy'd his hands in blood.

The direful deed perform'd, he went
  To ſhew his golden ſpoils,
When vengeful Juſtice, unawares,
  Surpris'd him in her toils.

Wild Robert ſeiz'd, at once was known,
  (No crape had hid his face)
Impriſon'd, tried, condemn'd to die!
  Soon run was Robert's race!

Since ſhort the time the laws allow
  To murderers doom'd to die,
How earneſt ſhou'd the ſuppliant wretch
  To heaven for mercy cry!

But he, alas! no mercy ſought,
  Tho' ſummon'd to his fate;
The Cart drew near the Gallows Tree,
  Where throng'd ſpectators wait.

Slow as he paſt no pious tongue
  Pour'd forth a pitying pray'r;
Abhorrence all who ſaw him felt,
  He, horror and deſpair.

And now the diſmal death-bell toll'd,
  The fatal cord was hung,
While ſudden, deep, and dreadful ſhrieks,
  Burſt forth amidſt the throng.

Hark! 'tis his mother's voice he hears!
  Deep horror ſhakes his frame;
'Tis rage and fury fill his breaſt,
  Not pity, love, or ſhame.

"One moment hold!" the mother cries,
  "His life one moment ſpare,
"One kiſs, my miſerable child,
  "My Robert, once ſo dear!"

Hence, cruel mother, hence, he ſaid,
  Oh! deaf to nature's cry;
Your's is the fault I liv'd abhorr'd
  And unlamented die.

You gave me life, but with it gave
  What made that life a curſe;
My ſins uncurb'd, my mind untaught,
  Soon grew from bad to worſe.

I thought that if I 'ſcap'd the ſtroke
  Of man's avenging rod,
All wou'd be well, and I might mock
  The vengeful pow'r of God.

My hands no honeſt trade were taught,
  My tongue no pious pray'r;
Uncheck'd I learnt to break the laws,
  To pilfer, lie, and ſwear.

The Sabbath bell, that toll'd to church,
  To me unheeded rung;
God's holy name and word I curs'd
  With my blaſpheming tongue,

No mercy now your ruin'd child
  Of heav'n can dare implore,
I mock'd at grace, and now I fear
  My day of grace is o'er.

Blame not the law which dooms your ſon,
  Compar'd with you 'tis mild;
'Tis you have ſentenc'd me to death,
  To hell have doom'd your child.

He ſpoke, and fixing faſt the cord,
  Reſign'd his guilty breath;
Down at his feet his mother fell,
  By conſcience ſtruck with death.

Ye parents, taught by this ſad tale,
  Avoid the path ſhe trod;
And teach your ſons in early years
  The fear and love of God.

So ſhall their days, tho' doom'd to toil,
  With peace and hope be bleſt;
And heav'n, when life's ſhort taſk is o'er,
  Receive their ſouls to reſt.

15. *The execution of Wild Robert: being a warning to all parents*, 1797.

With their superior craftsmanship and simple moral messages, the Religious Tract Society's broadsides aimed to undermine the appeal of vulgar broadsides. This one told its readers that Wild Robert's sorry end was his neglectful mother's fault. She collapses appropriately as she realizes the truth of the matter.

well as simple moral messages.[18] The Religious Tract Society pumped out this stuff for another couple of decades, diversifying output with heavily subsidized 24-page pamphlets like the laundered account of McLean the highwayman's execution, or the optimistic *Judgement and mercy: Being an account of several malefactors who appeared to die true penitents, and humble believers: Collected for the use of prisoners condemned to die.* These tales were priced at 8*s.* per hundred, with a 'considerable allowance to subscribers and booksellers'. For better or worse, some presumably found their way into condemned cells.[19] Backed by harassing magistrates, such enterprises checked the vulgar tide of street ballads, but they could not stem the tide of vulgar broadsides. The interesting question is whether they really needed to try.

The penny dreadfuls of the 1830s and 1840s did glorify villains both fictional and real; and some, with their lurid pornographic illustrations, can raise eyebrows even now. But for the most part it was the literal vulgarity of street literature—its independence of patronage and good linguistic manners—that offended polite commentators more than the contents did, for after the 1790s class distinctions rigidified and clearer demarcations were established between vulgar and polite language and literature.[20] The truth was that, so far as *contents and messages* were concerned, execution sheets were determinedly decorous. They always agreed that wickedness should meet its just deserts; that although lesser felons might be 'unfortunate', they were invariably 'deluded' and their punishment 'just'; that murderers, rapists, and sodomists were always 'monsters'; and that

> Of all the crimes recorded,
> In history from the first,
> The horrid crime of murder,
> It is the very worst.

The language of repentance, retribution, and warning was ubiquitous:

> So all young men take warning,
> By his untimely end,
> For blood for blood will be requited
> By the laws of God and man.[21]

[18] Their advertisement announced their sale in London and Bath 'and by all booksellers, newsmen, and hawkers in town and country'. Bulk purchase by shopkeepers and hawkers was discounted: 9*d.* for 25, 1*s.* 3*d.* for 50, 2*s.* 3*d.* for 100. See V. Neuburg, *Popular literature: A history and guide* (Harmondsworth, 1977), 255–9.

[19] BL tract, 1578/3683.

[20] O. Smith, *Politics of language, 1791–1819* (Oxford, 1984) ch. 1.

[21] Pitts's *Verses on the execution of William Corder, for the murder of Maria Martin* (1828): Madden Collection, 9.2.

By mid-century it was noted that 'the whole of the last dying speeches and confessions, trials and sentences, from whatever part of the country they come, run in the same form of quaint and circumstantial detail: appeals to Heaven, to young men, to young women, to Christians in general, and moral reflection. We have seldom met with one of a different character.'[22]

In these senses scaffold sheets were never as dangerous or as tacitly subversive as the canting ballads were. They never spoke about the subjection of weak people to powerful. The execution of traitors or rioters was never described in terms which played into radical hands. If they did hint at a criticism of the spectacle they battened on, it was usually through oblique allusion to the onlookers' pity, or through a profession of pity for condemned women (if young and pretty), or through the appropriation of opinions already safely expressed in parliamentary debates on capital laws. Thus a broadside like the one reproduced here, published by the London printer Gilpin in 1847, sniped with wonderful accuracy—but safely so, since abolitionism was in fashion:[23]

GRAND MORAL SPECTACLE!

Under the Authority of the Secretary of State for the Home Department

THIS DAY, SATURDAY, APRIL 17, 1847,

A YOUNG GIRL

SIXTEEN YEARS OF AGE

IS TO BE

PUBLICLY STRANGLED

IN FRONT OF THE

County Jail, Bury St Edmonds

SHE WILL APPEAR

Attended by a Minister of the Church of England,
clad in his Robes Canonical;

ALSO BY THE HANGMAN,

The Great Moral Teacher,

who after fastening her arms to her side, and putting a rope round her neck, will strike the scaffold from under her; and if the neck of the wretched victim be not by this shock broken, the said MORAL TEACHER will pull the legs of the miserable girl until by his weight and strength united he

STRANGLES HER

[22] Anon., 'Street ballads', *National Review*, 26 (1861), 405.
[23] BL broadside, 1888. c. 3, fo. 18.

This Exhibition (the admission to which is free,) is provided by a '*Christian* Legislature', for the instruction of a '*Christian People*:', and is intended to impress on the minds of the multitude an abhorrence of all cruelty, a love of mercy and kindness, and a reverence for human life!!!!

The very few extant Bristol broadsides to adopt mildly critical postures in earlier decades similarly aimed at safe targets sanctioned by parliamentary debates.[24] When Bonner, for example, put out his *True, full and particular account of the dreadful execution of Mary Jones, which took place in London some years ago, for stealing goods in a linen draper's shop, to the amount of only five shillings*, it coincided with the 1823 repeal of the death penalty for stealing privately in a shop. The crime had in fact been brought to the Commons' attention by Sir William Meredith in 1777; Bonner used Meredith's account verbatim, without attribution or dating.[25] Moreover, most of these 'critical' broadsides addressed the plights of young women, again always a safely chivalric posture to adopt. When a girl was convicted of infanticide in 1824, Bonner regretted that 'the laws operate so severely' against girls in her position: 'let him who is without fault cast the first stone at her'. When, also in Bristol, Shepherd detailed the *Life, trial, character and sufferings of Miss E. Clark . . . transported for life at the Old Bailey Sessions, April 1826*, she turned out to be the 'sixth daughter of a respectable gentleman named Clark, residing near Bristol'; she 'had received the best of education her parents could give her, and is, in every sense of the word, an accomplished girl'. Then followed a well-worn moral tale. Aged 17, she fell from virtue through 'seduction' and went to London, pregnant. There she became a servant and a prostitute, and was set up to pass a forged £5 note. At her twelve-hour trial 'her prepossessing countenance and becoming demeanour excited universal commiseration, and on the Judge passing sentence of death upon her, a shriek of sympathy echoed through the whole Court'. In prison she gave birth to a girl. Her sentence commuted to transportation for life, she begged leave to take her baby with her to Australia. Refused, she was driven into a 'frenzy of despair'. 'Her feelings we cannot more beautifully describe than in the following beautiful verses with which she addressed her infant a few hours before leaving this country':

[24] Ibid. 1880. c. 20, fos. 414, 380.

[25] Dickens was also to refer to the case in his preface to *Barnaby Rudge* (1841). Mary Jones, aged 19 and 'most remarkably handsome', committed the crime when her husband was pressed into the army, leaving her so indebted that she was turned on to the streets to beg with her two children. 'Her infant was suckling at her breast when taken to the place of execution.' She was hanged, the broadsheet made clear in Meredith's words, 'for the comfort and satisfaction of some shopkeepers in Ludgate Street'.

And when the dark thought of my fate shall awaken
The deep blush of shame on thy innocent cheek;
When by all, but the God of the orphan forsaken,
A home, and a father, in vain thou shalt seek;
I know that the base world will seek to deceive thee,
With falsehood like that which thy mother beguild:
Deserted and helpless—to whom can I leave thee?
O! God of the fatherless—pity my child!

There were a few other occasions when publishers safely dissociated themselves from the execution. In the *Account of the horrid execution of Samuel Brown, who was cut down whilst alive, and hanged a second time, on Saturday last, at York*, the behaviour of the drunken hangman was allowed to be 'revolting to every feeling of humanity'. The hanging of the Bristol rioters in 1832 was likewise a scene 'at which humanity shudders—four unfortunate individuals, standing upon the verge of eternity, the sole cause of which was habitual drunkenness'.[26] However, broadsides seldom showed any fuller compassion than this. In this Bristol case, Tindal CJ's elaborate demonstration of the 'enormity' of the riots was quoted without comment.

With these specialized exceptions, then, most sheets either described trials and executions with emotions well guarded, or else they moralized unashamedly. Some relished the details of 'the horrible and appalling spectacle witnessed on the fatal gallows', and their bathos could be resounding.

Some thousands of person did appear,
As to the stake she then drew near,

proclaimed the chap-book published when Christian Bowman was hanged and burnt for coining in 1789:

Think, think, how shocking was her fate,
By flames consumed in the street.[27]

Or, as a pregnant servant might say to the farmer's son in Wells who had seduced and then murdered her:

''Twas you my fond bosom who first did betray,
And by your enticement you led me astray:
To ruin my virtue was all you desired.'
She clamped her teeth, clos'd her eyes, and expired.

[26] BL broadside, 1880. c. 20. [27] Ibid. 1077. g. 36.

Or execution songs adopted the standardized tropes anciently required for effective street crying ('Come all you thoughtless young men, a warning take by me') and issued warnings no less dire than those favoured by Hannah More:

> Spectators all, both young and old,
> Who view this shocking sight,
> A warning take by their sad end,
> And steer your conduct right:
> Let no dishonest ways before
> E'er enter in your mind,
> Serve God in righteousness and truth,
> To virtue be inclined.[28]

The taste for forceful exemplification was sometimes yoked to implausible sentimentality—

> The Lord Chief Justice then passed sentence,
> While tears fell from his Lordship's eyes.
> Prisoner you have been convicted.
> You must prepare yourself to die!

—but invariably led to a resounding conclusion:

> Then while we act in an upright manner,
> With a conscience clear we need not fear.
> Let those that in bad paths are steering
> Think on the fate of Courvoisier.[29]

The felon might be allowed his heroism if he was as game as Robert Hartley was when he was hanged in Maidstone in 1828, but the moral judgements of authority would still be appropriated. Hartley 'behaved in a most hardened and impenitent state', his crimes were 'heinous', the exhortations of the chaplain 'pious'.[30] And for the most part condemnation was absolute, even of women convicted of infanticide: 'No darkness can cover a deed so foul from His eye who beholds all our ways, and we hope it will be a warning to all young females, not to forsake the paths of virtue.' When a woman was convicted in Shrewsbury in 1821 not only of infanticide but of then, unimaginably, putting the dead baby in a pie which she took to an unsuspecting baker to be baked, it was 'impossible to peruse without horror . . . this most cruel deed, which for depravity and wickedness, far exceeds all the annals of murders in the memory of man'—which was fair enough comment.

[28] Ibid. 1880. c. 20. [29] Ibid. 1880. c. 12; also Madden Collection, 11.604.

[30] BL broadside, 1888. c. 10.

Murderers got shortest shrift: 'here is another instance of the depravity of our nature, and the ills resulting from giving way to passion'; or 'This is one of the most unnatural and horrid murders we can remember. . . . We see, in this instance, a proof of the excellence of the Laws of our country, which will not screen the great man [in this case a surgeon] any more than the poor, but dispense justice to all'; of a Gloucester man who murdered his father, mother, and servant in 1825: 'God never suffers such monsters in wickedness to go unpunished' . . . and so on.[31] These stances were probably more overt in the provinces than in London, and more developed as time passed across the century 1750–1850 too: but Pitts's and Catnach's death-sheets can yield much the same harvest.[32]

3. *Purchasers*

It is important to know who determined this moral caution, customers or printers, if we are plausibly to link these attitudes to popular mentalities. It cannot be denied that printers had their say, especially in tense times when eyes must be kept on magistrates. Hone's acquittals in 1817 secured *de facto* immunity for most forms of street literature, but those who 'pattered' the broadsides in the street still found 'there was plenty of officers and constables ready to pull the fellows up, and . . . a beak that wanted to please the high dons, would find some way of stopping them'.[33] Printers had their own prejudices too. Catnach thought that radicals were 'the scum of the country' and counted himself an old tory. And many of his provincial counterparts were evangelicals, to judge from the pieties of their output.[34]

Even so, customers' tastes appear to have been determining. Radical printers like Russell, Guest, and Watts in Birmingham, Wheeler and Livsey in Manchester, or Thomas Willey of Cheltenham, all active in radical or Chartist publishing between Peterloo and the 1840s, were not averse to putting out radical sheets and songs in years of high excitement; so that it is difficult to believe that they were ignorant of the bloody code's political implications or uncritical of it.[35] None the less,

[31] BL broadside, 1880. c. 20.

[32] On most religious or political matters, broadsides were cautious: Anglicans, Quakers, Baptists, and Methodists would all be referred to favourably, and economic antagonisms were rarely emphasized: Neuburg, *Popular literature*, 129–30, 137, 142.

[33] E. P. Thompson, *The making of the English working class* (1965), 722 n., citing one of Mayhew's patterers.

[34] Hindley, *Catnach*, 409.

[35] See R. Palmer, *The sound of history: Songs and social comment* (Oxford, 1988), 13–14, and R. Palmer (ed.), *A touch on the times: Songs of social change 1770–1914* (1974), for examples.

when it came to execution sheets, their output was not significantly different in tone from those produced by Catnach. For some two hundred years, though they changed in ambition, quality, and quantity, execution sheets changed little in their moral postures. People might be induced to buy more sheets in new elaborations; but they must be given the moral certainties which they apparently needed if the world was to retain ethical meaning. So yet another kind of mentality is reflected here. But whose? Before we jump to conclusions, we must note next that the buyers of broadsides were more diverse than is often recognized.

Unexpectedly, plebeians were not the only purchasers of broadsides, or even in some areas the dominant ones. If it was true that most street literature was 'written by persons of the class to whom they are addressed', it is clear that we are dealing with a clientele at least as literate as the hacks employed by printers.[36] Mayhew learnt that 'tradespeople' were the readiest buyers in some parts of London. And although in London by mid-century 'gentlefolks won't have anything to do with murders sold in the street; they've got other ways of seeing all about it',[37] the polite or would-be polite classes in the provinces were never so aloof.

A provincial genre flourished which differed markedly from the vulgar metropolitan model with its crude woodcuts and sententious verses. Tradesmen, farmers, and genteel folk bought sober and undecorated execution homilies which confirmed their interests and values, reflected on the condition of the times, or preached usefully to servants and dependants. Sheets for such people were abundant, their prose directed over the heads of the labouring classes, their commentary sometimes ambitious. Thus a late eighteenth-century Bristol sheet on the occasion of an infanticide trial ascribed the recent increase in child-murder to the influx into Bristol of males 'of dissolute principles', which added 'materially to the injury committed on the tender minds of our various females'; it also ascribed it to the rage for emulation and luxury. The interests and values of those addressed are very apparent:

> When an enquiry is made at a door, one receives an answer from a lady so well dressed, with her hair in buckles etc. that she is immediately addressed as the mistress of the house and the mortification is very great on finding that this lady is perhaps only a scullion, or under house maid. To provide dress our servants are obliged to use every stratagem to get money, and a general prostitution appears to have taken place among them, so much so, that it is dangerous

[36] 'Street ballads', 399. [37] Mayhew, i. 282–4, 234.

for a country girl to come into town for service, for as dress is the order of the day, she is hurried by her companions to emulation, which her wages not affording, she is quickly led on to ruin.[38]

Productions of this kind seem to have been most common in agricultural towns where smaller populations and lower literacy inhibited plebeian sales. In Ipswich the three rival printers Bransby, Dorkin, and Scoggins usually aimed at or presumed an informed readership with full access to respectable values.[39] When two murderers were executed in 1809, Bransby ascribed the recent decline in executions to 'the humane discretion of the executive power' which now put 'a more liberal construction on severe penal laws'. When two women hanged for infanticide five years later, Dorkin noted that there had been seven Ipswich executions in the past couple of years and insisted that children must be trained 'early in habits of industry, and a regular observance of the Lord's Day'. When another woman was hanged for infanticide in 1815, Dorkin thought that the people of Ipswich might as well not be 'living in a civilized state of society, but in one of the rude and barbarous ages of former times!' He mustered some pity for the girl: 'it is shocking to humanity to consider that these dreadful visitations of the law should so often be necessary: and we cannot but view with sorrow these heart-rending scenes of woe'—but 'necessary' these executions, in his view, certainly were. And when during the East Anglian rick-burnings of 1816 a labourer Joseph Bugg was hanged in Ipswich for arson, the writer felt that the law must make a terrible example, for arson aroused the 'indignation of every honest man': 'To pass over the circumstances that brought this unhappy man to his end, without comment, would in a great measure defeat the purpose for which publications of this sort are intended to produce, of deterring others from falling into the same error.' Then the text played directly into the interests of its apparently far from humble purchasers:

Most truly do we lament the state of our Manufacturing and Agricultural poor, but the times are equally hard for the employer or tradesman, and if the pretence of these lawless proceedings arise from the necessity of the advance in wages, want of employ, or for the want of food, the burning of barns, stacks, or working implements of their employers will only add to their distress. In no country in the world are the wants and necessities of the poor so liberally provided for as in England, and even now in its most distressed state, what appears hard to them, would in any part of the Continent be considered a luxury. . . . The laws are strong, and effectual for the protection of the peaceable inhabi-

[38] BL broadside, 1880. c. 20, fo. 154. [39] Ibid., CUP. 407. MM. 29.

tant:—malice and revenge for injuries, whether real or imaginary, cannot be allowed by any civilized state.

The tradition survived: an Ipswich execution for buggery in 1830 elicited a long essay advocating 'A radical cure for crime' through religious education. All this was material fit for purchase by masters and for consumption by servants.

So much having been said, however, most broadsides, in the bigger centres and London especially, were unambiguously directed at humble people: we are safe in presuming that their ethical and sentimental messages accorded with plebeian expectations. So much is obvious from their style, marketing, and reception. When in 1774 Boswell composed and had published a broadsheet to vindicate his condemned client John Reid, the printer 'undertook to get the case which I had drawn in John Reid's style printed and cried, to conciliate the lower populace'.[40] Moreover, 'the most striking characteristic of the first phase of the expansion of imaginative literature was the sheer volume of noise which accompanied it'.[41] On execution days in the 1780s, Place recalled, 'every ragged man, woman and child bawld [the "death-verses"] . . . Some blowd horns, and a continual clamour was kept up for some hours during the morning—and indeed till the middle of the afternoon. They all used the same words, and the same tone in chanting them.'[42] Hackman observed in 1777 that last dying speeches were cried on the streets as commonly as muffins and matches were (Plate 16). A halfpenny ballad singer was chanting one in a 'ridiculous tone' outside his window as he wrote, and familiarity had 'even annexed a kind of humour to the cry'.[43] Blind people were common singers of death-verses.[44] First (as Mayhew later recorded) they chanted their wares: "Ere yer hav' jest pa—rinted and pub—lish—ed, the last dying speech and con—fession of Thomas Blook—who was hex—e—cu—ted this morning hat the Hold Bailey for the 'orrid murder of his wife, and hof 'is hinnercent babbies—together with a correct copy of werses written by the hunfortinit man the night before he suffered.' Then they sang the verses in a nasal twang to psalm tunes, accompanying themselves with a fiddle if they had one.

Although the fiddling and twanging style was going out of fashion in

[40] W. K. Wimsatt and F. A. Pottle (eds.), *Boswell for the defence, 1769–1774* (1960), 321–2, 332.

[41] Vincent, *Literacy and popular culture*, 201; cf. Spufford, *Small books*, 68 and *passim*.

[42] Place papers, BL Add. MS 27826, fo. 100.

[43] Borrow, ii. 714.

[44] For a portrait of a blind chanter, accompanying himself on a fiddle of catgut drawn over a bladder or tea canister attached to a mop-stick, see J. T. Smith, *Vagabondiana, or Anecdotes of mendicant wanderers through the streets of London, with portraits of the most remarkable drawn from life* (1817), 45 and pl. facing.

16. Thomas Rowlandson, *Last dying speech and confession*, 1799.

Rowlandson's broadside seller is based on Hogarth's in Plate 26 below. Even the pocket-picking boy repeats Hogarth's point that new criminals are made as old ones are killed. Patterers like this were still at work in Mayhew's London, though a dying breed. They held hands to their ears the better to pitch their caterwauling 'chaunts'.

the 1820s (Mayhew's informants later found it quaint), the chanting tradition survived. Even though by the 1860s ballad singers were a disappearing race,[45] Mayhew in 1851 still found running patterers sweeping up and down the poorer streets as noisily as they could. 'The greater the noise . . . the better is the chance of sale.' The verses still had to be 'such as the patterers approve, as the chaunters can chaunt, the ballad-singers sing, and—above all—such as street-buyers will buy'.[46]

Migrant sellers from London competed in the provinces too, especially when notorious murderers were hanged, like Corder in Bury St Edmunds in 1828 or Rush in Norwich in 1849. At Corder's execution demand was so lively that, as one seller told Mayhew, 'I got a whole hatful of halfpence. . . . Why, I wouldn't even give 'em [discounts of] seven for sixpence—no, that I wouldn't. A gentleman's servant come out and wanted half a dozen for his master and one for himself in, and I wouldn't let him have no such thing.' The readiest sales were at the foot of the scaffold, as the speed of production implies:

I never works a last dying speech on any other than the day of execution—all the edge is taken off of it after that. The last dying speeches and executions are all printed the day before. They're always done on the Sunday, if the murderers are to be hung on the Monday. . . . The flying stationers goes with the papers in their pockets, and stand under the drop, and as soon as ever it falls, and long before the breath is out of the body, they begin bawling out.

The sheets also found buyers in the poorest and most remote villages. In the 1840s London sheet-sellers were reaching these places on their pack-horses just as chap-book dealers had for centuries past: 'I've been through Hertfordshire, Cambridgeshire, and Suffolk, along with [the murderers] George Frederick Manning and his wife—travelled from 800 to 1,000 miles with 'em.' Mayhew learnt how in a Norfolk village in 1849 two impoverished families had to club together to buy a penny broadsheet: but buy it they did. His informant saw a timeless vignette at night through an uncurtained cottage window—'eleven persons, young and old, gathered round a scanty fire':

An old man was reading, to an attentive audience, a broad-sheet of Rush's execution, which my informant had sold to him; he read by the fire-light; for the very poor in those villages, I was told, rarely lighted a candle on a spring

[45] 'Those two somewhat shabby companions, with voices of brazen twang . . . their hands filled with broad-sheets, their eyes keenly glancing round for every possible owner of a spare half-penny, and making the whole neighbourhood ring with their alternate lines and joint chorus of some unspeakable ditty, sung to a popular air, with variations on the spur of the moment,—alas, where are they gone?' 'Street ballads', 400.

[46] For most of the following references, see Mayhew, i. 214–35, 280–1, 301–4.

evening. . . . The scene must have been impressive, for it had evidently somewhat impressed the not very susceptible mind of my informant.

Despite a developing market for ballad slips, yards of song, and fictitious 'cocks' about elopements, seductions, assassinations, and suicides (these accounting for the bulk of printers' lists if not of sales), murder was the big pull. A rape afforded a Seven Dials balladeer 'great satisfaction', but 'a murder—an out-and-out murder—if well timed, is board, lodging and washing, with a feast of nectared sweets for many a day': 'A good murder will cut out the whole of them.' In mid-century London up to a hundred running patterers and a score of standing patterers and boardmen selling sheets at a penny or halfpenny a time could make twenty shillings a week if there was a good murder, a quarter of that when there was not. 'The more horrible we makes the affairs, the more sale we have.'

The Seven Dials board trade knew how to catch this taste too. Artists prepared water-coloured placards varnished in gum resin to sell to the patterers. They 'lay on the horrors . . . in the highest colours' possible: 'scarlet, light blue, orange—not yellow I was told, it ain't a good candle-light colour'. They 'must leave nothing to the imagination. Perspective and backgrounds are things of but minor consideration. Everything must be sacrificed for effect.' It is difficult for us now to conceive the lived immediacy of people's responses to images and detail of these kinds, but Mayhew had a good eye for such things:

One standing patterer, who worked a Mannings' board [before that couple's execution for murder in 1849] told me that the picture of Mrs. Manning, beautifully 'dressed for dinner' in black satin, with 'a low front', firing a pistol at Connor, who was 'washing himself', while Manning, in his shirt sleeves, looked on in evident alarm, was greatly admired, especially out of town. 'O, look at him a-washing hisself; he's a doing it so nattral, and ain't a-thinking he's a-going to be murdered. But was he really so ugly as that! Lor! such a beautiful woman to have to do with him.'

This firmly concreted imagining had its counterpart in broadsides' accounts of crimes. Topographical detail helped in the appropriation of sensation to oneself, rendering the horrible familiar and offering concrete reference points for gossip and conversation:

The victim of this sad affair
    Was found, as we may see,
Foully murder'd, hid beneath a stone,
    In Minerva Place, Bermondsey.[47]

[47] *Copy of verses, and apprehension of Mrs Manning on the murder of Mr O'Connor* (1849), Madden Collection, 11.71.

Thus the reading of pictures and narratives was direct and literal, as it had always been among the poor. John Clare 'firmly believed every page I read' in the sixpenny romances he cherished as a child, feeling that he 'possesd [*sic*] in these the chief learning and literature of the country'.[48]

4. *The gallows emblem*

The point need not be laboured that this trade reflected as much as it constructed a set of common ethical postures which all social classes shared—a fragile consensus (always capable of being broken, however, as particular cases dictated) that villainy, especially against others' bodies, must be punished. But it does not quite suffice to leave our assessment at that. Some mystery still attaches to the broadsheets' huge sales. The mystery lies in the fact that the sheets were repetitive and their moralizing intrusive and formulaic. Also they were far less sensational than critics made out. People were launched into eternity rather than strangled on ropes. And euphemisms veiled the most horrific crimes. The appeal for many purchasers could not have resided primarily in these censored and stylized narratives. Read half a dozen and you have read them all.

What drew the purchasers, then, was more likely the fact that execution sheets were totemic artefacts. They were symbolic substitutes for the experiences signified or the experiences watched. They were mementoes of events whose psychic significance was somehow worth reifying. Set up by powerful, remote, and broadly unintelligible agencies, these ritually inflicted deaths had not only to be accommodated within frameworks of moral meaning but also, somehow, to be tamed and possessed. So people bought execution broadsides as needfully as they stood on the scaffold's site after it was dismantled. Moreover, at the heart of this need were the woodcut illustrations. It is in the illustrations, so little attended to by historians, that the broadsheets' iconic meanings become explicit.

Apart from their loud and erratic typography, the most eye-catching features of execution sheets were their woodcut embellishments. These were 'all simply abominable', it was said in 1857, 'a full century behind the march of improvement' in graphic art.[49] Images were simple and undeveloped, their messages attenuated in comparison with those of religious and political propaganda (copperplate applying to those

[48] E. Robinson (ed.), *John Clare's autobiographical writings* (Oxford, 1983), 5.

[49] Smith, 'The press of the Seven Dials', 262.

services). Many execution cuts did service for generations. Publishers attached so little value to verisimilitude that some nineteenth-century sheets depicted men and women in Stuart costume; portraits of malefactors were cheerfully interchanged;[50] and the number and gender of scaffold victims would be clumsily superimposed on a standard scaffold image without regard for scale, and not always with reference to the real numbers hanged. Provincial publishers used well-worn Newgate images to illustrate local executions, and sometimes the compliment was returned. All execution woodcuts shared these and kindred defects, and nothing was done to improve them. Catnach used finer boxwood engravings for other kinds of broadsides but never for these. Why did none of this matter, either to printers or to their clientele? *Why were these rough-and-ready images for so long thought to suffice?*

Their deficiencies could be explained in terms of technological constraints upon the engraver and the expense of copperplate; or they could be explained in terms of the low status and pay of their artisan and sometimes apprentice engravers; or simply in terms of the undeveloped taste of their audiences. Apart from signboards, woodcuts had always been the cheapest and main source of imagery for a common people starved of pictures, and along with the ballads and tales they illustrated, they had long been pinned to cottage and alehouse walls as the chief source of 'learning, and often, no doubt, the delight of the vulgar'.[51] But this hunger for images does not explain their enduring simplicity long after Bewick pioneered boxwood engraving and the quality of cuts on song-sheets had so greatly improved. The execution blocks continued to suffice for other reasons.

Image-sophisticates now, we have lost ways of experiencing the impact of vernacular images of these kinds. But if the execution cuts are thought of as ideograms rather than as failed or clumsy representations, their meaning opens up at once. They had more than the mere quaintness ascribed to them today. Just as woodcuts of demons or monsters in chap-books and prints, drawing on the cultural codes of people who believed in magic, had long had the power to electrify their perusers, so, in their starkly simplified black and white contrasts, the figure dangling in grotesque disproportion, scaffold prints drew on an 'image magic'

[50] One supposed portrait of the Mannings' victim in 1849 derived from Lawrence's portrait of George IV: Mayhew, i. 284 and facing p. 278.

[51] T. Holcroft, *The memoirs of Thomas Holcroft*, cited by Shepard, *Pitts*, 29; cf. Watt, *Cheap print*, part II. As late as 1849 a satirical cut by John Leech depicts a gallows scene and a Greenacre broadside pinned over the mantlepiece of a working-class parlour: P. Anderson, *The printed image and the transformation of popular culture, 1790–1860* (Oxford, 1991), 25.

17. *Trial, conviction and execution of Samuel Fallows, on 14th April 1823 at Chester.* The potency of this most reduced of broadside woodcuts may paradoxically be enhanced by its poor craftsmanship. Accurate depiction mattered less to purchasers than did the possession of a totemic image, however rough, which might stand for the sensations which executions released.

which had fuelled the graphic imagination for centuries. They still retain a power to disturb for those who can see. If thinly for us, then deeply for their immediate audiences, the crudest representations in 'folk art' could carry something akin to a totemic meaning. If they named the object correctly, a few schematic indications would ensure their effectiveness, their power ensuing from an 'effective contagion' with the scene that was its source (Plate 17). Even today a distant memory of this survives in the children's word-game, hangman's noose.

These woodcuts contained no lessons. All they did was 'materialize a way of experiencing, bring a particular cast of mind out into the world of objects, where men look at it'.[52] So it is not surprising that popular icons on the subject of death were resilient and antique. In the 1830s some broadsheet woodcuts were still saturated in a medieval iconography of angels and demons, expressing the survival of a folk theology distant from official belief. Such images resonated with ancient associations, drawing on the medieval to early modern rhetoric of the emblem which had long furnished print culture with a repertoire of motifs. They drew on an image theory which refused to reduce the image to mere illustration; instead it granted it an efficacy of its own, investing it with its own affective charge and value.[53]

The ubiquity of the ideographical gallows can be set in a second framework. As it assimilated Hogarthian references, and as the

[52] C. Geertz, *Local knowledge: Further essays in interpretative anthropology* (New York, 1983), 99.

[53] R. Richardson, *Death, dissection and the destitute* (1987), 10–11; D. Freedberg, *The power of images: Studies in the history and theory of response* (Chicago, 1989), 1, 42, 112, 148; R. Chartier (ed.), *The culture of print: Power and the uses of print in early modern Europe* (1989), p. 6.

compliment was returned, the woodcut came to share a common attitude to the gallows with 'high' art. Although the popular woodcut was not designed to make a moral point as Hogarth's images were, the image of hanging remained iconic at both levels, and at both levels it was reciprocally sustained.

It is true that there were a few artists who sought realism when they depicted the scaffold. Rowlandson was one. Despite his irrepressibly comic line and cool detachment from his subjects, his gibbet drawings still shock. Two riders on horseback start back in horror before a ghastly corpse dangling in its iron cage; or men and women gaze up at four gibbeted corpses swaying in the wind (Plates 18, 19). Mary Evans's execution at York in 1799 is tautly sketched from the rear of the scaffold but from close quarters (Plate 20), the caricatured faces of parson and hangman crowding the terrified girl as the hangman adjusts the rope, her coffin waiting behind her. We can only guess why Rowlandson drew these pictures, noting the fascination with crime and death in all his work.[54] But it may still be significant for our argument that Rowlandson was almost alone in producing such images, and that the most realistic and coruscating depiction of any execution in our period was produced by a stranger to English taboos and visual language alike. Visiting England, Géricault drew a *Public hanging* in pencil and wash in 1820. Three men stand on a scaffold with hands tied and nooses around their necks. The top-hatted hangman hoods one figure, the central figure is hooded already, and the ordinary of Newgate preaches to the third, who ignores him, staring out at us with intense and darkly circled eyes. The picture is drawn from the scaffold's foot, the line fluid, the image fearsome (Plate 21).[55] No other depiction is as honest as this. At all levels of ambition, realistic as distinct from ideographic images of the scaffold were rare. This suggests either a failure of nerve or the power of taboo—that the emblem was a safer device, preserving decency and sustaining a distance from the horror. Or it suggests—more plausibly?—that in England the emblem had become a conventional language for

[54] Cf. the mocking skeletons of Rowlandson's *The English dance of death* (1815–16) and the punishment themes in the stocks of *Newbury Market* and *Dr Syntax attends the execution*: R. R. Wark, *Drawings by Thomas Rowlandson in the Huntington Collection* (San Marino, Calif., 1975), plates 243ff., 31, 322; R. R. Wark, *Rowlandson's drawings for the English dance of death* (California, 1966).

[55] This image is usually, but erroneously, assumed to depict the execution of the Cato Street conspirators. 5 men, not 3, were hanged and decapitated for treason then; Géricault would not have omitted the coffins and axe which were the centrepieces in the tableau (see Plates 33 and 34, Ch. 11, below). Géricault's earlier drawing of a decapitation in Rome (1816–17) was heroically stylized and passionless by comparison with this drawing: L. E. A. Eitner, *Géricault: His life and work* (1983), 223, 114; pl. 191, 200.

18. Thomas Rowlandson, *A gibbet* (1790s?).

Rowlandson was fascinated with death, scaffolds, and gibbets, though his irremediably comic line seldom did justice to their horror, so his attitude often seems cruel. But there is no doubting the intensity of his observation in the three drawings reproduced here and below. The sketch of Mary Evans (Plate 20) is especially unusual in its apparent sympathy for the condemned and in its contempt for the grotesques officiating at her death. Husband-poisoners could not commonly expect pity from a man.

19. Thomas Rowlandson, *Crowd by a gibbet* (1790s?).

this subject, with its own *sufficient* power. Why else would Thackeray in 1840 render his first glimpse outside Newgate of something 'awful to look at, which seizes the eye at once, and makes the heart beat', in terms of a starkly blocked image which needed no elaboration?

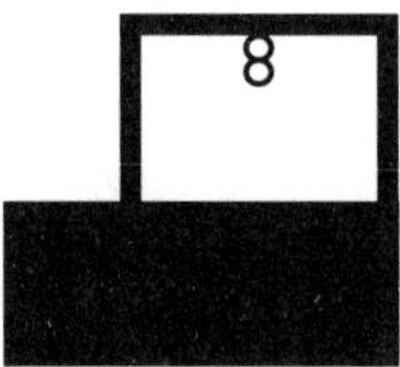

'There it stands', he wrote, 'black and ready, jutting out from a little door in the prison. As you see it, you feel a kind of dumb electric shock, which causes one to start a little, and give a sort of gasp for breath.'

The gallows emblem had never before been so ubiquitous in English graphic art as in the century before this sketch; nor in the nature of things could it be so thereafter. It would be too much to say that artists were haunted by it, for emblematic language was universal and the scaffold image one among many. Still, in copperplate as well as woodcut it

20. Thomas Rowlandson, *Mary Evans hung at York Augst. 10 1799 for poisoning her husband.*

21. Théodore Géricault, *A public hanging*, 1820.

Sketched while he was visiting London in 1820, Géricault's image confronts the terror of the condemned with unparalleled immediacy. It was never to the English taste to get as emotionally close to the subject as this. English depictions invariably diverted attention away from the scaffold's victims towards the crowd.

was so frequent from Hogarth's day all the way down to 1840 that it affirms the extent to which the scene loomed over the Hanoverian imagination, and what subliminally shared understandings attached to it across classes.

Some part of Hogarth's innovative achievement in and after the 1740s lay in his resort to a demotic visual language in order to structure his pictures around popular icons, proverbs, and chap-book moralities. As he went about London he would study walls for their graffiti and sign-boards for their devices—'with delight', he said, drawing upon them in pursuit of that 'pleasing labour of the mind to unfold mystery Allegory and Riddles'.[56] In pictures loaded with emblems, many turning on the visual violence of manacles, whips, prison bars, and pillories (no artist more beset by the prison than Hogarth), he used one image repeatedly: a gallows from which a corpse dangled.

This is scratched on the shutter of the Bridewell in *The harlot's progress* (1732) and on a wall in *The first stages of cruelty* (1751) (Plate 22). In plate 3 of *Marriage à la mode* a model of Tyburn's triple tree surmounts a cupboard containing a life-size anatomical figure at which a skeleton leers meaningfully. Someone holds up a broadsheet with a gallows emblem on it in *The polling*, one of his four prints of *An election* (1755–8). In *The invasion* (1756) the image of the French king drawn on a wall holds a gallows in its hand. *A just view of the British stage* (1724) is festooned with nooses, satirizing a pantomime on Jack Sheppard, executed that year. In the background of *Gin Lane* (1750–1) a suicide hangs by his neck in a ruined attic and a coffin hangs from an undertaker's sign-pole as a body would from a cross-beam. In *The idle 'prentice turn'd away, and sent to sea* (1747) a sailor points to a corpse hanging from a gallows on the distant shore, portending the apprentice's fate in the last of that series (Plate 22). Hogarth's last work, *Tailpiece, or Bathos* (1764), was saturated with the gallows' camouflaged representation, and in the distance a body hangs from a scaffold. In *The idle 'prentice* and *The harlot's progress* the emblem expressed the ineluctability of fate and retribution, the just deserts of the wicked. In *Tailpiece* it embraced the terrors of death and a life's defeat. And in *The stages of cruelty* it represented the pain which the powerful could deliver upon the weak.

Hogarth freed his successors to deploy these meanings repeatedly thereafter in print, and they did so energetically. Thomas Bewick

[56] R. Paulson, *The art of Hogarth* (1975), 30 ff., 64; J. Lindsay, *Hogarth: His art and his world* (1977), 20, 157.

22. Detail from Hogarth, *The idle 'prentice*, 1747.
The many references to the scaffold in Hogarth's works speak for its looming presence in the eighteenth century, as well as for its moral meanings. They also often hint at its ubiquity as an emblem scratched on urban walls (Plate 23).

mysteriously illustrated *British birds* and *Aesop's fables* with vignettes of gallows and gibbets in which carrion birds fly overhead and gleeful devils preside. When in *Aesop* he depicted the evil inherent in the scaffold in a deft little design of a devil swinging merrily on a gallows, did the evil attach to the law which hanged people there, to the felon who hanged, or to the object itself? It was not necessary to say (Plate 24).[57] With similar ambiguity, George Cruikshank consciously deployed Hogarthian motifs when he illustrated Ainsworth's 'sort of Hogarthian novel', *Jack Sheppard*, as late as 1839. In one picture (in which Wood the carpenter offers to adopt the baby Jack Sheppard) the derelict walls are covered in graffiti, drawing the eye even more than the central characters do. Pinned between them is the last dying speech and confession of Jack's hanged father Tom. Below that, someone has scratched 'Paul Groves

[57] T. Bewick, *History of British birds* (2 vols., 1797–1804), and *The fables of Aesop, and others, with designs on wood* (orig. pub. 1818; 2nd edn., 1823); cf. the collections of *Vignettes* (Newcastle upon Tyne, 1827).

23. Detail from Hogarth, *The stages of cruelty*, 1751.

Cobbler hung himsel in this rum for luv off licker'. And below that again, a stick-man dangles from a stick-scaffold.[58]

It was in the work of Gillray and Cruikshank that Hogarth's legacy was most direct. Depending on their current paymasters, they used the

[58] In another of Cruikshank's illustrations for the book, *Jack Sheppard exhibits a vindictive character*, the room's walls are covered in broadsides and the shelves groan with objects, one of which is a small model of the Tyburn tree with a body suspended from it, labelled 'Jack Hall a Hanging' (he of the chimney-sweep ballad noted in the last chapter).

24. Thomas Bewick, *Gallows vignettes*, 1797–1818.
Fine boxwood engravings of these gibbet-obsessed kinds were scattered throughout Bewick's work. Far from commenting adversely on hanging law, they celebrated the scaffold's morbid and romantic associations, or simply stated that the scaffold was the devil's domain towards which evil-doers must come.

scaffold emblem as loyalists to speak of anarchy and revolution or as radicals to mount critiques of injustice. Cruikshank's *French conscripts marching to join the Grand Army* (1813) pass under a skeleton on a gallows whose arm points the way 'To Russia'.[59] In his *The age of reason or the world topsy-turvy exemplified in Tom Paines works!!* bishops and ministers dangle from a gallows supported on a guillotine.[60] In Gillray's boisterously anti-radical print of the *Middlesex election, 1804* dung flies through the air as the mob cheers Sir Francis Burdett's coach; a banner depicts Britannia bare-breasted being flogged by Governor Wall; and in the foreground a rat is hanged from a noose while someone holds up a

[59] R. A. Vogler, *Graphic works of George Cruikshank* (New York, 1979), pl. 4.

[60] D. George, *English political caricature, 1793–1832: A study of opinion and propaganda* (Oxford, 1959), 183.

right-angled pole from which a stick-figure dangles, comically awry.[61] In radical incarnation, Gillray's *New way to pay the National Debt* (1786) castigated George III's avarice by depicting the king receiving moneybags outside the Treasury accompanied by a military band and by beggars, one of them legless and armless on the ground. A tattered broadsheet on the wall announces 'the last dying speeches of fifty-four malefactors for robbing a hen roost', and on an elongated scaffold they hang in a line of corpses, like slaughtered hens themselves. Plastered over the sheet is another whose text begins 'God save the King'.[62] In Cruikshank's 1819–21 cuts for William Hone's pamphlets the gallows recurs inevitably. A Janus-faced clerical magistrate holds up a cross in one direction and gallows, whiplash, and shackles in the other (see Plate 39, p. 513 below); an Irish bloodhound goes for the throat of a prostrate female figure symbolizing the Union, the background filled with gallows and gibbets.[63]

In 1875, in a letter optimistically headed 'How I put a stop to hanging', the ageing Cruikshank recalled how on the morning of 16 December 1818 he had walked down Ludgate Hill:

> seeing a number of persons looking up the Old Bailey, I looked that way myself, and saw several human beings hanging on the gibbet opposite Newgate prison, and to my horror, two of these were women, and, upon enquiring what these women had been hung for, was informed that it was for passing forged one pound notes. The fact that a poor woman could be put to death for such a minor offence had a great effect upon me.

He went home and designed on copperplate a parody of a *Bank restriction note* (Plate 25). Hone published it at a shilling a time. It was a best seller, allegedly netting some £700 and requiring the engraving of a second plate after two or three thousand impressions had been taken from the first. 'When it appeared in [Hone's] shop windows it created a great sensation, and the people gathered around his house in such numbers that the Lord Mayor had to send the City police (of that day) to disperse the crowd.' The mock banknote was signed for the governor and company of the Bank of England by 'J. Ketch' the hangman. It was decorated with shackles and skulls, and the £-sign was formed by a rope with a noose framing a prison through which faces peered. The

[61] D. Hill (ed.), *The satirical etchings of James Gillray* (1976), pl. 6.

[62] Ibid., pl. 7 and fig. 7; Vogler, *Graphic works of Cruikshank*, fig. 4.

[63] Facsimiles in E. Rickword, *Radical squibs and loyal ripostes: Satirical pamphlets of the Regency period, 1819–1821, illustrated by George Cruikshank and others* (1971), 55, 289.

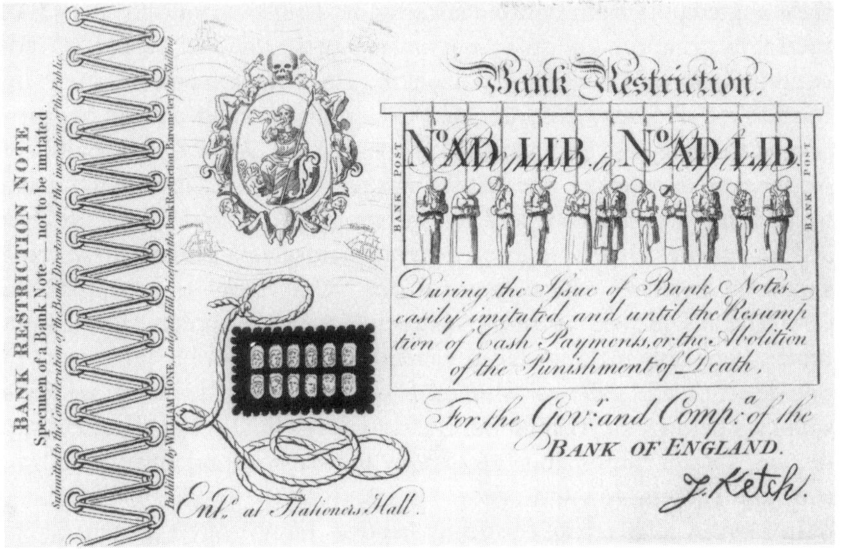

25. George Cruikshank, *The bank restriction note*, 1819.
A fifth of English executions between 1805 and 1818 were for forgery, mostly brought by the Bank of England for the forgery of banknotes. Published by William Hone and selling in many editions at a shilling a copy, Cruikshank's was the most virulent of all graphic attacks on Bank policy.

dominant motif was a line of eleven men and women, heads hooded and necks twisted, dangling in line from nooses.[64]

Cruikshank's wood engravings for Hone reduced the emblematic statement to essentials. But for the most part these and similar devices were deployed densely and almost incidentally in the copperplates of the caricature tradition, each one all but lost in the prints' detail. The density of reference required purchasers to read prints attentively, alert to hidden meanings and associations, and in this spirit they certainly were read.[65] But by whom? And how do these images in what, for want of a better term, may be called a 'higher' art connect with the vulgar tradition of the broadsheet woodcut?

Although prints of the quality of Hogarth's, Gillray's, and the early

[64] B. Jerrold, *The life of George Cruikshank* (1882), 90–4; F. W. Hackwood, *William Hone: His life and times* (1912), 198–205.

[65] 'Curiosity is implanted in all our minds,' as Hogarth observed, 'and a propensity to searching after, pursuing and surmounting, difficultys . . . enhances the pleasure of the pursuir and makes it a sport' [*sic*]: Lindsay, *Hogarth*, 20.

Cruikshank's were expensive and aimed at the politically literate, their accessibility to artisans and tradesmen should not be underestimated. Hogarth had deliberately directed some of his popular plates in the 1740s at 'the use and instruction of youth wherein every thing necessary to be known was to be made as intelligible as possible . . . [so that] the purchase of them became within the reach of those for whom they were chiefly intended'. The twelve plates of *Industry and idleness* sold at a shilling each, within the reach of master craftsmen, he hoped. Probably the originals did not penetrate far beyond London,[66] but the plates were republished cheaply many times, and the fact that even the original price need not inhibit popular sales is suggested by the success on the streets of Hone's and Cruikshank's banknote, also selling for a shilling. The more radical woodcuts of 1819 and 1831–2 sold more cheaply too, and were increasingly accessible in the unstamped penny newspaper: William Hone's 'little books' led the way in this dissemination.[67]

Those who could not buy could still see. Print-shop windows were beset by crowds 'of all ranks' gazing at *Industry and idleness* when it was first published,[68] just as decades later 'the blockade of people in front of [Hone's] house was usual on the appearance of every new pamphlet'.[69] Thackeray recalled the displays of print-shop windows in his youth ('Knight's, in Sweeting's Alley; Fairburn's in a court off Ludgate Hill; Hone's, in Fleet Street—bright, enchanted palaces') and how prints were perused:

> How we used to believe in them! to stray miles out of the way on holidays, in order to ponder for an hour before that delightful window in Sweeting's Alley! . . . There used to be a crowd round the window in those days of grinning, goodnatured mechanics, who spelt the songs, and spoke them out for the benefit of the company, and who received the points of humour with a general sympathising roar.[70]

'A well known source of gratis recreation to the unemployed is what is called "a picture-fuddle",' Charles Smith noted in 1857, 'when a party of idle hands will hunt up all the print-shops and picture-shops of a whole

[66] Even in the mid-19th cent. Hogarth's work 'remained unknown in the cottages and villages of the country': Smith, *The little world of London*, 242.

[67] Hackwood, *Hone*, 200; George, *Political caricature*, 249–50. In 1824 Hone claimed that his tracts showed 'what engraving on wood could effect in a popular way, and [excited] a taste for art in the more humble ranks of life, they created a new era in the history of publication. They are the parents of the present cheap literature, which extends to a sale of at least four hundred thousand copies every week': ibid. 186.

[68] Lindsay, *Hogarth*, 138–9.

[69] Hackwood, *Hone*, 203.

[70] Cited in Jerrold, *The life of Cruikshank*, 77–9.

26. William Hogarth, *The idle 'prentice executed at Tyburn*, 1747.
Gathered where Marble Arch is now, the Tyburn crowd celebrates the doomed apprentice's arrival by cart, accompanied by his Methodist preacher and his coffin; the ordinary of Newgate travels ahead aloofly by coach. The broadside-seller and the orange-seller, the bawds and the beggars, the boys fighting or picking pockets (thus embarking on Idle's career), were copied in many subsequent depictions. Moulding the very idea of the scaffold crowd over the ensuing century, the picture bore its share of responsibility for the enduring belief in the crowd's barbarism and criminality.

district, and spend perhaps the whole day in the contemplation of this gratuitous gallery.'[71]

Inevitably, Hogarth's images were appropriated in prints for the people.[72] The most enduring Hogarthian image in our context was his *Idle 'prentice executed at Tyburn*, plate 11 of *Industry and idleness* (Plate 26). William Gilpin observed in 1782 that he could 'not immediately recollect having seen a croud better managed, than Hogarth [had] managed' this one.[73] Although the image was picked up in the 1790s in Rowlandson's *An execution outside Newgate prison* (the same energetic crowd, the same orange-seller, broadside vendor, boys and dog), it reached its vulgarized apotheosis in artisan woodcuts of the 1820s and 1830s. Here it was often

[71] C. M. Smith, 'Amusements of the moneyless', in his *The little world of London*, 9–10.

[72] For an undated broadside visual quotation from *The idle 'prentice sent to sea*, see Lindley, *Woodblock engravers*, 18.

[73] W. Gilpin, *Observations on the River Wye* (orig. pub. 1782; 1973), 78.

27. *The trial and execution of Martin Clinch and Samuel Mackley for the wicked murder of Mr Fryer, in Islington Fields.*

This stylized woodcut was published by Pitts in the 1830s or 1840s (the hats suggest). It embellished many broadsides, including, as in this title, an account of two men's hanging in 1797—the representation of only one man not being held to matter.

reduced to a stylized and barely recognizable ideogram, the crowd roughly outlined, wooden, and static (Plate 27).[74] But in some more ambitious cuts, although Hogarth's scene is transposed to Newgate's exterior, the same characters appear—the orange-seller, the broadside-selling woman with her baby, the man and dog, and the boys brawling (Plate 28). And when the *Weekly Chronicle* in 1837 put out a near-full edition on Greenacre's execution, offering readers 'a scene in the Old Bailey, before the execution, engraved expressly for the *Weekly Chronicle*, by a distinguished artist', it rehearsed Hogarthian motifs religiously. The crowd was bigger, more respectable, and bonneted and top-hatted, and Newgate and the opposite houses were accurately drawn; but in the foreground the woman still sold oranges, a pocket was picked, boys

[74] For examples, see Hindley, ii, *passim*.

28. *The sorrowful and weeping lamentation of four lovely orphans left by J. Newton, a wealthy farmer, who was executed at Shrewsbury, on Monday last, for a cruel murder, on the body of his innocent wife and infant child*, 1823.

Hanged at Shrewsbury in 1823, John Newton was a habitual wife-beater who killed his wife in anger, watched by his terrified children. The ambitious woodcut illustrating the London printer Pitts's broadside on the case had no connection with the real Shrewsbury scene. With its Hogarthian figures, it was used to illustrate London executions too.

scuffled, and dogs fought (Plate 29).[75] In its many derivative forms, Hogarth's depiction must have been not the least source of élite anxieties about the scaffold crowd's energetic indisciplines, especially as the image got detached from the moralizing context of Hogarth's original cycle.

After the Peterloo massacre in August 1819, gallows symbolism began to be used almost exclusively in radical sheets. It enjoyed its last revival

[75] *Weekly Chronicle*, 7 May 1837. The image was used again for Cooper's execution in 1862: this 1862 version is reproduced without recognition of its origin in T. Gretton, *Murders and moralities: English catchpenny prints, 1800–1860* (1980), 56–3, and hence by T. W. Laqueur, 'Crowds, carnivals and the English state in English executions, 1604–1868', in A. L. Beier *et al.* (eds.), *The first modern society: Essays in honour of Lawrence Stone* (Cambridge, 1989), 334. In the 1862 version the scaffold is rudely superimposed on the print (Laqueur makes much of this), but in the original it was integral and to scale.

29. *A scene in the Old Bailey, immediately before the execution* [of Greenacre], 1837. 'Engraved expressly for the *Weekly Chronicle*, by a distinguished artist', this picture exaggerates the breadth of the Old Bailey, and fills the foreground with familiar Hogarthian figures. None the less the attempt at realism is unprecedented. The newspaper publication of such an image portended broadsides' gradual eclipse over the next thirty years.

in the turbulent years of the 1830 Swing uprising and the reform agitation thereafter. In *The blessing of new taxes!!!* by Lewis Marks (probably 1819), in which the regent and the ministers robbed John Bull of his last comforts, there were meaningful pictures on a background wall portraying the executions of Louis XVI and of Charles I and the bodies of the regent and Castlereagh hanging on a gibbet. In Heath's grim image of *Merry England* (1831), against a background of burning barns and with bayonets and soldiers' hats on the bottom margin, a heavy gallows looms, a noose hanging from it. On its beam a black crow perches, alongside a judge's wig crowned with the black cap (Plate 30). The duke of Wellington was *The real Swing!!* in a print of that name in the same year: and he swung on demons' tails between the arms of a heavy wooden gallows, to the acclaim of onlookers below.[76]

But the image was dying. The violence of Regency caricature was being replaced by milder comedies of family life and character; no longer censored or dangerous, the printmaker became jester or moral-

[76] George, *Political caricature*, 182, 242, 254, and pl. 91, 95.

30. William Heath, *Merry England*, 1831.

The use of the scaffold emblem to comment satirically on repressive law was in its last heroic age when Heath produced this carefully understated print for radical bookshops. It protests at the judicial killings and transportations which ended the Captain Swing disturbances in the agricultural south-east.

ist. Tales of criminal ignominy and retribution were increasingly depicted in visual narrative, their moral laboured, the economical eloquence of emblems diminished; and the future of graphic art henceforth lay in alliance with the printed word.[77] Later Victorians acknowledged that 'the present generation, examining these things, might wonder at the effect [the old prints] had upon the public mind'. They had to remind themselves that the effect had once been 'extraordinary'—that there had been 'a rush and a crush to get them'.[78] The demise of the scaffold image expressed also a growing squeamishness about the subject. It was slowly filtered out at politer levels of representation as polite people retreated as much from the contamination of the image as of the event, until we reach a *reductio ad absurdum* in Leech's depiction of the Mannings' execution for *Punch* in 1849, when the scaffold is wholly omitted, and the frame is filled by the crowd alone (see Plate 41, p. 604 below). Although the image lived on in broadsheet woodcut, right to the end of public execution indeed, in this as in other contexts the symbiotic relationship between élite and popular emblems was becoming tenuous, the old resonances lost, vibrant still among the vulgar, but repudiated by the polite.

People of all ranks whose senses might be numbed by the familiarity of hanging might still subliminally address the scaffold's meanings and *see* the ritual in the terms suggested by the pictorial devices. The image pervaded both the artistic and the popular imagination for a century, just as the looming image of the prison did from Hogarth and Gay on to Dickens. Furthermore, the uses Hogarth and his successors made of it suggest that gallows graffiti were chalked ubiquitously on urban walls—a demotic symbol which flourished independently, outside the woodcut. Nor was it always idly chalked. In mid-1790s Birmingham loyalist slogans gave way to chalked slogans calling for 'bread or blood', accompanied by gallows emblems. The image was also represented concretely—in the ten-foot-high gibbet plus effigy erected against a hated Middlesex JP in 1800, for example, which had its own cathartic meaning, about giving the élite a taste of their own medicine.[79] Humble people did not scratch images of judges in wigs or of courts in

[77] George, *Political caricature*, 257–60. See the later Cruikshank's moralizing series on *The bottle* (1847) or on *The folly of crime* (1844).

[78] Jerrold, *The life of Cruikshank*, 103.

[79] E. P. Thompson, 'The crime of anonymity', in D. Hay et al. (eds.), *Albion's fatal tree: Crime and society in eighteenth-century England* (1975), 281; E. P. Thompson, *Customs in common* (1991), pl. 6 and p. 481.

their pomp when they conjured up their images of justice or its opposite. The scaffold was the vehicle for the plebeian commentary.

Like all of us who reflect upon justice, humble and middling Hanoverians were buffeted by contrary assumptions as they contemplated hanging law. Images of justice were as multifaceted then as they are now. It is foolish to look for uniform postures, 'radical' or 'deferential' as the case may be. People steered a wavering course between tacitly ethical approval, sardonic and transgressive defiance, and mockery, sentimental anguish, or outright voyeurism. But what is clear is that those who assembled at the scaffold did not do so with the empty heads contemporaries ascribed to them. We have come a long way in these last four chapters from a closed sense merely of the crowd's ghoulishness, and can now admit resonances in their witnessing to which their enemies were blind. We are still not finished with this survey of plebeian texts; but it will be clear by now that their actions always made *sense*, even though at this remove in time, and even by the people themselves, precisely *what* sense can sometimes only be guessed at.

CHAPTER 6

THE PREROGATIVE OF MERCY AND THE PRACTICES OF DEFERENCE

1. *The home secretary's other papers*

SO FAR WE HAVE HEARD THE VOICES OF THE CONVICTED AND their families—indeed of the 'people'—only in the cries and gestures of the condemned on the scaffold or obliquely in the ballads and broadsides which commented on their plights. There remains a third set of texts, not in the public domain, which takes us behind the scenes observed hitherto, and closer even than ballads and broadsides do, to the attitudes and feelings of those convicted and doomed to face death or distant exile. They also show how capital law *worked*; and we use them extensively in later chapters.

Year by year in the early nineteenth century the Home Office accumulated in its overcrowded rooms a growing mountain of documents, now dirt-encrusted and tattered, whose survival historians have barely noticed. They were the petitions submitted by or on behalf of felons sentenced to death, transportation, or rotting years on the prison hulks. Sad documents all of them, they begged the judge, the home secretary, or the king for a mitigation of sentence under the royal prerogative of mercy. The dismal flood of appeals arrived on the home secretary's desk at a rate of some 1,300 a year between 1812 and 1822, increasing to nearly twice that rate by the later 1830s. The appeals occupied the centre ground of a judicial system whose very basis was the discretionary application or mitigation of penal pain. Along with their accompanying papers, sometimes extensive, sometimes brief, they have for the most part never been opened since they were first bundled and tied in ribbons nearly two centuries ago.[1]

[1] Appendix 1 discusses the arrangement, location, and coding of the petitions and their use in this book.

A phantasmagoric array of characters weaves its way through this vast archive. There are men and women innocently caught in the law's trammels, along with many rightly caught: poachers and pugilists; blackmailed homosexuals and impostors in sailor's clothing; street thieves, footpads, and highwaymen; penny-grinding clerks and occasional forgers; and discharged soldiers and sailors, one-legged or with head wounds from Waterloo or Trafalgar trepanned and wondrously sealed by tin plates. Over them judges, jurors, prosecutors, attorneys, and communities speak loquaciously. The archive is eloquent on lost practices too. It is full of letters penned hastily on the eve of execution and judges' or home secretaries' scribbles in response. Parchment crackles as the more ambitious petitions unfold. Memos are pencilled on scrap paper or penned on paper thickly handmade. Quill pens make many scripts spikily illegible; and everyone uses dashes instead of commas. Polite people's addresses are simple, the name of the addressee and the town usually sufficing. Surnames are staunchly English or Welsh, reminding one how cosmopolitan they have since become; while the spellings of the names of the poor often change disconcertingly within the records of single cases. Appellants' ages are negotiable too: few know quite when they were born, but rough guesses will do. The letters of the poor are poorly or phonetically spelt, betraying spoken accents wonderfully; the handwriting is laboured or scratchy even when dictated to others; and often crosses stand in for signatures. Then there are the faded pink ribbons which bind the bundles, and superadded a couple of centuries' ingrained dust and grime, the archival soil in which these people now lie quite buried.

In these papers we hear the voices of the common people. Accents apart, they are strange voices sometimes, like the poignantly antique language with which an abused Lancashire child cloaks her brutal experience: 'he pulled out his fie and for shame and put it into my fie and for shame'—and momentarily we are as much shocked by the Elizabethan remoteness of the image as by the crime it describes.[2] In some appeals attitudes were complex, subtexts ironic, deference expediential—not really felt at all. The more ambitious told elaborate tales of over-hasty trials and indifferent judges, perjured evidence, unexamined witnesses, unheard alibis, and misunderstood motivations. Sometimes affidavits and character testimonies, gathered by their local patrons, accompany them. Then petitions might be parchmented, scripts elaborate, literacy

[2] HO47.64/13.

high or legalistic, dossiers thick. These are the ones which are most useful for the deeper analysis of supporters' views undertaken in Chapter 16; but it is the briefer petitions, sent by felons or their families, which speak most directly from the plebeian or rural world, which are this chapter's concern.

Typically, these humble scripts knew what had to be said. They would announce their writers' confidence in the king's humanity and their enjoyment of the happiest reign known to mankind, then beg abjectly for the royal mercy. Many were bald and desperate pleas from people facing death or exile who had found no patrons to support them and could offer no evidence of character or of mitigating circumstance to move the king to mercy. Most were cringing and needful, like this classic of the kind which, quoted in full here, must stand for many:

> May it please your Most gracious Majesty:
> We the undersigned most unfortunate wretches and unhappy Subjects under Sentence of Death in your Majesty's gaol of Newgate, do most Humbly implore your Majesty's Clemency and Mercy to rescue us from the Dreadful and most awful fate which awaits us.—
>
> Sire, We are fully sensible of the great Crimes we have committed against the Laws of our Country yet do most fervently hope that your Most Gracious Majesty will be pleased to spare our lives and not thus cut us off without giving us an oppertunity [*sic*] of amending our lives and repenting of our former Transgressions.
>
> And may it please your Most Gracious Majesty as the Almighty Desireth not the Death of a Sinner, but rather that he may turn from his evil ways, and live, so do we most humbly trust in your Royal Majesty's well known Mercy and Humanity, may be extended towards us in our hour of Danger.
>
> O Sire, consider the lives of your Majesty's unhappy Subjects and do not cut us off from the face of the Earth amidst the lamentations of our Fond wives and unhappy offspring without one cheering hope of Atonement for the great offences which we have committed against the Laws of God and our Fellow Creatures.
>
> And may it please your Most Gracious Majesty, what ever Situation your Majesty may be pleased to place us in, we will with true humility endeavour to atone for the past and to show that Clemency has not been misapplied.
>
> O Sire, take our petition under your Merciful Consideration and for a long Continuance of your Majesty's prosperous and Happy Reign we your Most Humble petitioners As in duty bound will ever pray.[3]

This petition was signed collectively by thirty men huddled in Newgate's condemned cells in the winter of 1825–6. Half could only

[3] PC1.74.

scratch crosses against their names. Collective petitioning like this could not bode well for their futures. It meant that individuals had no grounds for a personal appeal. The King in Council, then, would send the youngest and least dangerous of those signing to the prison hulks and then as like as not to Van Diemen's Land or Botany Bay. How many of the rest would be left to hang depended on the Council's assessment of the perils of the time or on the royal whim; but that some would hang was certain. This particular petition was inscribed in a copperplate script whose curlicued elaborations matched the straining deference of its text, and almost certainly it was composed and written for the men by the keeper of Newgate or one of the sheriffs.

The vast majority of bundled dossiers before 1819 were at some time discarded. Only a small though useful proportion survive from before that year. From 1819 onwards all the dossiers were kept. Why was this? Nobody would ever ask to see the dossiers of the hanged or of the majority of the transported or imprisoned.[4] For most of them relegation to the archive had the finality of the graves many had just been put in. Probably their survival from 1819 onwards reflected the new-broom administrative ethic Robert Peel stood for when he took over the Home Office in 1822, a striving for a bureaucratic order less valued in previous generations, or else a readiness to deal more informedly with the debates on the criminal law which were now taking shape.

2. *The prerogative of mercy*

The sovereign's power to mitigate the sentences of judges through his prerogative of mercy was the chief and ancient emblem of his majesty as the Almighty's representative on earth. Mercy was likewise an emblem of his role as the bond of civil society, for the exchange between monarch and people was meant to be reciprocal. The prerogative blessed both 'him that gives and him that takes', as one formula put it in 1819. The felon who sought mercy from the king exalted the king by repaying him with gratitude and deference, and the return was a world in which right order was affirmed. The prerogative was 'the brightest jewel in the British crown, and the most precious of the rights of the people'.[5]

[4] Very occasionally, papers were dug out years after a first appeal to deal with a second appeal. E.g. James Empsom's sentence to 14 years' transportation was reported in Mar. 1824 (HO47.65). Sent to the hulks, he launched 2 further appeals for release in 1827 and 1829, based on the chaplain's reports on his orderly behaviour: hence further dossiers, HO17.49/7 and HO17.123/Yn30.

[5] E. Christian, *Charges delivered to grand juries in the Isle of Ely* (1819), 283.

By our period the prerogative worked through a well-oiled administration. In London and Middlesex (the Old Bailey's jurisdiction) it was exercised in the sovereign's person until Victoria's accession in 1837. King and council would sit after each Old Bailey session to determine the fates of those whom the judges or recorder had condemned there, ticking off those who would live or die with the sleazy insouciance which Chapter 20 will examine. Elsewhere in the provinces, meanwhile, the crown delegated decision-making directly to the judges or the secretary of state. In principle their recommendations were subject to the king's approval; but the mounting number of cases had long made his personal overview impractical. So when judges on assize circuits recognized mitigating circumstances in a case, or thought an excess of hangings impolitic, or were dissatisfied with verdicts, they pronounced the death sentences in court to make sure the prisoners were properly terrified, then, at the conclusion of their circuits, they would sign a circuit letter which recommended mercy, 'some favourable circumstance appearing in their respective cases'. When transportation to America was regularized in 1718, the king's acquiescence in these recommendations apparently became automatic, the secretary of state approving them routinely on his behalf.[6] The judges' control of the discretion was reinforced in 1823, when they were allowed merely to 'record' the death sentence in court and thus to commute death sentences on the spot. Either way, it was through these direct judicial channels rather than by petitioning that most condemned felons escaped the noose.

On the king's behalf, the home secretary dealt only with non-metropolitan appeals which sought mitigations beyond the judges' recommendations. In this task he was 'governed by no fixed rules or forms of procedure', as a much later Home Office memo nicely put it.[7] Sometimes he would set up his own enquiries into new evidence submitted by appellants; but in the end he would still send petitions, affidavits, or letters to the judge or magistrate who had tried the case; and usually he accepted their responses, even though he often nudged them towards the decision he desired.

There were famous difficulties in the way of this system's validation. With over two hundred capital crimes on the statute-books by 1820, the English code was infamously bloody. Most were property crimes added in the preceding century to protect property in a society in which

[6] Beattie, pp. 431–2. According to Radzinowicz, i. 112, until the end of William IV's reign judges' reprieves were routinely approved by the King in Council.

[7] HO45.9362/33391, internal memo on the prerogative of pardon, 1874.

market values were strengthening. To be sure, in practice most of these crimes were rarely encountered or tried; most prosecutions were under common law or much older statutes for 'standard' crimes; and many of the new statutes merely particularized offences which later generations would have embraced in one statute. So the list of capital crimes looked more appalling than it really was. None the less, nothing got over the truth that hangings were unusually numerous in England. Unfavourable comparisons with other countries were unavoidable (as the Introduction showed). English élites had to evolve a wonderful web of justification for this unique harshness, especially as criticisms of it multiplied outside their charmed world.

The system's most famous apologist was Archdeacon Paley. In 1785 his great work pointed out that a system which hanged so freely had libertarian effects. It guaranteed that cherished freedom from 'restraints . . . inspection, scrutiny, and control' which other nations had already and unwisely sacrificed to the despotisms which ruled them. Moreover, if more people were hanged in England than elsewhere, the reasons lay only in 'much liberty, great cities, and the want of a punishment, short of death, possessing a sufficient degree of terror'.[8] Paley agreed that the system seemed harsh, but justice and fairness in the popular understanding of the terms (as petitions were increasingly to define them by the 1820s, for example) were not his concern. The end of punishment was not justice, he insisted. It was deterrence—and deterrence through terror. However regrettably, much might have to be sacrificed to that great end. If by mischance an innocent person was sometimes hanged, the misfortune must be borne for the greater social good. But the victim in that case might take consolation from his being 'considered as falling for his country'.

Yet even at this point—and this was the key legitimation—all was not as bad as it seemed. The capital statutes 'were never meant to be carried into indiscriminate execution'. 'The legislature, when it establishes its last and highest sanctions, trusts to the benignity of the crown to relax their severity, as often as circumstances appear to palliate the offence.' Thanks to the royal prerogative of mercy, in other words, English people could be sure that 'few actually suffer death, whilst the dread and danger of it hang over the crimes of many'. (And, sure enough, in London in the year Paley wrote 54 per cent of the condemned were pardoned, rising to three-quarters in the decade following.[9])

[8] *Principles of moral and political philosophy* (1785 edn.), 541–2.

[9] Ibid. 527, 534; see also App. 2, below.

Nor was it forgotten that the mercy prerogative had deeper purposes, agreed upon by whig theorists as well as by tory. It not only softened harsh law; it also (as William Blackstone wrote) testified to the king's magnitude, and bound subject to monarch: 'These repeated acts of goodness coming immediately from his own hand, endear the Sovereign to his subjects, and contribute more than anything to root in their hearts that filial affection, and personal loyalty, which are the sure establishment of a prince.'[10] Blackstone might have added that the sovereign was not the only beneficiary of petitioners' gratitude. Petitioning was meant to reinforce the bonds of deference locally as well as nationally, to testify to the magnitude and benevolence of all who held power in the land. In helping mercy petitioners to address the king, gentlemen and gentry could collaborate in affirming the values to which petitioners had to defer, with success depending on petitioners' ability to meet the criteria through which élites identified the bases of good order. The 'fairness' of sentences determined élite responses to petitions in good degree: fairness was never lost sight of. But nor were the other variables lost sight of which determined the distribution of favours—petitioners' ages, sex, previous records, the numbers of their dependants, their likely cost to the parish, their characters, or the status of their patrons if there were any.[11] The system worked well enough to satisfy élites' belief in their own humanity and fairness.

The increasing criticisms of hanging, discretion, and the prerogative in the first thirty years of the nineteenth century actually ensured that the petitioning system entered its golden age in those decades. With attacks on its legitimacy mounting, neither crown nor executive could afford to discourage the process. Petitions were one of the few expressions of the link between sovereign and people which an unpopular regent or king could hope to enjoy in a time of mounting disaffection. Moreover, as trials, convictions, and death sentences multiplied after 1815, petitions had never been so necessary to support the legitimizing message that the harshness of English justice was tempered by mercy.

Petitioners had their own interests to serve in playing their roles to the full. Not only were there more convicted people to petition after 1815 or so; there were also rising expectations that hangings would have to be held ever more strictly in check if the land was not to be covered

[10] W. Blackstone, *Commentaries on the laws of England* (12th edn., 1793–5), iv. 397.

[11] See Chs. 18, 19; and, for an overview of these criteria, P. King, 'Decision-makers and decision-making in the English criminal law, 1750–1800', *Historical Journal*, 26/1 (1984).

in scaffolds. Political passions could also swell the urge to petition. When five Nottingham rioters were sentenced to death (and three hanged) in 1832, a petition for their reprieve was signed by 17,000 people in twenty-four hours.[12] The outcome of these pressures was that in the early nineteenth century petitions for mercy began to land on the home secretary's desk as never before. They flowed in not only from or on behalf of those left to hang but also from or on behalf of people whose death sentences judges had already commuted to transportation and who sought further mitigation. Some also came from those sentenced only to imprisonment at petty sessions for misdemeanours. Those under sentence of death in Newgate routinely petitioned the King in Council, and those under sentence of transportation nearly as routinely petitioned the home secretary.

Although petitions were not submitted by every convicted prisoner by the 1820s (the hopeless, the ignorant, and most petty offenders abstained), they were sent by a high proportion even of those who could not realistically hope for redress. In Shropshire, for example, a county with a low crime rate, seventeen people were capitally convicted in 1829: burglars, housebreakers, robbers, horse- or sheep-thieves, and one rapist. The only one of these who was hanged had attempted murder, and he made no appeal, his case hopeless. Fifteen of the seventeen had a sentence of death 'recorded' against them (and therefore knew that they were only to be transported); but this did not stop a third of them from petitioning. The rapist John Noden, whom the judge left to hang, was saved by an astonishing campaign against his execution organized by his home community, as we shall see in Chapter 17. Three of the non-capitally convicted petitioned too, against sentences to seven years' transportation.[13]

Most petitions were speedily processed. Not all were of a kind relevant to our study.[14] Many were 'to a considerable extent mere appeals *ad misericordiam*, which are not complied with', as a later Home Office memo put it bluntly.[15] These came in the form of letters from the wives of transported felons begging to be transported with their convicted husbands:

[12] J. Arnould, *Memoir of Thomas, first Lord Denman* (2 vols., 1873), i. 401.

[13] Criminal Registers, HO27.38; petitions traced in the Appeal Registers, HO19.

[14] By the 1820s nearly a third were in the form of standardized lists of prisoners on the prison hulks or in Millbank penitentiary who were thought worthy of release on grounds of good behaviour or illness. J. H. Capper, the secretary of the Home Office's criminal department, dealt with these on the recommendations of hulk or prison superintendents.

[15] HO45.9362/33391 (1874 internal memo).

In what Ever Light your Honer may Look at this your humble Petitioner begs your Honer will Pardon the Great Liberty taken in thus Addressing on the following Subject but beleave me sir it is indeed Distress and Truble in the extreem that as induced me. . . . I do crave mercy in the time of Truble and Do beg that my Gratious Soverign the king could be Pleased to grant me a Passadge to the Place where he [the husband] may have to go. . . . My adge is twenty one and have no children but are not sartaine of my Pragent State.

Often pleas like this were supported by overseers of the poor anxious to save on poor-rates by shipping off wives and children who would otherwise be without support. By the early 1830s they also gained credibility from the shortage of females in the transportation colonies.[16] But usually such missives were routinely answered in the negative by means of a printed form. Other pleas came from the parents of the transported, asking for information. One enclosed this letter home from a son in 1826, poignantly out of date on the identity of the home secretary:

We should be glad to know if you have hird any Moar a Bout our Sentence or are We both to be Hear for Life this is hart breaking to think of . . . o Mother how we are Scaterd through this World. . . . What a happiness it ould be to me if I could see you all once Moor but I Feer I never Shall as the Goverment hear is so Mutch altered and Little Liberty given unless you have Frinds in England to petishon for you to Lord Sidmith.

No response was elicited by this kind of thing either.[17]

What took up the home secretary's time were reasoned appeals against sentence from or on behalf of newly sentenced felons. Some ten to twelve thousand of these accumulated in the 1820s. In 1829, 1,018 petitions came from those convicted at provincial quarter sessions and assizes (a few from petty sessions); another 260 were from Londoners convicted at the Old Bailey, petitioning the King in Council. About a third had been sentenced to death. These appeals now made up a major part of the home secretary's business. Peel appropriately referred to his 'daily, I might say, hourly opportunities of witnessing the practical operation of the [capital] statutes which I am attempting to simplify and amend'. Petitions even chased him on his holidays.[18] By the 1820s the

[16] 'If your Lordship would grant a free passage for the family [consisting of the wife, two boys and four girls] (the number of females would be useful to the Colony), the parish will pay for their expenses to Portsmouth and will be benefited by the removal of them as the population is much more numerous than required for all necessary purposes.' PC1.77/I, 28 Jan. 1829. PC papers are full of these letters and pleas, increasing in numbers by the early 1830s.

[17] PC1.74/I: petition to Peel by Mary Lynch, 30 Mar. 1826. R. Hughes, *The fatal shore: a history of the transportation of convicts to Australia, 1787–1868* (1987), has used these PC letters.

[18] *The speeches of . . . Sir Robert Peel delivered in the House of Commons* (4 vols., 1853), i. 409. In parliamentary recess, while Peel was on vacation, mercy business was communicated to him by his secretary,

swelling appeal archive was the single bulkiest set of the Home Office's papers.[19]

Did petitions increase in ambition as well as in number? The answer is yes, if the growing elaboration of the appeals in the judges' reports is anything to go by. That the two volumes of these for 1787 and 1790 which were analysed by Peter King contained as many as 136 cases speaks for their relative brevity at that date.[20] By contrast, two thicker volumes for 1823 and 1827 contained a mere 27 appeals between them. It is true that some early petitions were as elaborate as the later. A Norwich campaign to pardon a housebreaker in 1784 was unusually ambitious in the number and energy of its well-placed appellants.[21] Famous too was the campaign for Dr Dodd in 1777: the petition for him was signed by 23,000 London householders on 29 yards of parchment. But most extant eighteenth-century appeals consisted of a single petition against sentence and the judge's scribbled comments on the case; only occasionally is there an accompanying letter or two.

By the 1820s there were still simple dossiers like this, the petitions rudely signed by relatives or friends of prisoners. But dossiers of several thousand words were now common, and some were complex enough to fill a small book. Thanks to the doubts which men attached to female victims' evidence, rape convictions always invited extended argument. The Noden rape case of 1829 leaves a dossier of 12,000 words, while the dossier on one John Pattern, condemned for rape at Warwick assizes in 1825, comprises 21,000 words. Pattern was hanged in the end, but the effort to save him was as extraordinary as was the Home Office's efforts to find the truth of his case. His file included the judge's transcript of

Dawson. Some of it was left to await Peel's return, and there were quiet periods: 'These are halcyon days for all under Secretaries,' Dawson wrote breezily to Peel in Sept. 1822, 'there being nothing with which they have to trouble their masters, and nothing to distress themselves. There is literally not a single letter of the slightest consequence today, and all applications in criminal cases for remission of punishment, which in fact are not many, I have reserved for your future consideration, being in Capper's opinion, cases which will well justify delay. . . . I hope you have good sport.' But 4 days later Dawson had to send 2 appeal dossiers on to Peel, and another day later petitions from 2 attempted rapists in Perth. Dawson added that the cases 'admit, I think, of an easy answer—no application for a mitigation of punishment, or for a remission of any part of it, has come from any other quarter than themselves—we have received no report in their favour. . . . I therefore anticipate your decision upon the case': 19, 23, 24 Sept. 1822, Peel papers, BL Add. MS 40351, fos. 103, 105.

[19] Petitioning for mercy increased in the 1830s and thereafter. By 1888 the Home Office was receiving annually some 25,000 papers, about 5,000 of them prisoners' petitions (G. R. Chadwick, 'Bureaucratic mercy: The Home Office and the treatment of capital cases in Victorian England', Ph.D. thesis (Rice University, 1989), 164 n.). But the main issue after *c.*1830 was the mitigation of transportation or prison rather than of death sentences (murderers' appeals usually being disallowed).

[20] See n. 11, above.

[21] HO47.1 (Matthew Barker's case), Sept. 1784. Cf. J. Johnstone (ed.), *The works of Samuel Parr* (8 vols., 1828), viii. 52–4.

the trial,[22] several petitions, a dozen affidavits sworn on Pattern's behalf, the reports of the home secretary's investigator sent to Birmingham to examine post-trial allegations and evidence, the judge's report, Home Office memoranda on the case, and a draft of Peel's decision to hang him after all.[23]

These contrasts between the eighteenth-century appeals and those of the early nineteenth century suggest that by the time Peel came to the Home Office dubious verdicts or harsh sentences were being subjected to closer public scrutiny and more elaborate argument than had been the case hitherto. Even rather obscure people were learning to challenge more dubious sentences with new confidence and sophistication. In the case of stronger appeals, the Home Office itself was reacting more conscientiously or cautiously too.

Yet most petitioners' efforts were in vain. The home secretary's or King in Council's scribbled response on most of them was either a curt 'nil', or when people were to hang, that bleaker euphemism, 'the law to take its course'. The infrequent grace of outright pardon was usually granted only when others confessed and could be convicted or when evidence of innocence was overwhelming.[24] Even when convictions were obviously dubious or sentences excessive, the luckiest petitioners could hope only for 'conditional' pardons. This meant they were let off hanging on condition of confinement to the penal hulks or transportation for life, or let off transportation on condition of imprisonment at home. Even those whose innocence was tacitly admitted were kept on the hulks for an exemplary period of half their sentence or so and then pardoned, as if the law's dignity must be sustained, its errors never openly admitted. There was never a question of compensation for wrongful imprisonment, let alone of apology, when free pardons were issued.[25]

[22] Until the establishment of the Court of Appeal in 1907, judges' notes on cases were the only official accounts of assize trials. When mercy appeals were made, early 19th-cent. judges submitted full notes to the Home Office only in major or involved cases; usually they only summarized the case. After the 1848 Crown Cases Reserved Act, it became customary for the full notes to be submitted both to fellow judges of the Queen's Bench and Exchequer and to the Home Office: Chadwick, 'Bureaucratic mercy', i. 103–4.

[23] HO47.68.

[24] Suspicion of post-trial affidavits and reconstituted alibis was enduring, judges always pointing out that it was easy to invent new evidence or alibis after trials were over. Examples in the judges' reports are multiple: see e.g. HO47.63/4 (Taylor's case *et al.*, 1822), and Ellis's case, discussed in Ch. 16.

[25] Thus the case of John Pledger, sentenced by Alexander J. at Essex assizes to 14 years' transportation on the grounds that he was one of a gang of about 20 men involved in night poaching. A good alibi convinced the landlord, Lord Braybrooke, of Pledger's innocence, and character testimony supported the appeal. This forced the judge to agree that Pledger might have been innocent and that he was 'not an unfit case for mercy'. Still, there was no question of retrial or of outright pardon. Peel simply

On the judges' advice, by far the majority of petitions were rejected outright, especially if the judge had already respited the death penalty at the completion of his circuit. Rejections seem to have been more frequent in London than in the provinces, for in smaller communities, where people knew each other and anxieties about crime were lower, patrons more often took up a petitioner's case, and support of this kind was hugely important in determining outcomes favourably. Even so, in 1829 some 87 per cent of petitions achieved nothing.[26] There was no question of a second trial, and despite mounting pressure for a more equitable appeal system by the 1830s, a formal appeal court was established only in 1907.[27] Yet still élites clung to the illusion that mercy was the brightest jewel in the royal crown, and they expected the nation to believe it. *Did* people believe it? What attitudes did the implacable processes of judging and punishing and pardoning elicit from those noisy and needful common folk who were the law's chief target?

3. *Deference?*

To get beneath petitions' rhetorical surfaces, recall first how urgently and needfully they were produced. They sound deferential, but this is not surprising. The first response of convicts and their families was desperation, followed by an anxious casting about for help and advice. A lot of energy, money, and anxiety went into appeals, and it was all intimidating for those unversed in the ways of the great world. Families could fall apart. 'The father nearly driven to despair, the mother at the point of death, and the wife raving mad and obliged to be confined'—thus the alleged consequences of a Gordon rioter's capital conviction in 1780: and why doubt that it was so?[28]

For most people, ambitious appeals and proceedings in error were out of the question. County clerks of peace charged £13. 10*s*. just for

scribbled on the dossier: 'To be sent to the Hulks as 7 year convict.' But the whig Home Office offered a free pardon a year later, when Braybrooke wrote that another man had confessed to the crime and that, with the gang's conviction, poaching was anyway now much diminished (HO47.75). For a similar compromise in punishment, see Anderson's and Morris's case, pp. 441–2 below.

[26] Computed from the Petition Registers' indication of the outcomes of each appeal catalogued: HO19.

[27] Meanwhile the restructuring in 1830 of the Court of Exchequer Chamber (the appeal court for the common lawcourts), the establishment in 1848 of the Court of Crown Cases Reserved (to deal with criminal cases), and the reform of the judicial machinery of the Privy Council in 1833 (to deal with appeals from colonial courts and ecclesiastical and Admiralty cases) hardly touched the plights of those who needed an appeal system most—convicted felons without means.

[28] SP37.21, fo. 159.

three copies of an indictment.[29] It was a rare case when four highwaymen convicted in Chester in 1822 employed a master extraordinary in Chancery to prepare a dossier remarkable for its exhaustive presentation and meticulously drawn maps: they must have had money to burn.[30] To have counsel mount an appeal to take to the twelve judges, as the banker-forger Fauntleroy did in 1824, you would need a fortune.[31] The best hope for poor felons was to enlist the sympathy of local professionals and gentlemen to mount campaigns for them. Sometimes a sympathetic defending counsel or attorney offered free or cut-price services. An attorney thus won the release of two Denbighshire labourers by citing law reports to prove the illegality of their fines and imprisonment with hard labour for a common assault.[32] When the attorney James Harmer tried to save Edward Harris from the gallows in 1825 (so convinced of his innocence that he published a book on the case), 'all that I received from the prisoner, or his friends, was £12, which was expended in employing two counsel, subpoenaing witnesses etc. . . . I have expended many pounds from my own pocket'—including a £50 reward for information.[33] The hopes of most prisoners condemned at the Old Bailey were 'nearly always' pinned on the help of the sheriffs or on the keeper of Newgate. These officials would spare 'neither time, nor trouble, nor money' to prove some mitigating fact, Wakefield observed: 'I have known Mr Wontner and his servants occupied incessantly for days and nights, in an attempt to prove the falsehood of evidence on which the prisoners had been ordered for execution.'[34]

But these were the lucky ones. People without resources, without patrons, sometimes without 'character', especially in the anonymity of the big city, had no choice but to purchase advice or go it alone. Those who turned to professional letter-writers or attorneys for their copper-plated petitions and fancy phrases might be impoverished by the process and still receive only thin assistance if they received assistance at all. Samuel Solomons, sentenced to death for housebreaking during the Gordon riots, not implausibly petitioned that witnesses would have confirmed his innocence had he been able to summon them, but 'alas,

[29] Cf. the solicitor James Harmer's request to Peel not to transport a convict until the man could raise money from friends for such an appeal. Peel agreed and the man was kept on the hulks pending the case: 7 Sept. 1829, PC1.77/I.

[30] HO47.63/4.

[31] HO17.87/I (Fauntleroy dossier, Qk42).

[32] HO47.64 (Roberts's and Ellis's case), Oct. 1823.

[33] J. Harmer, *The case of Edward Harris, who was executed at Newgate, for robbing and ill-treating Sarah Drew, investigated, and facts and arguments adduced, to prove his innocence* (1825), 86.

[34] E. G. Wakefield, *Facts relating to the punishment of death in the metropolis* (1831), in M. F. Lloyd Prichard (ed.), *The collected works of Edward Gibbon Wakefield* (1968), 242.

through the deception, ignorance or neglect of his Attorney they were not called although your Petitioners wife . . . sold and disposed of all her effects to raise money for the attorney'.[35] When John Kelley was sentenced in Surrey in 1822 to fourteen years' transportation for stealing a shawl, his distraught father was sent from one attorney to another, waiting hours for their attention, continually paying out fees and buying drink, until at last 'a sort of an attorney' charged him 2*s*. 6*d*. for writing half a sheet of paper:

> He was then informed that he must petition the judges, and which petition would cost 7/- to be drawn up. The father paid the man 3/- in advance, and was to pay 4/- when the writing was done, then £2 would be wanted to get the petition presented. Thus has this poor man been plundered and deceived, and all this business has been transacted at a public house, and these parties and their friends must all be treated with drink. . . . He has not only been out of employment in consequence, but has had his pockets drained.[36]

In view of these costs, most of the poor wrote their appeals unaided or called on neighbouring tradesmen to do it for them, announcing bleakly that 'in consequence of very humble circumstances' they were 'unable to obtain necessary professional aid'.[37] Male petitioners had some advantage over female, for in towns and cities they could sometimes draw on networks of past or present employers and workmates to add crosses or scratchy signatures to a petition. Women in service were more at the mercy of mistresses. No procession of fellow workers would ever proceed to the Home Office to ask for mercy for a woman, as one did for Samuel Wright in 1864.[38] Still, men had difficulties enough. When William Fenning tried to produce a written statement on behalf of his condemned daughter Eliza Fenning in 1815, 'being unable to do it himself, from the agitation of his mind, and his hand trembling very much', he went to the Pitt's Head tavern in Old Bailey and

> there asked the witness, John Wooddeson, to write for him, who, from the same causes, being equally incompetent, Fenning applied to another person in the room, a stranger, and asked him if he could write; he said he could, and then, upon the solicitation of Fenning, wrote [the dictated statement] in ink, on both sides of a scrap of paper, which Fenning gave him.[39]

[35] SP37.21, fo. 163.

[36] HO47.63/1: letter from the chairman of sessions to the Home Office, recommending mercy (punctuation changed). Peel agreed in this case.

[37] HO17.42 (docket Fw3).

[38] See p. 102.

[39] J. Watkins and W. Hone, *The important results of an elaborate investigation into the mysterious case of Elizabeth Fenning* (1815), 54.

Some petitioners had no choice but to address the Home Office in homely letters, as openly and simply as they would have spoken—rich material for historians of the interface between orality and literacy:[40]

Sir I ope you will not be long before you let Thos. Rigby of Blackden my Tennant to have his Libberty and I ope with the blessings of the almighty that he will have better conduct to be in bad company any more which I ope this will be awarning to him for ever, as he was as a hard a slave for his family as any Tennant I ever had before this happened and his wife as an Industeras a woman for her family for she as kept her Children all to school and with good Clothing and paid her rent ever since he went of and she as eight nise cows and Two Horses this is Two Cows more till she had wen he went of . . . T. Arden. Thomas Rigbys Lanlord[41]

Others were pitifully anxious about the risks of going it alone: 'My lord I have no advocate, no noble friend to plead for me, but a hope on your lordships commiseration, much have I suffered . . . I know your lordship could do every thing and can feel for the oppressed. . . . Pardon sir, the great liberty I take, but where to go but to the good and great.' Or: 'May it please your Lordship . . . to admit an apology for the liberty taken in thus addressing your Lordship and also for all errors this petition may contain we being ignorant in these matters we can only address your Lordship in language dictated by the heart.'[42] But still they ploughed bravely on, the fruits of autodidactic learning or cottage schooling strongly apparent, scripts cramped and uneven, styles and spellings guided by ear rather than by memory or eye:

Sir i have taken the liberty of riting to you in be half of my sone Joseph Bellefield as lying down [on the hulks] at woollege [Woolwich] to know if you whold be so kind as to inform me wat is sentense is he was convicted for a felony and was death recorded and i have recived a letter and he has informed me that he does not know what the court as been pleased to inflict on him there fore if you whold be so kind as to in form me a shall be humbly a bliged to you . . . it as allways been my practise to set be fore my children a good pattan and never . . . gave them any incourige ment to do hany boddy the least injery in the whorld and if hany thing could be done to save him from leaving is native country i should be very glad [if he could be sent] to the penny tensary at mill bank london and if it will not be to much trouble to you to in form

[40] For beginnings in the analysis of popular texts of these kinds, see O. Smith, *The politics of language, 1791–1819* (Oxford, 1984), D. Vincent, *Literacy and popular culture: England 1750–1914* (Cambridge, 1989), and C. Steedman, *The radical soldier's tale: John Pearman, 1819–1908* (1988), ch. 6.

[41] PC1.74 (26 June 1826).

[42] PC1.67 (4 June 1819); HO17.42 (Absolom's case, docket Fw3).

me if hany such a thing can be done I shall feel humbly a bliged to you . . . Sir i most humbly beg your pardon for this liberty of adressing my self to you.[43]

In ill-formed letters of their own, or through patrons, relatives, or friends, or through professional letter-writers or attorneys when they could afford them, or through prison governors, chaplains, or sheriffs if they could not afford them (relying on such people for their turns of phrase as much as they did for their pens, ink, and paper), the convicted and their families could sound very deferential indeed as they harnessed every conceivable rhetorical device to support their thin hopes. The urgent tropes of popular melodrama were much favoured, for in London these were usually people who knew the penny gaff and gothic romance as well as Drury Lane; theatrical gesture and speech inspired most of their communications.[44] The affluent shared this language too. 'Forgive Oh Most Gracious Sire a wretched and distracted Woman for presuming to approach your Royal person to supplicate Mercy to be extended to the unfortunate Henry Fauntleroy'—thus the wife of that most notorious of all forgers, hanged in 1824: 'Let not I beseech you the dreadful punishment of an ignominious death be inflicted on the husband of your supplicant, the father of the child, but spare oh mercifully spare the father's life, that disgrace may not be entailed on his innocent offspring and overwhelming misery inflicted on your petitioner.'[45] But the poor were almost as good at it as the rich. 'Oh! let me implore your Majesty in the sacred name of God for mercy, mercy, that great ornament and bright luminary of your Majesty's revered person,' wrote the father of a humble clerk Richard Gifford, also sentenced to death for forgery in 1829.[46] Those ignorant of these devices simply abased themselves before the king or his home secretary in the terms attorneys advised. 'We are very humble parties', two Denbighshire labourers announce in 1822, but 'have confidence in His Majesty's well known paternal solicitude for the meanest of his subjects'.[47] On the eve of 17-year-old Henry Hawkins's execution for burglary, his parents 'humbly pray to be admitted into your Lordships [Sidmouth's] presence, and to

[43] HO17.123 (docket Yn37). When Joseph Bellefield of Birmingham had the death sentence recorded against him for housebreaking in 1829, a substantial local petition made it clear that witnesses had attended the assizes for 2 days in order to offer evidence on his behalf, but had been ignored by the court and 'were obliged to return home for want of the necessary means to enable them to stay longer'. When Bellefield was sent to the Woolwich hulks to await 14 years' transportation, his father ('in a wretched . . . forlorn and helpless state of poverty') wrote to Peel in the terms cited; Peel scribbled 'Nil' on the dossier.

[44] M. Baer, *Theatre and disorder in late Georgian London* (Oxford, 1992), 11.

[45] HO17.87/I, Nov. 1824.

[46] HO17.54/I, Sept. 1829.

[47] HO47.64/26 (Roberts and Ellis).

be permitted on their knees to solicit your powerful interest with his Most Gracious Majesty, for a respite. . . . O my lord, do pray consider his tender age!'[48] 'On our knees for clemency', others write.[49] And on their knees we may imagine them really falling if lucky enough to be granted an audience with a Home Office clerk.

Some historians might wish to take the self-abasing postures of our texts at face value, and affirm that the deference here was 'real'. There is a hefty body of social historians of law and law-enforcement who are inclined to emphasize the more benign motifs in their country's history, particularly when relations of class and status are concerned, and who take a sanguine view of the respect which poorer people accorded to criminal law and its administrators in times past. Mistrustful of those who paint Hanoverian élites and law in darkest hues, they stress that the law was broadly just when all is said and done; it protected the weak as well as the strong, and was used by the weak as much as by the strong. It was a 'multi-use right available to *most* Englishmen', we are told; it was a 'resource available to and used by *almost* every layer in eighteenth-century society', another adds; 'the moral underpinnings' of the law must be allowed to have weighed as much as the 'prerogatives' of social and economic power did, we are elsewhere advised.[50] The common people's respect for and deference to the law, as to their betters, may be held to follow from these cheering dispensations.

Yet the law and humbler people's responses to it had numerous subtexts neatly evaded in these formulas, and some poorer people could read those texts more knowingly than historians do who think relationships of power rather tedious or negative things to attend to. In view of its costs, mystifications, and social biases, it would be odd if the poor and humble felt that the law was as much their law as is often fondly suggested. The very biases within the prosecution process indicate how little, structurally, the law need invite respect from those who were its prime targets. The words 'most' and 'almost' in the quotations above could do with stronger elucidation, one might feel.

[48] PC1.68 (4 July 1820). The boy was executed next day.

[49] Thus a burglar's parents in 1825, signing their petition with crosses: HO47.65.

[50] J. Brewer and J. Styles (eds.), *An ungovernable people: The English and their law in the 17th and 18th centuries* (1980), 20; J. Langbein, 'Albion's fatal flaws', *Past and Present*, 98 (1982), 101, 105; King, 'Decision-makers and decision-making', 33; C. Herrup, 'Law and morality in seventeenth-century England', *Past and Present*, 106 (1985), 105, 123. *So far as they go*, each of the cited propositions is true; it is only the balance of emphasis that may be debated, and what is omitted. Anyway, Brewer, Styles, and King helpfully qualify their positions, so I do not intend to parody them.

Here there are some well-worked research findings to repeat.[51] In late eighteenth-century Essex a third of the felony prosecutors at quarter sessions were farmers or yeomen, another third were tradesmen or artisans, while *only* a fifth to a sixth were labourers or husbandmen: was this last figure a high proportion or a low one? At Surrey quarter sessions, only 14 per cent of prosecutors were unskilled men, and only a quarter were such in the Black Country between 1836 and 1851. To some these figures suggest a consensual relationship to law; others, in view of the size of labouring populations, may think differently.

Add to this the fact that most prosecutions aimed *downwards* socially. The targets were rarely (were they ever?) social superiors, and they were not usually social equals. If a quarter of indictments in Essex were against artisans or lesser tradesmen, *two-thirds* were against labourers, servants, vagrants, and paupers. In the Black Country in 1836, 54 per cent of prosecutors' targets were manual workers. The social direction of prosecution in the cases covered by mercy petitions exhibited much the same bias. And even when humbler prosecutors did use the law, they were not necessarily imbued on that account with that 'acceptance of its basic legitimacy' which is so often and strangely insisted upon. Along with the middling classes, they could prosecute maliciously, to harass enemies, win rewards through false accusation, or exact reimbursements for the costs of prosecution—faults in the prosecution process with which the law's professional critics battled for nearly a century.[52] On the part of poor people, respect for the law in all these circumstances is the last thing to be assumed axiomatically.

That there were people whose deference to the great and their law was habitual and reflexive need not be denied. We do hear many felons in condemned cells, or on hulks, or in prisons acknowledging their sentences as 'just'—though to those with the literacy, status, and interest to record such opinions, of course. Some were forced to do so by parsonical brainwashing; some because there was no other way of making sense of their plight other than to accept the say-so of those who told them that they deserved what they got. Many believed it too, like the sadly defeated man whom G. J. Holyoake once saw sentenced at Gloucester to transportation for life:

[51] P. King, 'Crime, law and society in Essex, 1740–1820', D.Phil. thesis (Cambridge, 1984), 127, 180; Beattie, pp. 192–8; D. Philips, *Crime and authority in Victorian England* (1977), 125–9.

[52] D. Hay and F. Snyder, 'Using the criminal law', 50, and D. Hay, 'Prosecution and power: Malicious prosecution in the English courts, 1750–1850', 389–95, both in D. Hay and F. Snyder (eds.), *Policing and prosecution in Britain, 1750–1850* (Oxford, 1989).

When he heard the ferocious sentence, in genuine and awkward humbleness he made a rustic bow to the Bench, saying 'Thank you, my Lord.' Ignorance had never appeared to me before so frightful, slavish and blind. Unable to distinguish a deadly sentence passed upon him from a service done to him, he had been taught to bow to his pastors and masters, and he bowed alike when cursed as blessed.[53]

Deference to the law may also be anticipated among yeoman-to-middling kinds of people, or those who aspired to be like them, and among professionals and others with property, whom the law served best. When such people went astray their fawning language was not dissimulated.

But deference is a strange quality, its meanings never obvious. Doubtless everyone accommodates both a deferential and a non-deferential self at one and the same time, depending on contexts; but how is the 'reality' of deference to be taken for granted when it was offered by weaker people to stronger, under huge constraints? Deference should presumably make sense to the free self before it can be called 'real'; but when was a poor self, in that relationship, free? It is never clear that, when people sounded deferential in petitions, they were expressing those cognitive components of deference in which élites (and historians) put their trust. They were not acknowledging a reciprocal relationship contracted as between partners, the merciful duty and largesse of the great being met freely by the grateful respect of the humble. Sydney Smith was wiser and wittier than most. As he put it in 1826:

A prisoner who dislikes to undergo his sentence naturally addresses to those who can reverse it such arguments only as will produce, in the opinion of the referee, a pleasing effect. He does not therefore find fault with the established system of jurisprudence, but brings forward facts and arguments to prove his own innocence. . . . The way to get rid of a punishment is not . . . to say, 'You have no right to punish me in this manner,' but to say, 'I am innocent of the offence.' The fraudulent baker at Constantinople, who is about to be baked to death in his own oven, does not complain of the severity of baking bakers, but promises to use more flour and less fraud.[54]

In conditions of massive inequality, two quite different postures can be discerned. The first was the placatory capitulation of the hopelessly impotent in the face of great people's power—the voice of defeat and desperation, the fawning subservience of the oppressed, not the free deference envisaged by those who take language at face value. The

[53] G. J. Holyoake, *Sixty years of an agitator's life* (2 vols., 1893), i. 163.

[54] *Works* (2 vols., 1856), ii. 114.

second voice was ironic, calculative, and expediential, and, when thwarted, angry as well.

In the first instance, what recurrently astonishes in some texts is the writers' stoic forbearance as they told numbingly tragic stories without detectable anger about the penalties and mischances which judges and fate had thrown at them. Occasionally the self-abasement is *so* extreme that one suspects the operation of an elemental defence against the pains and humiliations which authority figures inflict: a defence in which victims identify with aggressors. In recent times this response has become familiar through reports on hostages' devotion to kidnappers, but it is known to psychoanalysts in many contexts, and we met it briefly even in the vehemence of execution crowds.[55] In the petitions it reveals itself in the most needful and defeated petitioners' struggle to establish between their superiors' values and their own a congruity at odds with their more healthily 'appropriate' feelings and interests.

The following letters, unbearably sad, provide a case in point. The first was written by a son incarcerated on a hulk to his anxious family; the second was written on receipt of the first, by the father to the Home Office. In both cases something deeper than cynical dissimulation or tactical self-censorship seems to be at work, especially when the first writer's age and the triviality of his crime are registered:

A son to his parents, from aboard the prison hulk Euryalis, Chatham, 15 April 1829

Dear Father and Mother I have taken The Earliest Opertunity of writing these few Lines to you Hopeing to find you in Good Health as it Leaves me at Present Thank God for it and I received your kind Letter witch you sent me and I was very Hapy to hear from you Dear Parents my poor Brother Solomon is no more he died last Monday weak at half past one in the morning and the Good Captain let me see Him on sunday when he was alive and he also let me go to his Funael and I thought that was very Good of him Dear Parents I hope that he is happyer than what I have or you Father and I hope that is in heaven and I hope that he is one of our Blessed Saviour Dear Parents Remember my Love to my Brother and Sisters and all enquiring Friends . . . and be so good as to answer this letter as soon as possible can and he died with a pain in his Chest and in a Decline Dear Parents if ever I do Return to you I will be A comfort to you I have no more to say at present from your unfortunate Son John Edwards and he was buried on the 10 of April no money nor Parcells allowed to be sent on Board

[55] 'Faced with an external threat', the subject may defend him- or herself 'either by appropriating the aggression itself, or else by physical or moral emulation of the aggressor, or again by adopting particular symbols of power by which the aggressor is designated': J. Laplanche and J.-B. Pontalis, *The language of psychoanalysis* (1983), 208–9.

If ever I get my own Liberty again
It will be a glorious hour—
It will warn me of a future time
To go a thieving no more
When I can to no parents fly
Nor to my native home
I shall curse the Day I went astray
To leave my mothers home.

The father, from Wolverhampton, to the Home Office, 3 May 1829

To the Right Honorable sir J H Cooper [Capper] Esquire Sir I take the Libberty of writting To you witch I hope I dount ofend you I had 2 Children Transported for taking a Chain out of a Cart in the Market place and they had chain again value 5/- Solomon Edwards was a 11 years of age John was 9 years of age and now sir Solomon died the 6 of April Sir I hope you will have mercy on his discontented and distressed Father and mother they are Bereft with The greatest of sorrow to think our Child Can not be forgiven as his years are but 12 But I hope and trust in god that you will have the goodness to for give him We hope witch will be a waren [warning] to his Body and to his sole pray pitty his father and mother as we are worn out with Trouble and we shall Be for Ever Bound to Pray for your goodness while Life remains your Humble and Petenaent servants John and Jane Edwards. They were transported [i.e. sent to the hulks ready for transportation] April 4 1826.

[*The father signs, the mother marks with a cross*][56]

We can never be categorical in our readings. The father might have spoken a different language in the tavern or at work. But in his craven address to the Home Office clerk, and in his son's reference to the hulk's 'good captain', and in the verses which the boy cites for the clarifying help they give him, we seem to witness their joint battle to align themselves *with* the system responsible for their tragedy, as if this was the only way in which they could make sense of the world, as if, in short, these were simply vulnerable and defeated people who needed the approval of their oppressors as the cowed child needs the approval of the parent. 'Deference'? Yes: but under what terms?

The second posture is easier to read. Fully internalized deference to superiors was probably most common among those enmeshed in client relationships on the land (sometimes in towns) and in the subordinations of the great estates or closed manorial villages. It cannot be assumed among those who had escaped that subordination—those increasing numbers whose free labour was either predatory or purchased piecemeal, who, when working at all, worked independently or

[56] PC1.77/I. The parents lived in Wolverhampton; there is no record of further action in this case.

in small units of production as yet undisciplined by the factory.[57] By the later eighteenth century the market had rendered such people quite literally free, though not in the sense lauded by élite ideologues. The more marginalized among them, rogues and vagabonds, were a chronic problem for those who yearned for the client-bonded order of things. If the connected middling classes and yeomen did a good deal of the prosecuting, these others were the people who largely had to face the judges and fill the prisons. It is inconceivable that they did not know how to adopt tactical postures in addressing those with power over them. At this level the role-playing between subordinate and dominant groups was theatrical, the display of deference reflecting only 'the *self*-portrait of dominant groups as they would have themselves seen', as it played prudentially into governing-class fantasies and self-images.

An example is the implausibly self-abasing petition of the Gordon rioter James Jackson in 1780, sentenced to death for helping in the riotous demolition of Newgate, its wording provided knowingly by the vicar of St John's, Clerkenwell: 'If the circumstances above related should have no weight with your Majesty, and it should be your Royal will that he should die, he will die with a conscience void of offence and his last prayers will be that God of his infinite mercy shower down blessings on your Majesty, your Royal Consort, and the whole of your illustrious Family.'[58] A king who could believe that a subject would die thus would believe anything; but then, this king might. In these conditions, with vicars and the like standing between petitioners and petitioned, the transcript through which plebeians sustained their private critiques of the great was usually hidden indeed. It was best heard when the public performance was ruptured *in extremis*, or in the gossip, jokes, festivals, and parables of the people which historians are only now beginning to study seriously.[59]

Whatever language the free plebeians overtly used in petitions, its purpose would be instrumental, its deference merely postured. In their needfulness most petitioners might be honest and well-meaning enough. They had no choice but to come clean, and often their stories were clean. But the most loosely connected would with ease invent alibis and persuade people to support them or testify to alibis after trials were over. 'To prove this by perjury is a common act of Newgate friend-

[57] E. P. Thompson, 'Patricians and plebeians', in *Customs in common* (1991), 36–42.

[58] SP37.21, fo. 159.

[59] J. C. Scott, *Domination and the arts of resistance: Hidden transcripts* (New Haven, Conn., 1990), 103 and *passim*.

ship,' Fielding wryly observed, and from his personal experience of Newgate Wakefield agreed: 'Almost every execution is preceded by perjury, having for its object to prevent the execution. In such cases the mass of people in London appear to think, that the crime of perjury is less than the crime of refusing to commit perjury for the prevention of death.'[60] Thus judges and home secretaries were not always being thick-skinned in suspecting that some petitions were perjured or forged—rural ones too. 'I understand a petition with some apparently respectable signatures has been presented in his favor,' the chairman of the Kent sessions warned the Home Office in 1826; 'but the names were attached to it without the knowledge of the individual. For the security of property in this neighbourhood I do trust that he will finally be sent out of the country.'[61] And it was also at this social level that the fragility of the deferential transcript would reveal itself fully when the law proved unjust or merciless at last.

Given the effectiveness of controls, such explosions from the hidden transcript hardly presaged resistance. But they did express an anger (in my judgement) both 'appropriate' and honest. We do not often hear these fierce voices, it is true. Usually only their delayed articulation is recorded, as in the Tyburn ballads, where anger is diluted into its sardonic after-echoes. Or we hear anger only muffled, after people had carried it back into daily living, burying it within themselves but making it a part of themselves too. Thus a wife who had to carry her anger back home with her after visiting her husband on the hulks at Chatham, believing him innocent: 'I found him in great distress drawing tymber on a cart like a horse he thinks it a very hard case that he is thus compelled to suffer . . . when he as done nothing but spoke the truth in defence of the laws of the nation. . . . There is no law for a poor man.'[62] She was one of thousands of people throughout the towns and villages of England who to their own satisfaction had learnt that the law was unjust and cruel. As their knowledge was diffused through neighbourly complaint and talk, it would have shaped the neighbours' views as much as or more than JPs' or parsons' homilies did.

But there were also cases when the anger exploded more immediately than this—and straight into our sources. The recorder of Exeter reported to the Home Office how one Samuel Giles betrayed the servile

[60] Wakefield, *Punishment of death*, 243.

[61] PC1.74/I (3 Feb. 1826).

[62] HO47.73 (Clifton's case), Feb. 1828. Clifton's wife asked the sympathetic local clergyman to whom she addressed this letter for a recommendation 'to see the Secretary of States myself I would try if I could beg my poor husbands freedom'. 'Nothing to be done,' Peel scribbled on the file.

role the law allotted him. Giles 'often extended his threats and hope of vengeance to the magistrates, and honoured me with a similar wish in court when I pronounced my sentence of seven years [transportation] on him. I had before imprisoned [him] for six months hard labour, once publicly whipped him, and he was once besides committed for a burglary.' His outbursts did Giles no good: after the latest of them he was kept permanently in irons.[63] But how many thousands of felons must have shouted their unrecorded defiance as he did, indeed as those condemned to die did whom we met in our first chapter? In cases like these it would be difficult to agree with E. P. Thompson that ruling-class control in our period was 'located primarily in a cultural hegemony, and only secondarily in an expression of . . . power'. To the contrary, we should rather agree that 'the tranquillity of social relations . . . depended not simply on attitudes, but on *power*'—pure and simple.[64]

Finally there were the anonymous letters. Some threatened vengeance for death sentences on others. 'You are a bloody rascal a bloody thief and a bloody murderer . . . and mark my words I shall not mind swinging for you . . . you bloody rascal,' someone wrote to Park J. when the forger Hunton hanged in 1828.[65] Or they came from victims themselves, like this beauty sent to a chairman of quarter sessions in 1829:

> Sir I am varry glad to in form you that whe are not agoin out of the Country this time but yo bluddy old roge you was determd to send us whether whe was gilty or not but whe shall not be long before whe are back and you shall pay for all you remember them three you sent for fourteen years at wakefield sessions but they will surprise you before the sumer is ore . . . you must keep a sharp look out both you and all the fals swearing crew for whe intend to have oure revenge you thought you could do everything in sending us out of the cuntry but you can do nothing at all i have found that out all reddy so you may go an ang yourself you old snot and all the bluddy crew.[66]

There were many such missives. Forwarding this one to the Home Office, the magistrate noted: 'I am so much accustomed to receive anonymous letters that in general I take no notice of them.'[67] The

[63] PC1.74 (Jan. 1826).

[64] Thompson, 'Patricians and plebeians', 43; D. Levine and K. Wrightson, *The making of an industrial society: Whickham, 1560–1765* (Oxford, 1991), 377.

[65] HO17.88/I.

[66] PC1.77 (Apr. 1829): 'It is a common opinion among the Thieves here', the chairman explained, 'that such of them as are sentenced to transportation at the Sessions are seldom if ever actually sent off, but that after being on the hulks for a short time they are discharged.'

[67] For the frequency of such letters in other contexts, see E. P. Thompson, 'The crime of anonymity', in D. Hay *et al.* (eds.), *Albion's fatal tree: Crime and society in eighteenth-century England* (1975).

magistrate affected contempt of course. It was important to dismiss such rebels as brutes, rogues, mavericks, madmen. But authorities knew better than really to dismiss such indications of turbulence beneath the ordered surface of things. Even polite petitions carried latent menace. The writers obeyed and respected the laws (their messages went), or if they did not, it was only because of bad example or pressing poverty; and they could use the languages and ostensibly accept the values of the lawmakers as they rehearsed their tribulations. Yet between the lines all petitioners unavoidably stated that life was not good for them and ought to be better. At this point they conveyed their subliminal *or else*: 'he who speaks of desperation to his Sovereign, threatens him.'[68] In anonymous letters like the one just quoted, the *or else* sounded out deafeningly. The magistrate certainly took note of the letter just cited. The home secretary did too. Peel ordered the warders of the hulk *Leviathan* to trace the writer by comparing convicts' handwriting in Sunday letters home, and to ensure that he was transported. The culprit did not oblige by writing, however; he was never traced. Not that it mattered. Most hulk convicts in that particular batch were transported anyway.

[68] A 17th-cent. source cited by Scott, who develops the point made here: *Domination and resistance*, 96.

PART III

THE LIMITS OF SENSIBILITY

CHAPTER 7

SYMPATHY AND THE POLITE CLASSES

Arguments and interpretations

1. *An argument*

WE LEAVE THE PLEBEIAN WORLD NOW. THE NEXT TWO PARTS of the book examine how the middle classes felt about the bloody code—or the polite and would-be polite classes, as we might more inclusively call them. The reference is chiefly to literary, professional, and commercial people, mainly in London, because there they left most records. But it sometimes has to include worldly-wise gentry, along with those trading and middling kinds who had an interest in displaying civility and good manners and making their opinions heard.

Within these groups, status and cultures varied by occupation, education, region, and religious denomination. Moreover, cross-fertilizations between their cultures and those of the common people were innumerable, so within this diversity boundaries were also hazy. This was no homogeneous class. None the less, it has always made some sense to talk about them collectively, and we have to do so in what follows. Although they had little direct leverage on parliamentary affairs before 1832, they often possessed local influence. They enjoyed increasing social and market power as producers and consumers, as well as prestige as writers, lawyers, or doctors. In their social lives they had long been distancing themselves from the vulgar to affirm their own achievements, standing, and civility, displaying 'manners' as one way of doing so. They had high investments in religious belief, or in debating the grounds of religious belief. And many had the access to the tools and concepts of classical education which common people lacked. All this shaped not only their styles of life but also what it was possible for them to think and consciously feel (and conversely what it was not possible for the vulgar to think and consciously feel). It also shaped unconscious lives. As they cultivated socially acceptable emotions, we shall see that the unseemly emotions they repressed could engage them in dilemmas and neuroses of which the vulgar might be innocent.

In old-fashioned narratives of progress these middle classes have always been regarded as main agents in the history of penal as of other reform. They are assumed to have stated most clearly the ethical and pragmatic desirability of abolishing discretionary and inhumane punishments, and to have carried most middle-class opinion with them. This view will need to be taken with pinches of salt in Part V of the book, but the grounds for it are not difficult to see. From the rationalist debating circles of the 1770s on to Romilly and Mackintosh, from John Howard on to Elizabeth Fry, deists, radicals, Quakers, and evangelicals of various persuasions did open penal reform to public debate. Many of these people were professionals determined to reform an *ancien régime* which excluded them from influence, or they were increasingly earnest believers moved by the hope of bringing felons to penitence. If they had little else in common, most none the less contemplated others' suffering in quite different ways from those of a century earlier. With opinions hatched in cultures of sensibility as well as of Enlightenment or of evangelical piety, they mainly agreed that sympathetic emotion was the bond of social and personal relationships, and that it was a mark of cultivation to empathize with suffering and strive to relieve it through benevolent works. The history of sensibility has a fair chance therefore of taking us to the core of what scaffold witnesses felt about scaffold deaths. Sensibility in theory had the power to erode that emotional dissociation from the terrified felon upon which the tolerance of all harsh punishment depends.

Some historians have argued that the diffusion of sensibility is a sufficient explanation for the mitigation of punishment across our long period. The starkest claim has been that among most European élites by the mid-eighteenth century 'a new threshold was reached in the amount of mutual identification human beings were capable of'—so much so that 'by the end of the eighteenth century some of the audience could feel the pain of delinquents on the scaffold'. This is unwisely said to have been the 'primary' cause of penal amelioration in ensuing decades.[1]

[1] P. Spierenburg, *The spectacle of suffering: Executions and the evolution of repression: From a preindustrial metropolis to the European experience* (Cambridge, 1984), 183–5, 187–96. Spierenburg follows Norbert Elias

McGowen puts the relationship more persuasively by showing how penal reformers generated an argument from their own experience and valuation of the feeling self. Sympathetically gifted themselves, they believed that social relationships might be healed by diffusing the same gift among the common people rather than by resort to coercive force—thus foreshadowing that long process of moral evangelization of the ensuing century. At odds with the pessimism of those who defended harsh bodily punishment on the grounds that the *plebs* could understand no other discipline, reformers believed they could displace penal violence by the gentler educative coercions of the prison, in which felons might learn their connectedness with others and identify themselves with society at large.[2] There is no doubt that many contemporary critics of capital law did find in sensibility, as others did in Enlightened or evangelical ideas, an oppositional *resource* with which to put the *ancien régime* on the defensive. It provided shared assumptions and vocabularies ('humanity', 'civilization') in terms of which action could be understood and directed. Sensibility certainly has to be part of the story of change.

That said, however, the following chapters keep a cautious distance from the notion that sensibility and sympathy self-evidently and conclusively undermined the rationale of old punishments. As the Introduction makes clear, political, economic, and bureaucratic explanations of penal change are more convincing than cultural explanations; and anyway among reformers harsh disciplinary and penitential imperatives were never negated. Then an additional truism presses upon the attention, central to the explorations in this part of the book: that the middle classes were not exempt from the murkier motives and impulses of humankind, so that not everything in their views of the world conduced tidily to those progressive narratives in which they themselves put their trust.

in linking sensibility shifts to state-formation (pp. 201–4), but his exemplifications of the 'threshold' are few and unconvincing. For a fine general review of this literature, see D. Garland, *Punishment and welfare: A study in social theory* (Oxford, 1990), ch. 10.

[2] R. McGowen, 'A powerful sympathy: Terror, the prison, and humanitarian reform in early nineteenth-century Britain', *Journal of British Studies*, 25 (1986), 313–17. Other historians argue similarly that an increasing antipathy to violence and cruelty marked an important point in a very long-term decline of violence in all social relationships. Beattie (p. 139) claims that this had profound effects on the law itself even before 1800: 'It is in the State's attitude toward violence that the new sensibilities are most evident', and 'the State's violence changed character as the opinion changed upon which it depended for its effectiveness'. For broader debate, see L. Stone, 'Interpersonal violence in English society, 1300–1980', *Past and Present*, 101 (1983), 22–33; J. A. Sharpe, 'The history of violence in England: Some observations', *Past and Present*, 108 (1985), 206–15 (and Stone's 'Rejoinder', ibid. 216–24); J. C. Cockburn, 'Patterns of violence in English society: Homicide in Kent, 1560–1985', *Past and Present*, 130 (1991), 70–106; A. Macfarlane, *The justice and the mare's ale* (1981).

In this latter connection, it should go without saying that the culture of sensibility only named desirable emotions (like sympathy), and that this was very different from guaranteeing the diffusion of those emotions, let alone from the taming of contradictory and more primitive feelings which in all of us refuse to lie completely low. Then as now most people occupied areas of confusion and contradiction where thoughts were mediated through murky feelings and feelings through murky thoughts, while sensibility pointed in many directions and forged many linkages, not all benign. Allied with gentility and evangelicalism, it as often advised the polite to keep their distances from the unseemly sights enacted on scaffolds as it advised sympathetic engagement with them. Moreover, the recommendations of sympathy had to overcome polite observers' social distance from, incomprehension of, and distaste for the social and sometimes political 'others' who chiefly hung on nooses—namely the tatterdemalion poor, or the traitors, responses to whose deaths we explore in Chapters 10 and 11. Most intriguingly, the recommendations of sympathy had to be reconciled with unregulated feelings about hanging which were much more elemental and vehement than any which sensibility could conceive or contain. All this explains some of the mystifying responses to the scaffold which we shall duly survey. It also explains why most middle-class élites were far less clear about the desirability of abolishing or mitigating the scaffold's penalties than their own and others' progressive narratives fondly imply. It does not go amiss to be wary of their finer words. 'The characteristics of a mentality are registered in acts as much as and more than in statements.'[3]

2. *Sympathy and sensibility*

Sensibility and sympathy, clearly, are central to our argument: let us first note some of their meanings and expressions. 'A sensibility of heart . . . fits a man for being easily moved, and for readily catching, as by infection, any passion,' it was said in 1759. This got to the point deftly. The ability to feel spontaneous emotion and to communicate with others through 'exclamations, tender tones, fond tears, / And all the graceful drapery' of pity and feeling (Hannah More) became a principle of civility and a mark of moral authority. It is pushing it to claim that around 1760 'everything changed in Western mentality: attitudes to life, marriage, the family and the sacred'. Continuities can always be estab-

[3] M. Vovelle, *Ideologies and mentalities*, trans. E. O'Flaherty (1990), 65.

lished.[4] But it is clear that the nature of *conscious* sociability altered substantially. This was a moment when among the polite and would-be polite instinctual passion was a little tamed and the urge to bond with fellow beings more consciously valued. Moreover, sensibility's cultural purchase was extensive. To catalogue its expressions is to trivialize them, but a weighty literature acknowledges its impact on many fronts.[5]

First and foremost, the culture of sensibility was rooted in new scientific understandings of body and mind, and particularly in the notion that sensation was the basis of thought. The 'nerves, or instruments of sensation', it was declared in 1733, 'are like keys which, being struck on or touch'd, convey the sound and harmony to [the] sentient principle'. Hartley's materialist psychology in *Observations on man* (1749) was influential here, drawing as it did on the associationist nerve theory of Locke and others. The book popularized the view that individuals were not only highly susceptible to external stimuli but also discerned 'truth' through them: sight, hearing, and touch were essential mediators between mind, body, and world. It followed that if you controlled the sensations determining character you might improve character itself. When Joseph Priestley popularized Hartley's work in 1775, he did not know whether it did more 'to enlighten the mind, or improve the heart; it affects both in so super-eminent a degree'.[6] Thanks to developing understandings of these kinds, 'humanitarian narratives' evolved in the eighteenth century, acknowledging the impact of sensations experienced in common by victim and witness alike, and so reshaping perceptions of suffering. From novels like *Clarissa* through to the affecting biographies contained in medical autopsies, a narrative form developed which came to speak 'in extraordinarily detailed fashion about the pains and deaths of ordinary people in such a way as to . . . connect the actions of its readers with the suffering of its subjects'.[7] The expressions

[4] Ibid. 80, 199. Cf. L. Pollock's riposte to historians who argue for the rise of a 'modern' affective view of childhood, that 'there have been very few changes in parental care and child life from the sixteenth to the nineteenth century in the home': *Forgotten children: Parent–child relations from 1500 to 1900* (Cambridge, 1983), 268 and *passim*.

[5] A. Gerard, *An essay on taste* (1759), 86; J. Dwyer, *Virtuous discourse: Sensibility and community in late eighteenth-century Scotland* (Edinburgh, 1987), 55–7; G. J. Barker-Benfield, *The culture of sensibility: Sex and society in eighteenth-century Britain* (Chicago, 1992), ch. 1.

[6] Barker-Benfield, *Culture of sensibility*, 6; A. J. Van Sant, *Eighteenth-century sensibility and the novel: The senses in social context* (Cambridge, 1993), 30, 56, 54; J. Priestley, *Memoirs of Dr Joseph Priestley (written by himself)* (orig. pub. 1806–7; 4th edn., 1833), 13.

[7] P. Brooks, *Reading the plot: Design and intention in narrative* (Oxford, 1984), 30–4, 268–9; T. W. Laqueur, 'Bodies, details, and the humanitarian narrative', in L. Hunt (ed.), *The new cultural history* (Berkeley, Calif., 1989), 177; Van Sant, *Sensibility and the novel*.

of kindly emotion in the rise of public philanthropy which these narratives mediated are well documented, as are the stirrings of conscience about slavery and the insane.

Sensibility might not quite have moved literature 'from a reptilian Classicism, all cold and dry reason, to a mammalian Romanticism, all warm and wet feeling'.[8] But emotion came to underpin aesthetic judgement and appreciation of the sublime. Of art 'a man, who is destitute of sensibility of heart, must be a very imperfect judge'.[9] Writing came to be valued as a process rather than as a product, the result being, as Dr Johnson remarked, that anybody could write like Ossian if he would abandon his mind to it, and that you would hang yourself if you read Richardson only for the plot. Inspired by epistolary novels of feeling, diaries and letters after 1740 similarly evinced an increasing interest in the exploration of the self and its emotions. The literary valuation of nature as if it had human qualities and feelings was transmuted in Romanticism into an identification with nature's turbulent moods.[10]

In evangelicalism and 'new dissent' religious feeling was expressed in a more passionately felt spirituality. Attitudes to death changed too. Even among humbler people it became more private than hitherto, the site of intimate weeping rather than of public display. Feelings about death became more secularized, more pathos-laden, more preoccupied with death's beauty. Tears also became a way of communing with others and with the self, demonstrating 'a new relationship with others regulated by emotional identification'—a new 'obligation of compassion and consolation'. 'The propensity to weep over strangers or humanitarian principles . . . was an affirmation of a new sensitivity to unhappiness and pain.' 'What is generosity, clemency, humanity, if it is not pity applied to the weak, to the guilty, or to the human race in general?' Rousseau cried.[11]

Then there was what Byron helpfully defined as the shift 'from cunt to cant', as decorum, privacy, and shame came to enshroud the bodily and sexual functions, while attitudes to women and children were

[8] Northrop Frye, 'Towards defining an age of sensibility', *English Literary History*, 23 (1956), 145–8, 150.

[9] Gerard, *Essay on taste*, 88.

[10] H. Blodgett, *Centuries of female days: Englishwomen's private diaries* (New Brunswick, 1988), 22–3, 40; M. Andrews, *The search for the picturesque: Landscape aesthetics and tourism in Britain, 1760–1800* (Aldershot, 1989), ch. 3.

[11] J. McManners, 'Death and the French historians', in J. Whaley (ed.), *Mirrors of mortality: Studies in the social history of death* (1981), esp. 119–20; A. Vincent-Buffault, *The history of tears: Sensibility and sentimentality in France* (1991), 17, 36–7.

affirmed within newly emphatic domestic ideologies.[12] The sentimentalizing of nature and the anthropomorphizing of animals testified to another 'profound shift in sensibilities' which worked upon all social levels, as Keith Thomas has shown.[13] The appeal to emotions posed new dilemmas for political arguments too. Sensibility could be harnessed to both conservative and radical arguments, dispute then turning on whether the emotions ruling each case were 'authentic', as the battle between Burke and Paine showed.[14]

Historians of culture are not always good at explaining the causes of major cultural shifts of this order. They often imply that cultural practices merely 'build upon' or react against earlier practices, as if those sequences were unproblematic. Sensibility certainly drew on new physiological understandings of body and mind achieved in the previous century and on the valuation of sensation which ensued; it was *taught*, we shall see. But its purchase depended also on the assertiveness of new social groupings for whom sensibility's messages served covert social functions.

Contemporaries had no doubt about its peculiarly specific social location. Commentators from Adam Smith on to Francis Place were convinced that a softening of social relationships resulted from the expansion of commerce and manufactures. 'Let us not forget that the trading and commercial interests, by enlarging the wants of mankind, have diffused the principle of benevolence,' Samuel Parr noted.[15] Even if it is an old joke that the middle classes have always been rising, this is the contextual explanation with which we can still live best.

The reference is to the familiar fact that by the early eighteenth century the culture of gentry and aristocracy was challenged by the middle people in commerce and professions. Their self-consciousness had been sharpened within the new affluence and sociability of the provincial town and of London alike. After the Wilkesite agitations it had been sharpened in a developing history of political confrontation too.[16] It has

[12] R. Porter, 'Mixed feelings: The enlightenment and sexuality', in P.-G. Bouce (ed.), *Sexuality in eighteenth-century Britain* (Manchester, 1982), 21; G. Vigarello, *Concepts of cleanliness: Changing attitudes in France since the Middle Ages* (Cambridge, 1985), 100–4; L. Stone, *The family, sex and marriage in England, 1500–1800* (1977); L. Davidoff and C. Hall, *Family fortunes: Men and women of the English middle class, 1780–1850* (1987).

[13] K. Thomas, *Man and the natural world: Changing attitudes in England, 1500–1800* (1983), 15; J. Passmore, 'The treatment of animals', *Journal of the History of Ideas*, 36 (1975), 195–218.

[14] S. Cox, 'Sensibility as argument', in S. M. Conger (ed.), *Sensibility in transformation: Creative resistance to sentiment from the Augustans to the Romantics* (1990), 63, 66.

[15] Cited by McGowen, 'A powerful sympathy', 321.

[16] P. Earle, *The making of the English middle class: Business, society and family life in London, 1660–1730* (1989), 10, 333; P. Borsay, *The English urban renaissance* (Oxford, 1989), pt. 4.

been claimed that as middle-class numbers and wealth grew after 1660, they merely emulated the manners and consumption patterns of superiors, and so bound 'the top half or more of the nation by means of an homogenized culture of gentility that left élite hegemony unaffected'.[17] Yet in the present context bourgeois sensibility had a reactive, differentiating, and status-affirming meaning which this view ignores. It entailed a *dis*sociation from aristocracy and gentry as much as it did from the common people. It was a way of differentiating the feeling bourgeois self not only from the unfeeling mob but also from an arrogant and exclusive aristocracy.

As the century wore on, upwardly mobile and educated professional, literary, and mercantile families, and some higher tradespeople too, had every interest in announcing their hard-won and not always secure status by elaborating their own cultural positions as well as consumption patterns. The effect was that their deference to and envy of the great had a hard critical edge to it. Élites, as Hume put it, not of birth but of virtuous attainment,[18] many discarded aristocratic and gentry mores as energetically as they did plebeian—so much so that one historian claims that the Hanoverian aristocracy became 'a tool of an increasingly dictatorial bourgeoisie' as capital competed with land as a basis of power.[19] New penal attitudes accommodated unambiguous hostility to aristocratic values. The capital code was 'a relic of Norman policy', it was declared in parliament in 1790, appropriate only to 'the days of Gothic tyranny and ferocity of manners'. 'At this period of civilization and refinement, a milder mode of punishment would be more adequate,' another writer put it some years earlier.[20] Thus articulating their own notions of sociability and justice, these people generated a provincial 'urban renaissance' and a heady tradition of anti-aristocratic dissent which, fused with commercial interest, survived deep into the nineteenth century. These are the material contexts of the 'huge, mysterious swing in sensibility' which is our concern.[21]

[17] L. and J. D. F. Stone, *An open élite? England, 1540–1880* (Oxford, 1984), 286–9, 409.

[18] D. Hume, 'Of the middle station of life' (1742), in *Essays moral, political, and literary*, ed. T. H. Green and T. H. Grose (2 vols., 1882; repr. 1964), ii. 375–80.

[19] P. Langford, *Public life and the propertied Englishman, 1689–1798* (Oxford, 1991).

[20] PH xxviii (1790), c. 782 ff.; W. Smith, *Observations on the laws relative to debtors and felons* (1777), 6, cited in McGowen, 'A powerful sympathy', 321.

[21] We should note another powerful interpretation too complex to incorporate here. T. L. Haskell discerns in 18th-cent. humanitarian motivations and actions less a purposeful self-interest than the ethical outworking of a new personality type matured by the disciplines and opportunities of market capitalism. Attending to the remote consequences of their acts, thinking causally, committed to 'new levels of scrupulosity in the fulfilment of ethical maxims', and acquiring new techniques for interventionist action, some businessmen (Haskell argues) were driven to extrapolate from the ethical systems learnt

None of this denies that there had to be teachers to diffuse this oppositional sensibility, and of these there were many. Not all, admittedly, spoke with one voice. The philosophical world, Mackintosh later observed, 'has been almost entirely divided into two sects,—the partisans of selfishness . . . and the advocates of Benevolence, who have generally contended that the reality of Disinterestedness depends on its being a *primary principle*'.[22] Some altogether denied the validity of sensibility as the basis of morality. Bentham did, for example. He preferred morality to emanate from enlightened self-interest. For the same reason, James Mill regarded 'as an aberration of the moral standard of modern times . . . the great stress laid upon feeling'.

Additionally, many came to doubt the sincerity of those who affected sensibility. Carlyle and Dickens were only the most prominent in the next century to decry the 'mawkish sentimentality' and 'diseased sympathy' for the criminal which they believed ensued from these postures. There was certainly much false posturing and many promiscuous effusions of emotion on criminal as on other subjects.[23] 'Virtue requires habit and resolution of mind as well as delicacy of sentiment,' Adam Smith had to announce sternly, and it was not without reason that Southey later parodied 'the sentimental classes, persons of ardent or morbid sensibility'. His attack was developed thereafter in innumerable conservative denunciations of 'that false cant of humanity by which *liberal* writers of the day perpetually endeavour to enlist public sympathy on behalf of the violators of law against its just inforcers'.[24] As has been well put, sensibility only too often 'signified revolution, promised freedom, threatened subversion, and became convention'.[25]

None the less, as late as the 1830s the reaction against Benthamism

in the market a new moral universe in which *not* helping suffering strangers might become an unbearable act, and in which helping them was also now technically feasible. Haskell's is an outstanding exegesis of a mentality of market capitalism: T. L. Haskell, 'Capitalism and the origins of the humanitarian sensibility': I, *American Historical Review*, 90/2 (Apr. 1985), 339–61; II, ibid. 90/3 (June 1985), 547–66.

[22] *The progress of ethical philosophy* (1851 edn.), 120.

[23] Take the journalist William Jerdan recalling his visiting a man awaiting execution for forgery in Newgate around 1811–14: 'My spirits were excited almost to distraction by the interview, and I wrung the poor fellow's hand, perhaps for the last time, and rushed from the dismal place. To my horror I found the doors bolted and egress denied; I screamed for the gaoler, but no one answered my call; reason had not time to exercise its influence, and after another fruitless effort with hand and voice, I tried to grasp the iron fastenings and fell down, as if shot, perfectly insensible upon the pavement.' Perhaps Jerdan's 'nerves did not recover the stroke for several weeks'; but equally he might have been reading too many overheated novels: *The autobiography of William Jerdan* (1852), i. 130–1.

[24] R. Williams, *Keywords: A vocabulary of culture and society* (1983) 280–3; C. E. Dodd, 'Facts relating to the punishment of death', *Quarterly Review*, 47 (1832), 197, cited in McGowen, 'A powerful sympathy', 318.

[25] Barker-Benfield, *Culture of sensibility*, p. xvii.

still drew on the belief that the organ of virtue was the sensibility rather than the conscience, and that benevolence—'the love of *loving*', as J. S. Mill put it—was the moral ideal.[26] The principle of sympathy had been valued in moral philosophy long before this. The 'new' view of others' miseries, and of the desirability of connecting with them emotionally and sensationally, was rooted in the associational psychology of the eighteenth century. It changed the recommended nature of social relationships fundamentally, and it shapes modern philanthropic agenda to this day.

Countering the pessimism of Hobbesian psychology, seventeenth-century writers like Henry More had already regarded sympathy and compassion as intrinsic to man's nature. More thought the passions and affections so much a part of the natural law that commiseration and compassion were natural checks to injustice: 'to take away the life of an innocent man, is so monstrous a crime, as tears the very bowels of nature, and forces sighs from the breasts of all men.' In 1711 Shaftesbury similarly defined what it was to be human in terms of the 'natural' capacity for sympathy. Human cruelty was 'unnatural', he thought; and he condemned those who took 'unnatural and inhuman delight in beholding torments, and in viewing distress, calamity, blood, massacre and destruction, with a peculiar joy and pleasure': 'to delight in the torture and pain of other creatures indifferently . . . to feed as it were on death, and be entertained with dying agonies; this has nothing in it accountable in the way of self-interest or private good . . . but is wholly and absolutely unnatural, as it is horrid and miserable.' From Shaftesbury on to Berkeley, Butler, and Hutcheson, it was understood that vice and selfish passion might be surmounted by natural benevolence.[27] Hume joined in. He also saw sympathy as a natural principle: 'Nature has preserv'd a great resemblance among all human creatures, and . . . we never remark any passion or principle in others, of which, in some degree or other, we may not find a parallel in ourselves.' The passions of others, therefore, might not only be 'imagined' but also experienced and felt, thus to become 'the very passion itself, and produce an equal emotion'. In political and judicial relations, he continued, we 'annex the idea of virtue to justice, and of vice to injustice', thus sharing the 'uneasiness' of the victims of injustice. Even 'when the injustice is so distant from us, as no way to affect our interest, it still displeases us;

[26] W. E. Houghton, *The Victorian frame of mind* (New Haven, Conn., 1957), 264–8.

[27] N. Fiering, 'Irresistible compassion: An aspect of eighteenth-century sympathy and humanitarianism', *Journal of the History of Ideas*, 37 (1976), 200–2.

because we consider it as prejudicial to human society, and pernicious to every one that approaches the persons guilty of it. We partake of their uneasiness by *sympathy*.' Hume later put greater stress on utility and 'advantage' in the processes which preserve 'peace and order among mankind', but he still insisted that the benevolent check to rational self-interest ('a feeling for the happiness of mankind, and a resentment of their misery') was primary in human nature and the origin of morality, the terms in which a general advantage might best be understood.[28]

Adam Smith's *Theory of moral sentiments* (1759) similarly identified sympathy as the source of benevolence and moral judgement:

> How selfish soever man may be supposed, there are evidently some principles in his nature, which interest him in the fortune of others. . . . Of this kind is pity or compassion, the emotion which we feel for the misery of others, when we either see it, or are made to conceive it in a very lively manner. That we often derive sorrow from the sorrow of others, is a matter of fact too obvious to require any instance to prove it.

What Smith contributed was the recognition that imaginative projection was the vehicle through which compassion was released. Here indeed is a key idea informing arguments in this book which are still to come: 'By the imagination we place ourselves in his [the sufferer's] situation, we conceive ourselves enduring all the same torments, we enter as it were into his body, and become in some measure the same person with him, and thence form some idea of his sensations.' It was, he argued, because each individual accommodates an imaginative and intuitive self which permits identification with those who suffer that self-interest can be subordinated to the interests of others. Striking here is Smith's insistence on the bodily organization of sympathy ('the mimical powers connected with our bodily frame', as Dugald Stewart later put it in developing his 'law of sympathetic imitation')—an associational process which located the capacity for identification with others in the natural, albeit involuntary and unconscious realm: 'When we see a stroke aimed, and just ready to fall upon the leg or arm of another person, we naturally shrink and draw back our own leg or our own arm; and when it does fall, we feel it in some measure, and are hurt by it as well as the sufferer.'[29]

[28] D. Hume, *Treatise of human nature* (1739–40); cf. J. Mullan, *Sentiment and sociability: The language of feeling in the eighteenth century* (Oxford, 1988), 29–42.

[29] A. Smith, *The theory of moral sentiments*, ed. D. D. Raphael and A. L. Macfie (Oxford, 1976), 60–1; Fiering, 'Irresistible compassion', 211; D. Stewart, *Elements of the philosophy of human mind* (1827) in *Works*, ed. W. Hamilton (1854), iii. 117, 129.

How did these precepts affect behaviour and feeling among the polite classes? As the Introduction noted, we should nowadays insist that emotional repertoires are neither 'natural' nor common to all people in all cultures and times. Statements about a 'feeling' are cognitive interpretations of a sensation, and in that sense feelings like pity, compassion, sympathy, etc. are learnt. But, if so, this only endorses the importance of the moral philosophers and novelists of Hume's generation, for it was from them that the new emotional repertoires were disseminated. Through the novels of sentiment and on again through many essays on taste and manners, the project upon which they laboured entered polite drawing-rooms. There, in principle, the conscientious would become practised in those forms of identification which acknowledged kinship with others, while even false posturing could reinforce a cultural predisposition. If culture had any bearing on changing attitudes to harsh punishment, it could only be through the broadening purchase of this principle among the polite classes, as part of the sanctioned, the habitual, the unspoken. We come now, however, to the complications which beset this promising story.

3. *Impeded narratives and resistant mentalities*

To imply that in the eighteenth century the polite classes were converted into amiably sympathetic beings is like arguing that they crossed a 'threshold' in their ability to identify with others; and as inept. Continuities in history always give the lie to assumed discontinuities. Moreover, people seldom fully absorb the lessons which the social world flings at them. All messages have to compete with collective memories, attitudes, and values bequeathed by past generations, as well as with individuals' compulsions, defences, and interests in the present. These things can cut across the principles so finely in question. They explain why most of us have difficulty in separating what we wish to see from what we actually see, and why most take comfort from established patterns of living to save 'strenuous thinking', anxiety, and conflict.[30]

These legacies from the past and projections in the present also explain why attitudes to punishment in the real world cannot be subsumed within a simple explanatory schema. A comprehensive survey would have to accommodate not only the sympathetic minority who thought that being entertained by dying agonies was unnatural, but also

[30] P. Gay, *Freud for historians* (1985), 133, 139, and ch. 5.

(say) the unexpectedly John Bullish animadversions of John Wilkes, who in 1784

> thought the happiest results followed from the severity of our penal law. It accustomed men to a contempt of death, though it never held out to them any very cruel spectacle; and he thought that much of the courage of Englishmen, and of their humanity too, might be traced to the nature of our capital punishments, and to their being so often exhibited to the people.[31]

It would also have to accommodate this laconic recording in a merchant's diary in 1824: '4 men condemned at [the] last assizes hang'd this morng. Self at home day and eveng, my cold continued very indiff[eren]t'.[32] Neither utterance finds a place in conventional narratives of penal reform, yet Wilkes's may reflect prevailing attitudes more accurately than the more attended-to critiques of the law which gainsaid it; while the merchant's preoccupations probably capture the quotidian mentality of busy businessmen better than any number of Quaker petitions will do. Moreover, many people in our period were as indifferent to the scaffold or as numbed by habituation to it as we all may be when confronted by repeated reports of distant horrors. Others were helped by knowing that the scaffold was necessary (order and property depending on it), or that retributive justice enacted the divine will. Then add the fact attended to in Chapters 10 and 11, that social or political distances between polite witnesses and scaffold victims were usually too wide for individuals' sympathetic imaginations to negotiate them successfully, and it is not difficult to agree that in most contexts sympathy for cruelly punished felons was only too easily compromised.

The impeded chronologies of penal reform bear witness to this. There was no inevitability in the movements from attacks on the bloody code in the 1770s to the code's dismantling in the 1830s, from Howard's recommendations of prison reform or Bentham's of a panopticon prison to the building of Pentonville prison in 1840–2, or from Colquhoun's advocacy of organized policing to Peel's metropolitan police of 1829. Over half a century intervened before aspects of the reformist programmes of the 1770s and 1780s began to be implemented, and it was hard battling all the way. Even as late as the 1820s many people did not think that major penal change was inevitable—which is why Peel's criminal law reforms in that decade aimed not to end the capital code but to reconstitute it by removing its inessentials.

[31] Romilly, i. 84.

[32] J. Fiske (ed.), *The Oakes diaries: Business, politics and the family in Bury St Edmunds 1778–1827* (2 vols., Woodbridge, 1991), ii. 291.

Straightforward political reasons helped to account for the long delays: the counter-revolutionary backlash of the 1790s, the defence of the constitution in disturbed times thereafter, and tory control of the executive until 1830. So did the shortage of resources and the weakness of bureaucracies by which alternative secondary punishments might be introduced. The delay might also have reflected something more difficult to pin down but more pertinent to the chapters immediately following: we follow French historians of 'mentality' here. Forward-looking ideas will always be constrained by what Braudel called the prisons of the *longue durée*, or by what Labrousse called the history of 'resistances'. The reference is to the inertias embedded in habits, attitudes, feelings, and ceremonies which new ideas can shift only slowly if they can shift them at all. Powerful resistances to change, these are the structures which 'get in the way of history', Braudel wrote: 'as hindrances [they] stand as limits . . . beyond which man and his experiences cannot go'.[33]

This part of the book is concerned with resistances, at the half-conscious or unconscious levels, of precisely these kinds. They alone do not explain the delays in penal reform: political and bureaucratic conditions determined chronologies more decisively. But by looking to the legacies of mentalities, we may mark quite precisely the limits of what culture could achieve, and of sensibility in particular. For all its implicitly psychoanalytical burden, the argument which these chapters develop is a simple one.[34]

The first premiss (taken up again in the next chapter) is that people were (and are) capable of the most elemental excitement—and thence

[33] The resistances Braudel had in mind were geographical, biological, and economic; but he acknowledged that 'mental frameworks too can form prisons of the *longue durée*': F. Braudel, 'History and the social sciences', in *On history*, trans. S. Matthews (1980), 31, 35. 'Ideological representations share with all systems of values a heaviness, an inertia, since their framework is made up of traditions': G. Duby, 'Ideologies in social history', in J. Le Goff and P. Nora (eds.), *Constructing the past: Essays in historical methodology* (Cambridge, 1985), 153.

[34] The dangers of 'psychohistory' have been much rehearsed by its enemies, with good reason, in respect of some practices and dogmatic practitioners. But some objections have been advanced so passionately that they invite analysis by the procedures they condemn. Despite shortcomings, the psychoanalytical narrative remains a major tool of understanding, for we have nothing else with which to address its central issues. Some very odd attitudes would remain unexplained in this book if we failed to note that punishment taps into primal wishes and fantasies; or how those wishes and fantasies are suppressed in the interests of decorum; or the distortions of feeling which ensue when forbidden wishes erupt none the less. The danger of anachronistic ascription is inherent in any writing about past motives, not only this form. To interpret motivation narrowly in terms of 'interests', as historians habitually do, itself smacks of a reductionism more simple-minded than any which may be held against historians with a fuller sense of the psyche's adaptational processes.

curiosity—about what happened on scaffolds. This has been claimed in the discussion of the scaffold crowd in Chapter 2; but the point applies to polite people as well. Dickens knew this when he noted

> the strange fascination which everything connected with [capital] punishment or the object of it, possesses for tens of thousands of decent, virtuous, well-conducted people, who are quite unable to resist the published portraits, letters, anecdotes, smilings, snuff-takings, &c &c &c of the bloodiest and most unnatural scoundrel with the gallows before him.

It is in our '*secret nature*', he added, 'to have *a dark and dreadful* interest' in the subject.[35] That puts it well. In all of us a dark interest about scaffold deaths may be as unavoidable as curiosity about sexuality, to which death may stand opposed but, by that token, be yoked in the most intimate relationship. Indeed, a Freudian would say of the scaffold that Eros and Thanatos did battle there. The scaffold tapped into primal excitement and curiosity about death, aggression, and destructiveness, while in the crowd's enjoyment of its own unanimity and in its support for the condemned, it might also bizarrely have been an arena of love.[36] In some of its eighteenth-century expressions sensibility tacitly sanctioned this excited curiosity. It encouraged people to practise their sympathetic responses by visiting felons in condemned cells or by attending executions, just as they were encouraged to witness and pity others' misery in the cells of Bedlam. Large numbers of comfortable people of both sexes were thus permitted to be as openly curious about hanging as their ruder seventeenth-century forebears had been (and as the crowd always was). Thus sanctioned, polite people might even have gone to hangings more frequently and zealously than Pepys's generation did. We are at the mercy of scant seventeenth-century recording; but certain ironic paradoxes cannot escape notice. This sympathetic age was not only that in which executions reached peaks unprecedented since 1530–1630; it might also—so far as cross-class attendance was concerned—have been the public spectacle's heyday.

[35] P. Collins, *Dickens and crime* (1964), 229 (my emphases).

[36] Thus *Civilization and its discontents* (1969 edn.), 55–6, 58–9: 'Besides the [libidinous] instincts to preserve living substance and to join it into ever larger units, there must exist another, contrary instinct seeking to dissolve those units and to bring them back to their primeval, inorganic state. That is to say, as well as Eros, there was an instinct of death . . . of aggressiveness or destructiveness. . . . Even where it emerges without any sexual purpose, in the blindest fury of destructiveness, we cannot fail to recognise that the satisfaction of the instinct is accompanied by an extraordinarily high degree of narcissistic enjoyment, owing to its presenting the ego with a fulfilment of the latter's old wishes for omnipotence. The instinct of destruction . . . must, when it is directed towards objects, provide the ego with the satisfaction of its vital needs and with control over nature. . . . It constitutes the greatest impediment to civilization. . . . [Civilization] must present the struggle between Eros and Death.'

The second stage of the argument is covered in Chapter 9. It acknowledges that some of the middle classes continued this curious witnessing until the very end of public executions in 1868 (for none of the processes which concern us operated universally or at uniform paces for all people). None the less there was a significant shift in polite society towards the end of the eighteenth century, and certainly into the nineteenth. Taboos began to encase the scaffold, thus altering its relationship with civility irretrievably. Sensibility lost some of its philosophically sympathetic edge as it 'became convention' and was subsumed reflexively within the codes of bourgeois decorum, often in alliance with evangelical earnestness. As this happened, the old curiosity became disreputable. People were advised to keep personal and emotional distances between themselves and all unseemly spectacles. Distaste for any association with hanging was projected on to the crowd with enhanced energy: the scapegoated crowd became the target of contempt, not the hanging itself.

The third stage of the argument allows that despite its deepening social sanctions, decorum made exorbitant demands upon many people. Curiosity about what happened on scaffolds was difficult to suppress. What had to be elaborated were systems of unconscious defence to tame the drives which claimed socially inappropriate gratification. These defences came to constitute the socialized, conscious, and judgemental self.[37] The more self-controlled the people of feeling, the more their primal interests obliged them to set up defences against the scaffold's full meanings and pain. They became squeamish, for example, or affected ignorance of what happened.

Sometimes the defences worked; but sometimes not. Repressed curiosity would then emerge embarrassingly, subjecting men and sometimes women to strange conflicts and anxieties. In these conditions, curiosity could seldom be expressed openly. Not unlike a repressed sexuality, it would manifest itself obliquely in fantasies or regressions, or be overlaid in anxiety, shame, and guilt. It is not difficult to concede in such exemplars that the demands of the social world were the foundation of anxiety, neurosis, and conflict—of civilization's discontents. 'Civilization is built upon a renunciation of instinct', Freud wrote: but 'serious disorders will ensue' if the renunciation is inefficiently achieved.[38]

[37] Anna Freud, *The ego and the mechanisms of defence* (Madison, Conn., 1966), 42–4. For a guide to terms, see C. Rycroft, *A critical dictionary of psychoanalysis* (1968), and R. Harré and R. Lamb (eds.), *The dictionary of personality and social psychology* (Oxford, 1986), 195.

[38] *Civilization and its discontents*, 34.

The many confusions about penal pain among polite Hanoverians are not now to be left off the record. From a humanitarian point of view it is not a cheering depiction which follows. We find few hearts bleeding for the hanging man or woman, or spirits indignant at the law's rough justice. We eavesdrop on the private susurrations of people who largely accepted things as they were, or who accommodated their witnessing in ways remote from those which compassion recommended. At this level, the history of feeling about execution is more often about congested, inward-turning, or blocked emotion than it is of sympathies openly released. This helps to explain why, regardless of their sensibility or civility, it was not obvious to most people before the 1830s that capital punishment for relatively trivial crimes was an inhumane way of dealing with crime and disorder. Most did not know how to deal clear-headedly with the scaffold's grim tableaux. What they saw, they saw through a glass darkly.

CHAPTER 8

WATCHING FROM CURIOSITY

1. *The unembarrassed witness*

DID POLITE PEOPLE EVER FEEL AN UNABASHED EXCITEMENT AND curiosity when they attended executions? Could they ever have attended without disgust or guilt overlaying those simpler feelings? Is it thinkable that readers of *Pamela* or *Sense and sensibility* were moved by the same impulses as the vulgar crowd? If so, what within their secret natures accounted for that 'strange fascination' and 'dark and dreadful interest' which Dickens discerned whenever the scaffold was in view? When did it cease to be permissible to contemplate executions in these open ways?

The supposition which shapes our answers was stated in the last chapter. It is that fascination with what happened on the scaffold was *primal*. It was common to all people regardless of social standing. At its most self-regarding, the fascination spoke for pleasure that it was others who died, and it expressed an elemental curiosity about how they died. 'I never saw a man hanged but I thought I could behave better than he did,' James Boswell declared;[1] or as a *New Grub Street* character admitted when he read of a Newgate hanging, 'There's a clear satisfaction in knowing that it is not oneself.' Provided that identification with the victim was checked, a fantasy about one's own relative longevity might be savoured as the other's fate was settled before one's eyes. Add to that the gratification of watching a life extinguished which law or morality had defined as deserving of extinction, and add again the crowd's self-loving solidarity as it watched—and the 'pleasures' of witnessing become more intelligible, even to those who would prefer not to admit such proclivities in themselves or their fellow creatures.

A modern critic asks our opening questions too, and gives her own

[1] F. Brady and F. A. Pottle, *Boswell in search of a wife, 1766–1769* (1957), 141.

supportive answer. Elisabeth Bronfen speculates about artists' strange compulsion to depict the death of beautiful women, and our pleasure in such images: 'How can we delight at, be fascinated, morally educated, emotionally elevated and psychologically reassured in our sense of self by virtue of the depiction of a horrible event in the life of another, which we would not have inflicted on ourselves?' She replies that these depictions

> delight because we are confronted with death, yet it is the death of the other. We experience death by proxy. In the aesthetic enactment, we have a situation impossible in life, namely that we die with another and return to the living. Even as we are forced to acknowledge the ubiquitous presence of death in life, our belief in our own immortality is confirmed. There is death, but it is not my own.

She adds that the artistic representation of death expresses a displaced anxiety about death, and a desire for death as well. It expresses 'something that is so dangerous to the health of the psyche that it must be repressed and yet so strong in its desire for articulation that it can't be'. In a 'gesture of compromise' the artist deals with the danger by representing death in 'the body of another person and at another site'. And since the visible corpse can be understood to stand for something else, what is literally represented may not be fully seen at all: wherein safety lies.

How this analysis might apply to scaffold witnessing need not be laboured. Scaffold deaths offered witnesses a similar affirmation of the self, the same admixtures of danger, pleasure, and desire, and the same quality of not quite seeing the fully actualized horror. The tableaux on scaffolds were not pictorial, but witnesses could still see them as *representations*. Enacted upon stages, hangings also signified something other than themselves as their reality was filtered through ascribed meanings; they could just as efficiently be dissociated from the witness's own death as pictures were. They might be emblematic of one's own death. But thankfully and cheeringly they were not *it*.[2]

There are many examples of open curiosity about scaffold death before the age of sensibility named curiosity 'philosophically', and so made it self-conscious—before decorum and civility rendered it shame-

[2] E. Bronfen, *Over her dead body: Death, femininity and the aesthetic* (Manchester, 1992), pp. x–xi. Arlette Farge also notes at French executions 'the paradoxical and contradictory coexistence between the fear and horror of death and a real taste for it'. Mass attendances at executions indicate attitudes which were the reverse of indifferent, she insists. Nor are these attendances to be ascribed to past people's familiarity with death, familiarity never being the same as indifference: *Fragile lives: Violence, power and solidarity in eighteenth-century Paris*, trans. C. Shelton (Cambridge, 1993), 185–7.

ful too. Go back to the seventeenth century in search of a baseline, and we find people enough who indulged curiosity without the least shame or pity. Pepys was only the most famous of them. He was shaken by a hanging only once, and then because the victim was a semi-gentleman with whom he identified (identification always being the enemy of equable dissociation). 'A comely-looked man he was,' he wrote of Colonel John Turner, whose hanging for burglary he watched in 1664. Pepys's curiosity was such that he paid a shilling for the privilege of standing 'in great pain' on the wheel of a cart, and with the rest of the crowd he had to wait patiently while Turner delayed the proceedings 'by long discourses and prayers one after another, in hopes of a reprieve'. But Turner 'kept his countenance to the end—I was sorry to see him'—so much so that after the man 'was flung off the lather [ladder] in his cloak', Pepys went home to his dinner 'all in a sweat'. What signifies is that this was the only such expression in the whole Diary: no sympathy otherwise, or shame. Pepys had watched many other executions without anxiety (he had sent his wife to book his place at this one), and he was to watch many more. Normally, his witnessing was buried without comment in the details of daily business:

> My Lord not being up, I went out to Charing Cross to see Maj.-Gen. Harrison hanged, drawn, and quartered . . . he looking as cheerfully as any man could do in that condition. He was presently cut down and his head and his heart shown to the people, at which there was great shouts of joy. . . . From thence to my Lord's and took [colleagues] to the Sun tavern and did give them some oysters . . .

Or:

> At the corner shop, a drapers, I stood and did see Barkestead, Okey, and Corbet drawne toward the gallows at Tiburne; and there they were hanged and quarterd. They all looked very cheerfully. But I hear they all die defending what they did to the King to be just—which is very strange. So to the office. And then home to dinner.

When he did comment, his reactions were squeamish, an emotion which blocks both sympathy and identification. On a rural excursion he and a lady companion 'rode under the man that hangs upon Shooter's hill; and a filthy sight it was to see how his flesh is shrunk to his bones'. Likewise, he gazed with curiosity upon the 'kidnys, ureters, yard, stones, and semenary vessels' of 'a lusty fellow, a seaman that was hanged for a robbery', who was dissected and lectured upon at Surgeon's Hall. He found it 'a very unpleasant sight'. None the less, 'still

I did touch the dead body with my bare hand'—to see how it felt. And his curiosity as to why gentlemen sentenced to death favoured a silken noose ended with a piece of wishful thinking which his own observations of the hanging man's struggles should have contradicted:

> [The silken noose] being soft and slick, it doth slip close and kills, that is, strangles presently; whereas a stiff one doth not come so close together and so the party may live the longer before killed. But all the Doctors at table conclude that there is no pain at all in hanging, for that it doth stop the circulation of the blood and so stops all sense and motion in an instant.[3]

Pepys was not the only seventeenth-century witness of this ilk. John Evelyn had the same curiosity. In 1682 he noted how before Colonel Vratz was hanged, he told a friend 'that he did not value dying of a rush, and hoped and believed God would deal with him like a gentleman'. After his hanging, Evelyn went to see the corpse of the 'obstinate creature': 'the flesh was florid, soft, and full, as if the person were only sleeping.'[4]

Curiosity about execution as unselfconscious as this never waned in succeeding generations. In later chapters we do not encounter an obliteration of old impulses, only their suppression in some people, and then at no consistent pace. Indeed these oddly innocent impulses to see the worst and to relish the ensuing sensations doubtless propelled most scaffold audiences, rich and poor, all the way down to 1868, however rationalized they became, and however officially repudiated and deplored.

In the eighteenth century we meet the impulses at all social levels. Farmers' wives might not have attended executions themselves because of distances, or perhaps because status differentials were tighter in rural communities than in London. But they made up for their absence by noting execution details in diaries and commonplace books with a curiosity as candid as Pepys's. News, rumour, and sensation had to be caught, pinned down, transcribed, by people starved of these things. 'Sir Ferrers executed, Monday 5 May 1760, for shooting his steward was hanged in his wedding clothes which were white cloth embroidered with silver, and white silk stockings,' Mary Grose confided to her diary on the Isle of Wight: 'When he mounted the stage he knelt for about three minutes with the clergyman, whom it was said to be his brother, then was executed where he hanged a full hour, was then put in a shell

[3] R. Latham and W. Matthews (eds.), *The diary of Samuel Pepys* (11 vols., 1970–83), v. 23, iii. 67, ii. 72–3, iv. 60.

[4] *Diary of John Evelyn*, ed. E. S. De Beer (5 vols., Oxford, 1955), iv. 276, 24 Mar. 1682.

not undressed and carried to Surgeon Hall.'[5] The diary of Mary Hardy, a Norfolk farmer's wife, included lengthy summaries of crimes, hangings, and gibbetings 'from the News Paper'. In 1775 she did go to an execution, of a highway robber in Norwich, noting afterwards only that 'he appeared very penitent & devout after his condemnation and at the place of execution'. In York that year she went to look at the grisly exhibits in the castle—the 'knife and fork used in quartering the rebels taken in the late Rebellion', 'a cord that one Smith condemned for sheepstealing hanged himself with in the condemned hole', and the skull of the man murdered by the notorious Eugene Aram in 1759. Two days later she visited Aram's gibbet, still standing on Thistle Hill.[6] Recording all this in her diary, she withheld commentary.

None of this witnessing was sadistic or cruel. Perhaps it related to the old sense that life and death were parts of a continuum along which all must move, the same unspoken understandings evinced here which allowed the felon to sit on his coffin, clad in his shroud, *en route* to the Tyburn gallows. Each witnessing could just as well be said to express the directness of people at ease with a curiosity which they saw no reason to repudiate.

What qualities of disingenuousness had to be marshalled to achieve such uncompromising witnessing? The best answer may take us to the young, for among children the scaffold's horror was seen freshly, and it was also unambivalently affirmed. It does not take dreadful stories like that of the 16-year-old in 1790 who, having watched a Bristol execution, hanged himself that day, his mind 'unhinged',[7] to make us realize that responses to the scaffold of the young were as immediate as any we might encounter.

Victorian critics of the scaffold crowd were often disgusted by the thought that plebeian parents took their offspring to executions. 'Mothers held up their babes to see, / Who spread their hands, and crow'd for glee', Coventry Patmore brightly versified in 1845. Yet not long before, middle-class parents' resort to scaffold and gibbet by way of example and punishment had been matter-of-course it seems. Hogarth's prints of the good and idle apprentices, the idle one ending stickily at Tyburn, were at the earlier and softer end of the devices wheeled into this service. Their morality pervaded eighteenth-century children's

[5] A. Crane, *The Kirkland papers, 1753–1869* (1990), 37.

[6] B. Cozens-Hardy (ed.), *Mary Hardy's diary* (Norfolk Record Society, 1968), 15, 17, 23–4.

[7] *The Times*, 20 May 1790.

books: 'every good boy was to ride in his coach, and be a lord mayor; and every bad boy was to be hung, or eaten by lions,' Leigh Hunt recalled of his boyhood reading.[8] In the middle range were the school visits: a Methodist schoolmaster of Reading took his pupils to watch 'improving scenes' at the gallows; a school holiday was given when a girl was hanged at Horsham in 1824.[9] At the hardest end, children could be flogged after being taken to executions 'that they might remember the example they had seen'.[10] In Mrs Sherwood's *History of the Fairchild family: or The child's manual: Being a collection of stories calculated to show the importance and effect of a religious education* (1818) the authoress has the good paterfamilias Fairchild punish his squabbling 6- to 9-year-old children by whipping their hands, depriving them of food, and taking them to see a gibbeted body, gruesomely described. Under the gibbet the children endure a homily on the wickedness of quarrelling and have to join their father in prayer. The book went through many editions, and in the 1900s some readers could still recall how it had terrified them as children.[11] If parents did not wield these horrific texts, nurses would. The young Samuel Romilly was frightened by his nurse's 'stories of devils, witches, and apparitions', of 'murders and acts of cruelty':

> The prints, which I found in the lives of the martyrs and the Newgate Calendar, have cost me many sleepless nights. My dreams too were disturbed by the hideous images which haunted my imagination by day. I thought myself present at executions, murders, and scenes of blood; and I have often lain in bed agitated by my terrors, equally afraid of remaining awake in the dark, and of falling asleep to encounter the horrors of my dreams.[12]

The barrister William Ballantine (1812–87) also recalled a nurse who 'helped to fill my soul with terrors' by tales of murder:

> The relation . . . of any horrible crime used to produce a most painful effect upon me . . . and I particularly remember hearing talked about the murder of a Mrs Donathy . . . and it is wonderful, after a long lapse of years, how vivid my recollection of it is. . . . Such memories remain—perhaps, also, their consequences. . . . Unheard and unpitied, I many a time cried myself to sleep.[13]

Ballantine's barrister father took him to the Old Bailey to watch the trial of Bishop and Williams, the burkers. William Page Wood, later Lord Chancellor Hatherley, was exposed by his father, the radical alderman, to the same spectacle: 'I once saw thirty or forty . . . condemned, some

[8] *Autobiography*, ed. E. Blunden (Oxford, 1928), 70–1.

[9] H. Potter, *Hanging in judgment: Religion and the death penalty in England* (1993), 22.

[10] J. P. Grosley, *A tour to London* (1772), i. 172–3.

[11] *Notes and Queries*, 10th ser., 2 (1904), 454–5.

[12] Romilly, i. 11–12.

[13] W. Ballantine, *Some experiences of a barrister's life* (5th edn., 1882), 3.

of whom were making grimaces at the judge, whilst others we expected to be left for execution were deeply distressed. The scenes were sometimes most painful.' (They also, however, 'gave me an early inclination for that profession to which my father, when I was about fourteen, told me that I was destined'.)[14] The antiquarian William Kelly's father took him as a boy to see Cook's gibbeted body in 1832, 'he telling us that he did so because he had no doubt it would be the last time such an event would ever take place in England'. The adult remembered this fifty years later.[15]

The adult seldom knew what to do with such memories, but what is striking in these cases is how the child's voice echoed in the concrete detailing and Anglo-Saxon vocabularies of the adult recalling. Charles Knight the publisher remembered being taken in 1804 to see two gibbeted bodies on Hounslow Heath when he was barely 4: 'The chains rattled; the iron plates scarcely held the gibbet together; the rags of the highwaymen displayed their horrible skeletons. That was a holiday sight for a schoolboy, sixty years ago!'[16] Another old man remembered how in the 1780s he was taken to watch a woman burnt in Winchester for treason: 'I sat on my father's shoulder, and saw them bring her and the marine to the field. They fixed her neck by a rope to the stake, and then set fire to the faggots, and burnt her.'[17] Gathorne Hardy and his classmates at Shrewsbury School between 1827 and 1833 were not allowed to watch executions at the castle opposite, but once they were let out too early and flocked to see a hanging body, never to be forgotten: 'in a smock frock with a cap over his head [he] looked like a mere sack'.[18] Wordsworth recollected his boyhood awe when he stumbled on a place 'where in former times a murderer had been hung in iron chains':

> The gibbet mast had mouldered down, the bones
> And iron case were gone; but on the turf,
> Hard by, soon after that fell deed was wrought
> Some unknown hand had carved the murderer's name.
> The monumental letters were inscribed
> In times long past; but still, from year to year,
> By superstition of the neighbourhood,

[14] W. Ballantine, *Some experiences of a barrister's life* (5th edn., 1882), 17; W. R. W. Stephens, *A memoir of the Rt. Hon. William Page Wood, baron Hatherley* (1883), 19.

[15] *Notes and Queries*, 6th ser., 8 (1883), 394.

[16] C. Knight, *Passages of a working life during half a century, with a prelude of early reminiscences* (2 vols., 1864–5), i. 40–1.

[17] *Notes and Queries*, 2 (1850), 6.

[18] Local people waited for the body to be cut down, so they could touch the dead man's hand to cure warts: A. E. Gathorne Hardy (ed.), *Gathorne Hardy, first earl of Cranbrook: A memoir* (1910), i. 22–3.

The grass is clear'd away, and to this hour
The characters are fresh and visible:
A casual glance had shown them, and, I fled
Faltering and faint, and ignorant of the road.[19]

Although most adults left only shallow records of childhood witnessings, some (like Ballantine) knew that they had 'consequences'. Wordsworth ascribed to his some part of his early moral awakening. Romilly could not have been remote from his childhood memory when he asked the Commons in 1813 if there was any 'father in this assembly who would wish his child to be present at such a sad scene' as an execution.[20] When, as home secretary, Gathorne Hardy introduced the Bill for hiding the scaffold in 1868, he told parliament how 'objectionable and *horrible*' they were (this is the third use here of the childish word). Usually the adult would recall such memories to affirm the barbarity of times past and the progress of the present. The adult was rare who continued to see as clearly as the undefended child might see. But some did retain that gift, none more so than Byron.

In 1812 Byron went to watch the hanging of John Bellingham, the deranged tradesman who had shot Spencer Perceval in the Commons. Here the suppression of detail and the comic linkage he established hint at a symptomatic blockage, centred in anxiety: 'after sitting up all night, I saw Bellingham launched into eternity, and at three the same day I saw [Lady Caroline Lamb] launched into the country.'[21] But the suppressed effect of his witnessing was released five years later in a horrific letter describing an execution he watched (through opera-glasses!) in Rome:[22]

The day before I left Rome I saw three robbers guillotined—the ceremony—including the *masqued* priests—the half-naked executioners—the bandaged criminals—the black Christ & his banner—the scaffold—the soldiery—the slow procession—& the quick rattle and heavy fall of the axe—the splash of the blood—& the ghastliness of the exposed heads—is altogether more impressive than the vulgar and ungentlemanly dirty 'new drop' & dog-like agony of infliction upon the sufferers of the English sentence. Two of these men—behaved calmly enough—but the first of the three—died with great terror and reluctance—which was very horrible—he would not lie down—then his neck was too large for the aperture—and the priest was

[19] Wordsworth, *The prelude, or Growth of a poet's mind*, ed. E. de Selincourt, rev. H. Darbishire (Oxford, 1959), bk. xii (1850), ll. 235–47.

[20] PD1 xxvii (1813), c. 108, cited by M. Wiener, *Reconstructing the criminal: Culture, law, and policy in England, 1830–1914* (Cambridge, 1990), 94.

[21] L. Marchand (ed.), *Byron's letters and journals*, ii. (1973), 177; and T. Moore, *Letters and journals of Lord Byron: With notices of his life* (2 vols., 1830), i. 357.

[22] Marchand (ed.), *Byron's letters*, v. (1976), 229–30.

obliged to drown his exclamations by still louder exhortations—the head was off before the eye could trace the blow—but from an attempt to draw back the head—notwithstanding it was held forward by the hair—the first head was cut off close to the ears—the other two were taken off more cleanly;—it is better than the Oriental way—& (I should think) than the axe of our ancestors.—The pain seems little—& yet the effect to the spectator—& the preparation to the criminal—is very striking and chilling.

The first [execution] turned me quite hot and thirsty—& made me shake so that I could hardly hold the opera-glass (I was close—but *was determined to see—as one should see every thing once—with attention*) the second and third (which shows how dreadfully soon things grow indifferent) I am ashamed to say had no effect on me—as a horror—though I would have saved them if I could.[23]

It hardly suffices to respond to this text by observing primly that Byron was 'irresistibly drawn by the macabre and the horrific'. Rather (as has been better said), voyeurism was here being turned into a form of honest witness, avoiding cant and testing courage and manhood. One could not flinch at these realities; one needed to watch another's death in order to experience it vicariously, as one would one's own.[24] As Byron did this, however, we note that initially he too saw things as the vulnerable child might, before the adult mind began to intellectualize and censor. He saw the first if not the second and third execution in all its detailed specificity: in the child's word again, as '*horrible*'.

2. *A sanctioned curiosity*

The old (or primal) curiosity about scaffold horrors was never wholly retracted. Byron proves its longevity, and so do many witnesses much later. But after the mid-eighteenth century rationalizations began to encase the witnessing. Curiosity began to be justified by new physiological understandings that one had to *witness* others' suffering if one were to experience them by proxy and achieve the compassion which prompted benevolent responses. As in this way curiosity became a valued element in the sympathetic sensibility, it acquired an alibi which it retained into the 1830s, though by then not without challenge. The alibi might even have prompted polite Hanoverians in the third quarter of

[23] My emphases. Byron's relationship to the sight is like Géricault's when he watched and drew the decapitation of a Roman criminal in 1816–17 with 'a steadiness of heart in the face of an almost unbearable reality'. His quest for a charge of emotional energy for his work also explains his painting decaying and dismembered limbs in his studio in preparation for his *Raft of Medusa* (1819). Delacroix thought these sketches a 'truly sublime' demonstration of the way in which art transforms the odious: L. E. A. Eitner, *Géricault: His life and work* (1983), 180–4.

[24] L. A. Marchand, *Byron: A biography* (3 vols., 1957), ii. 694; F. Raphael, *Byron* (1982), 67, 96, 133.

the eighteenth century to attend executions and condemned cells more zealously than hitherto. Meditations upon and records of attendances became more copious than they had been at the beginning of the eighteenth century, let alone in Pepys's day, though this may only be a trick of recording.

We meet curiosity's rationalization in James Boswell, for example, when he explained the 'philosophical manner' in which he visited hangings: to such a man undecorated curiosity would no longer do. Sharing with Johnson a sense that death was 'immediately an evil', he made it clear that executions provided him with occasion for the doleful contemplation of his own mortality.[25] He said that his interest in Hackman's execution centred on the latter's 'prayer for the mercy of heaven'; it was after watching hangings, too, that he would meditate on human fallibility and free will. As he deeply said to Johnson after watching fifteen men hang outside Newgate one day in 1784 (fully accepting the law's entitlement to teach such lessons): 'I was sure that human life was not machinery, that is to say, a chain of fatality planned and directed by the Supreme Being, as it had in it so much wickedness and misery. . . . Were it machinery it would be better than it is in these respects, though less noble, as not being a system of moral government.'[26] Death was necessary for any man to ponder who 'directs his thoughts seriously towards futurity', he wrote elsewhere, for 'death is the most aweful object before every man'. From this it followed

> that I feel an irresistible impulse to be present at every execution, as I there behold the various effects of the near approach of death. . . . I cannot but mention in justification of myself, from a charge of cruelty in having gone so much formerly to see executions, that the curiosity which impels people to be present at such affecting scenes, is certainly *a proof of sensibility, not of callousness*.[27]

There was some special pleading here: Boswell knew how his attendances might be regarded by people more genteel than he. On the other hand he also knew he stood on firm ground. His generation took it for granted that information was transmitted to the mind through the vibration of 'nerves' (anatomy and physiology had taught them this); hence sensations communicated through sight, hearing, and touch were indispensable mediators between mind, body, and world. On this view,

[25] C. Weis and F. A. Pottle (eds.), *Boswell in extremes, 1776–8* (1971), 285–6, 291; J. Hagstrum, 'On Dr Johnson's fear of death', *English Literary History* (14), 308–19.

[26] *Life of Johnson*, ed. G. B. Hill and L. F. Powell (6 vols., 1934–50), iii. 383–4, iv. 329.

[27] Boswell thought that the fact 'that the greatest proportion of spectators is composed of women' proved his point: James Boswell, *The hypochondriack* (Mar. 1778), ed. M. Bailey (2 vols., Stanford, Calif., 1982), ii. 282.

the witnessing of others' suffering almost became a moral obligation, for just as the nerves of a dissected creature would be stimulated by touch, so appropriate stimuli would stir human benevolence. This was not a new idea. As Mandeville had written: 'To see people executed for crimes, if it is a great way off, moves us but little, in comparison to what it does when we are near enough to see the motion of the soul in their eyes, observe their fears and agonies, and are able to read the pangs of every feature in the face.' Whether at the scaffold, inside the condemned cell, in prostitutes' refuges, or at the Foundling Hospital, it was upon concrete witnessing that the practices of sympathy depended. As another writer urged: 'Attend these hospitals; examine the mournful cases that offer; see what pitiable objects appear; such dismal spectacles as would pierce the hardest heart: 'tis not in human nature to be insensible of so much human misery.' The scaffold too was a stage upon which human nature could be observed in the 'common course of the world'. As Mackenzie's *Man of feeling* (1771) put it: 'we delight in observing the effects of the stronger passions . . . for we are all philosophers in this respect; and it is perhaps among the spectators at Tyburn that the most genuine are to be found.'[28]

On the part of polite ladies and gentlemen by mid-century, these theories sanctioned four kinds of scaffold witnessing—none, however, innocent of primal impulses, whatever the cultural validations ostensibly attached to it. The most frequent practice was attendance at hangings, by some men many times more than once. The next was the visiting of the condemned. The third and most specialized (because expensive) was the collection of and contemplation of images of the hanged. The fourth was attendance at the public anatomization of executed bodies. Each survived as more or less acceptable pastimes until the 1830s at least, even though by then codes of decorum and evangelical earnestness were labelling them disreputable and presiding over the conflicts attended to in the next chapter.

In the first instances, at scaffolds or in condemned cells, some of misery's witnesses meditated conscientiously enough upon the meaning of what they contemplated; but they were probably rare cases. After Wordsworth and Coleridge attended the trial of John Hatfield in 1803, Dorothy persuaded Coleridge to visit Hatfield in his condemned cell, such visits still being possible. Fascinated, he found Hatfield '*vain*, a

[28] G. J. Barker-Benfield, *The culture of sensibility: Sex and society in eighteenth-century Britain* (Chicago, 1992), 6; A. J. Van Sant, *Eighteenth-century sensibility and the novel: The senses in social context* (Cambridge, 1993), 30, 56, 54.

hypocrite. It is not by mere Thought, I can understand this man.'[29] More often, however, 'alibi' is a cynically appropriate word in these contexts, for curiosity of the most elementary kind could shield behind these sanctions, and many without sacrifice of reputation were able to share kinship with Pepys in enjoying sensations of which they made no sentimental use at all. Henry Angelo hired a Paris window to watch a man broken on the wheel in 1775, a spectacle which in his memoirs he described vividly but without outward feeling. Everyone 'shuddered' at the dreadful shrieks, he wrote; but he still hurried off to the place de Grève to watch another execution—'*curious* to see the ceremony'. Again, just as Walpole in 1760 had justified his detailed description of Lord Ferrers's execution on the cheerful grounds that 'the man, the manners of the country, the justice of so great and curious a nation, all to me seem striking', so it was '*curious hankering*' that impelled Angelo to watch the executions and decapitations of the Cato Street conspirators in 1820: but he recorded no meditation upon the horror beneath his Newgate window.[30] Gentlemen and grandees like the poet-scholar Thomas Warton, the wit George Selwyn, or the duke of Montagu attended executions obsessively in the mid- to later eighteenth century. Selwyn went to Paris to see the would-be regicide Damiens pulled apart by horses, hacked by knives, and burnt in the place de Grève in 1757; and Montagu, Selwyn, and their circle were jokingly called 'the Hanging Committee' by their friends. An unnamed aristocrat, possibly Montagu, attended hangings in disguise. Henry Matthews watched a public decapitation in Rome in 1818 as phlegmatically as gentlemen on grand tours had watched such scenes two centuries before. Unlike Byron, he was apparently undisturbed by the spectacle, wondering only whether the startled expression of the face suggested a continuation of consciousness after decapitation.[31] James Curtis in the 1820s 'had been present at every execution in the metropolis and its neighbourhood for the last quarter of a century. For many years he had not only heard the condemned sermons preached in Newgate, but spent many hours in the gloomy cells with the persons who had been executed in London during that period'—with Fauntleroy, for example, and Corder the

[29] D. V. Erdman (ed.), *The collected works of Samuel Taylor Coleridge: Essays on his times in the* Morning Post *and the* Courier (1978), i. 416.

[30] Walpole, iii. 311; H. Angelo, *Reminiscences* (2 vols., 1828), ii. 181–2.

[31] Angelo, *Reminiscences*, ii. 362–5, 473–4; H. Matthews, *Diary of an invalid, being the journal of a tour . . . in the years 1817, 1818, and 1819* (3rd end., 1820), 265. Matthews's question is still debated. A letter in the *Guardian*, 5 Mar. 1993, thinks that the decapitated head was conscious of its fate for 20 seconds, so long as blood continues to feed the cortex.

Red Barn murderer, about whom he wrote the most comprehensive book.[32]

The second sanctioned practice was prison visiting, though not at all in the exhortatory spirit which evangelicals later brought to the condemned cells. 'You can't conceive the ridiculous rage there is of going to Newgate,' Walpole wrote when over several days some three thousand people crowded the condemned cells to gawp at the manacled gentleman-highwayman McLean before he hanged in 1750. Visiting on this scale is not (I think) recorded in earlier generations, though again we are at the mercy of inadequate sources. The modern fine lady 'weeps if but a handsome thief is hung', Soame Jenyns's poem put it that year; and over McLean they wept in buckets: 'some of the brightest eyes were . . . in tears', Walpole recorded. Here the *Beggar's Opera* bore more responsibility than the novels of sentiment did, for there was much affectation, which Walpole found droll. Lady Caroline Petersham and Miss Ashe were 'the chief personages who have been to comfort and weep over' McLean, he wrote; and the ladies' real interest in that once dangerous but now manacled and tamed manhood was not hidden to him: 'I call them Polly and Lucy and ask them if they did not sing "There I stand like a Turk with his doxies around",' he added, referring to Macheath's mistresses in the opera and Macheath's song before Tyburn. Nobody asked McLean and other felons what they thought about these inundations. Walpole was considerate: 'As I conclude he will suffer, and wish him no ill, I don't care to have his ideas, and am almost single in not having been to see him.'[33]

Thirdly, there was the commissioning of portraits of the condemned and the collection of relics. The duke of Roxburgh collected malefactors' portraits as he collected books. Sir James Thornhill painted a Newgate portrait of Jack Sheppard in 1724, Hogarth one of the murderess Sarah Malcolm in 1733 and another of Lord Lovat in 1746 on the eve of his execution, and Nathaniel Dance etched the murderess Elizabeth Brownrigg in 1767. Ever fashionable, Boswell was moved by 'a very curious whim' to commission a portrait of his condemned client John Reid in 1774. He insisted on having Reid's 'picture done while *under sentence of death*' the better to capture the saturated emotion of the moment.[34] Many of these portraits claimed pseudo-scientific alibis of their own—to explore theories of physiognomy popularized by Le Brun or later

[32] C. Hindley, *The life and times of James Catnach* (1878), 188.

[33] Walpole, ii. 230, 218–19 and n.

[34] W. K. Wimsatt and F. A. Pottle (eds.), *Boswell for the defence, 1769–1774* (1960), 296.

Lavater, for example. Contemplating Sarah Malcolm's countenance, Hogarth saw only 'by this woman's features, that she is capable of any wickedness'. Walpole purchased Hogarth's *Sarah Malcolm* and hung it without anxiety at Strawberry Hill.[35]

These practices lost some energy in later decades, but they were not likely to be wholly suspended in a century in which Madame Tussaud would travel to Roehampton to buy the murderer Daniel Good's clogs in 1842. Sir Thomas Lawrence drew the Ratcliffe Highway murderer, Williams, after his suicide in 1811; Robert Cruikshank drew Eliza Fenning in her condemned cell in 1815 (see Plate 37, p. 357 below); George Cruikshank drew George Matthews in his in 1819; and William Mulready drew the murderers Thurtell and Probert in 1824 and the Mannings in 1849. By the last date the taking of casts of the condemneds' heads had replaced portrait-collecting, this time with alibis in phrenology. Sir Thomas Lawrence twice requested permission to take a cast of Thurtell's skull before his hanging in 1824; casts of Thurtell's head and body were on public sale in the 1840s—the artist W. P. Frith used them in the academy in which he was trained—and Thurtell's effigy was exhibited in Madame Tussaud's well into the twentieth century, as we saw in Chapter 4. Newgate prison itself preserved shelf upon shelf of cast criminal heads; an *Illustrated London News* picture depicts an elegant bourgeois family of mother, father, and two daughters being shown them by the turnkey as late as 1873.[36]

Fourthly, and most graphically, the dissection of hanged bodies.[37] Some scepticism will be in order here, though since this practice evinced a fruitful empiricism stretching back to Vesalius, it would be silly to deny that there was more to it than the provision of just another alibi for curiosity. What anatomy had achieved both in medicine and in its deep effects upon cultural knowledge should not be underestimated. As one historian notes, by the eighteenth century 'knowledge of circulation and neural processes made internal function increasingly vivid and therefore increasingly available as a literal and metaphorical means of

[35] Walpole, ii. 104.

[36] Mary Cowling, *The artist as anthropologist: The representation of type and character in Victorian art* (Cambridge, 1989), 287, 302–3, and figs. 285, 291–6; Mayhew, i. 223. On George Matthews (his case was taken up by the solicitor James Harmer, as discussed in Ch. 16, below), see L. Binyon, *Catalogue of drawings in the British Museum* (1898–1907), i. 301.

[37] Until 1832 the only tolerated supply of corpses for research came from hangmen or body-snatchers (paupers thereafter). By the Murder Act of 1752, judges could order murderers' bodies for dissection as part of the sentence in court: P. Linebaugh, 'The Tyburn riot against the surgeons', in D. Hay *et al.* (eds.), *Albion's fatal tree: Crime and society in eighteenth-century England* (1975); R. Richardson, *Death, dissection, and the destitute* (1987).

describing interior experience'; whence anatomists' knowledge was hailed as 'the foundation of natural theology', and to learn the basics of the subject became the mark of the educated man and often woman.[38] Even so, it has to be said that in some anatomical displays a more primitive curiosity was never wholly gainsaid.

As public anatomy lessons became social events in many eighteenth-century towns, they often drew fashionable audiences with the lightest sense of what they observed. Even the chief practitioners were seldom innocent of displays which had less to do with the pursuit of knowledge than with medicine's chronic struggle for public esteem.[39] Hogarth was aware of their association with morbid or voyeuristic excitements when he depicted Tom Nero's dissection in *The four stages of cruelty* (1750–1): grotesque surgeons leer over a disembowelled murderer's corpse exempted from the reverence otherwise accorded to the body. Rowlandson knew it too, in his sketch of anatomists leering at a naked female corpse (see Plate 31, p. 265 below). And in the real world there were plenty of people like Dr Aldini, whom we can meet in his own account, getting to work on the corpse of a Newgate murderer in 1803. After shoving one electrode up Foster's rectum and another into his ear and then connecting his battery, he watched how the muscles got 'horribly contorted and the left eye actually opened'. As the shocks gave the corpse 'the appearance of reanimation', he concluded that 'the power of galvanism, as a stimulant, was stronger than any mechanical action whatever'. He was looking for a cure for asphyxiation.[40]

Equally mixed motives may be read in the anatomization of the Red Barn murderer William Corder after his hanging at Bury St Edmunds in 1828.[41] First his half-naked body was exposed to public view on a table in the Shire Hall. Since the county surgeon had already cut open and folded back the chest, it was not surprising that 'the anxiety of the people to gain admission to see the mangled body . . . was as intense as it had been at the trial and execution': ladies went as well as gentlemen. More serious work ensued before an audience of medical gentlemen and Cambridge students. As was usual by now, the ceremony began with the wiring of Corder's limbs to a battery to make them twitch. We

[38] Van Sant, *Sensibility and the novel*, 12; A. Desmond, *The politics of evolution: Morphology, medicine, and reform in radical London* (Chicago, 1989); Barker-Benfield, *Culture of sensibility*, ch. 1.

[39] P. Ariès, *The hour of our death* (1981), 364–9.

[40] J. Aldini, *Essai théorique et expérimental sur le galvanisme* (2 vols., Paris, 1804), ii. 38, 50–3; and his *General views of the application of Galvanism to medical purposes* (1819), 79–83.

[41] Anon., *The trial of William Corder for the murder of Maria Marten, in the Red Barn, at Polstead, including the matrimonial advertisement, and many other curious and important particulars* (3rd edn., 1828), p. 73; J. Curtis, *An authentic and faithful history of the mysterious murder of Maria Marten* (1828), 304, 312–15.

do not know what inferences were drawn from their twitching, because on this the report is silent. The ensuing voyage of discovery was then described by the surgeon, at first clinically enough:

The first step of dissection was to examine the parts of the sternum, and accurately to describe them to the gentlemen present, which, from the fine state of the subject, and his great muscularity, were well marked; the external and internal abdominal wings were exposed to view, as well as the fascias, &c. &c. . . . A quantity of serious fluid was effused into both sides of the chest, (about two or three ounces), and the lungs were gorged with blood.

But soon the older curiosity intervened. There ensued 'an interesting discussion' about the causes of death by hanging:

whether it was *suffocation* or *pressure* upon the spinal chord. From the circumstances of the chest and shoulders of Corder being observed *to heave* several minutes after the drop fell, it was generally admitted that death most probably took place from the latter cause. . . . It is to be regretted that the brain cannot be examined, as the determination of making a skeleton prevents any part of the bones being destroyed.

In lieu of the brain, you could always dissect personality. Phrenological examination duly revealed the organs of '*secretiveness, acquisitiveness, destructiveness, philoprogenitiveness*, and *imitativeness*'. It went without saying that 'the organs of *benevolence* and *veneration* are almost wanting' (just as it went without saying four years later that the murderer Cook's head was 'very curiously formed, the animal passions being strongly marked, while the nobler ones [were] quite contracted').[42] Then:

The bones of Corder having been cleared of the flesh, have been re-united by Mr S. Dalton, and the skeleton is now placed in the Suffolk General Hospital. A great portion of the skin has been tanned, and a gentleman connected with the hospital intends to have the Trial and Memoirs of Corder bound in it. The heart has been preserved in spirits.[43]

Perhaps some learning was achieved in this ritual, but science here was also of its time and culture, dehumanizing an already stigmatized body of particular vulnerability. *Pace* one interpretation of other medical autopsies, no 'humanitarian narrative' attached to this one.[44] The clinical report just cited is eloquent also in its wide dissemination into the lay culture, for it was published in the book of the trial and

[42] Mrs Lachlan, *The conversion . . . of James Cook* (1832), 249.

[43] *Bell's Life in London*, 24 May 1829, cited in Hindley, *Catnach*, 190.

[44] T. W. Laqueur, 'Bodies, details, and the humanitarian narrative', in L. Hunt (ed.), *The new cultural history* (Berkeley, Calif., 1989).

execution. Moreover, after Corder's dissection there was no scientific justification for exhibiting his pickled scalp (with one ear attached) at the shop of a leather-seller in Oxford Street.[45] And, while we are about it, what explains the fact that to this day the selfsame relic, along with the volume of the trial bound in his skin, is neatly exhibited in a glass case in the Bury St Edmunds museum—undeniably 'curious' spectacle though it is? Schoolchildren enjoy looking at it, and so did I.

It could be said that this chapter identifies cultures of extraordinary morbidity, since, to our self-censoring ways of feeling, the people encountered here look like obsessive and heartless voyeurs. Yet none felt an obligation to repudiate his or her basic impulses—as polite Victorians were going to. They had not yet learnt the social need to censor excited curiosity about the elemental associations execution unleashed, or the certain relishing which went with it. What later generations experienced as the psychic danger of unseemly spectacle was only beginning to be acknowledged by some. If neurosis is bred in repression, when acts inadmissible to consciousness are made unconscious, on this subject they were not repressed.

But by the turn of the century that world was fading. More tightly regulated emotions claimed authority as bourgeois affluence deepened, social anxieties intensified, and gentility and evangelicalism took hold. The old candid response was curiosity in its many guises. The new response allowed some imaginative identification with others' suffering, as benevolence advised; but it might equally be smothered in the earnestness of the newly pious and genteel. With curiosity outlawed, inner conflict was likely; and out of the conflict disgust, fear, denial, and shame might be hatched, or odder reactions besides. It was not that emotion was absent in those who denied their curiosity and turned away from watching. The contrary was true. You do not have to be psychoanalytically sophisticated to know that when polite people made jokes about hanging, or got erotic kicks from it, or professed not to know about it, they were not only distancing themselves from their more candid forebears: they were also blocking a process which had the power to disturb them profoundly.

[45] Rayner and Crook, v. 216.

CHAPTER 9

ANXIETY AND DEFENCE

1. *Decorum and repression*

IN 1813 THE NOVELIST MARIA EDGEWORTH PERTLY WROTE TO A friend that she was pleased to live in an age when newspapers 'give subjects of conversation in common to people in the most distant parts of the country'. She was particularly grateful that executions were one of the subjects, because execution reports exposed the variety of human character wonderfully. 'Do you recollect', she chatted gaily, 'our seeing in the news-paper an account of a man who poisoned several racehorses by putting arsenic in the water-troughs? Also an account of a horrid woman who made her servant boy murder her husband; and ordered him to go back and finish him as composedly as if she had told him to kill a pig?'

> When this man was brought out for execution, he was more cool and composed than any other person present. When the cord was to be put about his neck it was perceived that it was too short and that it could not well reach him. He pointed to the hassock on which he had been kneeling and made a sign to them to put it under his feet! The woman who murdered her husband was an extraordinary compound of insensibility and sensibility. While sentence was pronouncing upon her, she shewed no symptom of feeling, but from the moment she was condemned she would not let any creature see her face. She kept it constantly hid, even till she was executed. On hearing the sermon the evening before her execution, she wept and sobbed bitterly yet in the morning seemed quite insensible and eat [*sic*] a hearty breakfast.[1]

The old curiosity is alive and well in this letter. What had changed, however, is the relationship women like Edgeworth now bore to it, and also the new note being sounded which points a way forward. Audible in her excited chatter is an edge of anxiety; and it is fuelled, one feels, by approximation to a realm which is more dangerous and tabooed than

[1] C. Colvin (ed.), *Maria Edgeworth: Letters from England, 1813–1844* (Oxford, 1971), 41–2.

hitherto. The taboo is acted out in the fact that the locations of Edgeworth's witnessing had changed too. She depends on reported sensation rather than upon the direct witnessing of old. Where once ladies had been allowed to contemplate others' suffering in condemned cells or at Bedlam, the better to experience sympathy, their granddaughters were advised to distance themselves from such contaminations. Feminists like Mary Wollstonecraft in 1792 might in protest rehearse the old sympathetic argument: 'The world cannot be seen by an unmoved spectator; we must mix with the throng and feel as men feel, before we can judge of their feelings.' But she spoke against the tide. In some circles, already, the very name of Tyburn 'effectually kept at a distance those who assumed a character for decency as well as those who really possessed one'.[2]

There were continuities, as ever. As late as 1827 Macaulay echoed Horace Walpole's condescending amusement when the latter had described ladies' responses to the manacled highwayman McLean in 1750. Macaulay told his sister how ladies sitting in the courtroom he attended on circuit became excited when another handsome highwayman was tried: ' "Such a nice gentleman;—and if he did rob the folks, why, it is hard to hang such a pretty man for that. It is a shame, so it is, to hang a fellow-creature for a few sovereigns. They would not have such a thing to answer for, not if you would give them the whole world." Pretty creatures!'[3] But discontinuities intrude here too. Macaulay's ladies sat in court rather than visited the condemned cell as Walpole's did. Though still loosely sanctioned within the alliance of science and sensibility, curiosity was looking indelicate, thanks to new inhibitions and stricter codes of gentility.

Men felt the change too. In many there was a turning-away from the spectacle, or its conversion to other effects, as the gratification of curiosity began to cause more psychic troubles than it was worth. The rationalized curiosity which people like Boswell brought to the scaffold (and curiosity of simpler kinds too) was having to defend itself. To advanced people who had naturalized the body, moreover, the body-politic metaphors in terms of which violence against the diseased criminal limb had been anciently justified began to sound obsolete.[4] This focused

[2] Revd J. Richardson, *Recollections, political, literary, dramatic and miscellaneous, of the last half-century* (1856), i. 20.

[3] T. Pinney (ed.), *Selected letters of T. B. Macaulay* (Cambridge, 1982), 35.

[4] R. Porter, 'Bodies of thought: Thoughts about the body in eighteenth-century England', in J. H. Pittock and W. Wear (eds.), *Interpretation and cultural history* (1991), 92; R. McGowen, 'The body and punishment in eighteenth-century England', *Journal of Modern History*, 59 (Dec. 1987), 651–9.

attention on the hanged man's or woman's struggles in ways which separated later eighteenth-century visions of the punished body from those hitherto prevalent, and it made them more difficult to bear. By the 1800s evangelical earnestness and campaigns for the reformation of manners developed further taboos to protect the feelings, and squeamishness was an expression of them.

If we agree that, on this dread subject, curiosity was 'primally' tempting, all these responses must have entailed repression—'the precondition for the construction of symptoms', as we know.[5] Doubtless most people dealt with harsh punishment by appealing to its ethical and political soundness. But among others the need to *deny* what was involved in hanging—the choking, the kicking, the witnessed pain—intensified as it became more difficult *not* to think about the process in personalized and immanently sympathetic terms. The result was that the honest witnessing which someone like Byron still achieved became rare. The old curiosity began to be overlaid in shame, for example, or denied in squeamishness, 'learnt ignorance', gallows jokes, and odder manifestations besides. Responses like these check the easy generalizations historians still make about this generation's humanitarian inclinations.

Sir Joshua Reynolds's confusions sharply illustrate this era's dawning *shame* about the curiosity it brought to its own brutalities. In 1790 Boswell enticed Reynolds to go with him to watch Mrs Thrale's female servant hang for theft, along with four others. It happened that the poor woman recognized Reynolds from the scaffold, and she bowed to him, politely. It was not the poignancy of this that detained the great man, however. What concerned him next day was that newspapers found fault with his attending at all, itself a portent of taboos to come. 'It was natural, they said, for a man of Mr Boswell's character to visit such a scene, but in one of the elegant and refined mind of Sir Joshua Reynolds it was extraordinary.' This embarrassment elicited an awkward letter from Reynolds to Boswell which testifies very precisely to the battle propriety was beginning to wage with curiosity, and how it was resolved. In his tortuously protested apologia, with its questionable analogies and assumptions about the acceptability of a *polite* hanging, Reynolds's anxieties and shame are made wonderfully transparent:

> I am convinced it is a vulgar error, the opinion that [hanging] is so terrible a spectacle, or that it any way implies a hardness of heart or cruelty of

[5] Freud, *Works*, ed. J. Strachey *et al.*, xvi (1963), 294.

disposition, any more than such a disposition is implied in seeking delight from the representations of a tragedy. Such an execution as we saw, where there was no torture of the body or expression of agony of the mind, but where the criminals, on the contrary, appeared perfectly composed, without the least trembling, ready to speak and answer with civility and attention any question that was proposed, neither in a state of torpidity or insensibility, but grave and composed . . . [*here the sentence peters out*] I consider it is natural to desire to see such sights, and, if I may venture, to take delight in them, in order to stir and interest the mind, to give it some emotion, as moderate exercise is necessary for the body. . . . If the criminals had expressed great agony of mind, the spectators must infallibly sympathise; but so far was the fact from it, that you regard with admiration the serenity of their countenances and whole deportment.[6]

Among other habitual voyeurs at the feast we meet dawning confusion too. Boswell seldom hid his fascination with the scaffold from friends, but even he began to worry lest the world think it unhealthy. When he attended the execution of his friend the Revd James Hackman in 1779, newspaper reports that he had ridden to the scaffold in Hackman's coach elicited from him a flurry of denials and embarrassed explanations.[7] Similarly George Selwyn, wit, sinecurist, MP, and friend of Horace Walpole, was noted for his 'well known singularity, that he had a particular penchant for executions. Whether it arose from a principle of curiosity, or philosophy, it is perhaps difficult to determine; but so it was, that scarcely any great criminal was carried to the gallows, but George was a spectator.' When this obituary in the *Gentleman's Magazine* for 1791 repeated the story that Selwyn had gone to Paris to watch the would-be regicide Damiens tortured and disembowelled in 1757, it was another sign of changing times that a friend wrote to the journal to insist that Selwyn was 'one of the most tender and benevolent of hearts' who only once attended an execution (Damiens's), 'and that accidentally, from its lying in his way', and that his reputation for loving hangings derived only from a joke about that episode made by Lord Chesterfield. Selwyn's necrophiliac tastes were in fact well attested; but both the reputation and its denial speak for a sensitivity on the subject which Pepys would have found strange. Selwyn was said to have disguised himself in women's clothes at executions to avoid recognition, a conjunction of excitement, gender inversion, and shame which was symptomatic of anxiety, to put it mildly.[8]

[6] C. R. Leslie and T. Taylor, *Life and times of Sir Joshua Reynolds* (2 vols., 1865), ii. 588–9.

[7] Boswell, *Life of Johnson*, ed. G. B. Hill and L. F. Powell (6 vols., 1934–50), iii. 383–4, 532.

[8] *Gentleman's Magazine*, 1 (1791), 183, 299; J. H. Jesse, *George Selwyn and his contemporaries* (4 vols., 1843), i. 5–12; E. S. Roscoe and H. Clergue (eds.), *George Selwyn: His letters and his life* (1899), 16–18; S. P. Kerr, *George Selwyn and the wits* (1909), 122–30.

Some began to turn away from the spectacle because their imaginative empathy was strong. Commissioned by the duke of Roxburgh to sketch Governor Wall on the eve of his execution in 1802, J. T. Smith shrank from the task when he saw in the doomed man in his cell 'death's counterfeit, tall, shrivelled, and pale . . . his soul shot so piercingly through the port-holes of his head that the first glance of him nearly petrified me. I said in my heart, putting my pencil in my pocket, God forbid that I should disturb thy last moments!' After watching the hanging next morning Smith managed to have breakfast as usual, but 'with little or no appetite'.[9] In others, denials would multiply retrospectively as curiosity in youth had to be lied about. Both Hone and Cruikshank, evangelical and sober in middle age, would insist that they had seen those executions which influenced them in the 1810s only accidentally: 'an immense crowd . . . carried me along with them against my will', wrote Hone implausibly of his close witnessing of Eliza Fenning's execution in 1815.[10] Leap ahead to 1840, and 'the sight has left in my mind an extraordinary feeling of terror and shame', Thackeray writes after watching Courvoisier hang. Such was the transition in question here that Thackeray shut his eyes when Courvoisier fell.

Perhaps dawning anxieties were also projected into new forms of special pleading when the capital code was defended. In 1785 Paley advanced the political and moral justifications for capital punishment with unprecedented ambition, travelling a good distance from the uncomplicated vindictiveness of the early eighteenth-century pamphlet *Hanging not punishment enough*, or even from Fielding's writings on the subject. But his elaborate pleading was necessary because older certainties had become uneasy. Similarly it was not coincidental in this era that the grimier businesses of hanging began to be tidied up—with mechanized drops, the shifting of its location from Tyburn to Newgate in 1783, and eventually (the process was a long one) its concealment behind prison walls in 1868. It seems also to have become more necessary to insist on the 'otherness' of the hanged man or woman, as we shall see. Even while Hartleian materialism taught that all people were kin, the condemned were to be defined with increasing explicitness as social others (poor people and thieves) or political others (subversives or traitors) or psychological others (monsters, murderers); and you learnt not to look at them without shuddering. To see them as human was becoming helpfully difficult.

[9] J. T. Smith, *A book for a rainy day, or Recollections of the events of the years 1766–1833* (1845), ed. W. Whitten (1905), 179–80.

[10] See Ch. 13.

All this, to repeat, had costs. Some costs can be read in some of the darker fantasies into which once free-ranging but now repudiated curiosity took the newly inhibited men of the nineteenth century. There was, for example, the morbidity turned against the male self. Victorian doctors had themselves half-asphyxiated by assistants to find out what hanging was like, and showmen suspended themselves by the neck to display their control of the process. Some died, perhaps in ejaculatory experiments, like the 14-year-old Nottingham boy who after watching an execution wanted to 'know how hanging felt' and was found shortly afterwards hanging dead from a tree.[11] But the most revealing fantasies of the repressing male attached to the hanged woman.

To the male mind regressing freely into unchecked association, the bucking female body as it hanged could elicit obscene fantasies. We hear nothing of this from earlier centuries, because cultural or religious censorship was powerful, and such associations were repressed. The danger was still tacitly acknowledged, however, in the peculiar nature of the punishments inflicted on female felons. In early Europe women were often buried alive to avoid their exposure by hanging. Right to the end of the eighteenth century in England, similarly, traitresses were burnt rather than disembowelled. Blackstone said that this was out of the 'decency owing to their sex'. It was possibly a solution misconceived, since, as Michelet noted when he justified the French repudiation of the penalty of burning, 'the first flame to rise consumed the clothes, to reveal poor trembling nakedness'.[12]

By the end of the eighteenth century and into the nineteenth the secret gratifications some men might extract from female execution were cautiously admitted. Freud's notion that 'the regression of the libido without repression' is a source of 'perversion' has some point here.[13] Taboo and repression were intensifying so much that in some individuals they fractured, so that the shame of witnessing the dying or the dead female body became pleasurable in itself. Sadistic echoings were not accidental. Sadism itself was a product of this time, and of sensibility itself. It valued the release of emotion for its own sake, in the violence against and the shaming of another's body.[14]

Rowlandson's sardonic sketch (*c.*1780) of anatomists ogling and

[11] A. S. Taylor, *The principles and practice of medical jurisprudence* (4th edn., 1894), 35, 39; for another case, see *Examiner*, 17 Aug. 1831: 252.

[12] Camille Naish, *Death comes to the maiden: Sex and execution 1431–1933* (1991), 8, 82.

[13] *Works*, xvi. 344.

[14] N. Fiering, 'Irresistible compassion: An aspect of eighteenth-century sympathy and humanitarianism', *Journal of the History of Ideas*, 37 (1976), 213–14; P. Ariès, *The hour of our death* (1981), 370.

fondling the leg of a dissected and naked female corpse was an early comment on the fellowship which sexual prurience and death were henceforth to enjoy in male cultures (Plate 31).[15] In the Gothic novel or in the art of Fuseli and Etty death and macabre desire were likewise erotically bonded. The beautiful and desired corpse came to be encased in a morbidity which by the nineteenth century was in full flower. 'The death of a beautiful woman is, unquestionably, the most poetical subject in the world,' as Edgar Allan Poe wrote in 1846.[16] Dickens's disgust at public executions, even his capacity to enter empathetically into Fagin's despair in the condemned cell, did not stop his attending several

31. Thomas Rowlandson, *The dissection* c. 1775–82.

On the left of Rowlandson's drawing anatomists ogle a naked female corpse as they stretch out the leg. The limbs are only faintly sketched, as if Rowlandson sensed that this was dangerous ground. But his comment on the illicit impulses released by anatomization is clear.

[15] R. R. Wark, *Drawings by Thomas Rowlandson in the Huntington Collection* (San Marino, Calif., 1975), pl. 2; R. Paulson, *Rowlandson: A new interpretation* (1972), 17, 89.

[16] Ariès, *The hour of our death*, 374–95, and *Images of man and death* (Cambridge, Mass., 1985), 210–17; E. Bronfen, *Over her dead body: Death, femininity and the aesthetic* (Manchester, 1992), 59.

of the spectacles to reinforce his dislike of the watching crowd. He and four friends paid ten guineas for the use of a roof and back kitchen of a house overlooking the gallows upon which the Mannings were hanged in 1849, and though it was the crowd which shook him most, the image stayed with him of Mrs Manning's 'fine shape, so elaborately corseted and artfully dressed, that it was quite unchanged in its trim appearance, as it slowly swung from side to side'.[17] The 16-year-old Thomas Hardy in 1856 likewise watched a young murderess hanged outside Dorchester gaol. According to his second wife, the experience gave 'a tinge of bitterness and gloom to his life's work', and charged his fictional creation of Tess. In old age Hardy chiefly recalled 'what a fine figure she showed against the sky as she hung in the misty rain, and how the tight black silk gown set off her shape as she wheeled half-round and back'. Two years later Hardy watched from afar (through a telescope!) as another murderer hanged at Dorchester, using this image in *Desperate remedies*. Remarkable on this occasion, his biographer notes, was his 'almost total lack of horror at the hanging itself, and [at] the fate of the hanged man', absorbed as he was by his own reactions. In this capacity for narcissistic and eroticized projection, the bourgeois would continue to betray the older Adam, a being repudiated and besieged no doubt, but as alive as ever beneath veneers of respectability and right feeling.[18]

2. *Squeamishness and denial*

Dead people cannot be laid on couches; what profits it to diagnose them Vienna-style anyway? One reply is that historians project anachronistic and present-day constructions on to the past all the time: the past can hardly be seen otherwise. Another answer is that some projections are more fruitful than others; and when confronted with the strange behavioural patternings apparent in the present context, we may fairly apply one of the central claims in the psychoanalytical understanding of adaptational processes. This is that when the well-socialized conscience (the superego which Anna Freud termed the mischief-maker) is at odds with instinct, conflict tends to be controlled or modified by unconscious mechanisms of defence. Some defences are more efficient than others but none is without peril. All may distort the individual's relation to the

[17] P. Collins, *Dickens and crime* (2nd edn., 1964), 234–40.

[18] R. Gittings, *Young Thomas Hardy* (1975), 57–61. For the disconcerting view that the witnessing of a woman's execution was 'a male rite of passage, a violent, erotic experience that wrests a boy from childhood', see B. Kalikoff, 'The execution of Tess d'Urberville at Wintoncester', in W. B. Thesing (ed.), *Executions and the British experience from the 17th to the 20th century* (Jefferson, NC, 1990), 114.

reality perceived by other people. Moreover, most are ways of coming to terms with what is rather than of conceiving what might be: they are sources of 'resistance' again. Defences take many forms, but the two most easily discerned in relation to our subject in this period were denial and reaction formation (in gallows humour for instance).[19]

Denial is the easy one. In relation to execution, the most developed form of denial was *squeamishness*. Some historians interested in the relationship between punishment and sensibility have confused squeamishness with sympathy or empathy. Squeamishness was no less an expression of civility than these latter emotions were, and it has been witnessed in all eras, and not only ours, among élites who announced their status by lowering their thresholds of disgust. But the reaction must not be misunderstood. It was not an empathetic threshold that was lowered. Empathy and sympathy are democratic emotions, extending their generous warmth to all.[20] Squeamishness by contrast refuses to accept the pain which sympathetic engagement threatens. It denies material reality or others' emotions and blocks the echoes of these within the self. It is a colder, more distanced, more aesthetic emotion, defensively fastidious in the face of the rude and the unsightly.

Squeamishness was never better witnessed than in the hardening response to the gibbeting of murderers' corpses, a common spectacle on the commons and highways of eighteenth-century England. The 1752 Murder Act had regularized ancient practice by permitting judges to order murderers' gibbeting as part of the sentence in court. On Hounslow Heath up to a hundred gibbets were said to have stood in the 1770s, 'so that from whatever quarter the wind blew, it brought with it a

[19] *Repression* stops a forbidden wish, idea, or anxiety from becoming conscious. In *denial*, the conscious individual disowns a part of him- or herself. Through *projection*, that repudiated part can be attributed to others, who thus bear the individual's guilt. Through *regression*, the individual reverts to the gratification techniques of an earlier psychosexual stage to compensate for current anxiety. *Reaction formation* converts anxiety about aggression into its opposite. *Rationalization* entails the intellectual explanation provided for an emotionally determined response. *Sublimation* is said to be the most mature form of defence, when energy is diverted, sometimes highly productively, into other contexts. Through *introjection*, the self takes into itself what it perceives in the object of its desires and wishes. For elaboration on these (she discerns 10 defence mechanisms), see Anna Freud, *The ego and the mechanisms of defence* (Madison, Conn., 1966), esp. 42–4, 55, and *passim*.

[20] Empathy may be defined as 'the understanding and sharing of another person's emotional experience'. It is distinguished from sympathy ('feeling with') by expressing a stronger identification with the object of it: R. Harré and R. Lamb (eds.), *The dictionary of personality and social psychology* (Oxford, 1986), 91–2. *Empathy* in the present context is an anachronistic word, coined in English only in 1904 (*OED*) to translate the *Einfühlung* which, in German aesthetics, described the process of 'feeling into' the emotions expressed in aesthetic objects or natural scenes. For subsequent usages, see L. Wispé, 'History of the concept of empathy', in N. Eisenberg and J. Strayer (eds.), *Empathy and its development* (Cambridge, 1987), 17–37.

cadaverous and pestilential odour'.[21] Pepys had been disgusted by a gibbeted body long before this; but it remained a popular spectacle, a place of licensed *frisson*, and protests from squeamishness before 1800 are difficult to find. London was said to have been a deserted city on the Sunday following Lewis Avershaw's gibbeting on Wimbledon Common in 1795; for several months his decomposing body provided a favourite Sunday outing.[22] Southey recorded a gibbeting in Stourbridge in 1812 which drew 'more than 100,000 people . . . and a kind of wake continued for some weeks for ale and gingerbread'.[23]

Fastidious feeling was moving against the practice, however. Southey thought gibbeting an offence to 'public feeling, and public decency'. By the time Best J. ordered a murderer to be hanged in chains on the high road to Brigg in Lincolnshire in 1827, local inhabitants' petitions prevented it.[24] We get a sense of what was afoot here from a letter one William Sykes wrote to the home secretary to protest at the gibbeting of men executed under Admiralty jurisdiction along the banks of the Thames. He did pity 'the scare-crow remains of the poor wretches who long since expiated by death their crimes. . . . It is said that "persecution ceases in the grave". Let these poor remains find a grave, and the remembrance of their offences pass away.' But aesthetic repugnance was his dominant feeling. If the sight was a 'sad' one, it was more audibly 'a disgrace and nuisance', 'revolting, disgusting . . . dishonourable to the law's omnipotence, and discreditable to the administrators of the law'. It 'excites feelings of disgust in the breasts of numerous travellers to Ramsgate, Margate, France, the Netherlands, &c. &c. . . . I have heard many ladies anxiously inquire if the boats had passed the gibbets, and not until then would they come upon deck.'[25]

The two last men gibbeted, both in 1832, were a Jarrow collier, William Jobling, for murdering a colliery owner, and a bookbinder, James Cook, for a gruesome murder in Leicestershire. The vulgar still flocked to the spectacle. Twenty thousand people watched Cook's crumpled body hoisted in its cage on a gibbet thirty-three feet high, its

[21] R. Southey, *Letters from England* (orig. pub. 1807; 1951), 45.

[22] The Bow Street runner Townsend, examined by the 1816 select committee on the police of the metropolis, claimed the credit for catching Avershaw and for choosing the site of his gibbet, the only difficulty being that it was 'so near Lord Spencer's house'. But 'if there was a person ever went to see that man hanging, I am sure there was a hundred thousand'. Townsend thought that gibbeting was helpful in keeping the crime and the punishment permanently in people's minds: W. Hone, *The table book* (1827–8), ii. 150.

[23] Radzinowicz, i. 218–19 and n.

[24] Southey, *Letters from England*, 45; A. Hartshorn, *Hanging in chains* (1891), 107.

[25] HO44.14, fo. 87.

'face covered with a pitch plaster, and over it . . . the cap he suffered in': 'An immense concourse of spectators lined the road to the gibbet, and several ranting preachers were busily engaged the whole of the day, in singing, praying, and preaching. The saints were also diligently employed in giving tracts away, and the place throughout the day seemed like a fair.'[26] But times had changed. Jobling's body was removed from its gibbet by his fellow colliers and given a decent burial. Cook's body had to be removed pre-emptively by order of the home secretary. That spelt the end of the punishment. When in parliament in 1834 Ewart moved to abolish gibbeting, he called it 'an odious practice', and Lord Suffield agreed that it was 'unsuited to the present state of public feeling'. The most interesting implication in the speeches was that gibbeting was no longer understood. Its 'only effect', Lord Suffield declared, 'was that of scaring children, and brutalizing the minds of the people. It could produce no moral effect whatever.'[27]

In England as in western Europe most expressions of élite repugnance at execution turn out to be of these self-distancing as distinct from sympathetically self-engaging kinds.[28] George IV's often-proclaimed humanity on behalf of the condemned co-existed with a stunning insensitivity to their terror as they awaited the outcome of their pleas for mercy. He too pitied them ('poor men'); but his distaste was chiefly about the personal contamination entailed in an offensive ritual undertaken in his name.[29] This posture was common to all ensuing generations otherwise pleased with their humanity: capital punishment was becoming the more tolerable as it was removed from sight.

We shall see more fully in the last chapter that all this had obvious bearing on the abolition of public execution in 1868. Aversion to the scaffold crowd was the chief reason for abolition, but the higher squeamishness played its part too. At executions 'a new sheriff occasionally almost faints', it was reported in 1866. One sheriff declared that he had never seen a hanging before he was elected, and he decided never to

[26] Mrs Lachlan, *Narrative of the conversion . . . of James Cook, the murderer of Mrs Paas* (1832), 249–52; T. Frost, 'The last gibbet', in W. Andrews (ed.), *Bygone Leicestershire* (1892), 202–3.

[27] PD3 xxii (1834), c. 155, and PD3 xxiv, c. 1221–2.

[28] Stone scaffolds might well have been removed from public places in Holland, especially near fine houses, but only because they embarrassed residents of quality. Damiens's execution in Paris in 1757 made Louis XV spend the day in secluded melancholy and prayer; he none the less 'signed the sentence,—and then desired he might not be told when it was to be executed,—because it would hurt his feelings!' as Southey wryly exclaimed. On the same occasion, the comte Dufort de Cheverny absented himself and his family from Paris, insisting that, on his return, the execution should not be referred to: Southey, *Letters from England*, 220; J. McManners, *Death and the Enlightenment: Changing attitudes to death in eighteenth-century France* (Oxford, 1981), 383–4.

[29] See Ch. 20.

attend one without having sat out a likely victim's trial and hearing him confess the crime: 'If a woman had been hanged I should not have gone to the execution.'[30] In the event the squeamish defence was well achieved, and in many countries it still is. Hanging people behind prison walls was to serve the law's purpose adequately for almost a century after 1868. As Mill put it in 1836, it was an effect of civilization 'that the spectacle, and even the very idea, of pain, is kept more and more out of sight of those classes who enjoy in their fulness the benefits of civilization'.[31] Mill approved of this process, since he thought it indicated the amiability and humanity of the 'more opulent classes'. But this was sadly to misconstrue its purposes and effects.

Squeamishness was the denial invoked when directly evasive strategies failed and a dangerous reality insisted on claiming attention. But denial had its freest range when reality could be evaded altogether. Peter Gay has discerned in Victorian attitudes to sex the posture he terms 'learnt ignorance'—that 'ignorance unconsciously desired, informally imparted, and assiduously fostered' which served as an 'unplanned, if highly adaptive defence' against socially dangerous feelings.[32] By the nineteenth century there was an equivalently learnt ignorance about hanging people. Although they were far from exclusively adept at it, women reveal the strategy most clearly. Among women, feelings about the hanging body as about the sexual body were blocked with exceptional efficiency. Social prescription now firmly insisted that respectable women put a distance between themselves and the unsightly and improper.

It was possible to strike a compromise between convention and the old curiosity. This was what Maria Edgeworth achieved with her newspapers. It was also what the Norwich Quaker novelist Amelia Opie achieved when she released her fantasies in a wonderfully displaced fascination with assizes. As a girl in the 1770s she had been dazzled by the assize procession. When she found that ladies were allowed to attend trials or causes, she attended the civil court, and when once the judge let her sit on the bench next to him, a lifelong addiction began. 'All was new, exciting, and interesting.' Judges began to reserve a place for her and to admit her into their private rooms or coaches. She was still attending civil courts in her seventies, sitting there demurely without

[30] RC 1866: 149, 222.

[31] J. S. Mill, 'Civilization' (1836), in J. M. Robson (ed.), *Collected works of John Stuart Mill* (Toronto, 1977), xviii. 130–1.

[32] P. Gay, *The bourgeois experience, Victoria to Freud*, i. *The education of the senses* (1984), 280–1.

refreshment for two days in succession. The assize week had for her 'attraction of an intellectual kind', she wrote. It fed her 'love of excitement'. 'A court of justice may be likened to a stage', but 'the emotions we behold are real, not acted . . . [They] are nature itself!' The limits are met, however, in the ladylike distance she kept from criminal business. 'I could not bear the thought of hearing prisoners tried, as the punishment of death was then in all its force.' She went to a murder trial only when assured that there would be an acquittal on grounds of insanity—otherwise 'I should have fled instantly'. She never attended an execution. The ringing of the bells before a hanging horrified her.[33]

Direct association with execution had become rare for gentlewomen. Right until the end of public executions newspapers questioned the characters of the many women who did attend. As genteel people retreated into more select conventicles of leisure, the assizes and quarter sessions acquired a mainly social meaning.[34] Ladies who looked forward to assize concerts, plays, and balls to flesh out the fashionable year were determinedly innocent about the grim rituals these festivities preceded. People 'look upon these meetings as a sort of family party', a French observer noted in 1819.[35] 'On Monday our Assizes are to begin,' Lady Roche had written cheerily in 1783: 'I know not who is to be hanged, but the ladies are making great preparations for dressing and dancing.'[36]

County bigwigs, magistrates, and sheriffs also experienced quarter sessions and assizes as convivial occasions. They went to assize or sessions dinners dressed up in court dress—'blue velvet coat, with huge cut steel buttons, bay for the hair, embroidered waistcoat, black satin breeches and appendages to match', as the Gloucester magistrate the Revd Francis Witts noted; the 'custom of dining together and passing the evening in each other's company tends much to maintain . . . friendly harmony and mutual good intelligence'. Blood-curdling assize sermons gave these occasions their political alibi, Witts enjoying one in 1829 'against the pernicious liberalism of the times'. Occasionally these people might be caught unawares by the drama or pathos of trials, their

[33] C. Brightwell, *Memorials of the life of Amelia Opie, selected and arranged from her letters, diaries, and other manuscripts* (2nd edn., Norwich, 1854), 23–5, 337, 355, 357, 385.

[34] P. Borsay, *The English urban renaissance* (Oxford, 1989), 143–4, 155, and ch. 6.

[35] M. Cottu, *On the administration of the criminal code in England, and the spirit of the English government*, in *Pamphleteer*, 16/31 (1820), 7.

[36] Lady Roche to Scrope Bernard, 18 Apr. 1783: N. Higgins (ed.), *The Bernards of Abington and Nether Winchenden: A family history* (1904), iii. 53–4. The Salisbury assizes of 1763 were celebrated by 'two exceedingly good balls', at which the sheriff was in such spirits that afterwards he 'could not get to his lodgings [without] some supporters, and Lord Seymour Webb . . . walked into the dirtiest ditch . . . and came out the most dismal figure': Borsay, *English urban renaissance*, 155.

personal defences outflanked. After watching a murder trial and conviction in 1835, Witts had to walk about for half an hour before turning to other business 'to calm the agitation of my spirits'.[37] But for the most part the festive note continued to sound out in its noisy denial of reality. As Haydon the painter wrote in disgust on the day Daniel Good was executed in 1842:

the Whole Town is buried in dissipation: routs, Masques, balls, Fetes, & Idleness. They are half intoxicated, every body is cracked, nobody reads, reflects, or thinks, and in the midst of all this reeling splendor, a Murderer is hung, declaring his innocence to the last! It seems as if God let Mankind alone sometimes after having settled general principles.[38]

3. *Darwin's dream and gallows wit*

In 1838 Charles Darwin recorded a dream in his private notebook. Some have interpreted it as expressing his fears that commitment to the theory of natural selection would destroy his scientific reputation, or that his imminent marriage would diminish the energy he could give to science. These are the threats he is said to have escaped in the dream.[39] But this addresses only the dream's manifest content and ignores its central symbol. The dream bore anxieties which had nothing to do with career or marriage. It also bore the key defence against anxiety, which is of interest to us next. It took the form of 'wit':

Was witty in a dream in a confused manner. thought that a person was hung & came to life, & then made many jokes. about not having run away &c having faced death like a hero, & then I had some confused idea of showing scar behind (instead of front) (having changed hanging into his head cut off) as kind of wit, showing he had honourable wounds.—all this was kind of wit.—I changed I believe from hanging to head cut off [there was the feeling of banter & joking] because the whole train of Dr Monro experiment about hanging came before me showing impossibility of person recovering from hanging on account of blood. but all these ideas came one after other, without ever comparing them. I neither doubted them or *believed* them.—Believing consists in the comparison of ideas, connected with judgement.

[What is the Philosophy of Shame and Blushing?][40]

[37] D. Verey (ed.), *The diary of a Cotswold parson* (1978), 180, 34, 79, 107.

[38] W. B. Pope (ed.), *The diary of Benjamin Robert Haydon* (5 vols., Cambridge, Mass., 1963), v (24 May 1842).

[39] H. E. Gruber, *Darwin on man: A psychological study of scientific creativity* (1974), 43; R. Colp, 'Charles Darwin's dream of his double execution', *Journal of Psychohistory*, 13/3 (1986), 287–9; A. Desmond and J. Moore, *Darwin* (1991), 263–4.

[40] P. Barrett *et al.* (eds.), *Charles Darwin's notebooks, 1836–1844: Geology, transmutation of species, metaphysical enquiries* (Cambridge, 1987), 555–6.

The source of the execution image here is no mystery. Darwin was probably too squeamish ever to watch a real-life execution. But people were hanged opposite Shrewsbury School when he was a boy, and Cambridge was much excited by a hanging while he was an undergraduate there in 1829.[41] In particular the reality of hanging was forced upon him when, aged 16, he entered Edinburgh University and attended Alexander Monro's anatomy lectures. The lectures were 'as dull, as he [Monro] was himself', Darwin reported. The truth was, however, that he was something other than bored: he was disgusted. Monro presided over his students 'bloody and begrimed', he wrote: 'I dislike him & his lectures so much that I cannot speak with decency about them. He is so dirty in person & actions.'[42] The subject of anatomy also 'disgusted' Darwin, and he fled from operating theatres if he could.[43] And since Monro drew on the lecture notes of his father who had preceded him in the anatomy chair, Darwin was exposed to one of that father's more memorable preoccupations. In the 1770s the elder Monro had speculated about the possibility of resuscitating hanged bodies, and both father and son experimented on the subject.[44] Darwin's disgust was compounded by the fact that anatomy schools depended on murderers' hanged bodies. These associations provided material enough to fuel the gallows image in the dream a dozen years later, and in essence it was not a witty one at all. Why Darwin needed to keep his distance from it is no mystery either.

Darwin attributed his lifelong horror of others' pain to the 'humane influence' of his sisters. In fact the suffering encapsulated in the gallows image was his own, blocked since childhood. His mother had died appallingly of a tumour when he was 8. Her children thereafter never mentioned her name, and as an adult Darwin admitted he could barely remember her.[45] What he did remember was his displaced trauma a month after his mother's burial when he watched the burial of a soldier outside his school. It was this denial of her death and his transposition of grief on to another representation of it that was determining. From that period on he was always squeamish about blood, and horror stories

[41] Desmond and Moore, *Darwin*, 69.

[42] F. Burkhardt and S. Smith (eds.), *The correspondence of Charles Darwin*, i. *1821–1836* (Cambridge, 1985), 25.

[43] G. de Beer (ed.), *Autobiographies: Charles Darwin; T. H. Huxley* (Oxford, 1974), 4, 11, 13, 25.

[44] Boswell turned for help to the elder Monro when he planned to resuscitate his hanged client Reid in 1774: see Ch. 10. For the younger Monro's reflections on how hanged people died (by loss of air rather than by apoplexy, he decided), see A. Monro, *Essays and heads of lectures on anatomy [etc.] . . . by the late Alexander Monro, secundus, MD* (Edinburgh, 1840), pp. xliv–xlv, 97.

[45] Desmond and Moore, *Darwin*, 14.

terrified him. He acquired, he wrote, a 'keen instinct against death'. 'Fear must be simple instinctive feeling,' he noted shortly before the dream: 'I have awakened in the night being slightly unwell & felt so much afraid though my reason was laughing & told me there was nothing, & tried to seize hold of objects to be frightened at.'[46]

So Darwin's dream-defence against his fears becomes clear. After condensation into the execution symbol, disgust/fear was converted into a refusal to accept the hanging and into a sense of triumph in evading it. Most strikingly there is the subordination of disgust and fear to the defensive reaction which he experienced as '*wit*'. Evading the noose and then noting the absurdity of the evasion *amused* him. A repudiated fear or excitement was converted into its opposite. The fact that Darwin's free associations then immediately led him into speculation about shame and blushing was consciously related to his current psychological speculations, but this too suggests a congestion of feelings beyond consciousness.

Reaction formations were present in many of the defences against scaffold death noted hitherto, but gallows humour—'wit'—was among the most characteristic. Some contemporaries had a fair idea of its components. While watching Courvoisier hang in 1840, Thackeray found it 'curious that a murder is a great inspirer of jokes. . . . We all like to laugh and have our fling about it; there is a certain grim pleasure in the circumstance—a perpetual jingling antithesis between life and death.'[47] Charles Lamb in 1811 noted this too, though he took a literal view of it. He thought that the 'general levity' with which the gallows comedy was treated was elicited by 'the absurd posture into which a man is thrown who is condemned to dance . . . upon nothing'. Execution by beheading would never be a matter for jest, but hanging a man invited 'vulgar jokes':

> To see him whisking and wavering in the air . . . to behold the vacant carcase, from which the life is newly dislodged, shifting between earth and heaven, the sport of every gust: like a weathercock, serving to show from which point the wind blows; like a maukin, fit only to scare away birds; like a nest left to swing upon a bough when the bird is flown: these are uses to which we cannot without a mixture of spleen and contempt behold the human carcase reduced. We string up dogs, foxes, bats, moles, weasels. Man surely deserves a steadier death.

[46] Barrett *et al.* (eds.), *Darwin's notebooks*, 532; Colp, 'Darwin's dream', 284.

[47] W. M. Thackeray, 'Going to see a man hanged', *Fraser's Magazine*, 22 (1840), 150.

Absurdity was compounded by 'the senseless costume with which old prescription has thought fit to clothe the exit of malefactors in this country': 'Let a man do what he will to abstract from his imagination all idea of the whimsical, something of it will come across him when he contemplates the figure of a fellow-creature in the day-time (in however distressing a situation) in a night-cap.'[48] Hackman had felt the same when he watched Dr Dodd mount the scaffold in 1777:

> The preparation of the unhappy victim mixed something disagreeably ludicrous with the solemnity. The tenderest could not but feel it, though they might be sorry that they *did* feel it. The poor man's wig was to be taken off, and the night-cap brought for the purpose was too little, and could not be pulled on without force. . . . Every guinea in my pocket would I have given, that he had not worn a wig, or that (wearing one) the cap had been bigger.[49]

The basis of much humour in this and earlier eras is often thought to have been an uncensored relish for others' misfortunes. Hobbes had believed that laughter was born in the 'sudden glory' felt at the sufferings and infirmities of others.[50] Doubtless in some people gallows humour was still of this freely relished kind, as when an early Swiss observer at Tyburn thought that 'a man cannot very well forbear laughing to see these rogues set themselves off as heroes by an affectation of despising death', a thing which we do not find laughable at all. But Lamb's sense that *spleen and contempt* attached to the incongruity hints at the need to find a more complex reading in an age of deepening inhibition; for these terms acknowledged the tension, cruelty, and pity at the comedy's centre when its unbearable elements were displaced into laughter. Nowadays we might rephrase the insight but leave it intact. Humour is one 'among the great series of methods which the human mind has constructed to evade the compulsion to suffer'; it displaces the true feeling through 'the censorship of conscious thinking'.[51]

It is not difficult to discern the evasive functions of gallows wit among gentlefolk, nor the significance of its near-precise social location among those in whose interests gallows deaths were inflicted. Plebeians had their own gallows humour, but it confronted that machine

[48] 'On the inconveniences resulting from being hanged' (1811), in P. Fitzgerald (ed.), *The letters and writings of Charles Lamb* (1903), iv. 341–52.

[49] Borrow, v. 20.

[50] R. K. Simon, *The labyrinth of the comic: Theory and practice from Fielding to Freud* (Tallahassee, Fla., 1985), 130–1.

[51] *Jokes and their relation to the unconscious*, ed. J. Strachey (1960), 171, 228–35; cf. Freud, 'On humour', in *Works*, xxi. 162–3.

ironically, not facetiously.[52] As Tyburn ballads show, the knowing irony of plebeian wit was born of a more intimate experience of hanging. Plebeian irony was usually embedded within meaning-reversals,[53] as in the sardonic terms for hanging which Linebaugh notes:

> A hanging day was a 'hanging match', a 'collar day', the 'Sheriff's Ball', a 'hanging fair' or the Paddington Fair. To hang, like a dance, was 'to swing', to 'dance the Paddington frisk', 'to morris'. It was 'to go west', 'to ride up Holborn hill', 'to dangle in the Sheriff's picture frame', 'to cry cockles'. . . . Hanging was to have a wry mouth and a pissen pair of breeches; it was to loll out one's tongue at the company. A man hanged will piss when he cannot whistle.[54]

If the jollity of the smart sets in clubland was never as open as this, it was because they had more to be anxious about: the killings were for or by them. So they rang the changes on breezy bluffness or painful facetiousness, with a kind of nervous bantering in between. The over-protested insensitivity of the wit which ensued can be breathtaking. 'Harrington's porter was condemned yesterday,' a friend wrote jauntily to George Selwyn in 1757: 'Cadogan and I have already bespoken places at the Braziers, and I hope Parson Digby will come time enough to be of the party. I presume we shall have your honour's company, if your stomach is not too squeamish for a single serving.' Because Selwyn went to Paris for Damiens's execution instead, Henry St John wished him 'good sport at the place de Grève, to make up for losing the sight of so notorious a villain as Lady Harrington's porter'. He added that Selwyn's friends had had a full view of this man 'as he went to the gallows with a white cockade in his hat, as an emblem of his whole innocence'. 'The dog died game,' another of Selwyn's correspondents reported. He 'went in the cart in a blue and gold frock . . . and had a white cockade. He ate several oranges in his passage, inquired if his hearse was ready, and then . . . was launched into eternity.'[55] This breezy tradition still flourished sixty years later. When Henry Angelo booked his half-crown window in 1820 to watch the hanging and decapi-

[52] As in the colloquial term 'jarvey' for a hackney-coach driver, 'a compliement paid to the class in consequence of one of them named Jarvis having been hanged': W. Ballantine, *Some experiences of a barrister's life* (5th edn., 1882), 18.

[53] As in the one awful gallows joke that Freud cited: 'Well, this week's beginning nicely,' said the rogue on his way to execution: Freud, *Jokes*, 229.

[54] In Grose's *Classical dictionary of the vulgar tongue* (1785), only money and the sexual parts were given more synonyms than hanging: P. Linebaugh, 'The Tyburn riot against the surgeons', in D. Hay *et al.* (eds.), *Albion's fatal tree: Crime and society in eighteenth-century England* (1975), 66.

[55] Roscoe and Clergue (eds.), *Selwyn: Letters and life*, 17; Jesse, *Selwyn and his contemporaries*, i. 345, 354–5.

tation of the Cato Street conspirators next morning, the gentlemen already there welcomed him cordially—'Hail fellow well met! We'll rough it together as if in a Margate steam-boat'—and he played cards with them until breakfast at six in the morning.[56] In 1840 Monckton Milnes prepared for Courvoisier's execution by spending the previous night in his club cracking gallows jokes. And it was in such locales in the 1840s that the cant parodies of Tyburn balladry took hold:

> The faking boy to the crap is gone,
> At the nibbing chit you'll find him;
> The hempen cord they have girded on,
> And his elbows pinned behind him.
>
> . . . .
>
> The bolt it fell—a jerk, a strain!
> The sheriffs fled asunder;
> The faking boy ne'er spoke again,
> For they pulled his legs from under.[57]

It was all wonderfully insensitive, but helpfully so; for audible in many such exchanges was what was *not* said and could not be said, the history of a silence. It was in their key omissions that the pleasurable danger of these exchanges resided. The more closely utterance drew towards the unspeakable, the more likely it was to convert tension into facetiousness. Sir Peter Laurie caught the tone when he described the patented scaffold he introduced as sheriff in 1824:

> Between the time of my improvement in the hanging-machine, for which I received my Knighthood—(one man's meat, ye see, is another man's poison) being approved of, and its being made use of on the first malefactor, I never slept for terror. I had constantly before my mind the fate of the Regent Morton, who invented the Maiden, and was the first to suffer by it; of Dr Guillotine, who introduced his namesake into France, and was the first upon whom the experiment was tried. . . . Never did bride wish for the bridal machine with greater eagerness than I longed to see a man fairly hanged and dead on my improved stretcher. . . . I am now safe, and my mind is at rest. By being instrumental in the exaltation of others, I have exalted myself. The machine has not had much work since I have been Mayor. Had it not been for the squeamish humanity of the Parliament, Cootes [?] would to a dead certainty have exhibited the efficacy of my improvement.[58]

It is unlikely that these postures are found before the early eighteenth century. Their release depended on the erosion of cosmological under-

[56] H. Angelo, *Reminiscences* (2 vols., 1828), ii. 181–4.

[57] 'Flowers of hemp, or The Newgate garland', *Tait's Edinburgh Magazine*, 8 (1841), 223.

[58] P. Laurie, *Maxims of Sir Peter Laurie . . . lord mayor of London in the year 1833* (1833), 38.

standings of execution, but also on the fuller realization of the physical pain of the secularized body. In this sense they acknowledged the impact of sensibility backhandedly but clearly enough.

How long could such facetiousness last? Silly people continued their banter at the condemneds' expense throughout the nineteenth century and beyond, and empire's burdens sanctioned its more open release. The unspeakable Baden-Powell, ever-anxious 'to see the fun' when 'natives' were executed—and executing them himself when he could—provides ripe late examples.[59] But on the whole Victorian earnestness announced that this humorous mode was as unseemly as Christian prescription had declared it to be in earlier eras. So we hear less of it by the mid-nineteenth century. By then the fast set could be called to account for their vacuity. One of the Revd R. H. Barham's *Ingoldsby legends* wrote their epitaph parodically in 1840. The bored Lord Tomnoddy learns delightedly that there is to be a hanging next day:[60]

> to see a man swing
> At the end of a string,
> With his neck in a noose, will be quite a new thing!

Hiring a floor at the Magpie and Stump at Newgate, he invites his friends to a festive night in anticipation. They play comic games, cork their noses black, get drunk: but alas they fall asleep under the table, so they miss the moment of execution (feelingly described: 'God! 'tis a fearsome thing to see / That pale, wan man's mute agony,— / The glare of that wild, despairing eye, / Now bent on the crowd, now turn'd to the sky'). When they wake up,

> They stared at each other, as much as to say,
> 'Hollo! Hollo!
> Here's a rum Go!
> Why, Captain!—my Lord!—Here's the devil to pay!
>
> The fellow's been cut down and taken away!
>
> What's to be done?
> We've miss'd all the fun! . . .'
> What was to be done?—'twas perfectly plain
> That they could not well hang the man over again:
>
> What was to be done?—The man was dead!
> Naught could be done—naught could be said;
> So—my Lord Tomnoddy went home to bed!

59 T. Jeal, *Baden-Powell* (1989), 181–6 and pl. facing p. 292.

60 'The Execution: A sporting anecdote', in *The Ingoldsby legends, or Mirth and marvels* (orig. pub. 1840; 1875 edn., 3 vols.), ii. 20–6.

Thackeray's account of Courvoisier's execution (also 1840) suggests that this cameo was not heavily overdrawn. From their hired windows on that occasion 'the Mohawk crew' flung articles into the crowd and one squirted the plebeians below with brandy and water. 'Honest gentleman!' Thackeray wrote in scorn, 'high-bred aristocrat! genuine lover of humour and wit!'[61]

How can these many forms of defence be construed as 'resistances' to the mitigation of the capital code? If death was always fearsome, and hanging now potentially more so than ever, would not most people wish to diminish its psychic dangers by curtailing those avoidable deaths on scaffolds? How could they resist such curtailment? The simple answer is that in a society undefended by police or adequate secondary punishments, necessity stood in the way. But we have a further answer now. If 'defences' extruded the experience from conscious feeling, it became a matter of diminished urgency to demand the process's curtailment; it was possible to continue hanging people with relative equanimity. Add to that the defences to which we turn in the next chapters, mediated through the social and political distances between the hanged and the polite classes, and the recognition deepens that most polite witnesses viewed the gallows tableau through filters which hid its primal terror from them.

[61] Thackeray, 'Going to see a man hanged', 155.

CHAPTER 10

EXECUTING 'SOCIAL OTHERS'

1. *Social distance, hanging, and 'normality'*

AS THE PROMPTINGS OF SYMPATHY AND OF DECORUM COMBINED to muddy polite people's feelings about execution, means of evading engagement with the terror of the condemned became more elaborate, it seems. The previous chapter surveyed individuals' unconscious strategies of evasion. This chapter and the next are about the capacity of the scaffold's polite witnesses to economize in their responses to executions by means of social and political discriminations. They survey the constraints on sympathy which in a highly polarized society timelessly ensued from the social and hence emotional distances between the propertied classes and most victims of the law, these last so alien to élite understanding that it was next to impossible to identify with them.

Two reference groups carry this part of the argument. The first, in this chapter, were the tatterdemalion poor who were the gallows' chief victims.[1] Their exclusion from the sympathy extended to socially intelligible scaffold victims speaks for an economy of the emotions organized along lines of class. How far poor gallows victims were beyond the reach of the kinds of sympathy which might strive to *save* them can be inferred from the identities of those with whom polite people *did* feel identificatory pity. These were usually people rather like polite witnesses themselves. They were middling and once-respectable or quasi-respectable kinds of people, people who might die politely as it were. Or else they were emotionally accessible through other forms of

[1] We must take it as read here that throughout the 18th cent. and into the 19th, vast income inequalities, together with fluctuations in wages, work conditions, and food prices, and post-war unemployment, affected prosecution rates and the numbers hanged. For development of these themes, see V. A. C. Gatrell and T. B. Hadden, 'Criminal statistics and their interpretation', in E. A. Wrigley (ed.), *Nineteenth-century society: Essays in the use of quantitative methods for the study of social data* (Cambridge, 1972); D. Hay, 'War, dearth, and theft in the eighteenth century: The record of the English courts', *Past and Present*, 95 (1982), 117–60; and particularly P. Linebaugh's investigation of the economic origins and experiences of *The London hanged: Crime and civil society in the eighteenth century* (1991).

contiguity, like that between patron and client. No offence to Scottish sensibilities is intended by the co-option in this chapter of James Boswell as a witness to 'English' attitudes; that voluble diarist's discriminating responses to executions can hardly be evaded in this context, as we shall see. Also discussed are the discriminating responses of those who got upset about the Revd Dr Dodd's execution in 1777. The chapter ends by leaping on to Thackeray's very different response to a hanging in 1840. Here at last and extraordinarily we do encounter a man who transcended class boundaries by associating himself with the murderer Courvoisier's scaffold terror. This rare projection has to be explained by noting the impact on Thackeray of the political and material climate by then.

The second reference group were traitors, addressed in Chapter 11. Their exquisitely aggravated punishment (hanging, disembowelling until the 1790s, decapitation until 1820) purposefully defined them as alien beings beyond right-minded sympathy. *Frankenstein* had been published two years before the executions of the Cato Street conspirators of 1820: and here, it was duly said, were monsters indeed. Even so, the punishments inflicted on the bodies of these people were extraordinary for so polite an age. We must ask why they survived so long, and how cultivated witnesses could watch treacherous heads hacked off and dripping, and not vomit.

It bears repeating before we proceed that many people withheld their vomit for reasons we all know about. Some subordinated their feelings about capital punishment to the claims of hierarchy, property, and morality, and so felt nothing about it that we can detect. 'Nature exacts my tenderness, but the Law my rigour,' as Henry Fielding said. He allowed room for tenderness: '[The felon's] dread of a sudden and violent death . . . the importunities, cries, and tears of a tender wife, and affectionate children, who, though innocent, are to be reduced to misery and ruin by a strict adherence to justice . . . form an object which whoever can look upon without emotion, must have a very bad mind.' But, he had added approvingly, 'whoever by the force of reason can conquer that emotion must have a very strong' mind.[2] Sometimes, similarly, people felt rather jolly about the whole hanging business because they were the victims of crime. Robbed by a footpad, Thomas Rowlandson

[2] H. Fielding, 'An enquiry into the causes of the late increase of robbers' (1751), in *The Wesleyan edition of the works of Henry Fielding*, ed. M. R. Zirker, *An enquiry into the causes of the late increase of robbers and related writings* (Oxford, 1988), 164–5. M. Madan held the same view: *Thoughts on executive justice, with respect to our criminal laws, particularly on the circuits* (1785), 80–1.

the artist was 'mightily pleased' when he procured the conviction and execution of another robber: 'Though I got knocked down, and lost my watch and money, and did not find the thief, I have been the means of hanging *one* man. Come, that's doing something.'[3]

Above all there was the indifference to penal pain which came from too much familiarity with it. 'What a number of executions do we read in every week's paper!' a cleric exclaimed in 1780. 'Their frequency makes them less shocking and terrible, and a list of poor wretches launched into eternity is now read by many with as much indifference as if it were a list of Births or Marriages.'[4] James Hackman observed in 1777 that people became accustomed to the idea of hanging merely by hearing the repeated chant of broadside sellers in the London streets. 'All that strikes us, is, the ridiculous tone in which the halfpenny ballad-singer chants the requiem':

> How many men, how many women, how many young, and, as they fancy, tender females, with all their sensibilities about them, hear the sounds, by which at this moment I am disturbed, with as much indifference as they hear muffins and matches cried along the streets? The last dying speech and confession, birth, parentage, and education—Familiarity has even annexed a kind of humour to the cry. We forget that it always announces the death (and what a death) of one fellow being; sometimes of half a dozen, or even more. A lady talks with greater concern of cattle-day than of hanging-day.

It was likewise without emotion that Parson Woodforde recorded a hanging in Oxford in his undergraduate days, or noted the condemnation of eight prisoners at Thetford assizes: 'J^s^ Cliffen . . . was hanged on Thursday last at Norwich on Castle-Hill and behaved most daringly audacious. . . . [His] body was this day carried to Badley Moor and there hung in chains at one Corner of the said Moor.'[5] Along with others of his profession,[6] Woodforde's diary withheld personal engagement without difficulty. This was news merely, recorded as changes in the weather might be.

[3] H. Angelo, *Reminiscences* (2 vols., 1828), ii. 324–6.

[4] O. F. Christie (ed.), *The diary of the Revd William Jones of Broxbourne and Hoddesdon, 1777–1821* (1929), 79–80: 'I was present at the last Oxford Assize,' Jones continued, 'and saw a villain "the most daring and hardened the Devil ever possessed" receive sentence of death. By one of the papers now before me I learn that he is executed.'

[5] J. Beresford (ed.), *James Woodforde: The diary of a country parson* (abridged edn., Oxford, 1975), 6–7, 246 (6 and 23 Mar. 1761, 28 Mar. 1785).

[6] Cf. the insouciance of another cleric in his diary in 1754: 'a good for nothing farmer . . . was condemned to die for cow stealing . . . he unfortunately over reached himself, and was caught in the noose at last: as the halter generally cancels all debts, I must be contented to have mine of about fifty shillings, wiped off with the rest of my neighbours by the same sponge': D. Gibson (ed.), *A parson in the vale of the White Horse: George Woodward's letters from East Hendred, 1753–1761* (Gloucester, 1983), 60–1.

Indeed, among people like Woodforde hanging was bizarrely domesticated, so deeply was its normality accepted. 'Hangdog' is a term whose literal reference to an era of criminal animal-hangings is now lost,[7] but Woodforde's age gave it fresh meaning. He and other gentlemen killed pets by hanging them. 'My dog Pompey came home shot terribly, so bad that I had her hanged directly out of her Misery,' Woodforde recorded in his diary in 1777. And again in 1794: 'One of my Greyhounds Fly, got to Betty Cary's this morning and ran away with a Shoulder of Mutton undressed & eat it all up. . . . I had the Greyhound hanged in the Evening.' The Berkshire parson George Woodward recorded in 1759: 'Our parish . . . has lately been in great confusion, on account of a mad dog, who passed through and bit several [dogs]; but they have hanged them all.'[8] Bewick even illustrated the practice (Plate 32). Hanging is not an obvious or convenient way of killing domestic animals. Is the practice ever to be met with *after* the era of public hanging?

When hangings were thus normalized, it was difficult to feel much about them one way or the other if you only read about them in newspapers. Gentlefolk who endlessly complained about the desensitization of common people at the scaffold hardly realized their own. It took an imaginative effort or personal shock to realize that the concept of normality failed to connect with the fearsomeness of what happened. Hackman himself had been shaken out of his indifference only when he saw Dr Dodd hang in 1777. The spectacle had to be personalized before he saw it: 'How shocking that a man with whom I have eaten and drunk, should leave the world in such a manner! A manner which, from familiarity, has almost ceased to shock us, except when our attention is called to a Perreau or a Dodd.'[9] Ironically this was written two years before the wretched Hackman was himself hanged for shooting dead his beloved Miss Reay, the earl of Sandwich's mistress, as she emerged from Covent Garden theatre. Irony was compounded when the classical

[7] 'Hangdog': 'a despicable or degraded fellow fit only to hang a dog, or to be hanged like a dog' (OED). In early modern England animals guilty of killing sheep or involved in bestiality cases were sometimes punished by hanging: K. Thomas, *Man and the natural world: Changing attitudes in England 1500–1800* (Harmondsworth, 1984), 97–8; E. P. Evans, *The criminal prosecution and capital punishment of animals: The lost history of Europe's animal trials* (1906).

[8] Beresford (ed.), *Woodforde diary*, 136, 457, 542; Gibson, *A parson in the vale of the White Horse*, 115.

[9] The letters were published by Sir Herbert Croft as *Love and madness: A story too true, in a series of letters between parties whose names would, perhaps, be mentioned, were they less known or less lamented* (1780). Walpole doubted their authenticity but acknowledged that 'they enter into [Hackman's] character' (Walpole, vii. 338–9). Even if faked, they reflect a contemporary perception. They were reprinted in Borrow, v. 21.

32. Thomas Bewick, *Vignettes* (1797–1818).
The first vignette might be interpreted as a deft Hogarthian comment on cruelty to animals and humans were it not that Bewick's engravings were never so didactic. It might be better read as a neutral observation about the normality of hanging, even in putting down domestic animals. 'Late, but in earnest,' the suicide's motto reads in the second vignette. Hauntingly morbid, it assimilates hanging to a Romantic version of normality.

scholar Dr John Warner went comfortably enough to see Hackman's own body laid out for dissection: 'he is now a fine corpse at Surgeon's Hall . . . a genteel, well-made young fellow,' he reported; 'there has been a deal of butchery in the case'.[10]

Equable responses of these kinds are easily culled from polite diaries and letters in this era. But it was not because some kind of 'pre-modern callousness' ruled, or because hangings did not signify much in that age when all is said and done. Each hanging was loaded with significance for any who engaged with it. What helpfully checked contemporaries' experience of that significance was not only familiarity, but also the fact that between victims and witnesses huge social distances yawned.

2. *Boswell and the limits of sympathy*

In the nineteenth century expanding bureaucratic definitions and punitive stigma were to deepen the so-called criminal's separateness from social life, the label in good part helping to create the reality. But already in the eighteenth century, in London chiefly but in big towns too, the marginality and otherness of lawbreakers in general and of scaffold victims in particular needed no underscoring. The identifications upon

[10] Radzinowicz, i. 177 n.

which empathy depended had always been blocked at the hazy boundaries which separated élites from the vulgar.

It confirms the importance of social contiguity in releasing sympathy for the condemned that for some centuries the most discernible locus of compassion had lain, and still lay, not among the polite people but among middling people and the poor. Admittedly the point should not be overstated. A middling man on the make, like Francis Place, was quite capable of betraying a feckless brother-in-law by petitioning in 1799 for his life transportation to save him from a hypothetical future hanging (incurring underworld threats on his life for his pains). Place also delighted in his experiences as an Old Bailey juryman. For all his radicalism, he largely acquiesced in the law and its harsh ways.[11] Doubtless most people like him did too. None the less, on the part of the farmers, tradesmen, and middling kinds who usually sat on juries or who brought most prosecutions, there was a long-standing reluctance to prosecute, together with a readiness to resort to life-saving partial verdicts in capital property cases. When at a murderer's execution in 1582, for example, we are told that 'all (where of some were such as a man would have thought had never a teare to shed at such a sight, having viewed diverse the like and more lamentable spectacles) with wet eies beheld him',[12] we encounter something more than a merely cultivated empathy, and it was independent of the postures of sensibility too. Face-to-face connection with the offender or broadly congruent social standing mediated the imaginative and emotional projections which underpinned this behaviour. It was David Hume who put this point most sharply. Sympathy is engaged only when 'external objects acquire any particular relation to ourselves, and are associated and connected with us', he wrote. This principle found perfect expression in Hume's friend James Boswell. A man who under certain conditions could feel deeply, Boswell's sympathetic limits were determined pretty precisely by the social distances between himself and suffering others.

On the face of it, Boswell is not the most sympathetic of subjects in this book. He enjoyed hangings. At least, he went to as many as he could. 'I . . . am never absent from a public execution,' he wrote. He took friends along to watch, as he took Reynolds to see the hanging of Mrs Thrale's old servant.[13] Nor in later years was he unduly pained by

[11] M. Thale (ed.), *The autobiography of Francis Place* (Cambridge, 1972), 97, 134, 292.

[12] The chronicler Holinshed, cited in J. S. Cockburn and T. A. Green (eds.), *Twleve good men and true: The criminal trial jury in England, 1200–1800* (Princeton, NJ, 1988), 148. See p. 407 below.

[13] J. Boswell, *Life of Johnson*, ed. G. B. Hill and L. F. Powell (6 vols., 1934–50), ii. 93 n. 3; C. R. Leslie and T. Taylor, *Life and times of Sir Joshua Reynolds* (2 vols., 1865), ii. 588–9.

the spectacle. Perhaps it helped that Dr Johnson had once teased him by declaring that sympathy for others' distress was an affectation. This interchange occurred the day after Boswell saw four men hang at Tyburn in 1769:

JOHNSON 'Why, Sir, there is much noise made about it, but it is greatly exaggerated. No, Sir, we have a certain degree of feeling to prompt us to do good: more than that, Providence does not intend. It would be misery to no purpose.' BOSWELL 'But suppose now, Sir, that one of your intimate friends were apprehended for an offence for which he might be hanged.' JOHNSON 'I should do what I could to bail him, and give him any other assistance; but if he were once fairly hanged, I should not suffer.' BOSWELL 'Would you eat your dinner that day, Sir?' JOHNSON 'Yes, Sir, and eat it as if he were eating it with me. . . . Sir, that sympathetick feeling goes a very little way in depressing the mind.'[14]

In 1783, likewise, Boswell agreed 'perfectly' with Johnson's blunt lament for the abolition of the Tyburn procession. Executions now, Boswell agreed, had 'not nearly the effect which they formerly had', because magistrates 'had too much regard to their own ease'. He thought the ritual should be made more terrifying in order to enhance its deterrent effect.[15] Hence most horribly, in 1783, his recommendation of the 'best' capital punishment he knew, the form practised in 'Modern Rome':

The criminal is placed upon a scaffold, and the executioner knocks him on the head with a great iron hammer, then cuts his throat with a large knife, and lastly, hews him in pieces with an axe; in short, treats him exactly like an ox in the shambles. The spectators are struck with prodigious terrour; yet the poor wretch who is stunned into insensibility by the blow, does not actually suffer much.[16]

These representations of Boswell look appalling, but they are not to be read superficially. They concealed complex emotions and ideas. It was not for nothing that as a student at Glasgow in 1759–60 Boswell had attended Adam Smith's lectures upon which *Theory of moral sentiments* was based, nor that he continued to be a friend of Hume's. Admittedly he was often confused about his motives, which is one reason why he was so fascinated by them. When he argued with his friend Wood about

[14] *Life*, ii. 94. Johnson's position on capital punishment was better conveyed in his question 'Who knows whether this man is not less culpable than me?' Given the scaffold's failure to deter, he argued for its confinement to the most heinous crimes: 'The necessity of proportioning punishments to crimes' (1751), in W. J. Bate and A. B. Strauss (eds.), *Samuel Johnson:* The Rambler (3 vols., New Haven, Conn., 1969), ii. 241–7.

[15] *Life*, iv. 188–9.

[16] Boswell, *The Hypochondriack*, ed. M. Bailey (2 vols., Stanford, Calif., 1928), ii. 284. These essays first appeared in the *London Magazine*, 1777–83.

the soul, and Wood confessed that 'he seemed to have no formed principles upon the subject, but just had ideas, sometimes of one kind, sometimes of another, floating in his mind', Boswell might have said as much himself.[17] His 'philosophy' was sensation-based, not cognitive: but that after all only suited the case. Nobody took sympathetic principles more seriously or tested them more purposefully. An acute self-consciousness resulted, witnessed in a lifetime's exploration of himself in his journals. He thought himself 'a being very much consisting of feelings': 'my existence is chiefly conducted by the powers of fancy and sensation.'[18]

Sensibility took him into dangerous explorations. That many of them invited hostile interpretation he well knew. We have already noted his shame about his interest in executions, and how he tried to account for it 'in a philosophical manner'. But at best there was a brave determination in his quests as he followed them to their risky conclusions. This is how his recommendation of the Roman butchery should be construed. He disapproved of cruelty, insisting that criminals should receive 'as *mild punishments* as are consistent with terrour'. But terror (i.e. deterrence) *was* the chief end of punishment; he could conceive of no other end. It followed that punishment should allow no room for those superadded responses from sensibility which subverted that end—the responses he himself (and many in the crowd) knew: identification, pity, and sympathy. As things were, he implied, hanging was sufficiently bearable to watch to permit the suffering to be transferred from the scaffold to the sympathetic spectator, who would then forget the crime's magnitude and see the law as cruel. Better to convert execution into a hyperbolic Roman ritual (provided the victim was speedily killed) which would attack the senses as violently as it did the body. This would allow space *only* for terror, and none for the pity which at the scaffold should be allowed no place at all.[19]

Two episodes in Boswell's scaffold witnessings return us to this chapter's social point. The first hanging Boswell witnessed was in May 1763. Aged 22, new to London, and casting about for something to do, he took it into his head to visit Newgate on the eve of an execution. That familiar curiosity drove him there, sanctioned by the need to hone his sympathetic responses. In the prison courtyard, among 'a number of

[17] W. K. Wimsatt and F. A. Pottle (eds.), *Boswell for the defence, 1769–1774* (1960), 342.

[18] C. Ryskamp and F. A. Pottle (eds.), *Boswell: The ominous years, 1774–1776* (1963), 97–8.

[19] G. Turnbull, 'Boswell and sympathy: The trial and execution of John Reid', in G. Clingham (ed.), *New light on Boswell: Critical and historical essays* (Cambridge, 1991), 113.

strange blackguard beings with sad countenances, most of them being friends and acquaintances of those under sentence of death', he was lucky enough to see the morrow's victims dragging their shackles to their last service in the prison chapel. One was a big Irish woman, another a forger, each of whom he ignored. It was the highway robber Paul Lewis who caught his eye, for Lewis was a young blade rather like Boswell himself. He was 'a genteel, spirited young fellow . . . dressed in a white coat and blue silk vest and silver, with his air neatly queued and a silver-laced hat, smartly cocked. . . . He walked firmly and with a good air, with his chains rattling upon him, to the chapel.' Nicknamed 'the Captain' (he had served as a naval lieutenant) and the son of a clergyman, Lewis reminded Boswell of Macheath in the *Beggar's Opera*, a personage and a play Boswell greatly fancied. Clearly Lewis also reminded Boswell of himself, because he was unexpectedly shaken by this encounter. It lay on his mind the rest of the day 'like a black cloud', and he had a sleepless night. Although 'sensible that I would suffer much from it', he felt 'a sort of horrid eagerness' to witness Lewis hang at Tyburn next morning; so along he went.

In his witnessing, something unexpected happened to him. Sympathy was one thing; what ensued was another. He plunged into a depression which lasted several days. He was so disturbed by the hanging that he censored his account of it:

> I took Captain Temple with me, and he and I got upon a scaffold [platform] very near the fatal tree, so that we could clearly see all the dismal scene. There was a most prodigious crowd of spectators. I was most terribly shocked, and thrown into a very deep melancholy. . . . Gloomy terrors came upon me so much as night approached that I durst not stay by myself: so I went and had a bed (or rather half a one) from honest Erskine, which he most kindly gave me.

He was 'still in horror' the next day, and again had to spend the night with a friend. On the third morning he rose 'heavy, confused, and splenetic'. The following night 'I was still so haunted with frightful imaginations that I durst not lie by myself, but rose and sallied straight to Erskine, who really had compassion on me, and as before shared his bed with me.' He fought angrily with himself: 'I am too easily affected. It is a weakness of mine. I own it.' Over the following days he sought solace in friendship and 'dissipation', but his release came only six nights later through a bleak affirmation of life over death in one of his many purchased copulations. As he so often did, he used the wonderful conjuction of the little death to obliterate his sense of the larger one—its pleasure, however, tainted by self-loathing and guilt. In the Haymarket, he

> picked up a strong, jolly young damsel . . . [and] conducted her to Westminster Bridge, and then in armour complete did I engage her upon this noble edifice. The whim of doing it there with the Thames rolling below us amused me much. Yet after the brutal appetite was sated, I could not but despise myself for being united with such a low wretch.[20]

When Boswell first met Dr Johnson a week later he had recovered his composure, but five years later he remembered how he had been 'shocked to the greatest degree' by Lewis's hanging: 'I was in a manner convulsed with pity and terror, and for several days, but especially the nights after, I was in a very dismal situation.' Even in 1778 the memory induced 'an invincible horror' whenever he passed Newgate.[21] In short, a startling *identification* with Lewis had been achieved.

The second episode in which Boswell was shaken by his capacity for identification ensued in 1774. It centred on his relationship in Edinburgh with the condemned sheep-stealer John Reid. Some years earlier Reid had been the fledgeling barrister's first criminal client, and Boswell's defence then had secured Reid's acquittal. Now indicted for another sheep-theft, Reid was convicted in Edinburgh's high court of justiciary and sentenced to death. Boswell was unsettled by the failure of his defence on this occasion, and believed that Reid was convicted only on his past record. Vexation and distress catapulted Boswell into a seven-week-long obsession with Reid's plight which quite dominated his diary.[22] He drew up petitions to the king, wrote letters to the well-connected, lobbied the lord advocate, and published a popular broadside on Reid's behalf (significantly adopting the man's identity and language). Meanwhile he regularly visited Reid in the condemned cell to press him to admit to his guilt if he was guilty (Reid swore his innocence to the end), and to prepare him for the worst. Boswell felt very alive under these pressures, and monitored his own reactions narcissistically. He was amazed by his own 'firmness of mind while I talked with a man under sentence of death, without much emotion, but with solemnity and humanity'. And he preened himself on his 'wonderful' firmness during his heart-breaking interviews with Reid and Reid's wife.[23]

The oddest motif in this relationship turned on Boswell's plan to have Reid resuscitated after hanging. Edinburgh's professor of anatomy, Alexander Monro, told him that 'the thing might be done by heat and

[20] F. A. Pottle (ed.), *James Boswell's London journal, 1762–1763* (1950), 251–6: 3–9 May 1763.

[21] C. M. Weiss and F. A. Pottle (eds.), *Boswell in extremes, 1776–1778* (1970), 282.

[22] Wimsatt and Pottle (eds.), *Boswell for the defence*, after 1 Aug. 1774, *passim*, esp. 249–68, 292–355.

[23] Ibid. 276, 289.

rubbing to put the blood in motion . . . the best way was to cut a hole in the throat, in the trachea, and introduce a pipe'.[24] So Boswell pressed compliant surgeons into service, scoured Edinburgh for an inn-room or stable where the operation might be performed, and found one at last in the face of much hostility and at some cost to his dignity. In the event he allowed friends to dissuade him from the plan lest it perpetrate a futile cruelty towards a man now prepared to die, and undermine his own reputation. So Boswell spent the last hours comforting Reid and his family in the condemned cell. He attended the execution, and then hired a cart for Reid's wife to convey his corpse to his village, overriding local objections to its burial in the churchyard there.

For most of this time Boswell's hyperactivity had kept anxieties at bay. When his feelings did catch up with him, it was because a dreadful image caught him unawares. On the eve of execution he saw Reid dressing in the white linen clothes and black ribbons his wife had brought for his hanging: 'I was not much affected when I saw him this morning in his usual dress. But now he was all in white, with a high nightcap on, and he appeared much taller, and upon the whole struck me with a kind of tremor.' Later that evening 'gloom came upon me'. Boswell was 'so affrighted that I started every now and then and durst hardly rise from my chair at the fireside'. Even so, he philosophized determinedly about his feelings (he had recently dined with Hume): 'I had by sympathy sucked the dismal ideas of John Reid's situation, and as spirits or strong substance of any kind, when transferred to another body of a more delicate nature, will have much more influence than on the body from which it is transferred, so I suffered much more than John did.'[25]

For all the marvellous self-regard here, consistent with contemporary understandings of the social distribution of sensibility, it is clear in both Lewis's case and Reid's that Boswell had projected his personality into that of another being and experienced the other's pain as his own. His careful account of Reid's comportment betrayed a reverence before the death he had to witness and a sensitivity to suffering which for once cut

[24] Ibid. 304. This was the Monro whose son's lectures on the subject so disgusted Darwin as a student: see Ch. 9. Boswell and/or Monro might have been inspired by the formation in London in 1774 of the Humane Society, which aimed 'scientifically' to resuscitate drowned people. But the resuscitation of the hanged had long been a matter of experiment and folklore. Sir William Petty allegedly revived a hanged woman with symptoms of life by putting her to bed with another woman. 'Half-hanged Smith' was revived by medical bleeding in 1705 (and according to *Newgate calendars*, he later reported a near-death experience remarkably like those reported by people in similar situations nowadays); a surgeon allegedly revived a man in 1767 by incising the windpipe; Dr Dodd's body, after hanging in 1777, was taken to an undertaker's in Tottenham Court Road and placed in a hot bath to revive it (vainly, however): Griffiths, pp. 175–6.

[25] Wimsatt and Pottle (eds.), *Boswell for the defence*, 300.

across the boundaries of class. The fact that these are the first developed explorations of this response to execution in diary- or letter-writing may merely reflect the diary's evolution from its primary use hitherto in spiritual self-scrutiny.[26] Yet that itself affirms a sea change in the history of feeling and the view that an empathetic threshhold was crossed here which might connect meaningfully if not easily with the history of punishment.

We come now, however, to the limits to Boswell's empathy, most to the point in this argument. First, there was a telling omission in Boswell's record of Lewis's hanging in 1763. His diary was silent about the hanging alongside Lewis of the broker condemned for forgery and the woman Hannah Dagoe condemned for burglary.[27] Dagoe should have been particularly difficult to obliterate from the record. A strong, masculine-looking basket-woman of Covent Garden, 'the terror of her fellow prisoners', she had stabbed a witness against her. At Tyburn, shouting defiance in Irish brogue, she freed herself of her bonds, struggled with the executioner, then tore off her clothes and threw them to the crowd so that the hangman would not have them. Then, half-naked and with the noose around her neck, she covered her eyes with a handkerchief, threw herself from the cart, and broke her own neck.[28]

Given the saturated sensation of this frightful fracas, why did Boswell's diary not refer to it? Perhaps the manner of Dagoe's dying was a censored element in his depression: it was too horrible to record. It is more likely, however, that his silence simply expressed his emotional distance from Dagoe. In the Newgate courtyard he had noted her in passing only as a 'big unconcerned being'. Her vulgarity and her lack of 'concern' at her fate marked her as one with whom a gentleman could have no sympathetic connection. With Lewis, conversely, identification depended on a projection mediated through a fantasy of social likeness (just as in Reid's case it had depended on the relationship between barrister and client). Lewis was a relatively well-born *galant*, dressed dashingly and of Boswell's age, a fitting target for the projection of Boswell's own unease. In Lewis, Boswell confronted an emblem of his own mortality and guilt. So an honest pity was born: 'Poor fellow!' Boswell had written of Lewis's appearance in the Newgate courtyard: 'I really took a great concern for him, and wished to relieve him.' In each case, projective identification or real connection made empathy possible.

[26] P. Delany, *British autobiography in the seventeenth century* (1969), 63–4, 168.

[27] Anon., *A genuine account of the . . . life and transactions of J. Rice, broker, for forgery; P. Lewis, a highwayman; and Hannah Dagoe, for stealing goods out of a dwelling-house* (1763).

[28] Rayner and Crook, iv. 14–17, for accounts of Lewis and Dagoe.

In an era when most of the condemned were Dagoe-like rather than Lewis-like, these conditions were rarely achieved. It is not surprising that the two episodes recounted here were exceptional in Boswell's career. Give or take the occasional forger (and Revd Hackman), most other doomed people remained as remote from him as their forebears had been from Pepys. This distance enabled Boswell's feelings about those executed to settle into complaisance: 'Still, however, I persisted in attending [executions], and by degrees my sensibility abated; so that I can now see one with great composure.'[29] Nor is it surprising that Boswell's feelings about hanging were never focused within a generalized critique of capital punishment. In this he shared kinship with his class and time. Thus we confront among polite people the discriminating mental structures common to all who live in highly stratified societies. As Fielding said of the high and the low people in *Joseph Andrews*: 'so far from looking on each other as brethren in the Christian language, they seem scarce to regard each other as of the same species.'

3. *The death of Dr Dodd, 1777*

Boswell's were moments of private crisis. But our diagnosis is confirmed by a public crisis at this time, centred on that most significant hanging in the history of collective feeling—the Revd Dr Dodd's for forgery in 1777. Sometime chaplain to the king, a sentimental sermonizer widely known in polite circles, but notoriously disreputable in most of his financial transactions, Dodd's death sentence was for several months the sensation of polite London, generating a campaign for pardon unprecedented in its ambition, with Dr Johnson famously giving a lead.[30] What needs emphasis here was the public emotion the case released. In an age when public tears flowed freely, they flowed never so freely as now. 'The judges, the jury, the counsel, the spectators, all the world was bathed in tears,' an incredulous foreign observer recorded. Henry Angelo added that the case moved 'almost every bosom to sorrow or sympathy'—'The conversation in every circle was of "poor Dr Dodd!"' Talk in taverns and at Angelo's father's dinner table was domi-

[29] *Public Advertiser*, 26 Apr. 1768: republished in *London Magazine* (1783), 203.

[30] Some 23,000 people, led by the 'Gentlemen, merchants and traders of London, Westminster, and Southwark', signed the petition to the king: 'Such a list was never collected, perhaps, on behalf of any individual offender against the laws from the earliest period of society' (Angelo, *Reminiscences*, i. 459). One petition and a letter from Dodd survive in SP37.12, fo. 78. See G. Howson, *The macaroni parson: A life of the unfortunate Dr Dodd* (1973); and J. Burke, 'Crime and punishment in 1777: The execution of the Reverend Dr William Dodd and its impact upon his contemporaries', in W. B. Thesing (ed.), *Executions and the British experience from the 17th to the 20th century* (Jefferson, NC, 1990), 59–76.

nated by argument about the case, and on the eve of the execution Angelo's mother wept. At the Tyburn hanging (Angelo attended it) 'never did so general a sympathy prevail':

> Every visage expressed sadness; it appeared, indeed, a day of universal calamity. . . . Thousands sobbed aloud, and many women swooned at the sight. . . . [Dodd's] corpse-like appearance produced an awful picture of human woe. Tens of thousands of hats, which formed a black mass, as the coach advanced, were taken off simultaneously . . . [The crowd's] silence added to the awfulness of the scene.[31]

Radzinowicz claimed that Dodd's execution marked 'a definite stage in the crystallisation of public opinion' on the death penalty, and it is true that Sir William Meredith referred to the case in the Commons when he criticized the bloodthirstiness of English law in property cases. Radzinowicz also claims, however, that Dodd's was the first case 'to stir the public conscience'. Reactions to it were 'undoubtedly a symptom of the growth of humanitarianism', he says.[32] Alas, what the episode marked better was something less elevated. 'We live in an age in which humanity is the fashion,' a contemporary wrote of the case. Another thought so too: there is sometimes 'a sort of fashion in feeling, when sorrow, as it were, becomes a national epidemic'.[33] *Fashion* is apt in this connection. Far from being symptomatic of a growing humanitarianism, this was a moment of collective excitement, when sensibility coincided with a particular apprehension of social interest. That interest was of precisely the self-identificatory kind which had enabled Boswell to feel for Lewis.

For all his disrepute, Dodd was a man of sensibility. Moreover, he died as any of the polite classes would die (and as plebeians like Dagoe did not), that is, he died politely, full of 'painful apprehension of a public execution, attended with all the tragic, and yet disorderly, parade normal in this country', but also 'under a deep sense of sin . . . look[ing] to that Lord by whose merits alone sinners could be saved'.[34] It also helped that he was, Dr Johnson noted, 'the first clergyman of our church who has suffered public execution for immorality'. He 'was a

[31] D'Archenholz, *A picture of England* (1790), 145–6, cited by Radzinowicz, i. 460; Angelo, *Reminiscences*, i. 456, 460, 475. 'Are you not glad, Madam, there is an end of talking of poor Dr Dodd?' Walpole wrote after the execution: 'I felt excessively for him . . . for between the law and his friends, he suffered a thousand deaths': Walpole, vi. 449.

[32] Radzinowicz, i. 451, 468 n.

[33] J. Hawkins, *Life of Samuel Johnson* (1787), 521, cited by Radzinowicz, i. 468 n; Angelo, *Reminiscences*, i. 456.

[34] J. Villette, *A genuine account of the behaviour and dying words of William Dodd, LL D, who was executed at Tyburn for forgery* (1777), 12.

preacher remarkably diligent and persuasive', his mercy petition explained to the king, 'a principal Institutor of charitable societies . . . a man much to be reverenced for his abilities and usefulness'. Above all, it added, the execution of a cleric 'so active and conspicuous may have very pernicious effects on common minds and increase that disrespect for the clergy which is always productive of laxity both in principles and practice'.[35] In other words, Dodd's death was painful and dangerous because it challenged the right order of things; executing a clergyman turned the world upside down.

These were the considerations, remote from a generalized benevolence, out of which the polite conscience was constructed. For we note finally that there was no epidemic of feeling about the 15-year-old boy Joseph Harris who was hanged alongside Dodd. Harris had robbed a stage-coach of two half-guineas and a few shillings; Dodd had tried to get away with £4,300.[36] Dr Johnson and his friends never mentioned Harris, not a word. Nor was there an epidemic of feeling three years later when twenty-five Gordon rioters, four of them women, were hanged on gallows dispersed through the streets and squares of London. Tattered blackguards beyond sympathy and intelligibility, their terror could not easily be re-experienced within the feeling self. Anyway, their deaths made a political sense which Dodd's did not.

4. *Thackeray: The modern man of feeling?*

In quest now of significant change, we leap far forwards to 1840, to an episode which at first glance reminds us of Boswell's encounter with Lewis's hanging; yet in key respects it was an episode unthinkable in Boswell's day. Either we meet in Thackeray just another young man going to a hanging thoughtlessly and being shocked by it; or we meet a new and different kind of sensibility—a modern one, we might call it, by which we might mean one intelligible to us: inheriting the precepts of eighteenth-century sympathy certainly, but responding to a hanging *democratically* at last, through an identification with a humble victim regardless of his social status. Either that, or another way of interpreting the episode entirely. Might it be merely that in 1840 we enter a different material and political world, which offered peculiarly appropriate (though instable) conditions for the release of empathy across class boundaries? This is the explanation that works best, not least because,

[35] SP37.12, fo. 78.

[36] I. Gilmore, *Riot, risings and revolution: Governance and violence in 18th-century England* (1992), 177–8.

realistically, it offers another bleak reminder of the temporal fragility and contingency of humanitarian impulses.

One day in July 1840, aged 29, with a family to feed and his novels still to be written, Thackeray planned to make a little profit out of the murderer Courvoisier's hanging. Courvoisier was a French valet who had brutally murdered his aristocratic master Lord William Russell, and the sensation was compounded by the fact that Lord William was the uncle of Lord John, who as home secretary had presided over the virtual obliteration of the bloody penal code in 1837 and who still sat in the Cabinet. Courvoisier's modelled head was to be displayed at Madame Tussaud's into the twentieth century—a reliable index of fame and the old curiosity too (see Plate 13, p. 118 above). Thackeray lightly offered *Blackwood's Magazine* some articles containing 'as much fun and satire as I can muster', and he thought Courvoisier's execution would do for the first article: 'I'll go on purpose.' So, early on the fatal Newgate morning off he went with his friend Monckton Milnes, witlessly walking into a witnessing which shook him to the core.

Elsewhere in the book we have noted and quoted from Thackeray's close scrutiny of the crowd. Pertinent here in his article (duly written without fun or satire) is his scrutiny of himself. Days after the execution he was 'miserable', Thackeray told his friends. The spectacle 'weighs upon the mind, like cold plum pudding on the stomach, & as soon as I begin to write, I get melancholy'. 'It was a horrible sight indeed,' he wrote again, 'and I can't help mentioning it for the poor wretch's face will keep itself before my eyes, and the scene mixes itself up with all my occupations.'[37] This sounds very like Boswell after Lewis's hanging eighty years earlier. Like Boswell in Lewis's case, Thackeray had annexed Courvoisier's terror to himself, and terrified himself in doing so.

But if Boswell's reactions mark something of an opening in the history of identifications with the condemned criminal body, Thackeray's mark something of an end. There were three differences between Thackeray's and Boswell's reactions. First, Boswell's identification with Lewis had depended upon a fantasy of like social standing. Thackeray, however, played with no such fantasy: Courvoisier was a servant. Secondly, while Boswell censored what he saw when Lewis hanged, Thackeray's gaze was unflinching. He shut his eyes as Courvoisier

[37] G. N. Ray (ed.), *The letters and private papers of William Makepeace Thackeray* (4 vols., 1945), i. 451–5. Thackeray's articles were turned down by *Blackwood's*; 'Going to see a man hanged' was published in *Fraser's Magazine*, 22 (Aug. 1840), 150–8.

dropped, but till then he looked straight and minutely into the horror, recording it as Boswell was unable to do:

> His arms were tied in front of him. He opened his hands in a helpless kind of way, and clasped them once or twice together. He turned his head here and there, and looked about him for an instant with a wild, imploring look. His mouth was contracted into a sort of pitiful smile.

Thackeray sees Courvoisier here as a discrete and atomized body. He is seen free of the representations which had denied the individual in the early modern sinner: a merely vulnerable body like Thackeray's own. And finally there was Thackeray's exploding protest, quivering with emotion, we note, not coolly hatched in thought, and not conceding any credit, either, to the ethical and punitive principles which reconciled most people to capital law:

> The sight has left on my mind an extraordinary feeling of terror and shame. It seems to me that I have been abetting an act of frightful wickedness and violence performed by a set of men against one of their fellows. . . . I came away down Snow Hill that morning with a disgust for murder, but it was for the murder I saw done. . . . I feel myself ashamed and degraded at the brutal curiosity which took me to that brutal sight; . . . I pray to Almighty God to cause this disgraceful sin to pass from among us, and to cleanse our land of blood.

In all this it was indeed (as the Introduction suggested) as if a great divide had been crossed since Boswell's day. In past generations pity for the hanged and distaste for the hanging had always been available in some circles. But apart from Gibbon Wakefield on the boy John Bell in 1831 (a text offering another sign of changed times of course), pity for a common murderer and disgust at those who took his life had never been so nakedly expressed in print before. What released these emotions? Thackeray was a kind and generous man, and there is no reason not to give that its due. But something outside character—and more fleeting—was behind this impassioned but momentary protest of 1840 (since the posture was not sustained).

Two kinds of answer are to hand. On the long view, the conditions which enabled Thackeray to release anger, shame, and disgust were such that earlier generations could never have enjoyed. Not only was there a long process of cultural learning behind him, a sympathetic ground well prepared; but disgust at the penal cruelties of the old order was now sanctioned by material and bureaucratic shifts which promised a security old punishments could not provide. By the 1830s and 1840s a new form of disciplinary state was evolving. London, the boroughs, and the counties now either had or could have professional police forces if they wished; prosecutors were pushing up prosecution rates satisfactorily; penitentiaries were planned or a-building; and magistrates' courts were disciplining more people more speedily, cheaply, and lightly than before. Hangings began to look unmodern, and disgust at them appropriate.

The narrower view is that in 1840 Thackeray was swimming with a briefly flowing tide—along with others who, in that and immediately following years, supported total abolition of capital punishment: Dickens, Mill, Carlyle, Jerrold of *Punch*. This literary fashion thrived on a dramatically changed political climate. With Russell's capital statute repeals in 1837 breaking a long deadlock, executions had suddenly acquired scarcity value, so people thought about them as never before. Also influential was the abolitionist movement in parliament. Just before Courvoisier's hanging, ninety independent MPs from the professional or commercial classes supported Ewart's motion for total abolition, unthinkable before 1832. These contexts make it hardly surprising that Thackeray sounds like a 'modern' man. He wrote his article at a moment when to many the old penal ceremonies had quite suddenly ceased to make sense.

Yet it is a final comment on the fragile dependencies and the evanescence of passing fits of 'humanity' which were noted in the Introduction that the giddy moment did not last. Since hanging continued until 1964, Thackeray's protest was no portent of things to come either. Along with Dickens, Carlyle, Mill, and *Punch*, Thackeray later changed his mind, as the book's last chapter will show. By the end of the decade each of these men thought it was all right to hang murderers after all. 'Progress', they decided, would be served well enough by having the messy business hidden behind prison walls. So in the end it was squeamishness, not humanity, that won the day.

CHAPTER 11

EXECUTING TRAITORS

1. *Cato Street monsters*

1820 SAW THE LAST EXECUTIONS FOR HIGH TREASON IN THE OLD-fashioned way. In that year eight men were publicly hanged and decapitated. Five of them had led the plot laid in Cato Street to assassinate the Cabinet at dinner; they were executed before a Newgate crowd of 100,000 on 1 May. The other three were Scottish radicals who had attempted an abortive uprising, and they were executed in Stirling and Glasgow in August and September. In each case, after conventional hanging, their corpses were taken from their nooses and laid in coffins on the scaffold platform. One by one their heads were lifted out to rest on a block to enable the executioner to hack at their necks with a knife. Then, dripping gorily on to sawdust, each head was held up to the spectators, the executioner crying out its name. Only three decades earlier the bodies would also have been eviscerated to convey the magnitude of their offence a little more clearly.

There were no more such displays after 1820.[1] Treason prosecutions lapsed until the Chartist 1840s, and then the convicted were to be transported for life. This makes the decapitations of 1820 look anachronistic in retrospect, as if they were only delayed after-echoes of a Tudor punishment self-evidently ripe for abolition. But the oddity of what hap-

[1] The numbers executed for treason in the 18th cent. are imprecisely recorded, but patterns are clear. After the score or so Jacobite executions in the 1740s, the punishment all but lapsed until a score of Irish rebels were executed in 1794–8 and as many again in 1803. English radicals got off lightly, thanks to jury resistance and Hardy's acquittal in 1794 (Dr Watson and Thistlewood narrowly escaped conviction in 1817). But Despard and 6 fellow conspirators were executed in 1803, followed by 3 Pentrich insurgents in 1817, and finally, in 1820, by the Cato Street conspirators and the Scots. T. B. Howell (ed.), *A complete collection of state trials* (1816–18), and J. Macdonell (ed.), *Reports of state trials: New series*, cover most cases 'of constitutional interest', but are not comprehensive. For Irish and Jacobin trials in the 1790s, see E. P. Thompson, *The making of the English working class* (1965), C. Emsley, 'An aspect of Pitt's "Terror": Prosecutions for sedition during the 1790s', *Social History*, 6 (1981), 155–84, and the same author's 'Repression, terror and the rule of law in England during the decade of the French Revolution', *English Historical Review*, 100 (1985), 801–26.

pened in 1820 cannot be dismissed so cavalierly. These heads were hacked off, when all is said and done, at a time when English people loved saying that cruel and unusual punishments in other countries proved how backward foreigners were. So questions still stand. How did the polite classes stomach such a gruesome punishment in this enlightened country, not to say in this age of Jane Austen? Why did the punishment survive so long? What price civility, sympathy, and sensibility as people watched heads dripping blood on to sawdust?

With regard to 1820, it was not as if Arthur Thistlewood and his Cato Street friends or the Scots rebels had greatly threatened the social order. The Cato Street men had itched to 'plunge a dagger into the heart of a tyrant', and the Scots had planned an armed insurrection. But government spies easily broke open their plots; and since Liverpool's government needed to restore credit after Peterloo, it probably let the plots run in order to make political capital from them.[2] If so, it was a dangerous gamble. Brutal executions could easily have worsened the government's position by making martyrs. The decapitations of 1820 fuelled Scottish radicalism a decade later, for example.[3] Yet the most peculiar aspect of these episodes was that although there was some newspaper-muttering, there was little concerted outcry of disgust against them. Opinion was either cowed by the violence or approving of it.

Loyalist majorities always knew what they were supposed to think about these judicial killings. Executions in such excess made statements about the crown's offended power and the monstrosity of this highest of crimes. They preserved order against those whose defiance of the crown jeopardized order, hierarchy, and sovereignty itself. Loyalist pamphleteers seized on traitors' executions to insist that their fate, while 'it has excited Christian feeling on behalf of their precious souls, will be acknowledged in every loyal and patriotic bosom to be most justly merited'.[4]

Yet there was more to the nation's acquiescence in the punishment than pieties of this kind. The aggravated execution of traitors was so

[2] J. Stanhope, *The Cato Street conspiracy* (1962); Thompson, *Making of the English working class*, 700–10; D. Johnson, *Regency revolution: The case of Arthur Thistlewood* (Salisbury, 1974); I. Prothero, *Artisans and politics in early nineteenth-century London: John Gast and his times* (1979), 127–31; R. Sales, *English literature in history, 1780–1830: Pastoral and politics* (1983), ch. 7; I. McCalman, *Radical underworld: Prophets, revolutionaries and pornographers in London, 1795–1840* (Cambridge, 1988), 137–9, 145–8. Historians have given astonishingly little attention to the Scots.

[3] Anon., *An exposition of the spy system, pursued in Glasgow . . . 1816–20 . . . by a ten-pounder* (Glasgow, 1833).

[4] Anon., *The Cato Street conspiracy illustrated and improved . . . by a loyal subject of the British Empire* (Bristol, 1820), 3.

dangerously exciting and inflammatory a spectacle, so at odds with civility, potentially so counter-productive, and among polite people also so little commented on—that we may look for explanations for its survival to 1820 which lay below reasoned thought. One might have lain in the mystically constructed otherness of the traitor and the ancient cosmologies still subliminally associated with his punishment. Another was probably encapsulated in ancestral memories: the word 'treason' and the act of decapitation touched on associations reaching back past the Jacobites and Charles I to Elizabethan tableaux. Often illustrated, they were the stuff of chap-book and children's histories, and perceptions were filtered through them. It was difficult to see the immediate brutality of what happened—the blood, the sawed-at sinews, the heads held up by the hair—when what was unconsciously seen was an Elizabethan pageant re-enacted. We need a concrete case to begin with, however—so back to 1820 again. How and to whom the Cato Street conspirators' last hours and deaths were visually and verbally presented, how the authorities behaved, who prepared their deaths, in what moods the watchers watched, and how newspapers commented, are themes which illuminate both the locations and the limits of sympathy in this context with a clarity otherwise difficult to attain.

People differ in the ways in which they see and read great events. That there were both vulgar and polite ways of seeing what happened on that May Day morning of 1820 is suggested by two graphic depictions of the Cato Street executions. Each picture aimed at different social audiences and implied different understandings. Together they reinforce the irony encountered elsewhere: 'sympathy' with cruelly punished people was most readily released by popular audiences, not by the Jane Austen-reading élites in whose interests (by and large) this kind of justice was done.

The first depiction was one of several copperplate illustrations in the fifth volume of *The Newgate calendar improved*. The bulk of this tome reported the conspirators' trials verbatim; forty pages described their last hours and deaths. Cobbled together from newspapers and a shorthand transcription of the trial, the book aimed at those who could afford a memento of a great execution in an octavo half-leather edition liberally illustrated with portraits of the chief actors and locations in the drama.[5]

[5] G. T. Wilkinson, *An authentic history of the Cato-Street conspiracy; with the trials at large of the conspirators for high treason and murder . . .*, in *Newgate calendar improved*, v (2nd edn., 1820).

33. *The execution of the Cato Street conspirators, 1 May 1820*: from *An authentic history of the Cato-Street conspiracy*, 1820.

For plotting to assassinate the Cabinet, five conspirators were publicly hanged and decapitated, the last English traitors so dealt with. This and the next illustration—one 'polite', the other 'vulgar'—appeared in books each of which constructed or presumed in its readers quite contrary attitudes to the conspiracy and its bloody aftermath.

The illustration of the execution reflects its address to this market (Plate 33). It is oddly demure, massively understating the violence of the event. The background is dominated by the classical arch of Newgate's Debtor's Door, the foreground by the crowd. The crowd is unrealistically thin. The artist claims space chiefly for virtuoso characterizations of couples and bonneted women, their clothes rendered with fashion-plate care. The crowd is a polite one. It is not at all like the real crowd, which was kept far distant from the scaffold and supervised by soldiers and artillery pieces. The figures on the platform, framed by these motifs, are drawn by a clumsier hand and are less interesting than the foreground figures. But there is one shocking detail in the platform picture. The corpses of four of the hanged conspirators sit in a tidy line under the gallows awaiting transference to their coffins and the executioner's knife. Their heads are still hooded, their legs stretch out neatly before them, and their backs are kept erect by the nooses around their necks. Four living figures move about the platform, one approaching the line of corpses with a knife. To the left the headless body of the fifth conspirator can just be seen in its coffin; its truncated neck peeps over the rim. The executioner holds up its head to the crowd. The head drips a little blood, discreetly. But the scale reduces it to near insignificance, so the effect is not unduly alarming.

The second depiction is cruder and more explicit (Plate 34). For popular consumption, it is the fold-out woodcut frontispiece in a pamphlet which summarized the whole course of the trial and executions in thirty-eight pages for sixpence.[6] This publication does not aim at the polite people; it wants artisans' and shopkeepers' money; it wants to talk to the world from which the conspirators were drawn. In the Cambridge University Library copy the frontispiece woodcut is garishly water-coloured. Perspective is clumsy, but the detailing is lasciviously complete. 'Awful Execution of the Conspirators', its caption proclaims. Part of the crowd is drawn in close-up around the platform. There are hats mainly, and constables with tipstaffs ring the platform, but a few strong plebeian faces gawp upwards too, and cavalry on horses are lightly sketched in the background. The people are there, in other words; and so are the soldiers.

The scaffold platform dominates the picture, and nothing is sanitized here either. The hooded and seated corpses, arms bound, are again sustained by their nooses, but this time not neatly. They lurch awry and their heads plausibly loll. The executioner's assistant is master of the scene, all tousled hair and trousers loose. He holds up the ghastly head of Thistlewood by the hair. The head is shockingly large, out of proportion. Gouts of blood drip from it, water-coloured red. One of the five coffins on the platform is filled with Thistlewood's decapitated body, the block adjacent to the red-painted neck; an axe is propped against another coffin. Every figure is numbered and labelled, so nothing is missed. To the left, bearing a curved dissecting knife daubed in red, is

[6] Anon., *The trials of Arthur Thistlewood, James Ings, William Davidson, Richard Tidd, and John Thomas Brunt, for high treason* (2nd edn., 1820).

34. *Awful execution of the conspirators*: from *The trials of Arthur Thistlewood . . . for high treason*, 1820.

the masked figure of 'the surgeon who cut off the heads'. The hangman, squat and ugly, leans against a gallows post, wiping his hands and smirking at the corpses. The best-dressed man on the platform, top-hatted, cravatted, and holding a sheet of paper, is labelled 'the prompter', perhaps the sheriff: he it is who has instructed the executioner's assistant when to hold the head on high and to shout 'This is the head of Arthur Thistlewood the Traitor.'

This is not the detailed reality the distant crowd could have seen. It is the detail of the impossible close-up. But there is no question which of the audiences addressed by these two pictures is invited to see the violence of the event more clearly. There is no question which better conveys the event as news, and as a horror.

Each picture expresses a distinct ideological posture. The emotional reticence of the first supports the view sustained in its text. The message is clear. The monstrosity of the event adheres not to the punishment but to the traitors; the law retains dignity and meaning. The lives of these 'wretched and misguided men' were 'justly sacrificed to those violated laws, which form the bond of union, and the sinews of society in this highly-favoured land'. It has 'pleased Almighty Providence to protect the rulers of this country from the diabolical machinations of a set of lawless wretches, who sought to erect their own interest on murder, rapine, and treason. Their names are transmitted to posterity, branded with the most horrible crimes that disfigure their human nature.'[7] Thus although the pleasurable excitements of the spectacle are licensed in the picture (the blood drips and the hanged figures shockingly droop), the excitement is subordinated to a political message. The scene is not dominated by the arch of Newgate and soldiers are not omitted for nothing.

The popular text, meanwhile, cannot tell us what to feel: radical publishers fear prosecution and must be discreet. So, ostensibly, it allows us to be titillated, disgusted, horrified, moved, awestruck, as we choose. But just as in its accompanying picture the emotional attack on the reader is uncompromising and the law nakedly violent, so the text contrives to nudge readers towards an appropriate understanding. Although the text has to declare that the traitors were 'deluded' and their sentences 'just', the blame for their misdeeds is squarely placed on the government spy Edwards. It was he who 'fattened the conspirators for slaughter' by instigating their crime; and, mysteriously, he was not prosecuted. The conspirators were vulnerable to his blandishments only because some were 'inflamed by public occurrences—others starving from want'. It is against the spy who set them up and who escaped justice that 'natural and just indignation' should be directed. This indignation would 'almost swallow up the horror we must feel at the atrocities intended to be perpetrated by these miserable men'. The implication is that the law has not got it right. Radicals' suspicions were justified. Thistlewood had been a marked man ever since he had escaped a treason conviction in 1817. The conspirators fell into a government trap and the trial was stage-managed to secure exemplary convictions.

Despite their differing purposes and audiences, the two texts have one thing in common. Each describes the last hours of the condemned men and the execution in detail, the 'respectable' text more meticulously than the popular. From these and similar publications, we know more

[7] Wilkinson, *Authentic history*, 352, vii.

about the last hours of the Cato Street conspirators than about those of any other condemned men of the time. Our polite text devotes forty pages to their conduct, and, while its graphic depiction might be discreet, the text is explicit.

This is unexpected. What can explain the fact that a loyalist text attends so poignantly to the traitors' anguish in their last days and hours, as it was betrayed in word, gesture, and pathetic minutiae—in the letters they wrote to wives and children, for example? On the face of it, this looks like a humanitarian narrative, concentrating on the condemned men's pains to elicit *sympathy* for them. Here again we seem to meet that disconcertingly inconsistent mixture of humane consideration and cruelty which characterized the Tyburn procession or the lax disciplines of condemned highwaymen's last Newgate days.

But was this narrative humanitarian? The detailing of the last hours actually conveyed a message useful to loyalists. In the record of their defiance the conspirators were allowed to reveal their unnatural dispositions—as monstrous others, indeed, outside the reach of sympathy or civilized understanding. There were shocking and monstrous things to report. As deists, four 'repeated their disbelief in the divinity of Christ, and refused . . . to seek pardon of their offended Maker'. The clerical ordinary's ministrations were rejected. The condemned service had to be abandoned lest the doomed men mock it. Brunt used communion wine to drink the king's health. Ings wished that his body might be conveyed to the king 'and that his Majesty, or his cooks, might make turtle soup of it'. Beyond that, the pathetic narrative can as well be said to appeal to tastes older than the sympathetic—to old curiosity, in short. We are not meant to feel much pity for these men. Rather, we are expected to want to know how men *in extremis* comport themselves; how one *might* prepare to have one's head chopped off. The information which is given to settle these interesting questions is drained of humanitarian intent. All that happens is that the reporter surrenders to antique fascination with what ensued, to explore how character coped. We notice how we too surrender to this curiosity. How *did* they cope—and how would we?

They coped marvellously as it happened. When in the last minutes their irons were struck off, four of the men were quite composed. The butcher Ings, dressed in butcher's garb so that the hangman would not get his best clothes, kept singing 'Oh! give me death or liberty!' He turned to a turnkey: 'Well, Mr Davis, I am going to find out this great secret,' and he hoped God would be 'more merciful to us than they are here'. 'Aye, to be sure,' Brunt joined in, 'It is better to die free, than to live slaves!' Brunt spurned last efforts to make him repent: 'What have I done? I have done nothing! What should I ask pardon for?' 'Come my old cock-o'-wax,' Ings said to Tidd on the way to the scaffold, 'keep up your spirits; it all will soon be over.'

On the scaffold the men faltered when they saw the sawdust and the coffins. But Ings 'raised his pinioned hands, in the best way he could, and leaning forward with savage energy, roared out three distinct cheers to the people, in a voice of the most frightful and discordant hoarseness'. The reporter thought these were 'unnatural yells of desperation . . . the ravings of a disordered mind, or the ebullitions of an assumed courage'. Perhaps they were, for Tidd told Ings: 'There is no use in all this noise. We can die without making a noise.' As the men were hooded and noosed, Tidd and Ings instructed the executioner on how to tie the knot: 'Do it well—pull it tight.' 'We shall soon know the last great secret,' Thistlewood said. Brunt's last act was to take a pinch of snuff from a paper in his hand, pushing up his night-cap to gain access to his nose. The others sucked oranges. Their exchanges continued until the executioner dropped the trapdoor which held them. Then 'the agonies of death were exhibited to the view of the crowd in their most terrific form'. Brunt kicked and choked for five minutes on his noose, until the executioner pulled at his legs.

An hour later Thistlewood's body was cut down and put in its coffin. When the cap was removed, the 'engorged and purpled' face was revealed. The neck was pulled forwards to rest on the block at the coffin's side. The masked man approached with his curved knife and 'his mode of operation showed evidently that he was a surgeon'.[8] He sawed the head off and handed it to the assistant executioner. The assistant displayed it high, announcing that this was the head of Thistlewood the traitor. The body was then shoved back into the coffin and the head jammed in alongside. One by one the others were dealt with likewise. Knives got blunt or turned on vertebrae: the surgeon used three altogether. The executioner accidentally dropped the last head, Brunt's, into the sawdust and the crowd shrieked. 'The rush of blood at each decapitation was so great', a later source recalled, 'that the end of the scaffold had the aspect of a slaughter house.'[9]

[8] A medical student was likewise employed to decapitate the Scottish insurrectionists Hardie and Baird in Stirling in Sept. 1820: *An exposition of the spy system, pursued in Glasgow*, 233. Few decapitations went smoothly in this period; see the account of Despard's in 1803, Ch. 1.

[9] C. Hindley, *The life and times of James Catnach (late of Seven Dials), ballad monger* (1878), 92.

This reporting also covers the crowd. It tells us (mandatorily) that it 'manifested all the thoughtless levity of a common mob'. But as usual the evidence proves nothing of the sort. It proves that the crowd had opinions, and that these were feared.

Outside Newgate on the day before the execution 'thousands' milled about, anxious for news. Many stayed up all night watching workmen erect the scaffold by torchlight; more assembled at the barriers at five in the morning. Among them were groups who indulged in 'language disgraceful to themselves, and alarming to those who felt anxious for the peace of the metropolis'. What this meant was that they were the kind of people who constantly attended 'factious' and 'mischievous' meetings—Billingsgate and Shoreditch people, like the conspirators themselves.

Tumult and riot were expected. Placards were prepared overnight announcing that the Riot Act had been read and instructing the crowd to disperse, just in case. Foot and Life Guards were deployed in the prison and hidden in buildings around Newgate or posted in line at each end of the Old Bailey, and six light field-pieces of flying artillery were assembled near Blackfriars Bridge, remaining there until it was all over. The City Light Horse and other regiments were kept under arms in barracks in Gray's Inn Road. A civil force of 700 men from the City wards was held in readiness. The scaffold was ringed by constables, and barriers kept the crowd unusually distant from the platform (so that when Brunt tried to address the crowd they were too distant to hear).[10]

Although 100,000 people were said to have filled the streets, most would have seen little or nothing; they flowed around corners far away. Those who did see waited in a tense silence at first, broken by low murmurs and ripples of excitement as the sheriffs and gentlemen made their passing appearances. When the conspirators appeared on the platform, someone cried 'God bless you, Thistlewood!' and all hats were removed. But it was the decapitations rather than the hangings that brought the crowd to life. 'Shoot that — murderer,' they cried when the surgeon began his work, and they hissed and hooted the executioner who held up Thistlewood's head. Howls, groans, and hisses accompanied every stage of the butchery, and women shrieked and fainted. Muskets kept the peace.

Meanwhile the authorities had been frantically busy. The King in Council met for two hours on Saturday to decide which of the

[10] *Morning Chronicle*, 1 May 1820.

conspirators should hang on Monday and which be transported. The full Cabinet and the judges ticked off the names one by one, and then ordered the executions to ensue swiftly—'to render the spectacle more imposing'.[11] That evening the sheriffs broke the news to the men. The five condemned took it coolly; some of the reprieved broke down in relief. There was a great scurrying of messengers to and from the Home Office with military instructions. The sheriffs persuaded Sidmouth to dispense with the drawing of the conspirators on hurdles lest 'inconvenience' be caused. The governor and sheriffs fended off the radical Alderman Wood, who wanted to question Thistlewood about Edwards the spy: the condemned were not to be disturbed, for there was spy-work to hide. The wives and children came to bid the men farewell, and these meetings were deeply affecting.

The surveyor of public buildings spent Sunday organizing the erection of the scaffold and barriers, and the workmen hammered all night. Through it all, the Revd Cotton dithered vainly around the impenitent conspirators. Davidson broke down at last, but he asked for a Wesleyan minister who turned out to be a humble journeyman 'in a situation in life not well adapted to reveal the holy tenets of salvation to a dying man'. So Cotton got a job to do on one conspirator at least. On the morning of the execution the five wives petitioned the king for permission to bury the bodies, 'confident that all desire of further vengeance has ceased'. Friends applied for the bodies too. Disconcertingly, they proposed to exhibit the bodies to raise money for the surviving families. Both requests were refused by the Home Office: the bodies were to be buried in quicklime inside the prison. Then there was the toing and froing in the final hour: the last breakfast, the arrival of sundry noblemen to breakfast with the governor and to watch the preparations in the press-yard where the manacles and shackles were removed, the last offers of communion . . .

Almost the only detail not included in the published report is the official assessment afterwards of how it all went, a gem of its kind. It was a 'quiet' ceremony, the lord mayor informed Sidmouth later that day:

> There was by far the most dissatisfaction expressed at the decapitation, but it was but momentary—on the whole there has seldom been a more tranquil Execution witnessed—the Troops were so dispersed that in whichever way the populace approached the Old Bailey they must be seen in force to deter any attempts at rescue—the mob are beginning slowly to disperse and unless any-

[11] *Morning Post*, 2 May 1820.

thing fresh were to occur I think it will be unnecessary to trouble your Lordship with any further account of this unpleasant business.[12]

Also excluded from the published report is the memento which John Adolphus, defence counsel, sent to Lord Liverpool. He got all the conspirators to provide posterity with examples of their handwriting: old curiosity again. Accompanied by an engraved card on which Adolphus offered his compliments, scraps of paper which the prisoners obligingly inscribed with their last thoughts survive in Liverpool's papers (Plate 35). 'Sir, I Ham a very Bad Hand at Righting,' writes Richard Tidd. Thistlewood's was defiant:

> Oh! what a mine of mischief is a Statesman!
> Ye furies, whirlwinds, and ye treacherous rocks,
> Ye ministers of death, devouring fire,
> Convulsive earthquake, and plague-tainted air,
> All you are merciful and mild to him.
>
> Newgate, 27th April, 1820. ARTHUR THISTLEWOOD.[13]

What finally of the polite spectators to which our polite publication scarcely attends? All we are told is that the windows opposite the scaffold were packed, the best of them having fetched three guineas apiece. Many thousands would keep their distance from an event like this. That respectable radical Samuel Bamford 'would not have gone [to the Cato Street executions] on any account; and such places were the very last at which persons of our description should be seen'. This was wishful thinking, however. His friends went despite his disapproval—'*merely . . . from curiosity, to see how such things were done*', as they said in that old and revealing language. Curiosity brought Henry Angelo also, booking his cheaper half-crown window overnight and playing cards with its jovial occupants until morning. Bamford records that afterwards 'the execution was the subject of conversation in every place, and I soon heard as perforce I must, the particulars of the disgusting transaction'.[14] Particulars, yes: but of feelings we hear little.

Perhaps some of the fine people felt the pain of the event within themselves. If so, the spasm was brief. Princess Lieven wrote to Metternich an hour after the executions that they saddened her: 'I feel pity for these poor human beings, for these aberrations of mind and

[12] HO44.6, fos. 135–6; Sales, *English literature in history*, 162–3.

[13] Liverpool papers, BL Add. MS 38284; E. Henderson (ed.), *Recollections of the public career and private life of the late John Adolphus* (1871), 112.

[14] S. Bamford, *Passages in the life of a radical* (orig. pub. 1884; Oxford), 311–12.

tho in Newgate Close Confin'd
No fears Alarm the Noble mind
tho death itself Apears in View
Daunts not the soul Sincerely true

Let S——h: And his Base Colleagues
Cajole And Plot their Dark intrigues
Still Each Brittons Last words Shall be
Oh give me Death or Liberty

Newgate 27 April 1820
John Thomas Brunt

He That Answ^r^eth A Matter before he
heareth it. it is a folly and a Shame Upon him
Thou Shalt not Oppress a Stranger in a
Strange Land ~
Thou Shalt not Pervert the Judgement
of a Stranger. W Davidson

Oh! what a mine of mischief is a statesman!
Ye furies, whirlwinds and ye treacherous rocks,
Ye ministers of death, devouring fire,
Convulsive earthquake, and plague tainted air,
all you are merciful and mild to him.

Arthur Thistlewood

Newgate 27th April 1
1820.

Sir I Ham a very Bad Hand at Righting Richard Tidd

if my life is destroyed in this conspiracy I shall consider I ham a Murderd man the reason is I was not the Inventer of this conspiracy & I should never known the party if Mr Edward had not come to my house & got a quainted with me & it was through him I was drawn a one side in my distrest & unfortunate situation because I could not keep my Wife & children

James Ings

35. *The Cato Street conspirators' last autographs*, 1820.

The Cato Street conspirators' counsel collected samples of handwriting from each conspirator shortly before his hanging or transportation, and sent facsimiles to interested parties, including Lord Liverpool, 'one of the [conspiracy's] intended victims'. Reproduced here are the mottoes of the five who were hanged and decapitated. Thistlewood and Brunt trumpet defiance; Ings pleads poverty and manipulation by the government spy Edwards; the Methodist Davidson offers copy-book aphorisms in copy-book script; and Tidd meekly apologizes for his near-illiteracy.

imagination.'[15] Mrs Arbuthnot's brother Cecil Fane also went from 'curiosity'—that word again—wishing 'very much to see how they w^{d} behave' since he had never been to an execution before. He at least was a rare spirit whose feelings overcame him. 'When [the conspirators] were tied up,' Mrs Arbuthnot reported, 'he felt so nervous & in fact felt so much more than they themselves did that he retired into a corner of the room and hid himself that he might not see the drop fall' (here again the Boswellian conviction that the man of sensibility felt more than the punishment's vulgar victims possibly could). Significantly, Fane's squeamishness 'excited great contempt in the people who were in the room with him'. Mrs Arbuthnot herself briskly refused to feel sorry for the conspirators: 'One really ought to thank God that the world is rid of such monsters.' The most typical reaction was probably

[15] Stanhope, *Cato Street conspiracy*, 140–1.

that of the pretty young lady 'who kept her eyes fixed on it all the time &, when they had hung a few seconds, exclaimed, "There's two on them not dead yet"!!'[16]

Press reactions were dictated by political alignments. Over the preceding weeks the radical press had been contemptuous of the show trial but had to tread carefully. Few if any newspapers frontally addressed the psychodrama of the decapitation. The whig–radical *Morning Herald* was virtually alone in denouncing 'the barbarous and inhuman butchery of the lifeless corpse' as 'the offspring of irrational barbarian vengeance . . . a horrid anomaly in the escutcheon of a humane and civilised nation, . . . a relic of that heathenish spirit of revenge, so contrary to the genius of the Christian faith'. How could it be supported 'by men who can think and feel'? This is the language we should expect to hear in 1820. Its appeal to civility and progress still suggests something other than an imaginative identification with the men's terror, however. For the rest, next to nothing on the affective level. The *Morning Chronicle* briefly hoped never again to witness such a spectacle, but it was chiefly disappointed 'at seeing several women among the spectators'. The tory *Morning Post* was pointedly silent on the violence of the executions. Chiefly it eulogized the ministers whom the Cato Street 'monsters' had planned to slaughter. It castigated opposition newspapers which implied that ministerial injustice and folly had provoked the plot, and stressed the crowd's acquiescence and the unheroic grimaces of the doomed men. All the polite papers without exception deplored the 'thoughtless levity' of the mob, although it was the mob whom the spectacle upset most directly, not the polite people. This strange mistake was repeated all the way down to 1868, and in this book's last chapter it will need attention.

None of the reactions noted above was peculiar to 1820. In 1817 at Derby 6,000 people watched the deaths of the labourers Brandreth, Turner, and Ludlam. They had led the abortive Pentrich uprising. Contemporary accounts are less lavish than the Cato Street accounts, for this was far from London. But some newspaper detailings were precise.[17]

Here again reports told how the scaffold was furnished with a block of wood, sacks of sawdust to catch the blood, two axes, two sharp

[16] F. Bamford and the duke of Wellington (eds.), *The journal of Mrs Arbuthnot, 1820–1832* (2 vols., 1950), i. 15–16.

[17] For the following account, see *Examiner*, 9 Nov., *The Times*, 8 Nov. 1817; Radzinowicz, i. 226–7. For the Pentrich uprising, see Thompson, *Making of the English working class*, 659–69.

knives, a basket to hold the heads. Before the execution the men were ritually dragged round the gaol yard on a hurdle. Some nodded to watching prisoners, others shut their eyes; some prisoners in the yard 'screamed and wept most bitterly'. The three men were led to the gallows, they prayed, the trap dropped, and they hanged for half an hour. What is precisely noted is that Brandreth's beard looked 'very frightful from underneath the white cap', and Ludlam was 'repeatedly convulsed after he had been thrown off'. Then a masked coalminer chopped at Brandreth's neck but made a sorry job of it. His assistant had to hack through the sinews with a knife. The executioner held up the head to the crowd, crying three times, 'Behold the head of the traitor, Jeremiah Brandreth.' The other two corpses were dealt with in the same fashion.

Reactions? In this vignette, as at the Cato Street deaths, only the crowd reveals feeling. At Brandreth's decapitation

> there was a burst of horror from the crowd. . . . The instant the head was exhibited, there was a tremendous shriek set up, and they ran violently in all directions, as if under the impulse of sudden phrenzy. Those that resumed their station groaned and hooted. The javelin-men and constables were all in motion, and a few dragoons, who had been stationed at both ends of the street, drew nearer with drawn swords. But all became immediately calm. Very few of the immense multitude now remained, and these looked quietly on while the heads of Turner and Ludlam were successively exhibited. . . . The heads and bodies were then thrown into the coffins, and all spectators dispersed.

Watched by soldiers, the crowd was to a degree intimidated, but its disgust was not directed against the 'traitors'. Most knew that they had been betrayed by the government spy Oliver, and most left the scene at the first decapitation. From the élites, however, near-silence once more. One cleric recalled that his father 'came home sick and faint'; but there was only one full expression of sympathy for the men, and that was obscure at the time. The poet Shelley contrasted the silence which accompanied the Pentrich men's deaths with the hypocritical outpourings at the recent death of Princess Charlotte:

> These men were shut up in a horrible dungeon for many months, with the fear of a hideous death and of everlasting hell thrust before their eyes; and at last were brought to the scaffold and hung. They too had domestic affections and were remarkable for the exercise of public virtues. Perhaps their low station permitted the growth of those affections in a degree not consistent with a more exalted rank. They had sons, and brothers, and sisters, and fathers, who loved them, it should seem, more than Princess Charlotte could be loved by

those whom the regulations of her rank had held in perpetual estrangement from her. . . . What these sufferers felt shall not be said.

Shelley's response was effortfully empathetic, but his observation that few others perceived the victims in these terms remains his article's most eloquent point.[18]

So back to the opening questions: among people of sensibility what filters blocked imaginative engagement with bloody inflictions of this order? Granted the tense climate of the times and the understanding that the executions were expedient, we might still wonder whether the punishments were tolerated because unconscious assumptions and images helped determine the ways in which they were seen. It is difficult to deny this likelihood. The massive degradation of the traitor's body drew unambiguously on the body metaphors of past times, not of the age of reason or sensibility. It was Blackstone who had written that the traitor was to be 'exterminated as a monster and a bane to human society'.[19]

Furthermore, the traitor's kinship with the past traitors of English history was conveyed within the symbolic labels attached to him. Here were the same masked executioner, the same coffin, the same block, the same shrouded scaffold as had attended the deaths of Elizabethans or Jacobites—and the same holding up of a *head* as was illustrated in every woodblock on Charles I. The butcheries of 1817 and 1820 bore echoes of these kinds, subliminally licensing the *frissons* which came with excess. Except in the case of sensitive spirits like Cecil Fane, and perhaps of those who felt so much that they had to deny their feelings (their silence speaking, therefore), what was felt was the excitement appropriate to the witnessing of a 'historic' moment with monstrous others at its centre. The pain of these others might well be conscientiously attended to in narratives composed after the event: but what about during it? The pretty young lady at Mrs Arbuthnot's window sailed close to the wind of social disapproval in watching the Cato Street hangings at all. But she had an acceptable alibi for the release of gratifications she would have denied in everyday life. Although we cannot know how she slept that night, in that moment of sharp *frisson* which Mrs Arbuthnot noted she could not have felt that traitors' deaths had the slightest connection with her own.

[18] Shelley's pamphlet was published anonymously in 1817 as '*We pity the plumage, but forget the dying bird': An address to the people on the death of the Princess Charlotte, by the Hermit of Marlow* (reprinted in D. L. Clark (ed.), *Shelley's prose, or A trumpet of prophesy* (1988), 162–9).

[19] W. Blackstone, *Commentaries on the laws of England* (4 vols., 9th edn., 1783), iv. 380.

2. *Punishing treason*

Edgar Wind once suggested that exaggerated cruelty in early European punishments spoke for the legacy of prehistoric sacramental processes in which the quasi-deified criminal acted as the community's scapegoat and redeemer. If so, we may meet deep resistances in mentality here which possibly extended across vaster chronologies than those which historians usually explore:

> The very shape of the wheel and the cross, the very act of quartering, point to ideas of a cosmic order, and would be senseless but for a victim sacrificed for a cosmic purpose. It is probably more a sign of human inertia than of human wickedness that these practices were continued long after they had lost their meaning. A medieval executioner who broke a man on the wheel did not know that he was repeating the form of a ritual by which his ancestors had sacrificed a god.[20]

The point gains force from the fact that a medieval cosmology continued to shape some understandings of execution well into the nineteenth century, and never more so than in the execution of traitors. In the residual power of this symbolism mental continuities were enshrined which take us distant from interests of state.

The old language was explicit. Since it was in the traitor's body that felonious or treasonable thought was conceived, each bodily part bore its own responsibility for aspects of the crime and was punished accordingly. The body which had committed a crime against sovereignty—and thus against nature—must be infinitely degraded.[21] This corporeal symbolism was classically expounded by Sir Edward Coke in his Gunpowder Plot judgement of 1606. For Coke the meaning of the attack on the traitor's body lay in the fact 'that it is the physic of state and government, to let out corrupt blood from the heart'. It followed that the traitor should be drawn to execution across bare ground:

> as being not worthy any more to tread upon the face of the earth whereof he was made: also for that he hath been retrograde to nature, therefore is he drawn backward at a horse-tail. And whereas God hath made the head of man the highest and most supreme part . . . he must be drawn with his head declining downward, and lying so near the ground as may be, being thought unfit to take the benefit of the common air.

Then he should be 'strangled, being hanged up by the neck between heaven and earth, as deemed unworthy of both, or either'. Cut down

[20] 'The criminal-god', *Journal of the Warburg Institute*, 1 (1937–8), 244.

[21] J. Bellamy, *The Tudor law of treason: An introduction* (1979), 204.

alive, his privy parts were to be cut off and burnt before him, 'as being unworthily begotten, and unfit to leave any generation after him. His bowels and inlay'd parts taken out and burnt, who inwardly had conceived and harboured in his heart such horrible treason. After, to have his head cut off, which had imagined the mischief.' Having been quartered, the body was to be set on high 'to the view and detestation of men, and to become a prey for the fowls of the air'.[22] Women convicted of treason (usually of coining or of the petty treason of murdering husbands) were to be burnt alive. According to Blackstone, this was 'out of consideration for their sex', avoiding the bodily exposure which quartering would necessitate. But it was still the treacherous *body* that was the target of retribution.

What of this language could possibly survive in the period 1790–1820? Significant understandings, we shall suggest shortly. It is fair comment that most people act in terms of 'ready-made ideas, commonplaces and intellectual bric-a-brac, the remnants of cultures and mentalities belonging to different times and different places'. 'Knowledge' may be embedded in rituals, gestures, or collective remembering, quite at odds with intellectual coherence; it may not even be conscious.[23]

But before that is admitted, it would be silly not to notice first that the old understandings were *diluted* as time passed. Even in Elizabethan times mitigations of Coke's sentence were customary. Traitors were drawn to the scaffold on hurdles to save them from death or unconsciousness; others were strangled before disembowelling or burning; aristocrats were decapitated immediately by the king's grace. In the eighteenth century the classic punishment was further eroded as the fixation on the treacherous body was displaced by a concern with mind. Judges were still statutorily bound to pronounce the main elements in Coke's sentence for male traitors. Colonel Despard and his friends were still told in 1803 that they were to be

> severally drawn on an hurdle to the place of execution, and there be severally hanged by the neck, but not until you are dead, but that you be severally taken down again, and that whilst you are yet alive, your bowels be taken out and burnt before your faces, and that your bodies be divided each into four quarters, and your heads and quarters be at the King's disposal.[24]

[22] T. B. and T. J. Howell, *Complete collection of state trials* (33 vols., 1809–26), ii. 184.

[23] J. Le Goff, 'Mentalities: A history of ambiguities', in Le Goff and P. Nora (eds.), *Constructing the past* (Cambridge, 1985), 169. Cf. R. Chartier, *Cultural history: Between practices and representations*, trans. G. Cochrane (1988), 28; A. Burguière, 'Demography', in Le Goff and Nora (eds.), *Constructing the past*, 110–11.

[24] *State trials*, xxvii. 528.

But in practice 'all discretion [was] transferred to the executioner'.[25] He decided whether or not to strangle the victims before the butchery began, or to omit the disembowelling, judges sanctioning both merciful acts. Thus although the Jacobites of 1745 were sentenced to be half-hanged, disembowelled while alive, decapitated and quartered, and their heads to be spiked over Temple Bar, the executioner killed them after their symbolic half-hanging and before their evisceration: he thumped their chests and cut their throats.[26] There was only one *post mortem* disembowelling after this, when in 1781 the Tyburn hangman made a long incision in the breast of the French spy La Motte, cut out his heart, and threw it to the flames—but this after La Motte had hanged for fifty-seven minutes.[27] Thereafter the sovereign remitted this part of the sentence as an act of prerogative. And in England though not apparently in Ireland there were no more displays of spiked heads,[28] and decapitation was also discarded for the male treason of coining. Just as men were hanged to death before evisceration, so female traitors too were strangled before burning. The last woman burnt alive for petty treason (murdering her husband) was Catherine Hayes in 1726, but this was only because the executioner bungled the strangulation when the fire was prematurely lit and he burnt his fingers.[29] Women's treasonable bodies continued to be burnt *post mortem* until 1789, but that was the last of it. When Sir Benjamin Hammett invited the House to 'go with him in the cause of humanity', denouncing the punishment as a 'dreadful' sentence, a disgrace to the statutes, and one of 'the savage remains of Norman policy', parliament abolished the burning of women for treason in May 1790.[30]

One could say, then, that in the execution if not the pronouncement of sentence the penal process against traitors became more respectful of the body in the eighteenth century. By the 1790s we seem to be faced only with vestiges of the old sentence, even with a loss of the old corporeal understandings. Despard and his fellows in 1803 and Brandreth and his fellows in 1817 were solemnly put on sledges harnessed to horses and

[25] S. Romilly, *Observations on the criminal law of England, as it relates to capital punishment . . .* (1810), note F.

[26] *State trials*, xviii. 351 (Townley's execution).

[27] *Gentleman's Magazine*, 51 (1781), 341–2.

[28] The decapitated heads of 2 Irish rebels, Colclough and Harvey, were stuck on pikes over the Wexford market and sessions houses in 1798: Borrow, v. 397.

[29] William Guest, executed at Tyburn in 1767 for the high treason of clipping coin, was drawn to the scaffold on a sledge to mark the treason, but then only hanged. When Amy Hutchinson was executed at Ely in 1750 for murdering her husband, she too was simply hanged: Rayner and Crook, iv. 51, iii. 30–40, 184–6.

[30] PH xxviii (1790), c. 782 ff.

drawn on ritualized circuits of the gaol yards before they were made to mount the scaffold; but the symbolism was hollow by now. Sentenced to be drawn in this fashion and knowing that when all was said and done he would be decapitated only after his death, an Irish rebel protested at the mummery of it: 'Ah! what need of all this cookery?'[31]

A dilution of intended meanings was also evident in the fact that the deaths of traitors were seldom reported in the outraged terms which their aggravated humiliation was meant to invite. Their final comportment and words had always been scrutinized for indications of penitence, but by the later eighteenth century the revelation of character was the central interest. As in the descriptions of the Cato Street conspirators' last hours, readers were invited not so much to offer pity as to share the quality of *affect* achieved in the spectacle. This was how the Irish rebel James Coigly's death was described when he was hanged, decapitated, and buried beneath the gallows on Pennington Heath in 1798:

> While the cap was drawing over his face, he made signs to his friend to approach nearer, and, bowing, dropped his handkerchief at his feet. That most trying time between when the cap was drawn over his face and the falling of the platform, was longer than usual; during all this time, no trepidation could be discovered in his limbs or muscles. His lips moved, and his hands were lifted up in prayer, to the last moment. He died apparently with little suffering.
>
> The spectators behaved with great respect towards him. When he declared his innocence, a buz [*sic*] of applause ran through the multitude, and there was even some clapping of hands. Toward the close of his address, many of the spectators wept, and some of the soldiers were unable to repress their tears.[32]

Given the emotional intimacy achieved in this kind of narrative, in view of the crowd's reactions too, and since many by now shared Romilly's conviction that the old sentence and what remained of its execution was 'cruel and disgusting', the apparently unstoppable dilution of the treason sentence could be made to fit the requirements of a progressive and humane narrative very easily indeed. It could find a place in the history of the 'civilizing process' itself.

Yet finally we retract a little. That the horrors of the treason sentence survived up to 1820 and continued to blunt the sensibilities of polite

[31] B. Montagu, *Debate in the House of Commons, April 5, 1813, upon Sir Samuel Romilly's Bill on the punishment for high treason* (1813), 60. Radzinowicz (i. 225 n.), citing an anecdote recollected in 1880, attributes to Despard the cry as the hurdle confronted him: 'Ha, ha! what nonsensical mummery is this?' This must be a garbling of the original Irish story. Hanbury Price, condemned for the treason of coining, was drawn to the Newgate scaffold on a hurdle as late as 1828 (*Ann. Reg.* 1828: 64).

[32] *State trials*, xxvii. 252.

audiences are social facts as significant as are those of long-term dilution and demise. To dismiss those decapitations as anachronisms begs a large question. In an era when, after the conviction of Despard and his co-conspirators in 1803, the Cabinet had to be solemnly called 'at the request of the Lord Chancellor to consider what advice should be given to His Majesty respecting the disposal of the heads of the prisoners',[33] it might be clear that beneath the pressures making for change, older mentalities still flourished.

Unexamined assumptions from past times certainly blocked parliament's ability to imagine the treason punishment's end. Romilly found this when he tried to remove its aggravated elements in 1813–14 by having the traitor only hanged and his body 'put at the King's disposal'. If one MP, Frankland, talked of 'mischievous attempts to unsettle the public opinion with respect to the enormity of these atrocious offences', it was the attorney-general, Garrow, who defended the full penalty on the grounds that it had the sanction of centuries. To be sure, it was also argued that the treason sentence now entailed no more pain than hanging did. The punishment, 'though in appearance theoretically severe, is in reality practically humane', Frankland pointed out: the body was dead.[34] But this statement merely confirmed how much the 'theoretical' severity (the symbolic element in other words) continued to matter in the conservative mind. The old decapitating ritual would 'excite disgust against the crime', Frankland also said; it would induce terror, Garrow added. Moreover, treason must be the highest crime known to law: it alone could sow 'unutterable misery and confusion . . . deluging

[33] Pelham papers, BL Add. MS 33122, fo. 116.

[34] Those who had defended the *post mortem* burning of women had likewise claimed, as Loughborough did in 1786, that though the spectacle was 'attended with circumstances of horror . . . no greater degree of personal pain was sustained, the criminal being always strangled before the flames were suffered to approach the body': Radzinowicz, i. 478. The burning of women until 1790 for the petty treason of husband-murder or the treason of coining itself indicates how difficult it was for sympathetic imaginations to see to the terrifying heart of these penal relics. The hanging and burning in Mar. 1789 of the last woman executed for coining, Christian Bowman, was noted with astonishing restraint—first in a broadside which only moralized at her expense, and then in the *Gentleman's Magazine*'s bald record that 'the woman for coining was brought out after the rest were turned off, and fixed to a stake and burnt, being first strangled by the stool being taken from under her'. In parliament Hammett might castigate the stake as a remnant of Norman policy; but his key point was the pragmatic one that 'the shocking punishment did not prevent the crime'. The *Universal Daily Register* condemned the punishment because it betrayed male obligations to womankind, because it degraded the watching mob, and—with wonderful bathos—because the smoke from the fire caused a public nuisance, making ill people in the neighbourhood even iller. See *The life and death of Christian Bowman, alias Murphy* (1789) (partially reproduced in J. Ashton, *Chapbooks of the eighteenth century* (orig. pub. 1887; 1989), 453–4); *Gentleman's Magazine*, 59/1 (1789), 272; *Universal Daily Register*, 24–7 June 1788.

the country in blood and slaughter, in plunder and devastation'.[35] In this context at least, if in no other, those who defended the *ancien régime* wanted proportional punishment to apply. How could it not be proper, Garrow asked in 1813, to attach a 'greater disgrace' to the murder of a virtuous king than to the murder of the meanest subject; how otherwise could 'authority . . . be protected from the violence of the subject?'

Reformers mustered heavy irony against these arguments, but to no effect. 'To throw the bowels of an offender into his face, one of the safeguards of the British Constitution!!' Ponsonby snorted. 'You may cut out the heart of a sufferer and hold it up to the view of the populace and you may imagine that you serve the community,' Romilly declared; 'but the real effect of such scenes is to torture the compassionate and to harden the obdurate.'[36] Romilly reintroduced his Bill in 1814 and it was then passed in modified form, but only by accommodating Ellenborough's and Eldon's insistence that the sentence for the quartering of the traitor's body be retained even if it was unlikely to be enacted, and Yorke's amendment that insisted on beheading after the hanging.[37] Small gain therefore. In future, traitors' fates were to be no different from Despard's. They were still to be sentenced as Brandreth and Thistlewood and their colleagues were sentenced: 'to be drawn on a hurdle to the place of execution, and there be severally hanged by the neck until you be dead—and that afterwards, your heads shall be severed from your bodies, and your bodies divided into four quarters shall be disposed of as His Majesty shall direct'.[38]

Parliament's resistance to the abolition of the aggravated sentence could be construed as a reasoned response to social disorder. This was no time for gratuitous mitigations of the law; the lessons of the French Revolution could not be denied. Also the truth seemed self-evident that 'the great and important principles of this constitution are the best, the wisest, and the freest, that the sun ever yet saw'. Still, Romilly's contempt for those who diluted his Bill shows that the difference between rationalization and rationality was not hidden to sceptical minds: 'I am not so unacquainted with the nature of prejudice as not to have observed that it strikes deep root; that it flourishes in all soils and spreads its branches in every direction.'[39] He was right to see that ratio-

[35] The lord president pronouncing sentence at the Stirling treason trials in 1820: see C. J. Green, *Trials for high treason in Scotland, under a special commission held at Stirling, Glasgow etc. in 1820* (3 vols., Edinburgh, 1825), iii. 540.

[36] Montagu, *Debate on the punishment for high treason*, 3–4, 6, 19, 41.

[37] Radzinowicz, i. 519–20.

[38] *State trials*, xxxii. 1393.

[39] Romilly, iii. 100; Montagu, *Debate on the punishment for high treason*, 41.

nalization drew on deeper needs than reasoned interest. It was not obvious to reason that the ritual elements in traitors' punishment increased popular disgust against treason. We cannot know how deeply the intended messages were accepted by watching crowds or newspaper readers: doubtless disgust was often transferred unconsciously from the law's brutality to the law's victims. What we do know is that such disgust as was *audible* in the crowd was directed at the sentence rather than at the victims, and that martyrs were made. It can only have been, therefore, that judges' and tories' defence of the treason punishment was based unconsciously upon a respect for the power of symbols which was part of their ancestral inheritance. Their conviction that popular audiences would be cowed by the power of symbolic aggravations was a projection of their own submission to that power.

As for the poise of those high people who watched the old Elizabethan display from the Newgate windows in 1820, it is undeniable that the executions were conducted in their interests and that this helped them. But they were also watching a tableau whose sanctioned meanings were rooted in their sense of their own history, and in inherited visual memories as well. Even in 1820 the empathetic imagination had to do battle with legacies of these kinds.

PART IV

PUBLIC OPINION

CHAPTER 12

OPINION AND EMOTION

1. *Inventing public opinion*

A PENAL REFORM NARRATIVE WHICH ASSUMES THAT MOST OF THE middle-to-middling classes became ever more indignant about harsh punishment in the early nineteenth century looks problematic now. For most people the strangulation of felons on scaffolds always signified something other than itself, it seems. Not much in the grimy business was seen as we can afford to see it at our safe distance. Few saw or heard the terrified urinations and ululations on the platform. Few saw how felons' plight might raise questions about how strong people subjected weak. And if some did see the pity of it all they usually turned away. Hierarchy was ordained and order and property must be protected somehow.

So where did 'public opinion' lie in the murky attitudinal brew surveyed in Part III? Where were the early nineteenth-century agents who felt the cruelty strongly enough to want to change things—not because they were inefficient but because they were cruel? We discern how artificial a construct 'public opinion' was in this as it is in all eras. Rare individuals did confront scaffold cruelties; but the full range of feelings and views in the opinionated classes could never be accommodated within the opinion which it was in reformers' interest to identify. When reformers did appeal to public opinion, they appealed to a construct which they had in some degree invented.

Yet constructs do signify; illusions do too; and so does social learning, as people cultivate feelings socially modelled by others which connect plausibly with their perceptions of the material world and their standing in it. So those 'teachers' who constructed new models have still to be accounted for. If by and after the 1810s small groups of reformers contrived the illusion that right-minded people held execution in abhorrence, this was a contrivance of consequence. Public opinion alone would not end the bloody code; no tory politicians in the 1820s would defer openly to it. But the notion that there was an opinion was increasingly difficult to ignore. With the explosion of debates after 1815 on a range of issues (penal reform far from the most prominent), and aided by a more vocal newspaper press, people came to believe not only in the existence of 'public opinion' but also in its inherently virtuous dispositions: so that more and more were induced to lend support to their conceptions of it. The naming of such a construct can bring it into being when the naming is energetic and purposeful enough.

If opinion was mobilized, however, it was not exclusively through argument. We may be sensationalists enough ourselves to agree with the eighteenth century's holistic view of 'thought'. Calculations about harsh law's inefficiency and counter-productiveness were important, but so was the appeal to emotion. It is not a history of ideas or interests that follows in this part therefore, but studies illuminating how ideas were energized by feeling. 'Reform', 'justice', 'fairness', 'benevolence', 'progress', and 'humanity' were key elements in the ostensibly reasoned attack on *ancien régime* law, but it was through the emotional resonances associated with these images that opinion would be chiefly touched.

A narrative summary may be helpful at this point. Criticisms of the capital code were audible in the mid-eighteenth century. They came from great men—Johnson or Fielding—and small, even from crime-reporters, it has been shown.[1] But the really giddy time for law reformers was the 1770s and 1780s. With Wilkes's triumphs fresh in mind and enlightened thought the rage, optimism itself appeared reasonable as privilege looked set to tremble before reason's clarion call. Reform was on the agenda, and penal reform was part of it. Reform '*will* take place', Capel Lofft proclaimed in 1786, establishing then a characteristic association of these years: 'The people of this country will remember, that unequal severity in the laws is always either consequent [on] or preparatory to despotism in the constitution.'[2]

In 1770 Sir William Meredith moved for a parliamentary committee to consider more proportionate punishments. A further enquiry was proposed in 1787 as the loss of the American colonies and the ensuing transportation crisis galvanized debate.[3] Both initiatives fell flat, but parliament was at least exposed to arguments for proportion and certainty

[1] P. Rawlings, *Drunks, whores, and idle apprentices: Criminal biographies of the eighteenth century* (1992), 25–6.

[2] Capel Lofft, appendix to J. Jebb, *Thoughts on the construction and polity of prisons* (1786), 36–7.

[3] Radzinowicz, i, chs. 12, 13.

in punishment, to unfavourable comparisons with the penal regimes of enlightened Europe, and to the language of humanity, sympathy, and feeling as well. Howard's and Hanway's exposure of prison conditions meshed with Blackstone's, Eden's, and Bentham's critiques of the law to shape debate for decades yet. Extra-parliamentary opinion also moved that way. As Newgate executions reached their awesome peak in the 1780s the *Universal Daily Register* waved the banner of penal reform with a fervour which would have disconcerted its later readers, when it became *The Times*:

> Reformer, take your example from the King of Prussia, and the Emperor of Germany. . . . If the executions in all the cities on the continent of Europe, for the last twelve months, were summed up together, they would be found to fall far short of those that will take place in London in one day only this week. . . . Why are not the Judges called upon by the Crown to attempt at least some system for the preventing of such sanguinary sacrifices—are we more infamous and corrupt than our neighbours? Does not the coronation oath bind the executive power, as far as conscience can be bound, to administer *justice* in *mercy*? . . . The poor thief pinched by want, or pressed by necessity, should not suffer equally with notorious offenders, and cruel murderers.

Punitive and disciplinary urges were not far below the surface of this efflorescence, it has to be said. Some mitigations of punishment might be recommended for lesser offences, and a few prisons were tidied up as Howard recommended. But the *Register* was not the only organ to suggest that punishment for atrocious crime should be made more terrible rather than less (by secret and private hangings, for example); Bentham thought the same (and so, we saw, did Boswell). It was rationality that was the goal, not kindness. Those not hanged should be harnessed to slave labour, the *Register* thought. Seventy thousand people must have hanged since 1685, it massively overestimated: 'What a shame! when it is so well known that we have not a harbour, nor a river, complete in the whole kingdom. Are there besides, no hills to reform, no waste lands to cultivate, no stone to saw, no mines to work? If there are, why hang men in the prime of their lives?'[4]

Reformist optimism about the imminence of penal change was silenced after this brief flowering in the 1780s. For some twenty years after 1789 there was a hiatus in the penal reform debate. The anti-revolutionary backlash ensured that Paley (never Bentham) became the textbook to which politicians, clergymen, and universities turned. By the time Romilly returned the issue to the parliamentary agenda in

[4] *Universal Daily Register*, 12, 31 Jan., 9, 13 May, 5 Oct., 7 Dec. 1785.

1808 reformers' sights had to be set lower than they had been. Still, his attack on selected capital statutes brought Quakers into the cause, and when in 1819 the Commons agreed to set up a select committee to consider parts of the capital law it began to look at last as if 'opinion' really was gathering slow momentum. How little was achieved in the 1820s, however, later chapters make clear.

The innovators in this early story were rationalists, radicals, and lawyers of various persuasions—less hagiographized but more active than the Quakers and evangelicals who have usually been given the credit. It was a smaller world then, and most of these constituencies knew each other. (It is still surprising to find that Gibbon Wakefield's grandmother was Elizabeth Fry's aunt, however.) So while the distinctions which follow are convenient, they are also overemphatic.

The rationalists attacked the capital code's ineffectiveness, not its cruelty. It failed to deter, they said, and its viciousness corrupted the people. It was too inflexible and limited to protect property, they added. And because punishments were excessive, they lost the support of prosecutors and juries whose endorsement of the law was essential. These men—lawyers and doctors mainly—were in some guises dangerously advanced for their times. Deists or unitarians, they said startling things. 'One reason why I cannot think that death ought so carefully to be avoided among human punishments, is that I do not think death the greatest of evils': thus Romilly in younger years. He was not thinking of the rewards of the afterlife here. Romilly had escaped childhood indoctrinations through his father's attendance at a French chapel where little more than 'a kind of homage [was] paid to the faith of his [Huguenot] ancestors'. His comment rather reflects that collapse of literal belief in eternal hell-fire which enabled many in the English Enlightenment to regard death as a natural metamorphosis to be accepted, even welcomed.[5]

What most moved them against an indiscriminate death penalty were beliefs or assumptions which drew remotely on Locke's view of the mind at birth as a *tabula rasa* and more immediately on David Hartley's *Observations on man* (1749). An associational psychology was the basis of their argument with 'irrational' punishments. If people acquired their ideas and personality from the bodily associations imprinted in the course of social living, criminals might best be

[5] Romilly, i. 14–15, 278; R. Porter, 'Death and the doctors in Georgian England', in R. Houlbrooke (ed.), *Death, ritual and bereavement* (1989), 84.

reclaimed through the application of measured pains or rewards within reformative penal systems. This recognition of the felon's capacity for guilt and reform eroded old understandings. As McGowen has put it, the criminal body, hitherto a medium for expressing concern for the health and integrity of the social body and state, became a locus of individual pain and rights.[6] Novelists from Defoe to Goldsmith refracted these views too, their narratives not the least important of the writings which stressed 'the power of confinement to reshape personality'.[7] In these terms the metaphorical, sovereignty-affirming, and theological structures in terms of which the criminal's bodily pains had hitherto been justified were slowly subverted.

Lawyers' involvement in these networks was crucial. They included not only Bentham, Blackstone, Eden, and Romilly, but also the workaday attorneys and barristers of the 1810s, 1820s, and 1830s, and by the 1840s abolitionist MPs like Ewart. Trained legal minds like these (not all of them practising) were the most effective advocates of reform since they were uniquely qualified to challenge the *ancien régime* through the axioms of justice itself.[8] Bentham has received most attention, but he was taken seriously only late in this story, his influence otherwise confined to small and advanced circles. Not only did he publish much of his work in French, but it was as late as the 1820s that the *Westminster Review* publicized his ideas and whig lawyers like Denman and Brougham began to laud him.[9] His name has overshadowed lesser but cumulatively more influential practitioners whose numbers, professionalism, literary output, and journals were expanding well before this. In their quest for more controlled trial processes and evidential rules, and in their increasingly energetic interventions in petitioning for mercy, we shall meet in Chapter 16 a peculiarly informed constituency. More than others, it had the intellectual capital to put the old system on the defensive where it mattered most—in courtrooms rather than in private studies.

'Radicals' were at work next, opposing oligarchical government and institutions, their opposition to the capital code part of a larger stance embracing the desire for parliamentary reform or the admission of

[6] R. McGowen, 'The body and punishment in eighteenth-century England', *Journal of Modern History*, 59 (Dec. 1987), 651–79.

[7] J. Bender, *Imagining the penitentiary: Fiction and the architecture of mind in eighteenth-century England* (Chicago, 1987), 1.

[8] D. Lieberman, *The province of legislation determined: Legal theory in eighteenth-century Britain* (Cambridge, 1989), ch. 10.

[9] For Denman's review of Bentham's work on evidence (which Bentham published in French), see *Edinburgh Review*, 40 (1824), 169–207.

dissenters' rights, or entailing opposition to all the workings of oligarchy, war-making included. Some were middling to humble men: bookseller-publishers like Hone or attorneys like Harmer, hovering on the fringes of popular radicalism, who knew well how to appeal to 'the people'. Exploiting cases of rough justice to expose oligarchical law at its most brutal, their highly significant statement was that the law should be 'just' in its processes as well as its punishments. The justice they appealed to was not that contrived in Paleyite terms by judges and the great people, or the calculatedly utilitarian justice of Benthamites. It was couched in terms of a commonsensical 'fairness', with proportion at its centre—the same fairness which informed petitions for mercy in the 1820s from people remote from central debates, as we shall see.

Finally, bringing the language of redemption to the reform cause and sustaining it into the 1830s and beyond, there were the evangelicals and Quakers: prudential, causally thoughtful, ethically high-minded, and familiar with the techniques of interventionist action through engagement in worldly business. Through campaigns of conversion and moralization, many sought to avert the moral cataclysm immanent in the diffusion of vice, disbelief, and Jacobin ideas, guided by belief in 'the corruption of human nature, the atonement of the Saviour and the sanctifying influence of the Holy Spirit'. Howard, Hanway, Paul, and others had anticipated their role in the 1770s and 1780s, but it was only in and after the 1810s, with Romilly's bandwagon already moving, that Quakers climbed upon it. Late joiners though they were, by the 1820s they had substantially shifted the vocabulary of reform.

When historians address large changes of this order, their emphases are shaped by hindsight, so there is an unrealistically predictive element in the summary just given, as there may be in what follows. The people attended to here did not dominate political culture or speak for large minorities, let alone majorities. None the less, between 1800 and 1830 they managed to construct an oppositional opinion about the capital code, and we must see something of how it was done.

Chapter 13 recounts two stories which reveal the complex mixtures of feeling and thought which energized radical indignation at harsh justice; wronged *women* were at their centre. Chapters 14 and 15 address the extent (but also, notably, the limits) of the contributions of Quakers and other pious people. Chapter 16 discusses the diffusion of the 'commonsensical' views of justice revealed when middle and lawyerly kinds of people appealed for felons whom they thought wrongly convicted or too harshly sentenced. Finally the microhistory in Chapter 17 confronts us with the kinds of opinion which a local community could refract, using the appeal process to reconcile the values of the community (misogynist, as it happens in this case) with those of the formal law.

The common motif in these studies is the *affective* as much as the rational basis of opinion, and the rest of this short introductory chapter explores this further. In varying degrees, all the constituencies mentioned above accommodated and reflected the sensibility which had softened polite classes' relationships to the social world for a century past. We have seen how strong the defences against sympathetic prescriptions could be; and many in the reform camps—Benthamite utilitarians particularly—kept aloof from them. Yet despite much ambivalence, the valuation of feeling, benevolence, and sympathy as bonds between individuals and within society, and the endorsement of empathetic identification with others' misery as a valued posture, could not but subvert that dissociation of the self from others' pain upon which tolerance of harsh punishments in part depends. Even if the direct attack on harsh law was seldom couched in these terms, most of the key constituencies had been touched by the cultures of sympathy, and, however rhetorically, they yoked the language of sympathy and humanity to their cause. It is the chief argument of this part of the book that in many critics these affective promptings determined the expression of thought itself.

2. *Acting from feelings*

None who assailed the capital code in parliament between 1770 and 1819 could afford to work overtly from sentiment. All had to protect themselves against the die-hard charge that an undiscriminating sympathy smacked of affectation. Romilly and his supporters admired Howard and consorted easily with evangelicals while not sharing their beliefs; but they argued from Beccaria and Bentham. They deployed empirical evidence to prove the irrationalities, disproportions, and counterproductiveness of the more offensive capital statutes, claiming that they failed to deter the criminals in question, and inhibited prosecutors from prosecuting on capital charges or juries from convicting on the full charge when doing so would hang a man. Only secondarily did they tend to refer to the cruelty of punishments, and hardly at all to the felon's moral reformation and salvation.

It does not do to read root-and-branch reformist intentions too far back. The reformers' initial ambitions were modest. Romilly's earliest decision to act against capital punishment was muted to a degree. In 1807 he recalled how, twenty-three years earlier, when a young barrister on circuit, 'some instances of judicial injustice . . . made a deep impression' on him. He resolved to 'remove that severity in our law, which has arisen from no intention of the Legislature, but altogether from accidental circumstances'. The reference here was a modest one, to the depreciation of money: people were hanged for stealing goods that were worth less now than when the capital statute was passed. He intended only to fix the sums 'according to the difference between the then and the present value of money', thus, he hoped, to end 'that shameful trifling with oaths, those pious perjuries (as Blackstone somewhere calls them), by which juries are humanely induced to find things not to be worth a tenth part of what is notoriously their value'.[10]

Higher ambitions than these would have been unrealistic. Without hope of office while the prince of Wales withheld favour, it was sensible for reformers to attack narrow targets. Although they welcomed the support of anything that could be called public opinion (their arguments depending on some real or imagined confirmation in that arena), it was the parliamentary vote that mattered. Romilly's successes in dismantling a few capital statutes between 1808 and his suicide in 1818 depended on the support of whigs of the Holland and Lansdowne connection. In such arenas feeling had its place, but a cool pragmatism counted for more if tory die-hards like Ellenborough or Eldon were to be swayed.

That said, however, among the whig–radicals there were still men trained in the arts of sympathy. Most had risen from the commercial or professional classes. Most had read Rousseau as well as Beccaria and Bentham. Some had consorted with or been on the edges of the constitutional reform clubs of the 1780s, and as young men they had welcomed the French Revolution with zeal. Furthermore, if one touchstone of sensibility was a respect for and tactical deployment of the 'humanitarian narratives' through which the pathos of others' suffering might be apprised—then even the coolest of reformers deployed the device shamelessly. Sir William Meredith in 1777 had waxed angry at the injustice of executing poor Mary Jones for shoplifting, her baby at her breast—a shocking story which echoed into the broadsides of the 1820s (we met it in Chapter 5) and eventually into *Barnaby Rudge*. Forty years later Sir James Mackintosh made a similar cry about another judi-

[10] Romilly, ii. 229–30.

cial cruelty: the decision to put a poor Irish woman to trial on the very day set for her husband's execution. Although Mackintosh coolly preened himself after his parliamentary speech for having made 'a forcible statement' with 'some pathetic passages' (quickly forgetting the woman herself), there is no doubt that he was practised in the needed tones of pathos.[11] And among reformers outside parliament the demands of the feeling heart were more freely released. Whether channelled by faith or by the vexation common to ascendant groups marginalized by dominant cultures, actively sympathetic virtue could become an oppositional emblem and cause, anger and indignation a sustaining source of energy.

Indeed, actively sympathetic virtue or benevolence were watchwords of the age. Benevolence was 'the base of private contentment and of social good', Dr Jebb announced in 1772.[12] Only active virtue opened the path to happiness, Samuel Parr added in 1789: 'Jesus connects the love of our neighbour with the love of God,' and 'piety is defective and unavailing without active virtue.'[13] God's purpose was 'the happiness of his creatures', the Quaker William Allen's journal the *Philanthropist* chimed in in 1811. The journal enjoined 'the duty and pleasure of cultivating benevolent dispositions', for

> it is the duty, as well as the interest of [God's] rational beings, to co-operate with him, in producing this desirable end, by mutual endeavours to promote the comfort of each other, and to sweeten with sympathy, those bitter cups, which are the portion of many in this probationary stage of being. . . . Virtue and active benevolence [will gratify] the best feelings of [the] heart . . . [and] all, even the poorest, may render material assistance in meliorating the condition of man.[14]

This was again the language of sensibility, placing the linkage of the 'rational' with the 'sympathetic' at the centre of the interpretation. Benevolence was enlightenment itself: the heart spoke to the mind; the mind was guided by the heart.

Among reformers expressions of the linkage were multiple. Compassionate hearts were worn on sleeves, and the language of sensation was ubiquitous. In Romilly reasoned argument against capital statutes would be yoked to an uncompromising condemnation of

[11] R. J. Mackintosh (ed.), *Memoirs of the life of . . . Sir James Mackintosh* (2 vols., 1835), ii. 340–1.

[12] J. Jebb, *The excellency of the spirit of benevolence: A sermon preached before the University of Cambridge . . . December 28, 1772* (Cambridge, 1773).

[13] 'On benevolence' (1789), in J. Johnstone (ed.), *The works of Samuel Parr, LL D* (8 vols., 1828), vi. 468.

[14] The opening essay, 'On the duty and pleasure of cultivating benevolent dispositions', *Philanthropist*, 1 (1811), 1–2.

public execution as 'a disgusting spectacle . . . shocking to humanity'. After an anti-slavery meeting in 1818 which heard 'horrible accounts of cruelty by masters to their slaves', Mackintosh reported that he had 'never observed any man so deeply and violently affected by the recital of cruelty as Romilly'.[15] Cruelty stirred them at other levels, too. In 1809 Romilly supported the Bill for the prevention of cruelty to animals brought in by his friend Erskine. The Bill was a measure of humanity, he told parliament: cruelty towards the dumb creation was barbarous, and led also to cruelty towards human beings. This was not rhetoric; such perceptions were lived out in daily life. Romilly confirmed this in an amused account of Erskine's relationship to the animal world. Self-parodying here, earnest elsewhere, such manifestations memorably convey the range of these circles' *kind-heartedness*:

> [Erskine] has always had several favourite animals, to whom he has been much attached . . . a favourite dog, which he used to bring when he was at the Bar, to all his consultations; another favourite dog, which, at the time when he was Lord Chancellor, he himself rescued in the streets from some boys who were about to kill him . . .; a favourite goose which followed him wherever he walked about his grounds; a favourite mackaw, and other dumb favourites without number. He told us now [at dinner] that he had got two favourite leeches. He had been blooded by them last Autumn . . .; they had saved his life, and he had brought them with him to town; and had ever since kept them in a glass; had himself every day given them fresh water; and had formed a friendship with them. He said he was sure they both knew him, and were grateful to him. He had given them different names, Home and Cline (the names of two celebrated surgeons), their dispositions being quite different.[16]

3. *Chivalry and the female victim*

No subject better brought to light the relationships between benevolence, sympathy, and oppositional thought than that of the wronged woman, particularly if she was of inseminable age and fetchingly vulnerable to male wiles. When reformers from Meredith on to Mackintosh told their affecting stories of rough justice, wronged women were usually central to them. Nor is it accidental that wronged women were the subject of two of the most extraordinary appeal campaigns of the early nineteenth century, examined in the next chapter. As men transformed the wrongly judged or punished woman into a victimized icon, even the

[15] Romilly in the Commons, 1813, cited by M. Wiener, *Reconstructing the criminal: Culture, law, and policy in England, 1830–1914* (Cambridge, 1990), 94; Mackintosh (ed.), *Memoirs of Sir James Mackintosh*, ii. 345.

[16] Romilly, ii. 288, 235.

most reasonable among them might betray the erotic fantasies which they projected upon their often implausible creations.

The story of the so-called Keswick Impostor John Hatfield and of his female victim shows how much the Romantic sensibility had to answer for in this connection. In 1803 Hatfield was hanged in Keswick for forgery and for impersonating an MP, after a long career of criminal impostures and pretended marriages. What drew most attention to his case was the fact that with false promises of marriage and under an assumed identity, he had seduced one Mary Robinson, an innkeeper's daughter in remotest Buttermere, dumping her after he had had his way. She stands as an archetype of the fantasies at issue here. Coleridge, who had once glimpsed the real Mary on a Lakeland tour, knew her as 'rather gap-toothed, and somewhat pock-fretten', he remembered. Worse, she was aged 30, and unmarried, and might be thought needful. But none of this stopped Coleridge's transforming Mary into a child of nature deluded and betrayed by a Romantic villain of darkest hue, 'her former character for modesty, virtue, and good sense' deepening the pathos of her plight. As Coleridge mooned over 'her exquisite elegance, and the becoming manner in which she is used to fillet her beautiful long hair', he so infected Wordsworth with his fantasy that Mary found her place in *The prelude* (VII. 316–59); and thence the fantasy spread to 'counterfeit Hatfields and Marys in abundance—in farces, melodramas, and novels, most of them with little recourse to the tale of real life'; it helped the force of these that her age was reduced to a sweeter 18. In such flabby fictions she reappears to this day.[17]

By conventional standards Mary should have paid the price for her fall. Her character was destroyed. But Romanticism could cut subversively across domestic doctrines. Seduction and loss of virtue were almost the conditions of her accommodation within male forgiveness, so long as she could be presented as the victim of her own passionate nature and of the man who had tapped into it. 'The acute sensibility peculiar to women' deserved male compassion, after all. Inducing as it did 'sudden excesses', 'unmerited attachments', and 'groundless discontent', not to mention 'weakness', 'pusillanimity', and 'feebleness of character', it was an affliction which men must help the weaker sex to bear.[18]

[17] R. Holmes, *Coleridge: Early visions* (1989), 338–41; D. V. Erdman (ed.), *The collected works of Samuel Taylor Coleridge: Essays on his times in the* Morning Post *and the* Courier (3 vols., 1978), i. 357–8, 403–16; Anon., *The life of John Hatfield, commonly called the Keswick Imposter, with an account of his trial and execution for forgery; also his marriage with 'Mary of Buttermere', illustrated with her portrait* (Carlisle, 1846).

[18] T. Gisborne, *An enquiry into the duties of the female sex* (7th edn., 1806), 34.

What was going on in a case like this need not be laboured. The weaker sex became a vehicle through which men registered their potency, benevolence, and chivalric selves. *Causes célèbres* like Abraham Thornton's rape and murder of Mary Ashford in Birmingham in 1817 were harnessed to warn women to be virtuous but also to submit to male protection.[19] As, tacitly, this view came to embrace criminal women too, it could not but shift attitudes to the law which doomed them. A woman led astray by her nature or by a designing villain and then condemned undermined belief in justice itself. The law might itself come to stand for male villainy pitted against the tremulous needfulness of the female heart. At this point the fantasies of the Romantic sensibility might lead the most reason-loving of men into passionate opposition to a justice capable of such wrong. Chivalric compassion could expose harsh law to an attack which was difficult to counter simply because *it was not reasoned at all.*

There had been little sign of these dispositions in the early eighteenth century, even though for centuries fewer women than men had been prosecuted, proportionately more acquitted, proportionately more convicted on lesser charges than those named in the indictment, and proportionately more pardoned if condemned. One historian has discerned in this 'an often instinctive chivalry, or if you like embarrassment' about the punishment of women; but this view is itself a little naïve.[20] What the differences better expressed was the perception that women posed less of a threat to lives, property, and order than men did.[21] Criminal women were not exempted from harsh treatment in centuries past. They were allowed a capacity for wrongdoing equal to men's and treated accordingly. In early *Newgate calendars* and ordinary's *Accounts* female criminals were mostly assumed to have been led into evil ways by their own depravity or taste for luxury, the chivalric ideal having little purchase here. They received no better treatment than men in prison or on the scaffold, sometimes worse.

Well into the nineteenth century, the atrociously criminal woman was still deemed a monster, as Esther Hibner was in 1829. Infanticides were still to be hanged as examples to others; disloyally thieving servants must hang too. But as broadsides have shown us, a deepening sadness was allowed to attach to these cases. Women whose sufferings ensued from misguided sexual passion or loyalty to powerful men became subjects of sympathy, and male villainy became the active prin-

[19] Anna Clark, *Women's silence, men's violence: Sexual assault in England, 1770–1848* (1987), ch. 7.

[20] G. Elton, in J. S. Cockburn (ed.), *Crime in England 1550–1800* (1977), 13.

[21] Beattie, p. 439.

ciple in such stories. In these conditions women's physical punishment became a delicate matter. Anxiety about executing women, about burning their bodies for coining or for murdering husbands (the certainty of male abuse apparent in the murders), or about whipping them bare-backed and hence bare-breasted in public, now tended to be activated by the sense that even at their worst women were creatures to be pitied and protected from themselves, and perhaps revered, like all women from whom men were born. It was in the 1780s that the tensions between the pain customarily delivered upon the criminal woman and the chivalric ideal seem to have reached breaking-point.

The execution of Phoebe Harris in June 1786 might have been a formative moment in this process. She was convicted at the Old Bailey for the treasonable crime of coining silver. On the morning of her death six men were hanged conventionally on the scaffold before her. Then two sheriff's officers conducted her from the Newgate door to a wooden stake in the middle of Newgate Street. There she was made to stand on a stool while a noose was put round her neck and attached to an iron bolt driven into the top of the stake. The ordinary of Newgate prayed with her, the stool was removed, and, with her body convulsing a foot or so off the ground, she choked noisily to death over several minutes. Half an hour later the executioner chained the corpse to the stake and heaped two cart-loads of faggots around it to which he set light. Bits of the body were visible in the fire two hours later. This death was watched by some twenty thousand people.

In earlier decades such spectacles had passed with little comment. Now the *Universal Daily Register* found Harris's execution 'disgraceful', 'inhuman', and 'shamefully indelicate and shocking'. The sentence was unjust to women, it announced: 'Why should the law in this species of offence inflict a severer punishment upon a woman than upon a man?' Admittedly the paper rehearsed other objections: the terror of the punishment hardened the hearts of the people; the watching mob 'breathed nothing but execration upon the police and magistrates; and after the unhappy victim was consumed, amused themselves with kicking about her ashes'. But when Margaret Sullivan was similarly hanged and burnt two years later, the paper pulled out the chivalric stops without inhibition—with deliberate exaggeration of Sullivan's pain, too (she was not burnt alive):

> There is something inhuman in burning a woman, for what only subjects a man to hanging. Human nature should shudder at the idea. Must not mankind laugh at our lay speeches against African slavery, and our fine sentiments on

Indian cruelties, when in the very eye of the Sovereign we roast a female fellow creature alive, for putting a pennyworth of quicksilver on a half-pennyworth of brass? . . . There is no place in London, where the circumstances of the torturing barbarity, on the body of the unhappy female yesterday roasted alive by the sentence of the law, is not talked about with horror. Shame on it, thus to attack the female sex, who by being the weaker body, are more liable to error, and less entitled to severity.

Parliament itself agreed with this indictment. The burning of women was abolished in 1790.[22]

Thenceforth uneasiness about the excessive punishment of women increased. The fame of Elizabeth Fry's ministrations to Newgate women was one manifestation of it. Coleridge's explosion in 1811 when he read a newspaper account of how a woman was publicly whipped for a misdemeanour was another. To whip a woman, he fulminated, was not only at odds with the nation's 'progressive refinement and increased tenderness of private and domestic feelings (in which we are doubtless superior to our ancestors)'; it was also to injure her 'in the first sources and primary impulses of female worth'. Not least did it degrade the man who delivered the punishment:

Good God! how is it possible, that man, *born of woman*, could go through the office? . . . Never let it be forgotten, that . . . *the* woman is still *woman*, and however she may have debased herself, yet that we should still shew some respect, still feel some reverence, if not for her sake, yet in awe to that Being, who saw good to stamp in her his own image, and forebade it ever, in this life at least, to be utterly erased.[23]

Coleridge also spoke to his times. The public whipping of women ceased in 1817 and private whipping was restricted to men in 1820 (to be abolished in 1862).

We are ready now for those two appeal stories already mentioned. They suggest that, in appropriate and mainly metropolitan conditions, it was around the plight of wronged women that 'opinion' against harsh law would be mobilized most effectively.

[22] Broadside in Hindley, ii. 177; *Gentleman's Magazine*, 56/1 (1786), 524; Radzinowicz, i. 211–12; *Universal Daily Register*, 22, 23, 27 June 1786, and 24, 25, 26, 27 June 1788; PH xxviii (1790), c. 782 ff.

[23] Erdman (ed.), *Works of Coleridge*, ii. 139–41.

CHAPTER 13

WRONGED WOMEN

The Stories of Sarah Lloyd and Eliza Fenning

1. *Introduction*

IN THE EARLY NINETEENTH CENTURY TWO CAMPAIGNS AGAINST the rough justice and execution delivered upon young women are of extraordinary interest for any student of attitudes to and feelings about capital law at this time. The first was set in Bury St Edmunds in 1800 when a servant girl, Sarah Lloyd, was hanged for theft from her mistress's house. The second was set in London in 1815: another servant girl, Eliza Fenning, was hanged for attempting to murder her master's household by poisoning their dumplings with arsenic. The campaigns to save each woman or rehabilitate her reputation were unprecedented. Nobody had made such fusses about the hanging of a common person before, or at least fusses which left such records. The campaigns suggest a readiness to challenge criminal justice which had not been easy to focus in the eighteenth century, and they portended many challenges to come. Each campaign testifies not only to the incorporation of long-standing critiques of *ancien régime* law, but also the close relationship between thought and emotion in those most concerned. Each was mobilized by engagement with *narratives* of injustice, with real victims at their centres. If opinion was to be moved at all, it would be as much by stories like these as by Beccarian or Benthamite argument.

The campaign to save Sarah Lloyd was extravagant to a degree. Its chief protagonist was the deistical gentleman barrister Capel Lofft, JP, one-time associate of Wyville, Price, Fox, Godwin, Priestley, Clarkson, Howard, and Hazlitt—a man of enlightened sensibility in whom kind-heartedness was neither an affectation nor a fashionable contagion but a passion to which political and social self-interest was subordinated to the point of love, obsession, and magnificent folly. Lofft's and Sarah Lloyd's story was a romance which takes us to the heart of our concerns: the indissolubility of passion and reason in this era. The 1815 London campaign speaks similarly. Like the campaign for Lloyd, Eliza Fenning's bore the imprint of men who combined a righteous and passionate anger with purposeful opposition to what they saw as a corrupt judicial and moral order.

The campaigns also bear a complex relationship to the meanings of public opinion, on which we say more at the end of the chapter. The Bury campaign had no effect on opinion. Few people in London heard about the case; fewer would have cared. It was remote and ahead of its time. Only when Romilly opened the parliamentary debates on capital law in 1808 did public receptivity change. Dramatically change it did, however. When Basil Montagu had tried to publish his collections of observations on the punishment of death not long before 1808, his publisher told him 'there was not sufficient interest upon the subject'. In 1812, with Romilly's parliamentary ball rolling, he was able to publish the *Opinions* in three expensive volumes. The London campaign in 1815 benefited from this sea change. It was a more closely orchestrated affair, provoking an outcry which publicists like William Hone turned to brilliantly propagandist effect. It was aimed at popular to middling audiences but was long remembered above those levels.

Finally of course, both Lloyd and Fenning were female, young, pretty, wrongly judged, and wrongly hanged. They were also (according to key constructions) the victims of designing men. Neither was so beautiful in her life as in her dying. By this date the most rational of men might melt a little when tragedies so romantically charged touched them.

2. *Capel Lofft and the execution of Sarah Lloyd, 1800*

We begin the first story with a riveting text, Capel Lofft's own. It was written after the appeal for Sarah Lloyd's life had failed. Lofft published it as tribute and requiem in the *Monthly Magazine*. It was written elegiacally in the tones of love, the weaker body of woman here turned into icon:

> I have reason to think that she was not quite nineteen. She was rather low in stature, of a pale complexion, to which anxiety and near seven months' imprisonment had given a yellowish tint. Naturally she appears to have been fair, as when she coloured, the colour naturally diffused itself. Her countenance was very pleasing, of a meek and modest expression, perfectly characteristic of a

mild and affectionate temper. She had large eyes and eyelids, a short and well-formed nose, an open forehead, of a grand and ingenuous character, and very regular and pleasing features; her hair darkish brown, and her eyebrows rather darker than her hair: she had an uncommon and unaffected sweetness in her voice and manner. She seemed to be above impatience or discontent, fear or ostentation, exempt from selfish emotion, but attentive with pure sympathy to those whom her state, and the affecting singularity of her case, and her uniformly admirable behaviour, interested in her behalf.

In a later letter Lofft gave an account of Sarah Lloyd's death. She died as sweetly as she had lived:

There was no platform, nor anything in a common degree suitable to supply the want of one; yet this very young and wholly uneducated woman, naturally of a very tender disposition, and, from her mild and amiable temper, accustomed to be treated as their child in the families in which she had lived, and who consequently had not learned fortitude from experience of either danger or hardship . . . appeared with a serenity that seemed more than human; and when she gave the signal, there was a recollected gracefulness and sublimity in her manner that struck every heart, and is above words or idea. . . . After she had been suspended more than a minute, her hands were twice evenly and gently raised, and gradually let to fall without the least appearance of convulsive or involuntary motion, in a manner which could hardly be mistaken, when interpreted, as designed to signify content and resignation. At all events, independently of this circumstance, which was noticed by many, her whole conduct evidently showed, from this temper of mind, a composed, and even cheerful submission to the views and will of heaven; a most unaffected submission entirely becoming her age, sex and situation.[1]

The law's view of Lloyd was more prosaic. She was an illiterate servant girl of Hadleigh near Ipswich, aged between 18 and 23 (attributed ages of the illiterate often varied widely). She had neither been taught the Lord's Prayer nor was she chaste. She had a sexual relationship with one Joseph Clarke, a plumber and glazier of low repute. He was the son of a respectable tradesman with influence behind him, and he had promised her marriage, she said. Lloyd was a servant in the house of a kind mistress. One night in October 1799 she robbed her employer. She then set fire to her house, and escaped unambitiously to her mother's home nearby. She was arrested with ease and put into Bury gaol, to wait for the Bury assizes in March 1800.

In the event the arson was not tried and of the burglary she was acquitted. But she was convicted of a capital larceny to the value of forty shillings, since some of the stolen items were found at her

[1] Borrow, v. 412–17; Rayner and Crook, iv. 239–41.

mother's house. She was condemned to death. The sentence might have been commuted had it not been for the arson, had she not also betrayed the bond between mistress and servant—'a crime that must be punished', Sir Nash Grose declared when he sentenced her in court. As the judge later told Lofft:

> there is a superior mercy due to all inhabitants of houses, and masters of families, with which the pardon or even respite of this unfortunate woman's sentence seems to me incompatible. Such is my abhorrence and idea of her crime as it affects the publick, that if Clarke or any other accomplice had been convicted I [should still be] of opinion that for example's sake her life ought not to have been spared.[2]

A pamphlet published locally after the execution endorsed this view:

> To masters and mistresses of houses, [Lloyd's crime] showed the necessity, not merely of watching the conduct, but of improving the morals and instructing the minds of their domestics in religious truths. To servants, and more particularly to female servants, it points out the importance of sobriety, chastity, fidelity, honesty, obedience. . . . She first fell a victim to seduction. She lost her virtue, her sense of rectitude, and her life.[3]

There was a petition for mercy, but the judge reported adversely and Portland the home secretary saw no need to put it to the king. Sarah Lloyd was set to die on 9 April.

In the case so far there was absolutely nothing out of the ordinary. Hundreds of such cases ended on gallows without fuss. Certainly the wronged mistress recommended Lloyd to mercy, stating that her behaviour until the crime had been 'proper and decent'; the rector of Hadleigh, the Revd Hay Drummond, circulated his own petition for others to sign. But this was standard too. Prosecutors often recommended mercy to avert communal criticism, while among rectors *noblesse* obliged.

What transforms this story, then, was an accident: Lloyd's confession to a prison visitor that she had been seduced by her lover, and the communication of the tale to Lofft. The tale fitted plausible and familiar sequences, and by this date nobody who had read a sentimental novel would doubt what to feel about it. While courting her for three months under the promise of marriage, Joseph Clarke had 'debauched' her.

[2] This account is based chiefly on 'Sarah Lloyd's case, miscellaneous correspondence' (E1/20.6.1); and 'Copies of letters concerning Sarah Lloyd', in the Common-place Book of Sir Thomas Cullum (Ac. 317/1): Suffolk County RO (Bury St Edmunds). It is not clear how those parts of the dossier which properly belong in the Home Office papers found their way here.

[3] Anon., *Trial of Joseph Clarke, the younger, and Sarah Lloyd, at the assizes, held at Bury St Edmunds, March 20, 1800* (Ipswich, 1800), 10.

Thereafter he came repeatedly to her bed as the household slept. She trusted and loved him. On the night of the crime, talking of marriage, Clarke induced her to steal her mistress's pockets and watch. In weakness and love she did so. He set fire to a wood-house to cover their tracks, and advised her to escape to her mother's house. If arrested, she was to say that two passing soldiers had set her up to the crime.

Thus a designing villain had been found. The Revd Drummond's petition for the 'poor unfortunate deluded girl' got the story rolling. Lloyd was Clarke's helpless instrument, it declared; if time were allowed, he might be forced to acknowledge his part in the crime and diminish her guilt. 'There never was a case where the most respectable persons of every description in the County of Suffolk . . . were more zealously interested in behalf of such an unfortunate woman, or indeed in any place whatever.' The petition was widely signed by local gentry. Drummond wrote to his brother the earl of Kinnoul to ask him to lay it before Portland for presentation to the king.

The case might still have passed on to its natural conclusion had Drummond not then turned for help to Capel Lofft. Lofft had served on the grand jury which had passed the indictments against her and he had attended the trial. To involve a man of Lofft's quixotic character and principles in a case like this was like exposing a distressed maiden to a knight errant with a wooden leg and a fair deal to prove.

Lofft was born in 1751 the son of the duchess of Marlborough's private secretary. As a young barrister and literary figure, he had moved quickly into advanced metropolitan circles, rising to the challenge of stirring times with the zeal of the advanced democrat. In 1775 he had joined anti-American War radicals and in 1780, with rational Dissenters like Jebb and Hollis, he helped to found Cartwright's Society for Constitutional Information, publishing his own advocacy of parliaments elected annually on a person- rather than property-based franchise, and associating with Horne Tooke and later Paine. He offered his services to the Anti-Slavery Society in 1777–8.[4] Most revealingly he wrote the introduction and appendix to John Jebb's *Thoughts on the construction and polity of prisons* in 1786. Here Lofft vented himself of his contempt for the 'vindictive jealousy' of penal laws which guarded 'the amusements of the great and wealthy' as if they were 'the very existence of society'. Penalties were disproportionate, erratic, irrational, and too often

[4] J. Fiske (ed.), *The Oakes diaries: Business, politics and the family in Bury St Edmunds 1778–1827* (2 vols., Woodbridge, 1991), i. 116 n.

applied to innocent and helpless sufferers. More than a ninth of those executed were killed for crimes 'far from implying irretrievable malignity'—picking pockets, horse-theft, and Sarah Lloyd's crime-to-be, theft from houses.[5] This publication was prophetic.

Prophetic too was the passion that the French Revolution unleashed in him. With William Godwin and others, he subscribed in 1789 to the Revolution Society's declaration 'That all civil and political authority is derived from the people; That the abuse of power justifies resistance; That the right of private judgement, liberty of conscience, trial by jury, the freedom of the press, and the freedom of election, might ever be held sacred and inviolable.' He wrote one of several attacks on Burke's *Reflections*—a good one, preceding Paine in this. He chaired meetings of the Friends of Parliamentary Reform, and in the mid-1790s as one of the network of anti-loyalists headed by Wyvill made a bad name for himself among tory gentlemen in Suffolk by moving addresses against the war with France at county meetings.[6] On top of this he published poetry of considerable ambition if not achievement, books on law and commercial affairs, and in 1791–4 he enlarged into four volumes Sir Geoffrey Gilbert's *The law of evidence*. This was a man who knew his law.[7]

He was also well regarded at this date. When Boswell and Johnson dined with Lofft in 1784, Boswell forgave him his whiggery, found his mind 'full of learning and knowledge and much liberality', and referred to 'this little David of the popular spirit'.[8] Lofft was at the heart of that caucus of 'literary caballers, and intriguing philosophers . . . political theologians, and theological politicians' (Burke's terms) who had seized the issue of law reform as zealously as they had that of constitutional reform and put themselves at the forefront of progressive radicalism—

[5] 'The laws as they are, will not be executed with that constancy which is indispensable to ensure their observance. Indeed, certainty of punishment does appear essential to sound policy: but reasonableness in the kind and degree of punishment allotted to offences must first be established, and the community must be satisfied it is, before that certainty can produce the proper effect, or be reconciled to justice and humanity': Lofft, introduction and appendix to J. Jebb, *Thoughts on the construction and polity of prisons* (1786), pp. vi, x, 34–5, 95.

[6] Fiske, *Oakes diaries*, i. 115, 131, 331–2.

[7] C. Lofft: *Elements of universal law* (1779); *Eudosia, a poem on the universe* (1781); *Observations on . . . parliament* (1783); *Abstract of the history and proceedings of the Revolution Society* (1789); *History of the Corporation and Test Acts* (1790); *Parliamentary reform: General meeting . . .* (1790); *Remarks on the letter of the Rt. Hon. Edmund Burke, concerning the revolution in France* (1790); *A vindication of the short history of the Corporation and Test Acts* (1790); etc.

[8] C. Bonwick, *English radicals and the American Revolution* (Chapel Hill, NC, 1977), 91; J. E. Cookson, *The Friends of Peace: Anti-war liberalism in England, 1793–1815* (1982), 14, 125, 137, 154; Boswell, *Life of Johnson*, ed. G. B. Hill and L. F. Powell (6 vols., 1934–50), iv. 278; M. Thale (ed.), *Selections from the papers of the London Corresponding Society, 1792–99* (Cambridge, 1983), 5.

until events in France and counter-revolution and government prosecution in England forced them into silence.

There had, however, been something a little forced in Lofft's zeal. And there was a flaw somewhere. He could never be a leader of men. The flaw turned on what his son later called a 'profuseness of nervous sensibility'. He was cursed by 'a weak and struggling voice, an extreme short-sightedness, and a person altogether, both in face and figure, singularly odd and diminutive'. He dressed badly, had to stand on chairs to be heard in public, and then his voice squeaked.[9] With the collapse of radical optimism in the 1790s, Lofft turned in on himself. Retiring to Suffolk, he acquired a bleakly realistic acknowledgement of what William Godwin called 'things as they are' (he finally accepted Sarah Lloyd's execution with the phrase 'since it must be'). He entered 'a class, now almost extinct, whose world was in their library', his son recalled: 'His love of literature was excessive. He existed for nothing else.' Increasingly dishevelled in dress, he lacked 'worldly prudence'. 'He walked uprightly in conscience, but in the ways of the world he wanted something to uphold him. . . . Embarrassment came upon him like an armed man.' This was to put it mildly.

Still, he was not the only one to capitulate to the times and retire from the great fight. At least he did not betray his early beliefs. He retained a vivid understanding of what Godwin again termed the 'modes of domestic and unrecorded despotism by which man becomes the destroyer of men'. So when Sarah Lloyd's case was drawn to his attention he was only too ready to hear it. What ensued had some inevitability about it. The servant girl was going to tap into a sensibility well trained in Enlightenment humanism but conscious of its recent defeats as well. He could hardly help making a victim like her the vehicle of his own frustrated ambition and anger. What was not expected was that he all but fell in love with her. In what he saw as her astonishing poise in adversity he recognized his own antithesis and complement. He was a man with no poise at all.

Lofft's first meeting with Lloyd in the condemned cell in Bury changed his view of the case. He had not seen her in person until this prison meeting, his letters reveal: in court she had been concealed by the crowd. Now he found every reason to accept that 'considering her age and sex', she must have been in Clarke's power from the first. Clarke

[9] For a contemporary depiction, see Fiske, *Oakes diaries*, i. 331.

was 'by far the greater criminal of the two', he declared. From every subsequent encounter with Lloyd, Lofft emerged with more elaborate versions of her history, relayed to any who might be enlisted in her cause. In these many exchanges by letter we trace a deepening obsession.

First, he learnt, on the night of the crime Lloyd had been much moved by Clarke's offer of marriage, but her integrity was such that she told him that she would not marry him until she had quitted her mistress's service, as agreed, on Michaelmas Day. Secondly, Lloyd told Lofft that she was pregnant and that Clarke had known this. This news touched off Lofft's most impassioned projections.[10] It was an important gloss on a tale of victimization in which he already believed devoutly. It also suggested a motive. Lofft discerned the 'incredible and horrid likelihood' that the designing villain had proposed the robbery in order to set Lloyd up for the gallows and to free himself of an unwanted marriage. This added a panicking urgency to his endeavours. The news had just come that Portland had dismissed the first petition, not thinking it worth laying before the king. The day of execution was set for the morrow, 9 April.

On the evening of the 8th, Lofft dined with Drummond and the prison chaplain and governor to hatch a desperate plan. Hoping that the news of Lloyd's pregnancy might yet sway the judge in her favour, they petitioned the under-sheriff for a respite of execution. Without the judge's or home secretary's authority this was improper; but Lofft cited Blackstone to prove that judges ordered execution only at 'a convenient time'. Persuaded that execution day was only 'a day of custom', the sheriff agreed to allow Sarah Lloyd a few more days to live. The several hundred people who turned up next morning to watch her die did so in vain.

Lofft then wrote to Portland to pre-empt the likely attack from that high quarter. 'Entirely without hope of life' and truly repentant, Sarah Lloyd was 'the deluded and miserable victim of a man who was under every duty to have protected her'. The court had not been told this. The king should be informed of the new circumstances forthwith. Lofft also wrote to the recent prime minister and old reformer Grafton, whom he knew from the giddy reform days. The girl had acted only out of a misplaced duty and honour, he insisted; hitherto virgin, she had known no

[10] The letters do not explain why, if pregnant, she chose to delay marriage, or where the signs of the baby were now, 6 months later. If she were pregnant, she could not have been hanged. It must be assumed that she miscarried or lied.

man other than Clarke. 'Ensnared and driven by him into her fate', her sexual surrender made her feel like his wife. Had this been known in court, she would surely have been pardoned. Meanwhile she was 'composed, resigned and prepared: wonderfully so . . . hardly any human being can be a greater object of mercy than she now'. If he were to be rebuked for the impropriety of securing a delay of execution, he added, 'all that I could suffer on this account weighs with me much less than what this poor repentant young creature must suffer if after this forbearance on the part of the Sheriff which permitted her life [it] is still not possible [for her] to be saved. This thought is indeed horror and torture.'

Lofft was right to anticipate trouble. Within two days the king's messenger arrived with two coruscating rebukes from Portland. The first dressed down the sheriff for breaching his duties and ordered the execution to proceed on 22 April. The second fumed that Lofft and Drummond had induced the sheriff to encroach upon the king's prerogative: 'The Throne is the only fountain of mercy, and it is of the utmost importance to the safety and happiness of the publick that no functionary should be suffered to transgress the precise bonds of the official duty which the law has assigned to them.' It added that 'the law must be implicitly obeyed'. 'The great object of punishment is example,' Portland proclaimed, and this case 'must be one of those which calls the loudest for being marked and branded with the most rigorous hand of the law as a warning and terror to misdeeds of a similar tendency'. Drummond was so agitated by this grand rebuke that he was unable to hold his pen and had to dictate his craven apology. By contrast, Lofft's response to Portland's letter was rather magnificent—suicidally so.

He held his hand for a few more days. News had come that Grafton would sign the second petition and with Kinnoul would press Portland to reconsider. Lofft wrote gratefully and garrulously to Kinnoul, rehearsing the minutiae of the case and again extolling Lloyd's 'modesty, clearness, and simplicity [in] all she says; her mild and placid countenance; and the evenness and consistency, the calm fortitude and modest preparation of her whole deportment in these most singular and most trying circumstances'. But the days passed and Portland remained silent. On the eve of execution Lofft knew the game was up. He composed two letters. The first was to the *Monthly Magazine*, denouncing the prejudicial accounts of Sarah Lloyd's crime circulated in the national press and rehearsing his own version.[11] The second was an

[11] Rayner and Crook, iv. 237–8.

incoherent but heroically defiant ten-page letter hastily scribbled to Portland, which the home secretary was unlikely to read closely if he read it at all. It rang the changes on Sarah Lloyd's 'modest, ingenuous mild countenance—speech and manners touchingly simple—no anxiety to justify herself, much less to accuse or blame any one', and it reviewed her relationship with Clarke. It impugned the judge's disinclination to try her for the arson, since examination of that charge would have revealed Clarke's villainy and saved her from the noose. And it made magisterially clear what Lofft thought of Portland himself and of the justice which had failed Sarah Lloyd so grievously:

> This is not any longer an application to your sympathy, to your official interposition, or even to the Royal Mercy. By the time you will be reading this (if indeed you will read it) Sarah Lloyd . . . will [have passed] beyond the limits of human power. . . . Your Lordship and the judge will I presume now enjoy the praise of having been the firm guardians of law and justice: the protectors . . . of property, of houses, and their inhabitants from robbers, from assassins and mid-night incendiaries. I am for the present very differently regarded—yet your love of the public welfare and security and respect for the safety and comfort of individuals I will dare to say cannot exceed mine. And pardon me, my Lord Duke, I would not change [our] situations.

Lofft's last defiance was to announce that out of 'sympathy and respect' he would attend Lloyd to the last moment, in order to 'soften those sufferings which, with far other intention, I have so much encreas'd':

> I thank Heaven, I have never yet attended an execution. . . . With respect to the pain to my own feelings I may doubt whether I should not prefer suffering this death myself to being the spectator. . . . [If I did not go] my imagination would present it to me as if I were a spectator: and with reproach to myself for not rendering this little, but last comfort.

If Portland's well-publicized rebuke had not already undermined his credit in county society, Lofft's behaviour at the scaffold now did. An extraordinary spectacle ensued. In pouring rain on the morning of 23 April Sarah Lloyd, with her arms pinioned, was conveyed the mile from Bury prison to the scaffold. Astonishingly Lofft sat in the cart with her, comforting her as she sheltered under his umbrella. He mounted the scaffold with her, and stood by her facing the crowd. Then he spoke for fifteen minutes, denouncing the home secretary for refusing to lay new evidence before the king. As he eulogized Sarah Lloyd's innocent and sweet temper she wept. The crowd did too. According to the *Bury Post*, 'scarcely a dry eye [was] present', and though this was an unfortunate phrase in those climatic conditions, the collective emotion was real. The hangman faltered as he prepared for the fatal moment, but Sarah Lloyd, according to Lofft, calmly assisted him by holding back her hair as he placed the noose round her neck. Her poise was astonishing as she hanged.

Sarah Lloyd's tragedy was over; Lofft's was not. He next took her funeral in hand. Press reports of that event are brief and uninformative. A local diarist, tellingly indifferent, notes: 'Sarah Loyd buried today. Mr Lofts & a vast concourse of canting people attended her Funera[l].'[12] This was no great matter in polite society; people turned away, embarrassed. We get the best sense of what happened from a Home Office informer. The times were dangerous in Suffolk. Wheat was high at £6 a quarter, the common people surly; spying eyes were watchful, Lofft a marked man.

Lofft decided on eight in the evening for the burial, but turned up with the body in the abbey burial ground an hour later, in the dark. A thousand people had assembled, and Lofft told them how Sarah Lloyd's mother had tried to hang herself when she learnt that the appeal had failed. He reminded them that 'the unfortunate sufferer now about to be interred was, until one fatal moment of seduction, surprise and horror [*this the reporting spy's sarcastic interjection*], respectable herself as she also came of a family which he pronounced to be truly respectable; for respectability was not to be confined to rank or riches, but is applicable to every person who gains his livelihood by honest industry.' (The spy was unimpressed. 'I am assured that the girl's father is a sot who cared not whether his daughter was hanged or not, and her mother had brought her up in ignorance even of the Lord's prayer.')

When Lofft then complained to the crowd that the authorities had denied her a burial service, the Bury preacher Mills contradicted him. The service had not been denied, because it had never been requested. 'Down with the parson!' people cried, and 'Hear Lofft, hear Lofft!' The crowd's further cries were carefully noted: 'The rich have everything, the poor have nothing'; 'The farmers are rich'; 'Lofft is the poor man's friend'; 'We will do what Lofft bids us'. Lofft was followed to his inn 'with shouts of applause and the honourable appellation of the "Christian", whilst poor Mills, who had voluntarily offered and read the service, was scouted and hissed home to his house. *Ainsi va la monde.*'

Two months later a tombstone was erected in the abbey yard; it stands there still. It gives the seducer Clarke his deserts at last, but

[12] Fiske, *Oakes diaries*, i. 389.

chiefly it reasserts the execution's legitimacy. Legitimation was likewise the purpose of the engraving and verses published by the Bury printer Birchinall (Plate 36). Sarah is shown in mob-cap and servant's dress. She weeps fetchingly into a handkerchief. She leans on her tombstone and points to its inscription. Four verses rub the moral home.[13] There was no chivalric forgiveness here. Lofft was furious. He protested fiercely in the press that she had already paid her dues to the law and that there was no reason for her errors thus to be 'drawn from their dread abode'. 'Least of all should a crime of the highest atrocity [arson] have been imputed by this tablet of reproach, a crime neither proved, nor brought to trial against her, nor I believe provable.' But he was beaten of course. The final unfolding showed how badly.

At both the execution and the funeral Lofft's enemies had listened to his harangues with some delight. His anti-war stances were not forgotten by loyalists. There were axes to grind, and excuses now to grind them. Two of Lofft's brother JPs sent their accounts of the execution to the Home Office. Describing how Lofft 'harangued the populace in terms not the most respectful to the executive government of the country', one left it to Portland to decide how to obviate 'the pernicious effects of such a conduct in a popular and active magistrate over the minds of the lower ranks of the community'. The other had no doubt what was needed:

> Probably you will know that Mr Lofft was some years violently democratic and that he continues to be very disaffected and hostile to government. I can confidently add that his Brethren of the Bench are heartily tired of his company and would be glad to get rid of him, and that the generality of people here (some of his own partisans excepted) agree on thinking that he well deserves to be removed. . . . He certainly is a dangerous man to be trusted with any means of doing mischief.

When Lofft compounded his offences by speaking at the funeral, one informer regretted that no mischief had ensued: Lofft might have been prosecuted then. As it was, 'a man, whose conduct so manifestly tends to defeat the purpose of executive justice and lessen the influence of a necessary example, is wholly unfit to be in any shape trusted with the administration of justice':

> Should he be struck out of the Commission, I am satisfied that his fellow justices will feel themselves relieved from a most vexatious embarrassment in their business; all the better class of people (including many of his partisans)

[13] E1/20.6.2.

36. *Sarah Lloyd of Bury St Edmunds*, 1800.

Hanged for stealing from her mistress, the servant girl Sarah Lloyd was the target of many chivalric male fantasies. But this engraving shows that some took a less charitable view of her, using her tale to warn girls in her station of the perils of seduction. Her gravestone, here accurately depicted, still stands intact in the abbey churchyard.

will think him rightly served; and the ignorant and ill-judging multitude will consider him as punished for his humanity and religion, will execrate the measure for a few days and then forget it.

On 6 May Portland submitted to the lord chancellor his view that Lofft's actions rendered him unfit to continue as a JP, and within a month Lofft was a magistrate no more. In the *Bury Post* Lofft replied to this humiliation with curt defiance. He was, he wrote, prouder of his removal from the bench than he was of his appointment. He tried to fight back a little,[14] but soon fell silent.

In ensuing years he withdrew from the society in which, as his son put it, 'he could not live but to his ruin'. He kept up some kind of end in Bury society, practising occasionally as defending counsel at the local assizes,[15] but his rare forays into metropolitan politics fell flat and old allies deserted him as his eccentricities grew. When among erstwhile London friends in 1809 he proposed Fox's memory, Crabb Robinson thought him pathetic: he 'implored attention, but all in vain'. He apparently joined the Society for the Diffusion of Knowledge upon the Punishment of Death after 1811 (and back in Bury he was heard toasting Romilly, 'reformer of penal laws'), but he made no mark there.[16] When he learnt in 1815 that the 'great Napoleon' was to be sent to St Helena (news 'almost overwhelming to me, though long accustomed to suffer much, and to expect every thing'), he wrote to the *Morning Chronicle* to propose that Napoleon should claim habeas corpus. (By odd coincidence his letter appeared in an issue of the paper which commented on Eliza Fenning's fate.) Charles Lamb asked Wordsworth: 'Have you seen a curious letter in Morn Chron by C.L. the Genius of Absurdity respecting Bonaparte's suing out his Habeas Corpus. . . . The man is his own Moon. He has no need of ascending into that gentle planet for mild influences.' And not long after, Dorothy Wordsworth said that she would 'pity his wife if it were not her own fault that she married him knowing what he was'. Secluded in their decaying Suffolk manor, his

[14] Some days later he wrote to Samuel Parr to ask for information on a Norfolk appeal in which Parr had been involved in 1784, when, as in Lloyd's case, the sheriff had unilaterally ordered a last-minute reprieve to allow an appeal to go forward. Those were 'better times than the present', he wrote, for any petition for life 'would at least have been suffered to reach the King': J. Johnson (ed.), *The works of Samuel Parr, LLD* (8 vols., 1828), viii. 52–4. This is referred to in Ch. 6; its papers survive in HO47.1 (Matthew Barker's case).

[15] Fiske, *Oakes diaries*, ii. 11, 17–18, 139.

[16] E. J. Morley (ed.), *Henry Crabbe Robinson on books and their writers* (3 vols., 1938), iii. 845; J. A. Hone, *For the cause of truth: Radicalism in London, 1796–1821* (Oxford, 1982), 146, 175, 248; Fiske, *Oakes diaries*, i. 147.

son recalled that a shadow descended on the family. Lofft dedicated his later years to classical studies, poetry, and travel. He died in 1824.[17]

So one of the advanced radicals of those blissful dawning days of the 1780s came to rest in defeat. In this he was akin to that other though fictional victim of powerful men's schemings and his own passion for justice, Godwin's *Caleb Williams* (1794). 'My life has for several years been a theatre of calamity,' that novel begins. Had the book been written a few years later one might think Lofft's story its model. Lofft too embodied the man of sensibility defeated both by his own impulsive character and by powers whose legitimacy he had always challenged. He achieved some magnificence. He showed what one man of feeling was capable of, and that not all need be self-serving evasion in the face of execution horrors. But there was also a kind of lunacy in what he attempted; and 'empathy' is a word hardly adequate to the splendour of the case. Identifying body and soul with the victim-woman in Sarah Lloyd upon whom he had projected his long frustrations, and then in love with the creature he made of her, Lofft acted out his own Godwinian tragedy and paid his price, as she paid hers.

3. *The beatification of Eliza Fenning, 1815*

Our second, very different, and much more influential story can also be introduced with a text. It is Johnny Pitts's halfpenny ballad-slip, sold on the streets, *Lines on the death of Eliza Fenning, executed at Newgate, July 26, 1815*:

My aching heart with pity bled,
    When poor Eliza! cloth'd in white;
At Newgate drop't her lovely head,
    And clos'd her eyes in endless night.

The weeping crowd all mournful stood,
    And many a plaintive sigh was heard,
And 'ere the awful scene they view'd,
    Each hop'd her life might yet be spar'd.

Poor hopeless maid! for poison try'd!
    Which caus'd her sad untimely death,
She vow'd that innocent she died,
    And seal'd it with her parting breath.

[17] *Morning Chronicle*, 2 Aug. 1815; N. Roe, *Wordsworth and Coleridge: The radical years* (Oxford, 1988), 15, 29–30; E. de Selincourt, *The letters of William and Dorothy Wordsworth* (Oxford, 1970), iii/2: 267–8; E. W. Marrs (ed.), *The letters of Charles and Mary Anne Lamb* (Cornell, Wis., 1978), iii. 125, 174; Capel Lofft, the younger, *Self-formation, or The history of an individual mind* (2 vols., 1837), i. 3–9.

A tender youth's devoted heart,
Had been to fair Eliza given,
But death resolv'd this pair should part,
And angels bore her soul to heaven.[18]

Clearly some myth-making was under way, among the populace as well as the polite. Although this ballad carried a stock woodcut depicting lovers in seventeenth-century dress, there is no proof of the existence of Eliza Fenning's beloved other than the fact that she died in clothes said to have been laid aside for her wedding.

We hear Eliza Fenning's voice more directly than Sarah Lloyd's, even if it was guided by mentors who taught her her phrases.[19] From her cell in the summer of 1815 she allegedly wrote many letters, and thirty were published. Publications of these kinds had many precedents,[20] but these were effusive gems of their kind, redolent of the high-flown tropes of popular melodrama. Thus to her spiritual counsellor: 'believe me, cruel and pitiable is my forlorn situation, but yet this trouble may be for some divine purpose to bring me to myself.' She bade 'farewell to my dear, unhappy, affectionate parents, whose breaking hearts cut my tortured breast, but God bless them and give them consolation amidst the awful scene of their oppressing woe'. 'I am murdered,' she wrote to them; 'but what a pleasing consolation within this tortured breast to suffer innocently.' And to her spiritual friend again: 'It is hard and pitiable, and indeed distressing beyond description, to suffer such an ignominious death, innocent as I am.' To a friend she sent a lock of her hair. She also wrote a mercy petition, addressing Home Secretary Sidmouth's 'well known goodness and mercy, which has repeatedly been extended to many miserable creatures under calamaties like myself': 'I most solemnly declare to a just God, whom I must meet, and my blessed Redeemer, at the great and grand tribunal when the secrets of all hearts will be known. Innocence induces me to solicit a fuller examination. I am the only child [surviving] of ten, and to be taken off for such an ignominious crime strikes me and my dear parents with horror.' On the

[18] Madden Broadside Collection, Cambridge University Library, 8/466.

[19] J. Watkins and W. Hone, *The important results of an elaborate investigaton into the mysterious case of Elizabeth Fenning* . . . (1815), 90–1. The following account is drawn from this source and the other pamphlets noted below; also from the *Examiner*, 16 Apr., 30 July, 13, 20, 22 Aug., 3, 18 Sept., 1, 15, 22 Oct., 5 Nov. 1815; *The Times* and *Morning Chronicle*, 31 Mar., 12 Apr., 27 July, 1 and 3 Aug.; Borrow, vi. 143–54 (for the fullest account, along with letters); Rayner and Crook, v. 159–64; and OBSP, Apr. 1815: 220–3. Fenning's appeal papers do not survive in the Home Office papers.

[20] Cf. *Miss Mary Blandy's own account . . . containing . . . original letters . . . whilst under sentence of death* . . . (1752).

scaffold too; asked by the Revd Cotton and her attending Methodist minister if she would speak, she declared: 'Before the just and almighty God, and by the faith of the Holy Sacrament I have taken, I am innocent of the offence on which I am charged.' Then she added *sotto voce*: 'I hope God will forgive me, and make manifest the transaction in the course of the day.' This last phrase hinted at another's soon-to-be-revealed guilt. The meanings of executions depended on statements like this. Every ear strained to hear them.

The appearance and demeanour of the condemned constructed meanings too. Eliza appeared at the Debtor's Door with two men to be hanged alongside her. Of the men we hear nothing in the reports, except that one of them wished her well and told her they should all three soon be happy. She was the figure people saw. She came out with the hangman's rope wound around her waist. She was dressed in 'a white muslin worked gown, and a worked muslin cap, bound with white satin riband; she wore a white riband round her waist, and pale lilac boots laced in front'. She appeared 'very interesting', reports said. She maintained a 'uniform firmness' throughout. She objected to the customary tying of the handkerchief round her eyes, refusing the hangman's dirty offering. 'My dear, it must be done,' the ordinary told her. She asked to be forewarned of the fall of the drop, but this too 'was what was never done', the ordinary said: 'it saved a great deal of pain to the convict'. She expired 'in resignation' and 'without a struggle'. As with Sarah Lloyd, this peaceful passing was not forgotten either. It was assumed to mean inner peace. 'The most heartrending sensations pervaded the minds of the thousands who watched the dreadful scene.' How the two men died with her we do not hear.

The crowd outside Newgate was bigger than any since the hangings of Holloway and Haggerty in 1807, it was said (45,000 attended then). No execution since Dodd's aroused as much interest as this one, according to the *Annual Register*. Rumour and gossip ensured that few were ignorant of what was going on. Most left the scene satisfied that something strange and wrong had occurred. As the *Examiner* put it: 'It is difficult to conceive what possible motive she had, if really guilty, to rush into the presence of her omniscient Creator, whose penetration no sophistry can elude, nor no art mislead, with a lie in her mouth.' The crowd smelled cover-up, malicious prosecution, wrongful conviction, cruel hanging—excuses for angry mayhem. On that and following evenings several hundred of the rough people assembled outside the prosecutor Turner's house in Chancery Lane crying 'Pull it down'; they were stopped only by police. Meanwhile a public subscription was opened for the Fenning parents. They had a bill to pay:

> For Elizabeth Fenning
> 1815
> July 26th. Executioner's fees, &c, striping }
> use of shell } £0. 14. 6.
> Settled. C. Gale, Junior.[21]

After her death, the publisher John Fairburn put out a purported portrait of Fenning, in fact one used earlier to depict the duke of York's mistress Mary Ann Clarke: the face is sweetly pretty. Next William Hone published a portrait drawn from life by Robert Cruikshank in the condemned cell (Plate 37). Here she is plainer, gazing vacantly ahead, a book or Bible in her hands, the fullness of her breasts spectacularly emphasized.[22] Stories circulated about her last hours—how she forgave all who gave evidence against her; how before her hanging she washed herself very carefully, in particular her feet; how she gave locks of hair to her attendants; how she worried that the time left was too short for prayer; how she kissed her hand to the prisoners at their barred windows as she left for the press-yard. The *Day* made the key point: 'Is it possible that this offence should have been committed by such a person—by a woman—a woman not engaged in the long habitual contemplation, and commission of revolting crimes—a child—a daughter, but recently sportive in all the guiltlessness of infancy?'[23]

Eliza's body was put on public display in her parents' lodgings, and for four days people queued to see it. 'She lay in her coffin seemingly as in a sweet sleep, with a smile on her countenance', and 'no part of her body changed colour in the least for three days after execution'.[24] Her funeral procession down Lamb's Conduit Street and her burial near the Foundling Hospital, as we have seen elsewhere, was a pageant of extraordinary ambition, ten thousand people attending it, many weeping.[25] She was as close to beatification as Londoners would ever bring a wronged maiden.

[21] *Examiner*, 6 Aug. 1815.

[22] F. W. Hackwood, *William Hone: His life and times* (1912), facing p. 99. Fairburn's print appeared in his *The case of Eliza Fenning, who was convicted on a charge of attempting to poison the family of Mr Turner, by mixing arsenic in yeast dumplings, containing her trial with the particulars of her execution, and funeral* (1815): its attribution as a depiction of Clarke comes from the BL catalogue.

[23] Watkins and Hone, *Important results*, 88–9; Hackwood, *Hone*, 99; press reports were summarized in Fairburn, *The case of Eliza Fenning*, 22.

[24] Fairburn, *The case of Eliza Fenning*, 38.

[25] See Ch. 2.

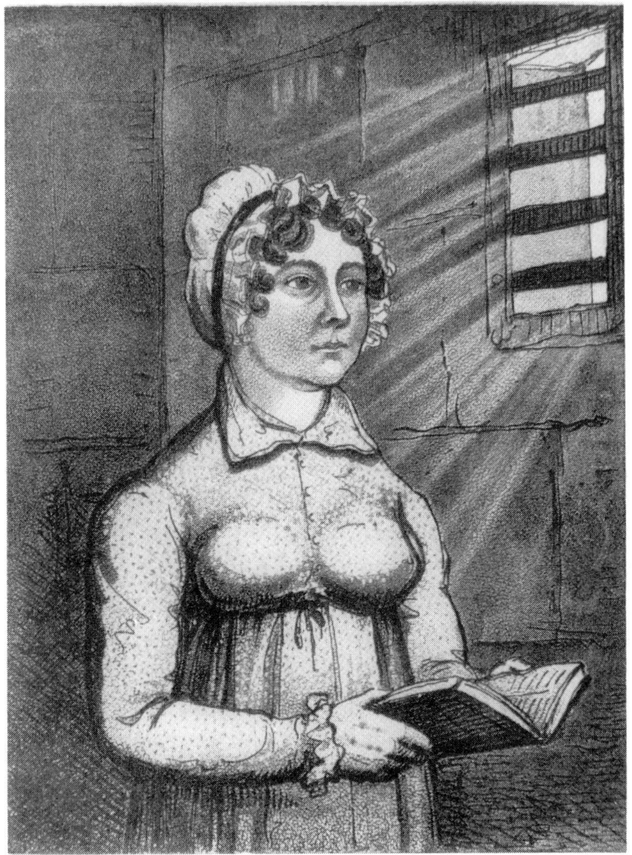

37. Robert Cruickshank, *Eliza Fenning*, 1815.

This engraving of Eliza Fenning reading her Bible in Newgate's condemned cell was published by William Hone to further his exposé of Old Bailey justice. Her guilt much in doubt, Fenning was turned into an icon of wronged womanhood by metropolitan radicals, and her case was remembered for decades after her death.

Eliza Fenning was common enough flesh and blood of course. Her oddest distinction was one which only we can notice. She was born in Hadleigh, Suffolk, in June 1793. This was where Sarah Lloyd was born and lived too. In so little a town the families would have known each other. The rest of her story is soon told. Fenning's father was a soldier. Discharged, he had migrated to London to become a potato-dealer with his brother in Red Lion Street off Holborn, while his wife took up a job with an upholsterer round the corner. Eliza learnt her letters at the dissenting Gate Street Charity School in Lincoln's Inn Fields.[26] She left school at 12 and entered service at 14. Now aged 22, she got employment as a cook in the household of Mr Turner, a law-stationer at the Holborn end of Chancery Lane. So everyone lived and worked within five minutes' walk of each other. Stories like this usually traversed narrow patches. (William Hone was not far away either. His bookshop was at the bottom of Chancery Lane across the road in Fleet Street, and a year later he opened a branch in Old Bailey.) Newgate and its scaffold was never more than a few minutes' walk away for all these people.

Six weeks after Eliza had joined them, the Turner family ate some yeast dumplings and fell violently ill, near to dying. She ate some dumplings too, and fell just as ill as they. The family's apothecary friend John Marshall filtered out the dumplings and found white crystals of arsenic in them. Since Eliza had cooked the things, she was whipped off to Hatton Gardens police office in no time at all, examined, and prosecuted for attempted murder. In April she was convicted and condemned to death at the Old Bailey by the recorder of London, Silvester. Sentenced, she 'instantly fell into a fit, screamed, and cried aloud most bitterly, and was carried from the dock in a state of insensibility'.[27]

The prosecution alleged that her motive in trying to kill the Turner family was revenge after a rebuke by her mistress for 'some indelicacy in her conduct': she had been caught visiting the Turners' apprentice lodgers by night, half-dressed. The Turners also found in her bedroom 'an infamous book, with a register on one of the pages, that explained various methods of procuring abortion'—so the opposition reported anyway. Visited in prison by the family, she vented her hatred of them, shouting 'insultingly' that they were responsible for the poisoning—perhaps with reason. At her execution sermon she became hysterical. Not the wilting violet of legend, then? We learn these things from the trial and from the Turners' apothecary friend, who published a book to advertise his skills at detecting arsenic—the only hostile publication among many on the case.[28] But her bile was understandable. Tried speedily and casually by a notorious recorder, her guilt was not beyond doubt.

[26] *Examiner*, 6 Aug. 1815.

[27] *The Times*, 12 Apr. 1815.

[28] J. Marshall, *Five cases of recovery from the effects of arsenic; with the methods so successfully employed for testing the white metallic oxides . . . to which are annexed many corroborating facts . . . relating to the guilt of Eliza Fenning* (1815).

The trial report occupies four brief pages in the *Old Bailey Sessions Papers*. It is a bleakly characteristic little record of rushed justice. Only prosecution evidence was heard; defence counsel Alley and Gurney were not allowed to sum up for Eliza (this by law), and their cross-examinations of the prosecution were cursory; Eliza's panicking protestations of innocence and her character witnesses were brushed aside. Silvester accepted the family's insistence that only she could have poisoned the dumplings. He noted the testimony of an apprentice in the house that she had advised him not to eat the dumplings. He directed the jury unhesitantly: 'Although we have nothing before us but circumstantial evidence, yet it often happens that circumstances are more conclusive than the most positive testimony.' How odd, he added sarcastically, that though the family was poisoned, Fenning had not rushed to their assistance: 'Gentlemen, if poison had been given even to a dog, one would suppose that common humanity would have prompted us to assist it in its agonies: here is the case of a master and a mistress being both poisoned, and no assistance was offered.' The fact that Eliza had been poisoned too and was writhing in her bedroom, declaring that she wished to die, was not noticed. The jury—not always a palladium of English liberty—returned their verdict of guilty after a few minutes' consultation.

Ensuing commentaries on the case found the required villain of the piece in the recorder, Silvester. Widely known as a randy reprobate—'Black Jack' he was called—he demanded sexual favours from any lady who came to beg him for mercy or justice.[29] He was also slapdash in his trying, though not alone in this: Old Bailey trials were usually conducted at breakneck pace—if the phrase be allowed—and prejudice against the prisoner usually ruled.[30] But his was a glaring prejudice and it did not escape notice. Although Romilly held aloof from the case, he noted in private the absence of motive and that the recorder had 'conceived a strong prejudice against the criminal', making 'some very unjust and unfounded observations' against her in his summing-up.[31]

Nor would the recorder let ill alone after sentence was delivered. When the prosecutor Turner was about to sign a petition for mercy, Silvester actually turned up at his house to advise against it lest suspicion be cast on another member of the family. Similarly, Silvester told the Quaker banker Corbyn Lloyd that the only reason he and others were inclined to interfere in Fenning's case was 'because she was a

[29] W. Jerdan, *Autobiography of William Jerdan* (1852), i. 131.

[30] See Ch. 19, below.

[31] Romilly, iii. 235–6.

pretty woman'. Most stunningly, it turned out that Silvester had deliberately suppressed a local chemist's revelation a few days before the execution that an unnamed member of the family (the son or an apprentice?) had lately been 'wild and deranged' and had threatened poisoning. It was to this that Eliza referred on the scaffold when she predicted that the transaction would soon become manifest. The recorder belittled the importance of the allegation when the case was taken to the King in Council, huge doubt though it cast on the conviction. Well might a 'respectable merchant' ask *The Times* 'by what accustomed usage or custom of our laws, or courts, or by what standards of justice' could such evidence be ignored.

There was other nasty work behind the scenes. The *Observer* and other papers spread reports on Fenning's disreputable character. They said that at the age of 12 she had been dismissed from her charity school 'for lying and lewd talk'; in prison she showed her 'amorous disposition' by sending love-letters to fellow prisoners; she said that she would tear the prosecutor's heart out if she could. Someone had handbills printed to announce that father Fenning was overheard telling his daughter to plead innocent lest he never walk upright in the streets again: a turnkey swore an affidavit to that effect, and Mr Fenning had to swear a counter-affidavit denying it. There was intimidation too. The *Public Ledger* refused to publish a letter for Fenning lest it be subject to legal harassment if it did. The *Observer* refused to advertise Hone's and Watson's book on the case.[32] Hone probably caught the point of all this rising excitement about the case when he noted that

> all the masters and mistresses of families, whose credulity or idleness rendered them proper subjects for alarums, were excessively devoted to the vociferous execration of the wickedness of servants. . . . Thus a sort of general cry was raised for the hanging of Eliza Fenning, as an example to all maidservants suspected, upon presumption of murderous intentions.

The night before the hanging the home secretary, Lord Chancellor Eldon, and Silvester met to review the representations made to save Fenning. They decided not to accept them.

We take it as axiomatic (for the grounds, see later chapters) that innocent people often hanged in this era. Better so, some thought, than let the guilty go free. Those wrongly hanged might be considered as dying for their country, as Paley had unforgettably said. Protests were rare.

[32] *Examiner*, 6 Aug., 22 Oct. 1815.

But times were changing. When Eliza Fenning hanged, the quality of justice was under parliamentary and newspaper debate. And new variables were beginning to matter.

First, it was not merely that Fenning was female, young, and pretty. She also sounded pious. She was Methodist, with energetic Methodist protectors. One whose anonymous letters to her were published promised her repeatedly that they would meet in heaven, which might not have cheered her. Her prison bible was supplied by St John's chapel, Walworth. The Revd Griffith Williams and members of Gate Street Chapel petitioned for her. Cruikshank's cartoon *Royal Methodists* included among its background images a placard referring to Dr Watkins's book *Vindication of Eliza Fenning*: it was subtitled 'a Methodistical Method to libel a judge and jury of the country—publish'd on speculation by one of that elect at the moderate price of *6s./6d.*'[33] She relied on Methodist ministers in prison and on the scaffold, not liking Newgate's clerical ordinary, Cotton, because, she said, 'he always urged her to confess herself guilty; and as she knew herself innocent, and he, whenever he saw her, treated her as if she was guilty, she could not bear to see him'. The funeral procession sang Methodist hymns.[34] So we know where the tutored piety of her well-publicized protestations came from.

Secondly, this was a climate in which newspapers mattered too. Letters on her case proliferated in the *Examiner, Observer, Morning Chronicle, British Press, Statesman*, and the *Day*. They were buried in columns given over to higher matters than these: London was jubilant after Waterloo. Still, press comment on a criminal conviction had not been vented on this scale before. The *Examiner* kept up a barrage of comment over several months; Hone's *Traveller* did the same. Some letters to these papers came from men who had visited her in her cell, announcing how she had 'wept her innocence'. Others pointed out worrying facts: a juryman had been deaf, for example, and could not have heard the evidence. Most noted that she could not have poisoned the dumplings, since she ate some herself. Others sent in character references from former employers. One published the allegedly amorous letters Fenning had sent to one of her fellow-condemned, in order to show that they were only polite and friendly, asking meekly whether he had petitioned for mercy.[35] Some (including two 'noble lords') had

[33] M. D. George and F. G. Stephens, *Catalogue of political and personal satires in the British Museum* (11 vols., 1870–1954), i. 597–9 (no. 12624).

[34] Watkins and Hone, *Important results*, 96.

[35] *Examiner*, 3 Sept. 1815.

mulled over the case as they read it in the papers; these said that they would petition the regent. The lawyer John Watkins later told Hone that he also had become interested in Fenning's fate from newspapers; he had petitioned as well. Fanned by newspapers, the gentle world was newly interested in the doings of courts. Trials were subjects of gossip, and the conundrums in puzzling cases like this engaged the idle and the benevolent mind. Moreover, in a year when City politics were fraught with aldermanic conflict, with evangelicals muscling in too, the doings of the Court of Common Council's recorder would never pass unnoticed.

Unfortunately the appeal archive on the case has not survived. Had we access to it, we might have traced a network of interest and comment spreading out in clubs and at dinners and embracing influential people. Newspapers report that a well-known solicitor (possibly the James Harmer whom we meet elsewhere; he also worked round the corner, in Hatton Gardens) told Sidmouth how dubious he thought the conviction was. 'An eminent Quaker' visited Eliza—William Forster or Thomas Buxton perhaps, both becoming concerned with Newgate conditions at this time. The Quaker banker Corbyn Lloyd visited Eliza and agreed to advertise for subscriptions for her parents after the hanging, and other Quaker banks joined the appeal. Basil Montagu got involved when he learned about the insane person in the Turner household who might have committed the crime: he wrote to the recorder for a respite. Even the regent's chaplain petitioned. Others applied to the City Corporation's shorthand-writer for access to his notes of the trial to confirm their suspicions about its conduct (and, thus encouraged, the shorthand-writer published his notes on his own account, in two editions no less).[36]

Finally, there was Eliza's own use of the press. Her mistress acknowledged that she could read and write well. Her letters, which Hone later published, were probably her own, though Hone admitted that they were corrected, punctuated, and edited. Their piety and finer tropes and those of her petition were doubtless pumped up for public effect by her Methodist friends, but they kept interest alive. 'How can I convince the world when brought in guilty at the bar of man?' she wrote to the *Examiner*: 'Yet there will be a grand and great day, when all must stand before the tribunal of God.' Thus the making of a romantic tragedy was under way before the hanging. So was a debate with justice in its sightlines.

[36] J. Sibly, *Circumstantial evidence: Report on the trial of Eliza Fenning* (2nd edn., 1819).

More was at stake in this story than quasi-erotic projection. Fenning's guilt was in doubt; the law's corruption was on the way to being exposed. So on questions like these there were knowing audiences to play to. Hard-bitten radicals whose strictures on the law already drew on a well-developed appreciation of 'old corruption' itched to seize their moment. Seize it some did.

The *Examiner* weighed in on the case in this spirit. Its editors Leigh and John Hunt knew what they were talking about. They had already been acquitted once for slandering the army on its savage military beatings. When Fenning's case broke, they had just been released from prison; Lord Ellenborough had sent them there for two years for the *Examiner*'s attack on the regent. Their experience was formative. In prison Leigh Hunt had watched the gallows being erected over the prison gate; he recalled a spectacle 'I shall never forget': 'a stout country girl, sitting in an absorbed manner, her eyes fixed on the fire. She was handsome, and had a little hectic spot in either cheek, the effect of some gnawing emotion. . . . She was there for the murder of her bastard child. . . . The gallows, on which she was executed, must have been brought out [and erected] within her hearing.'[37] Probably it was Hunt's pen which now denounced the *Examiner*'s rivals for being squeamish about 'exciting prejudices' against the law. He was not surprised by the tumults after Fenning's hanging, he wrote: 'In the conviction of that ill-fated girl circumstances did little, fact less, and situation everything,' and 'of direct evidence there was not an atom'. He denounced the 'smooth and specious fallacy' of 'the Protestant Jesuit, Paley, that confounder of the right with the profitable', that it was necessary for the innocent to suffer sometimes lest the guilty escape. 'To clothe law with all possible sanctity . . . [and] scarlet-clad oratory' had been 'the policy of every civilised nation, and none more so than our own'. 'What can we do more than Pilate did under similar circumstances?—wash our hands upon the *accident* of guilt or innocence, and go to dinner?'[38]

The other critical voices in this story were John Watkins and William Hone: it is from their huge analysis of the case that many of our details are drawn. Watkins had trained for the dissenting ministry in Bristol, turned evangelical, and come to London in 1794, there to live off the topographical and biographical essays which he pumped out regularly. Cruikshank's cartoon implies that he was 'methodistical' in 1815. He was a doctor of laws, so he knew what he was talking about in this

[37] Leigh Hunt, *Autobiography* (Oxford, 1928), 300.

[38] *Examiner*, 13 Aug. 1815.

story. 'Our criminal records are stained with too many instances of inconsiderate verdicts and inexorable sentences,' he petitioned the regent on Fenning's behalf. In letters to Hone he lamented that 'with all the abstract excellence of our laws, every facility is given to prosecution, while obstacles of all kinds are opposed to the manifestation of innocence.' It was strange that no judge had assisted in Eliza's trial, he observed. Under the recorder it was 'despatched in a manner, and within a space of time, little different from what occurs in an ordinary quarter sessions on minor offences'. The witnesses against Eliza had been drilled beforehand, he suspected: the prosecution case had been too neat for credibility.[39]

About Hone much more is known. Born in 1780, he was the dame-school-educated son of a devout law-stationer father who belonged to the Calvinistic Huntingdon Connection. Rebelling against his father's piety, he read Godwin and similar texts, turning atheistical (or at least Unitarian) and joining the London Corresponding Society when he was 16. Struggling upwards as a clerk, bookseller, and publisher by 1815, every halfpenny counting to feed his wife and family, he associated increasingly with better-heeled radicals like Basil Montagu and Richard Phillips, pulling other marginals like Cruikshank into his circle. In 1817 he achieved his apotheosis as the hero of radical London when he was tried by Ellenborough for publishing blasphemous parodies and acquitted thanks to his brilliant defence of himself. He made a modest fortune thereafter with his and Cruikshank's parodies of church, king, law, and ministers, until he saw the Redeemer's light in 1832 and died in 1841 the very model of the respectable publisher of items humorous, inoffensive, and engaging.

Hone's early involvement in Fenning's case is a reminder, if one is needed, that sympathetic temperaments were not cultivated exclusively by the salon classes. Of humble birth, literate but largely self-taught, youthfully secularized in belief, seeking wisdom in Paine or Godwin, upwardly aspiring and morally righteous, struggling in small business or clerical employment—many such people whose political consciousness was shaped in the 1780s and 1790s were fired to indignation by the sense that their own marginalization owed everything to oligarchy and privilege. At such social levels a *felt* instinct for justice flourished in hearts as in minds. In youth the poems of Goldsmith and Thompson had filled Hone's heart 'with rapturous joy'; his 'affections went forth to every

[39] Watkins, *Important results*, 157, 163, 171–3.

living thing'. He had read romances and novels 'incessantly', his desire for them 'insatiable'. And although he distanced himself from his father's piety, the model stayed with him of a man who 'was painfully affected by the sufferings of the poor; could he have willed it, every human being would have been in comfort and happiness'.[40] The struggling bookseller took the model to heart. His first publishing enterprise supported Montagu's exposure of asylum conditions in 1814, with Hone commissioning and publishing Cruikshank's etching of the lunatic William Norris ('riveted alive in Iron and for many years confined in that state, by chains 12 inches long, to an upright massive iron bar in a cell in Bethlam').[41] Altogether, when the Fenning case came to his attention a year later, his response to it was predictable:

> I was going down Newgate Street on some business of my own. I got into an immense crowd that carried me along with them against my will; at length I found myself under the gallows where Eliza Fenning was to be hanged. I had the greatest horror of witnessing an execution, and of this in particular; a young girl of whose guilt I had grave doubts. But I could not help myself; I was closely wedged in; she was brought out. I saw nothing, but I heard all. I heard her protesting her innocence—I heard the prayer—I could hear no more. I stopped my ears, and knew nothing else till I found myself in the dispersing crowd, and far from the dreadful spot.

Hone's recollection in the 1830s of this confrontation offers a nice example of retrospective shame. To have heard Fenning protesting her innocence in a crowd of 45,000 would have required Hone's being at the scaffold's foot; but in the 1830s he turned evangelical, so so old-fashionedly curious an encounter with a hanging could hardly be confessed to. Be that as it may, over the ensuing days he claims to have written (more likely he pirated) *The maid and the magpie, or Which was the thief?*, a parable drawn from the French melodrama about the circumstantial evidence which linked a guiltless maid with a magpie's theft. Getting George Cruikshank to illustrate it with 'a cut of a magpie hung by the neck to the gallows', Hone made some money to feed his hungry family at last: it was still in his lists at sixpence a copy in 1817.[42] The play also 'roused the public as to' Eliza's case, he claimed. In fact the public was aroused already, for other publishers had pointed the way. John Fairburn had published a shilling book on Fenning's case which

[40] Hackwood, *Hone*, 12, 47, 54, 58.

[41] Ibid. 95.

[42] Ibid. 99–101. Hone's version is not in BL. I. Pocock's version, *The maid and the magpie, or The fallacy of circumstantial evidence . . . first performed at the Theatre Royal, Covent Garden, on Friday, September 15th, 1815*, was published not by Hone but by John Miller: pirating was rife. For Cruikshank's illustration, see A. M. Cohn, *George Cruikshank, a catalogue raisonné* (1924), no. 525.

ran to forty pages and went through at least eight editions in 1815, following it with an addendum purporting to be Eliza's 'own narrative of circumstances which occurred in the family of Mr Turner' before the crime.[43] Other pamphlets followed: from the 'friend' who tended Eliza in Newgate,[44] and from the shorthand-writer who had reported the case for the *Old Bailey Sessions Papers*, in two editions.[45] There were several others besides.[46] Hone, in short, climbed on a bandwagon already in motion.

None the less, he it was who hammered home the hardest lessons in the Fenning case. Watkins heard about Hone's efforts, and contacted him. They embarked on research together. They put out their book a few months after the execution, its elaborate title trumpeting its importance:

The important results of an elaborate investigation into the mysterious case of Elizabeth Fenning: being a detail of extraordinary facts discovered since her execution, including the official report of her singular trial, now first published, and copious notes thereon: also numerous authentic documents; and argument on her case; a memorial to His Royal Highness the Prince Regent; and strictures on a late pamphlet of the prosecutor's apothecary. With thirty original letters, written by the unfortunate girl while in prison; and appendix, and an appropriate dedication.

It was an unprecedented production of its kind; 194 pages long, the appendices adding another 46 pages, costing 6*s*. 6*d*., and aiming at well-heeled readers, it handled evidence and arguments scrupulously and confronted the issues squarely. Its message was unambiguous. Fenning's, it declared, was 'one of the most extraordinary cases that ever happened in a civilised state', warning people 'to put little confidence in the reasonings of fallible magistrates, who have grown old in the ministration of death'. The law was only too easily 'converted into an engine of oppression, and an instrument of vengeance'. The law must perpetually be scrutinized lest people presume its incorruptibility; for 'amidst the habits of luxury, the cares of business, and the spirit of curiosity, ever inquisitive after novelties . . . abuses of the most flagrant nature, are too often suffered to pass unheeded or uncorrected,' leaving

[43] Fairburn, *The case of Eliza Fenning, and Elizabeth Fenning's own narrative . . . being a continuation of Fairburn's edition of the affecting case of Elizabeth Fenning* (1815).

[44] T. W. W[ansbrough], *An authentic narrative of the conduct of Eliza Fenning . . . till her execution . . . by the gentleman who attended her* (1815).

[45] Sibly, *Circumstantial evidence*.

[46] Anon., *The fullest report published on the trial of E. F. . . . with . . . particulars . . . respecting . . . her execution, etc.* (1815); Anon., *Further account of E. F., etc.* (1815); Anon., *The trial at large of E. F. . . . to which is added, an account of her execution, etc.* (?1815). None of these is in Cambridge University Library, and BL copies were destroyed by bombing.

the law 'in the hands of the crafty and vindictive'. It was clear in this case that Eliza's fate was determined by her employers' respectability ('by which is understood their opulence') as it was 'pitted against the humble poverty of their servant maid'.

In 1816 Hone followed the book with a discussion of *Four important trials at Kingston assizes*, one of them a poisoning case like Fenning's, which posed 'thirteen questions' to the law-writer Isaac Espinasse on the evidence which had hanged Eliza. It went through two editions. In the same year his account of the thief-takers' trials reminded its readers that 'all administration of law abounds in abuse, if unchecked and disregarded' and that it was 'in the nature of petty authority to intrigue, juggle, grasp, connive, and domineer':

> If the people are not moved into some indignation at the neglect of their *soi-disant* guardians, the healthful spirit of society is defunct, and the community is degenerating into a base rabble, similar to that which marked the declining Empire of Rome. On the other hand, if they both observe and despise . . . if the people at large should be thus convinced [of the extent of corruption in high places], it is barely possible that polite people may find the hanging of an innocent person now and then as something more than a bagatelle, and Nero cease to fiddle amidst burning Rome.[47]

4. *Residues and conclusions*

We know enough about the cross-currents which flow through all complex cultures to realize how easily dissident voices like Hone's or Hunt's may drown in competing clamours. We should not expect Fenning's story to be remembered for long, any more than Sarah Lloyd's had been. But in that case our expectation would be wrong. Dig a little more deeply and there are layers upon layers of remembering.

First, Romilly recalls the recorder's 'savage conduct' a year after the execution.[48] Next, there is a call for a second edition of Sibly's shorthand notes on the trial in 1819. Next again, a manuscript note in Dr Samuel Parr's hand survives buried in his library catalogues after his death in 1825. It is mysterious in its provenance, purpose, and explosive fury: why did he care so much? Who is referred to is not stated: 'I hold with the utmost confidence that Eliza Fenning was innocent and that the infernal malignity of her murderer is recorded for his punishment in a future world. He died of a debauch when he ought to have died by

[47] [W. Hone], *The whole four trials of the thief takers and their confederates . . . convicted . . . September 1816 of a horrible conspiracy to obtain blood money, and of felony and high treason* (1816), 71.

[48] Romilly, iii. 235–6.

the halter.—S. P.'[49] Move on to 1830, and Fenning's—presumably Methodist—friend Wansbrough, who had put out a pamphlet about her in 1815, urges Wakley's committee on coroners to remember the Fenning tragedy and ensure that coroners have medical expertise.[50] Leap on again, and a witness before the 1856 Lords committee on capital punishment is asked if he remembers the unseemly displays of Eliza Fenning's body after her hanging.[51] A year later Charles Phillips writes against capital punishment and asks: 'Who does not remember Liza Fenning?' He rhapsodizes on 'the fate of one so young, so fair, so innocent, cut down in early morn, with all life's brightness only at its dawn': 'little did it profit thee that a city mourned over thy early grave.'[52] Then on to the murder trial of Madeleine Smith in Glasgow in 1857. Smith's defence counsel refers to the Fenning case 'which, 42 years ago, was the subject of a division of opinion in every household in the land': he has heard that another had confessed to his guilt on his death-bed.[53] Next a small torrent of letters hits the newspapers, one from a Reverend J. H. Gurney. His uncle, the shorthand-writer, had told him how he heard from a Baptist minister called Upton that, in her last days, Fenning had confessed her guilt to him.[54] Move on to 1861 and *Blackwood's Magazine* summarizes the evidence for her guilt.[55] Another leap and there, amazingly, is Charles Dickens, telling one of his journalists in 1867 that:

> I believed the arsenic in Eliza Fenning's case to have been administered by the apprentice. I never was more convinced of anything in my life than of the girl's innocence, and I want words in which to express my indignation at the muddle-headed story of that parsonic blunderer [Upton] whose audacity and conceit distorted some words that fell from her in the last days of her baiting.

In Dickens's *All the Year Round* the journalist duly produces his story on Dickens's instructions: 'We are filled with a deep conviction that she was entirely innocent,' the author proclaims.[56] Next, in 1871 Charles Hindley recalls Fenning as a 'beauteous innocent creature', her gallows dress 'as spotless as her own purity', and her departing soul 'caught up by attending angels and carried to the heaven of eternal bliss'. He

[49] 'S. Parr', in W. T. Lowndes, *The bibliographers' manual of English literature* (1865), vi. 1787.

[50] T. W. Wansbrough, *Coroner: The following letter has just been addressed to the chairman of Mr Wakley's committee* (on coroners, respecting medical evidence at the trial of Eliza Fenning) (1830). Cf. W[ansbrough], *Authentic narrative of the conduct of Eliza Fenning.*

[51] SC 1856, vii, Q282.

[52] *Vacation thoughts on capital punishment* (4th edn., 1858), 88.

[53] F. T. Jesse, *Trial of Madeleine Smith* (Edinburgh, 1921), 272.

[54] *Ann. Reg.* 1857: 143.

[55] 'Judicial puzzles: Eliza Fenning', *Blackwood's Magazine*, 89 (1861), 236–44.

[56] M. Dickens and G. Hogarth (eds.), *The letters of Charles Dickens* (3 vols., 1882), iii. 240; 'Old stories re-told: Eliza Fenning (the danger of condemnation to death on circumstantial evidence alone)', in *All the Year Round*, 18 (June–Dec. 1867), 66–72.

knows it must have been so, for he tells us that the master's son confessed on his death-bed that he had ministered the poison in an act of revenge, when Fenning failed to submit to his embraces.[57] Does this unravel the mystery—including that reference of Samuel Parr's? Well, not according to the *Dictionary of National Biography*. Here (finally) Fenning gets one and a half full columns and yet another variant on her tale: 'the case of Eliza Fenning is remarkable as showing how powerful is a steady and consistent declaration of innocence on the part of a criminal to produce a general belief in it.' At the end of all this we still do not know who put the arsenic in the dumplings. However, all we need to register is that the story leaves residues, to put it mildly. It is not forgotten.

And so to a last suggestion, plausible in view of this long remembering. It is that Eliza Fenning's case helped a little to shift the terms of understanding about hanging law. Many other influences contributed to this outcome in the post-war years. With the country in turmoil, political trials numerous, the regent despised, radical satire constantly dripping, and traitors decapitated outside Newgate, radical iconography (Cruikshank's and Hone's not least) was awash with virulent images of hanging judges and clerical magistrates. At the populist level Paleyite rationalizations were not allowed to pass. Ellenborough was spat upon by the crowd when he left the court after Hone's triumphant acquittal in 1817. The fuss about Fenning was part of a louder anger in these years.

But it does look as if Hone and Watkins in 1815 had hit on a device of great influence in this accounting. To analyse a trial and conviction as soberly as they had analysed Fenning's was to strike the oligarchy's law on a vulnerable flank. It was not the high-flown principle of capital punishment that was addressed, as men like Romilly had to address it in parliament. What was addressed was the *casualness of justice* in terms a laity could understand. A precedent was set which others would follow. In 1807 James Harmer had tried just such an analysis when he published a book revealing the dubious evidence which had hanged Holloway and Haggerty, but he had watched his language carefully and the book made little impact it seems. When Harmer resorted to the device again, in 1819 and again in 1824 (see Chapter 16), his texts were as ambitious as Hone's. Beyond that, petitions for mercy on behalf of the harshly sentenced in the 1820s evinced a newly confident engagement not only

57 Hindley, i. 6–7 ('The heroes of the guillotine and gallows').

with the acceptability of capital punishment but also with the quality of trials and judgements. The Fenning case was not the least of the episodes to shift the terms of argument with harsh law after 1815.

Finally, we recall that in episodes like Lloyd's and Fenning's the critique of the law was mobilized less through reasoned argument about legal systems than through identification with the plights of women wrongly condemned, around whom luxuriant sentiment and sentimentality might flow. Despite our caution about the influences of humanitarian narratives, there is no denying that they could mediate the most energizing of projections. That both gentlemanly and popular radicals were capable of these projections—indeed, that the projections sustained their radicalism—this chapter has aimed to convey.

CHAPTER 14

PIETY AND BENEVOLENCE

1. *Belief, pain, and execution*

IT IS TEMPTING TO YIELD TO THE NOTION THAT EVANGELICAL philanthropists were those most qualified to intrude a feeling of benevolence into the long debates on punishment. In narratives of 'humanitarian' change they have always got a better press than radicals. 'Almost uniformly . . . on the side of right,' J. S. Mill said of them: the 'vast gain, not only to humanity, but to the ends of penal justice, we owe to the philanthropists'.[1] The titles of their biographies put them on the angels' side too: *Buxton the liberator*, *Fry the Quaker heroine* or *Lady of the prison*, *Clarkson the friend of slaves*, *Wilberforce the slaves' champion*, and so on. Such people were also vociferous in their own claims to represent opinion's progress and reform's advance. With their mounting attacks on plebeian and aristocratic vice, the slave trade, the treatment of the insane, and the inadequacies of popular education, the evangelical monopoly on virtue expanded enormously after 1790. So did their purchase on the nation's upper ten thousand. There were great differences between moderates and extremists; but by the 1820s evangelical doctrines are said to provide 'the most vivid insight into the "official mind" of the period', not to mention into the reflexive piety of many thousands of middle and middling people.[2] We know that they did much to mediate the long movement away from the stinking gaols of Howard's era towards the intended cleanliness and efficiencies of the penitentiary.[3] Must it not have been their equal achievement to untangle the defences against feeling for the executed felon which have impeded the pace of the humanitarian narrative so far?

[1] PD3 cxci (1868), c. 1048.

[2] B. Hilton, *The age of atonement: The influence of evangelicalism on social and economic thought, 1785–1865* (Oxford, 1985), 6, 8.

[3] M. Ignatieff, *A just measure of pain: The penitentiary in the industrial revolution, 1750–1850* (1978), 55, 66–7; R. Evans, *The fabrication of virtue: English prison architecture, 1750–1840* (Cambridge, 1982).

There are grounds for positive answers to this. Belief, especially Quaker belief, endowed people like Elizabeth Fry (the best and also a significantly exceptional exemplar) with an impressive determination and resilience, enabling them to confront cruelties without flinching, even if interests of class or control did shape outcomes. Nor were any propagandists more effective in their own causes than these who had honed their techniques in the anti-slavery agitations. Moreover, feeling was acknowledged and valued in evangelical faith, as was active engagement. By the end of this chapter—and in the next—their contribution must be given its due.

But there are also grounds for caution to be addressed first. One of them relates to the sources of this feeling engagement with others' distress. Was it 'faith' that achieved it? Or was it that the pious generated their sympathetic capacities by incorporating sensibilities which were as secular in origin as they were religious? Can the two be dissociated? Secondly, piety had many faces and agenda. The reference is not only to the divisions within evangelicalism, or to the authoritarian impulses historians have discerned in its penal ideas—the grim and less than humane disciplines of the penitentiary especially.[4] The reference is also to the fact that these pious people were not immune from the wishes, fantasies, and projections common to their time. We shall not find much opposition to hanging among them any more than among other people. Indeed, the fact that religious belief could *block* empathy for others' suffering would strengthen the claim that if they did feel that way, it was as likely to be due to the sympathetic values they had learnt with others of their class as it was to be due to 'faith'.

Many moderate evangelicals were as indifferent to penal issues as they were to other social questions. For decades yet, Hilton points out, they matched 'their *laissez-faire* or neutral conception of providence with a similar approach to the "Condition of England" ';[5] and most denominations defended conventional rules no less than the established church did. Few were disgusted by the horrors perpetrated in the name of law. None was as interested as radicals were in diagnosing those stark relationships of power which Hunt or Hone discerned in the Fenning case. Their political realism undermined that; but also they did not see the world that way. Those, Quakers mainly, who did think capital pun-

[4] The aim 'was not to punish less but to punish better; to punish with an attenuated severity perhaps, but in order to punish with more universality and necessity; to insert the power to punish more deeply into the social body': M. Foucault, *Discipline and punish: The birth of the prison*, trans. A. Sheridan (1979), 82.

[5] Hilton, *Atonement*, 16.

ishment an evil could still recommend a tactful quietism in the face of wrongful condemnation. In 1818 Elizabeth Fry tried to save the life of one Harriet Skelton, condemned for passing forged notes. Skelton was hanged, but what most upset Fry was that her efforts had offended the home secretary, Sidmouth. Humiliated by the great man's coolness, she had, she thought, 'too incautiously spoke[n] of some in power'. A decade later she advised her Ladies' Society for Visiting Prisons that 'much disadvantage' would 'generally accrue from endeavours . . . to procure the mitigation of the sentences of criminals':

> Deeply as we must deplore the baneful effects of the punishment of death, and painful as we must feel it to be, that our fellow-creatures, in whose welfare we are interested, should be prematurely plunged into an awful eternity; yet, while our laws continue as they are, unless they can bring forward decided facts in favor of the condemned, it is wiser for visiting ladies to be quiet, and to submit to decrees, which they cannot alter.[6]

Acceptance of capital punishment was also doctrinally rooted. On from the debates among early church fathers about its righteousness, Exodus always offered a conclusive text: 'life for life, eye for eye, tooth for tooth, hand for hand, foot for foot, burning for burning, wound for wound, stripe for stripe'. The thirty-seventh article of the Anglican Church still declares that 'The Laws of the Realm may punish Christian men with death, for heinous and grievous offences.' Admittedly there were tensions within eighteenth-century theology between the new psychology which saw human sympathy as a natural principle and traditional Christian teaching on hell and eternal punishment. Affective objections to eternal punishment became as potent as rationalist objections, for man could not be more compassionate than God.[7] None the less, from Burnet on to Butler, Anglican divines continued to insist that vengeance was a divine attribute of which civil magistrates must also partake if human institutions were to reflect God's ruling. Since within the divine order pain was an ordained consequence of sin, pain must be inflicted within a righteous human order too. Evangelicals did not deny these views. Believers must take their 'hints in the economy of public and private life, from the economy of Providence in the administration of the world', Hannah More wrote in 1813: 'We govern our country by laws emulative of those by which he governs his creatures.' Although

[6] F. Cresswell, *A memoir of Elizabeth Fry* (1856); E. Fry, *Observations on the visiting, superintending, and government of female prisoners* (1827), 24.

[7] N. Fiering, 'Irresistible compassion: An aspect of the eighteenth-century sympathy and humanitarianism', *Journal of the History of Ideas*, 37 (1976), 215.

the present penal laws were not ideal, they were part of that dispensation.[8]

The notion that the world was a school of suffering and that pain was the path to spiritual rebirth often became yet another way of *not* allowing the feeling self to experience the terror of those the law condemned. Those who believed in atonement by faith had to believe also that penal suffering might open the way to redemption and that what mattered was the soul, not the body. As like as not when evangelicals deplored capital punishment, it was not in protest at the bodily terror it inflicted but because hanging was an inefficient way of bringing felons to salvation. Thanks to the prerogative of mercy, execution was uncertain: felons' attention on the afterlife was distracted by the gamble. Overhasty execution did not help either. Elizabeth Fry worried that execution often 'cuts off the possibility of that amendment of life' which could be better produced by disciplinary and reformative 'punishments of a milder and less injurious nature'. Execution drove fewer 'into an effectual repentance and a saving faith' than a more extended suffering 'applied on the principles of a *wholesome discipline*'.[9] However (some felt), if those about to hang *could* be forced to repent in fear, they might as well hang, their reward imminent.

Should these views be construed as another 'defence' against 'anxiety' about scaffold killing? Hardly so, most historians would answer. What now looks like crass insensitivity when evangelicals faced the condemned expressed a providential view of the world which was older than that recommended by 'sympathy'. Evangelicals wanted the community of believers to be expanded and celebrated, and the pain of sin to be alleviated by the sinner's return to the fold, even at the point of death. But it did not follow that they were indifferent to the felon's fate. In 1819 the Quaker J. J. Gurney described his visiting a man about to be hanged as follows: 'I spent some little time with poor Belsham yesterday afternoon, and I may acknowledge, that *I was much comforted by my visit.*' The self-regard of the last phrase shocks now. But Gurney thought Belsham was about to enter the community of the saved, and that 'God was about to have that mercy upon him which men had refused'. There was something to celebrate. 'May the holy arm of Omnipotence be near

[8] Cited by Hilton, *Atonement*, 215. For fine general surveys, see R. McGowen, 'The changing face of God's justice: The debates over divine and human punishment in eighteenth-century England', *Criminal Justice History*, 9 (1986), 312–34, and his, '"He beareth not the sword in vain": Religion and the criminal law in eighteenth-century England', *Eighteenth-Century Studies*, 21 (1987), 192–211.

[9] Fry, *Observations*, 70, 72–4.

to support him in the moment of deepest trial! May God have mercy upon him through Jesus Christ!'[10]

On the other hand piety might be construed as 'defensive' if it is agreed that belief in bodily resurrection has been the main denial system through which most societies have kept the fear of death at bay. Even a Victorian—John Stuart Mill—recognized that this was one of the things that made religion 'useful'.[11] Conceived in this sense, the defensive function of faith might have been heightened in the later eighteenth century as anxieties about the legitimacy of execution increased: for increase they did of course. For centuries clerics had laboured to extract confessions and penitence from scaffold victims. If we take their words at face value, they did this to testify to the Lord's mercy, rather than just to legitimize the punishment and the authority which inflicted it (as some historians are too prone to think).[12] However venal they might be, most clerics probably believed what they said. But public words, pious or otherwise, always have hidden audiences and meanings. The speaker speaks to his own comfort too, affirming his own identity, his own place in hierarchy, his own relationship to authority and power. Should those structures shift a little, and should he be sensitive to the shift, the speaker's address to himself inevitably acquires new resonances and functions.

By the mid-eighteenth century the business of bringing felons to scaffold repentance was having to circumvent new difficulties. The legitimacy of the power to which the godly stood in intimate relationship was beginning to be questioned. Even discounting other subversively secular currents at work in that age (or the power of counter-attacks upon them), clerics had never before had to labour in a world in which doubts about criminal justice were levelled as they were now. 'Who knows whether this man is not less culpable than me?' Dr Johnson could ask anxiously in 1751: 'who can congratulate himself upon a life passed without some act more mischievous to the peace or prosperity of others than the theft of a piece of money?'[13] It was coming to be known that legal process was impatient, laws of evidence rough-hewn, trials rushed, judges indifferent. Too many victims for comfort died

[10] R. M. Bacon, *A memoir of the life of Edward, third Baron Suffield* (Norwich, 1838), 105.

[11] J. S. Mill, 'The utility of religion', in *Three essays on religion* (1885), 69–70.

[12] J. Sharpe, '"Last dying speeches": Religion, ideology and public execution in seventeenth century England', *Past and Present*, 107 (1985), 156, 163.

[13] 'The necessity of proportioning punishments to crimes' (1751): in W. J. Bate and A. B. Strauss (eds.), *Samuel Johnson:* The Rambler (3 vols., New Haven, Conn., 1969), ii. 241–7.

protesting their innocence or in unabashed defiance; and too many commentators were saying that this had meaning:

No country groans under an heavier load of Blood: Never did so many Martyrs die for the Laws and Liberties of a People, than have for the English Liberties and Laws: And in no nation have more people gone to the block and gallows, denying their being guilty of the crimes for which they were to suffer, and of late, the number of these innocent Victims have increased: a plain proof that either People are extremely obdurate, and believe no future state, or that the laws are defective, and the judges sometimes severe.[14]

In a world in which such risky things could be said, the ministrations of the devout to the condemned might well have to be more emphatic. Protesting a little too much, their ministrations shored up the authority which was challenged but also comforted ministers themselves as they accommodated themselves to their roles. An ever more compulsive belief in salvation could comfortingly diminish the obligation to confront the dangers inherent in the ramshackle processes of law and the pain it so often implausibly inflicted. For devout witnesses at the scaffold nothing helped to curb the anxious vapourings of empathy more than belief in the afterlife. The more compulsively belief in the felon's salvation through repentance was affirmed by some, the more intense, perhaps, the negated anxieties which underpinned the belief.

2. *Washing felons in the blood of the Lamb*

Say not, my fellow mortals, ''tis too late, my sins are too many and too aggravated.' The blood of Jesus, is the blood of him who was God manifest in the flesh. . . . As a fountain, it is now open, and cleanseth all from sin. God help you to come, and wash, and be clean. I speak now to the vilest among you, because the converting God of grace comes freely to the very chief of sinners.

It was in these terms that the Revd Henry Foster cheered up eight Newgate prisoners three days before they were hanged in 1776. His listeners were not, he reminded them, men 'whose works have been such, as have rendered you fit to live among men'. Luckily for them, however, God did not give his grace according to a life's works. It was belief that would save them. Believe they better had, for ''ere long, it is possible some of us must sink lower than the grave—low, as the bottomless pit'. He published his sermon afterwards for the benefit of one of the

[14] Anon., *The case of Mary Edmondson: By a gentleman of the law* (1759), 27.

departed men's widows and her four helpless orphans, which was something.[15]

What did God-fearing people like this achieve in the way of alleviating felons' terror? Granted their labours to improve prison conditions in ensuing decades, answers can cut both ways. It may be (if one can calculate such things) that the aggregate of suffering later endured by expanded multitudes in the solitude or silence of the Victorian penitentiary was greater than that previously endured by the relatively few who faced the scaffold or transportation boats. Men and women went mad there. Within the carceral city were abundantly witnessed the 'insidious leniencies, unavowable petty cruelties, small acts of cunning, calculated methods, techniques, sciences that permit the fabrication of the disciplinary individual'.[16] A historiography which stresses the viciousness of effects can score devastating points against optimistic narratives of 'reformist' intentions.

On the other hand it was thanks to men like the Revd Foster, even if not to him individually, that life in many prisons became healthier, and not merely more disciplined, after the revelations of Howard and his successors. Prisoners' grateful testimonies to prison reformers like Fry were not without meaning; and some historians can plausibly insist that the penitentiary which evangelicalism so strongly recommended expressed a spiritual ideal 'which place[d] the individual worth of the prisoner close to its centre'.[17] 'Humane' feeling was less central to the process of moralization than is implied here, but this is not to say that moralization was a mean enterprise when prisoners were otherwise treated like animals.

Whether moralization brought comfort to the condemned is another question entirely. Some defence against mortal terror was certainly on offer. They were told that their souls would live if only they believed. But was the defence accepted? Possibly. Many poor people had their own appreciations of an afterlife. Decades of evangelization had effect, even if messages were half-digested. They had deposited in many felons' minds 'the dangerous and most fallacious notion', Elizabeth Fry found, 'that the violent death which awaits them will [itself] serve as a full atonement for all their sins'.[18] This was wrong-headed since atonement

[15] H. Foster, *Grace displayed, and Saul converted: The substance of a sermon preached in the chapel of Newgate prison* (1777), 9, 13.

[16] Foucault, *Discipline and punish*, 308.

[17] W. J. Forsythe, *The reform of prisoners, 1830–1900* (1987), 228.

[18] Fry, *Observations*, 73. Fry is a main and authoritative source of these 'fallacious' beliefs: she reported them to similar effect to Romilly (Romilly, iii. 332–3), to T. F. Buxton (PD1 xxxix (1819), 822), and to Samuel Favell (Favell, *Speech on the propriety of revising the criminal code* (1819), 36).

required active contrition. Still, such people were at least softened targets as well as vulnerable ones. When exposed to a great preacher like Wesley their raw emotions could be manipulated as the minister chose. Preaching in Newgate in 1784 to forty-seven condemned prisoners, Wesley convinced himself that 'the power of the Lord was eminently present, and most of the prisoners were in tears. A few days after twenty of them died at once, five of whom died in peace.' He took the signs of penitence at face value: as well he might, for how else to cope with the collective killing of *twenty*? The tears seemed to speak for themselves, as did the hysterical cries of 'Welcome news! welcome news!' when the bellman came at midnight to tell them they were to die tomorrow, and the ecstatic hymn-singing during the last moments at Tyburn.[19]

Also there were always felons whose repentance was spectacular enough for them to be commemorated in exemplary pamphlets, their uplifting phrasings tidied up for consumption, their sticky endings efficiently harnessed to propagandize a cause greater than their own lives. Their cases were useful to those who would preach on (say) *God's indignation against sin, manifested in the chastisements of his people*—this in *a sermon preached on the occasion of the unhappy death of James Oldfield, who was executed at Tyburn, near York, on Saturday, the 28th day of April, 1770*. Sermons of this ilk multiplied in this half-century, most of them testifying to the successes of evangelical ministration in prison cells. 'I believe now my peace is made with God through Jesus Christ,' the condemned robber Matthew Lee supposedly told his minister on the eve of his execution in 1752:

> I experienced a great change in my soul, for while I was at prayer, last night, in great terror, apprehensive that I was going immediately body and soul to hell, I cried out so loud that my fellow prisoners in the next ward heard me. . . . But while I was in the midst of this horror and despair, I suddenly experienced a blessed and comfortable change; my mind was immediately calmed; I believed my sins were forgiven: the fear of hell was taken away: and I was so far from fearing death, that I was now more desirous to die than to live. . . . Jesus Christ hath washed away my sins in his own blood; and I am justified by his grace; my sins are covered with his righteousness, and he hath blotted them out of the book of his remembrance.

No matter that no footpad would speak so finely without editorial assistance: this case was valuable enough to be published in Dublin half

[19] N. Curnock (ed.), *The journal of the Revd John Wesley* (8 vols., n.d.), vii. 41, 32: cf. ii. 100, 334–6, 339, iii. 43, 381–7, vii. 230, 279.

a century later for sale at all Methodist preaching-houses in Ireland at 2*d*. a go. And who knows but that Lee did die happily? His reported cry to a relative at Tyburn was: 'My dear, don't vex yourself for me; for I shall be in heaven in two or three minutes.'[20]

But many supposed converts effused with their ministers only to impress prison authorities and win a pardon. The hysteria of others had less to do with belief than with the incontinent terror to which Lee's account bears witness. How shallowly most were converted was revealed among those who won last-minute reprieves. Even Wesley suspected a whiff of disingenuousness then: 'so often have I known [conversions and repentance] vanish away as soon as ever the expectation of death was removed'. Gibbon Wakefield later thought that only about half of those hanged died expecting happiness in another world, and then only because they had nothing else to believe in.[21] And lest half be thought a good tally, recall that this quota was achieved chiefly thanks to the vulnerability of the condemned and the refined cruelties applied in the course of constant clerical nagging which left even the most defiant with little chance of resistance. Hopefully not many achieved the refined cruelty of the Coventry prison chaplain who in 1849 held a condemned woman's hand over a candle flame for two minutes to give her 'a foretaste of her eternal fate'.[22] But some came close. When 15-year-old Richard Faulkner was condemned in Norwich for murdering a younger boy in 1810, he proved 'so shockingly depraved and hardened that after condemnation he repeatedly clenched his fist and threatened to murder the clergyman who attended the jail'. He had to be chained hand and feet in his dungeon. Faced with such recalcitrance, the attending clergy hit on the wheeze of dressing up another boy of the same age to look like the murdered one. When they led him into Faulkner's darkened cell, the apparition had its desired effect:

> [Faulkner] seemed so completely terrified that he trembled in every limb; cold drops of sweat profusely fell from him, and he was almost continuously in such a dreadful state of agitation that he entreated the clergymen to continue with him, and from that instant became as contrite a penitent as he had before been callous and insensible. In this happy transition he remained till his execution on Monday morning, having fully confessed his crime, and implored, by fervent prayer, the forgiveness of his sins from a merciful God![23]

[20] Anon., *Some account of the life and death of Matthew Lee, executed at Tyburn, October 11th, 1752* (Dublin, 1802), 6–8, 19.

[21] E. G. Wakefield, *Facts relating to the punishment of death in the metropolis* (1831), in M. F. Lloyd Prichard (ed.), *The collected works of Edward Gibbon Wakefield* (1968), 152–4.

[22] H. Potter, *Hanging in judgment: Religion and the death penalty in England* (1993), 48.

[23] Rayner and Crook, v. 63.

By the 1820s lay scepticism about these ingenious clerical devices was widespread. Later sceptics from Gibbon Wakefield on to Dickens and Carlyle mocked God's agents. Even discounting the tyrannies of new 'reformative' regimes, fashionable redemptionism could set vicious farces in motion, and comic rivalries as well. Prison authorities were often divided between high churchmen and evangelicals, as the aldermen in charge of Newgate were. Then prisons became playgrounds upon which denominational rivalries were played out with condemned souls as spiritual footballs. Neither side in these tussles bore convincing testimony to the notion that faith served their terrified targets better than it appeased ministers' own anxieties or served their quest for esteem. Neither side, soul-obsessed, thought the saving of *life* was their concern.

Wesleyan evangelists like Silas Todd had invaded Newgate in the mid-eighteenth century; Howard's work encouraged others to join him. But it was after Hannah More flooded prisons with salvationist pamphlets in the 1790s, and Fry had begun her task of cleansing the minds, souls, and bodies of Newgate women in 1817, that the prison world was opened to a veritable tide of ranters and methodisticals both discreet and loud, preparing the way for that revolution in the religious conduct of the penitentiary which was achieved by the 1830s. They had to be the right class of person to earn admission, it is true. Preachers 'in a situation in life not well adapted to reveal the holy tenets of salvation to a dying man' (thus a description of the tattered Wesleyan journeyman whom the Cato Street conspirator Davidson asked for before his hanging and decapitation in 1820) were turned away. But the rest came in unchecked: earnest Methodists like those who attended Eliza Fenning in 1815, or like Messrs Baker or Curtis, missing few of the eight annual opportunities which the Old Bailey offered for washing the condemned in the blood of the Lamb. No Newgate evangelists, however, are to be found among those denouncing the penalty of death for smallish property crimes. None is found among the petitioners for mercy, in all the thousands of appeals in the Home Office archives.

Ministering women came into prisons too—first through Elizabeth Fry's Association for the Improvement of the Female Prisoners in Newgate and then in expanding numbers across the country through her British Ladies' Society for Visiting Prisons. Glad as she was that women might now follow 'a crucified Lord, who went about DOING GOOD', Fry was aware of the dangers: 'Much depends on the spirit in which the visiter [*sic*] enters upon her work. It must be the spirit, not of judgment, but of mercy. She must not say in her heart, I am more holy than thou; but must rather keep in perpetual remembrance, that "*all* have sinned, and come short of the Glory of God." '[24] Some women followed her model conscientiously. Even the worst did their bit in setting up schools and embroidery classes as they spread the Word; they had bravery, or thick skins. But misogynist prison governors and chaplains always found female busybodies harder to take than male, and they had a point. A proselytizing mission was all that most women could reach for if they sought an independent role in that era. In time so many women wanted to do good that 'from forty to fifty carriages collected at the gate' of a prison on ladies' visiting-days.[25] For governors this must have been trying. Among many the benevolent impulse was as breathtakingly self-regarding as in Hannah More's fiction:

> Suppose now, Tom Saunders had not been put in prison, you and I could not have shown our kindness in getting him out; nor would poor Saunders himself have had an opportunity of exercising his own patience and submission under want and imprisonment. So you see one reason why God permits misery, is, that good men may have an opportunity of lessening it.[26]

When Sarah Martin, a spinster dressmaker in Great Yarmouth, passed the local gaol one day in 1810, she reflected how the inmates 'were shut out from society, whose rights they had violated, and how destitute they were of that spiritual instruction, which alone could meet their unhappy circumstance'. Thenceforth prisoners became an 'interesting field of occupation'. But in all her ministrations it never occurred to her to meditate upon the justice of the sentences which put people there. What impressed her was her own piety, and how her piety served *her*:

> seeing salvation, not in its commencement only, but from first to last, to be entirely of grace, I was made free! . . . The high assurance, that Christ was mine, and with Him 'all things', has never been withdrawn, but in all I have been called to resist, or conquer, or endure, or suffer, it has been a light from God not to be obscured, an ocean of comfort from the rock of my strength.[27]

The 'sufferings' of women like this inspired other women more than they did their captive audiences. When Miss Owston and Mrs Payne plied the Leicester murderer James Cook with exhortatory letters and tracts and visits before his hanging and gibbeting in 1832 (assuring him

[24] Fry, *Observations*, 2, 20–1.

[25] P. Priestley, *Victorian prison lives: English prison biography, 1830–1914* (1985), 105.

[26] H. More, *Stories for the middle ranks of society, and tales for the common people* (2 vols., 1818), i. 53.

[27] S. Martin, *A brief sketch of the life of the late Miss Sarah Martin of Great Yarmouth, with extracts from . . . her own prison journals* (Yarmouth, 1844), 6–7, 11.

that he deserved to suffer and surely would suffer, but only in the interests of eternal life and the Lord's greater glory should he repent in time), Cook succumbed to their persuasions—as well he might in view of his corpse's imminent elevation on a thirty-three-foot pole. But the ladies did nothing to stop his being hanged, anatomized, phrenologized, and gibbeted with a shaven skull covered in pitch-tar, even though the time was ripe for the cessation of these aggravations. A few critical noises would not have gone amiss, for as it happened Cook was the last man to be gibbeted in England.[28]

Of whatever sex—there were male counterparts enough—such people were hard for some to take then and are harder to take now. Read Mrs Lachlan's 355-page account of Cook's conversion (a healthy middle-class market for this kind of thing, expensively published and bound), and you drown in the surfeit of 'art's' and 'thee's' and 'thou's' and the gushing blood of the Lamb: that web of exhausted words and phrases which piety wove reflexively to obfuscate reality. For some these words were a way to understanding, sometimes of confronting; and the integrity of past manners must be allowed, since this was how such people saw the world. But that may be the very point. All generations contrive to hide others' pain from themselves through ingeniously differing filters. Evangelical belief could operate as efficiently to this effect as any other device.

3. *Ordinaries of Newgate*

There is nothing like the quotidian religiosity of an age for putting hagiographized subjects into perspective; so this section looks at the ministrations to Newgate wretches of the ordinaries of that great prison. These were the clerics in all England who had the most intimate experience of the condemned, and their relationships with their targets say more about the common potentialities of belief in this context than those of greater figures might do. These ministers will not subvert the view that faith could as easily mediate a detachment from suffering as an engagement with it. Their beliefs seem to have blocked that engagement quite healthily, in the interests of their own equanimity, that is to say.

It is not to the ordinaries that one would normally turn for evidence on the uses of piety. Venal, drunken, and the butt of London wits,

[28] Mrs Lachlan, *Narrative of the conversion (by the instrumentality of two ladies) of James Cook, the murderer of Mr Paas* (1832).

eighteenth-century ordinaries are memorable mainly for the *Accounts* of famous malefactors by which they supplemented their incomes.[29] Boswell thought well of John Villette (serving 1774–99), but Villette was not one who would need to defend his psyche against scaffold horrors. The story was told that when in 1778 a youth's execution was delayed by another's confession to the crime, Villette threatened the under-sheriff with prosecution for ordering a stay of execution without authority and urged the hangman to get on with the business: it was the public example that mattered, he said, not the guilt or innocence of the victim.[30] The Revd Dr Forde, ordinary after Villette until 1814, was a bit better, advocating a reformative prison regime though without taking steps to achieve one in his own domain. Consulted on the question by Bentham, he opined that executions 'answer no end whatsoever, either for punishment or example'. It was 'like the acting of a tragedy,' he wrote: 'a momentary tear of pity may be shed; but the next ribaldry obliterates the whole of the foregoing catastrophe'. But this only betrayed incomprehension of the condemneds' defences against terror and utter collapse. Forde showed his distaste for his duties as well as fears for his health by parading the Newgate cells with a nosegay of flowers under his chin to keep contagions at bay. He was contemptuous of the 'methodistical preachers' who outdid him in his tardy ministrations: 'It should not be expected of [the prisoners], with their habits, that they should be always crammed with preaching and prayers; they do not like it.' He was forced to resign when the aldermen's prison committee decided that 'he knows nothing of the state of morals in the prison; he never sees any of the prisoners in private . . . and does not go to the infirmary, for it is not in his instructions'.[31]

By 1814 the Court of Aldermen accommodated a quota of evangelicals. They tried to face up to the corruptions of the Newgate regime, goaded by radicals like Daniel Isaac Eaton and evangelicals like James Neild, whose critical views of Newgate and other prisons drew on Beccaria, Bentham, and Blackstone.[32] In 1814 a committee of aldermen set out on its own tour of English prisons. They saw the point of cleanliness, classification, sexual segregation, and the control of alcohol,

[29] P. Linebaugh, 'The ordinary of Newgate and his account', in J. S. Cockburn (ed.), *Crime in England, 1550–1800* (1977), 249 and *passim* for a full account of the 18th-cent. ordinaries' duties.

[30] Boswell, *Life of Johnson*, ed. R. W. Chapman (Oxford, 1980), 1319; M. d'Archenholz, *A picture of England* (Dublin, 1791), 149.

[31] *SC on Newgate*, PP 1813–14, iv. 249; the main source on Forde is B. Montagu, *An inquiry into the aspersions upon the late ordinary of Newgate* (1815), pp. x, 73–5, and app., pp. 7–12.

[32] D. I. Eaton, *Extortions and abuses of Newgate, exhibited in a memorial and explanation, presented to the lord mayor, February 15, 1813* (1813); J. Neild, *State of the prisons in England, Scotland, and Wales* (1812).

visitors, and girlfriends; but they concluded that not much could be done to make Newgate a model prison without rebuilding it across a site of thirty acres. So they contented themselves by obliging the keeper to draw up new rules for the prison, recommended that the ordinary be more diligent, and appointed the evangelical Revd Henry Samuel Cotton as a token of good intentions.[33]

Cotton was a fire-and-brimstone clergyman, but he retained many old manners of clerical life. The Cruikshank brothers' drawing of his rosy plump figure as he attended the condemned in the press-yard gives credence to the imprisoned free-thinker Haley's description of 'his fat, frowning, port-be-crimsoned countenance' (Plate 6, p. 66 above).[34] His 'rubicund visage betokened the enjoyment of the good things in life', the barrister Ballantine recalled. William Page Wood thought Cotton 'but too nearly resembled [the ordinary] who is depicted in Jonathan Wild'.[35] According to his enemies he was not above taking a bribe or two to save well-heeled felons from the noose.[36] He was an unimaginative man too. His official diary for the years 1823–5, a main source for the following, was a text of wondrous banality.[37] In it he contemplated his charges through a fog. He thought the death sentence had little effect on the condemneds' behaviour, in this echoing Forde's incomprehension of their despair. An Irish lad six months fresh in England, illiterate and without family, is 'quite as stupid as ever' and 'gives no rational account of [his crime], but persists that it occurred by accident'. Among other prisoners 'there is anything but seriousness; the majority . . . seem almost to have forgotten that they are under sentence of death, and

[33] *Report from the committee of aldermen, appointed to visit several gaols in England* (1816); London Corporation RO: Misc.249.5. Evangelical attacks on the Newgate regime continued: H. G. Bennet, 'A letter to the common council and livery of the City of London, on the abuses existing in Newgate; showing the necessity of an immediate reform in the management of that prison', in *Pamphleteer*, 11/277 (1818).

[34] For Haley, see *The Newgate Monthly Magazine, or Calendar of Men, Things and Opinions*, 1 (Sept. 1824–Aug. 1825), 140.

[35] When the Old Bailey was in session, 2 dinners were provided for the judges as they worked in relays, one at 3.00 and the other at 5.00; Cotton 'was most punctual in his attendance at both . . . and never affronted the company by abstinence at either': W. Ballantine, *Some of the barrister's life experiences* (5th edn., 1882), 65; W. R. W. Stephens, *A memoir of the Rt. Hon. W. Page Wood, Baron Hatherley* (2 vols., 1883), 19.

[36] 'Panting for gain . . . Dr Cotton has talked a great deal about charity, [but] that his charity begins at home is pretty evident,' Haley wrote in his *Newgate Monthly Magazine*, reporting that Cotton was willingly hoaxed by the news that a burglar awaiting trial was about to inherit £300: 'Out of pure Christian charity, our old parson provided him with counsel who, by some legal quirk, got rid of the capital part of the charge,' and then persuaded the recorder of London to sentence him for only 14 days.

[37] Newgate Visitors' Book, 1823–5 (London Corporation RO: 209C). 4th Geo. IV, c. 64, directed that, from 1 Sept. 1823, a book should be kept in every prison for submission to visiting justices, in which the chaplain and officers were to insert the date of visits to prisoners, with remarks. Cf. SC 1819: 61.

even appear to think my exhortations to them as inapplicable'. Yet others might appear 'quiet and orderly', but 'I understand that such is not their conduct out of my sight'. From others 'no serious impressions have been evinced'; 'many of them are lads of about 17 or 18 years old and from such an age little seriousness can be expected'. Another of the condemned said that 'he was a murdered man and that he was innocent of the crime, and that he had not had a fair trial'. It never occurred to Cotton that some of these people might be innocent, or angry, or numbed, or incredulous in the face of the short straws they had drawn.

Still, Cotton thinly did his best. He was indignant when he found boys of 8 and 10 in the prison for stealing pots, astonished when he found 'two female children of very tender years, under sentence of death, in the women's condemned cell, and [another] one for trial—and these facts are recorded in the year 1824 in the Metropolis of England'. 'Surely such a proceeding could never have been contemplated by the legislature,' he noted with unwarranted optimism. Moreover, he had much to bear. In the chapel, he noted, 'it would take ten turnkeys to maintain any order'. Prisoners were 'continuously laughing, yawning, coughing . . . and teasing other prisoners'; sometimes they urinated within sight of the lectern. At the pre-execution sermon, when the condemned squeezed into their own pew around an empty coffin, their boisterous parodies of the sermon could turn the chapel, as Linebaugh says, into 'the setting of two types of theatre: an official theatre of the Church and the courts versus a counter-theatre of the damned'. Officers were making almost £300 a year from spectators' fees. Turnkeys were bribed to allow whores, wives, and children to stay with prisoners overnight. In 1818 a visiting MP reported 'the grossest scenes in daylight', and another found Newgate's depravities 'proverbial': 'every man is visited by a woman . . . for the purposes of general prostitution.'[38] Until tamed by Elizabeth Fry and her ladies, the women's side was a hell-hole too. The governor dared not go there. Fry had to shield 'half-naked' women from visiting bishops; and they terrified Fry's first helpers. 'A den of wild beasts', 'half-naked women . . . of the most boisterous violence', one helper found them. Released to travel on open carts to the transport ships, women set up 'a sort of saturnalia or riot . . . breaking windows, furniture, or whatever came within their reach'.[39]

Cotton also had the recalcitrance of educated prisoners to deal

[38] W. J. Sheehan, 'Finding solace in eighteenth-century Newgate', in Cockburn (ed.), *Crime in England*, 235–6; Linebaugh, 'The ordinary of Newgate', 253–4.

[39] *Memoirs of the life of Elizabeth Fry by her daughters* (2 vols., 1847), i. 261.

with—enemies like Gibbon Wakefield, or anybody with opinions of his own for that matter. Over several months in 1824–5 the free-thinkers Haley, Perry, Hassell, and Campion (imprisoned for three years for blasphemy) smuggled out their exuberant exposures of the Newgate regime for the radical Richard Carlile to print as the *Newgate Monthly Magazine*. They put Cotton squarely in their sight-lines:

> Thou reverend pillar of the triple Tree,
> I would say Drop, for it was prop'd by thee;
> Thou Penny-Chronicler of hasty fate,
> Death's Analyst, Reformer of the State:
> Cut-throat of Texts, and Chaplain of the Halter,
> In whose sage presence Vice itself did falter.
>
> How many Criminals by thee assisted,
> Old Sam, have been most orthodoxly twisted?
>
> . . . .
>
> How oft hast thou set harden'd Rogues a squeaking,
> By urging the great Sin of Sabbath-breaking;
> And sav'd Delinquents from Old Nick's embraces,
> By flashing Fire and Brimstone in their Faces?

Finally, Cotton had something worse to cope with than impertinence and rebellion. We are not to forget that the prison was also the site of shrieks, weepings, and awesome collapse. This is how Cotton recorded the reception of the recorder's report on 25 May 1824:

> the men were assembled in the condemned room, where I communicated the dreadful intelligence . . . and an awful sight it was, to see two selected from this multitude before them all. . . . Hill exclaimed he was murdered, as three or four had been lately reprieved for the same offence [uttering bank-notes]. Osterby [a burglar] said he had not expected to suffer and exclaimed much against the officers who appeared against him. The remainder all fell on their knees and returned thanks to God and appeared thankful and serious for the moment. I then proceeded to the six women and communicated to them that they were all reprieved: this scene was most affecting and impressive—two were in fits, the rest wept incessantly, and it was long and with great exertion by the female attendants that any order could be restored.

The historian is always in danger of projecting his or her own responses on to past witnesses. We may guess quite wrongly at the dynamics involved in coping with such scenarios. Perhaps Cotton was just stupid or insensitive. Perhaps his social status blinded him to the pains of these others. But if on the other hand he is to be credited with some sympathetic energy, if he was a man of manners, dining with judges and barristers after all—he 'possessed a sort of dry humour', Serjeant Ballantine recalled, 'and I fancy was popular in the City'—then it is a fair guess that Cotton needed some transcendental resources to retain his emotional balance in confrontations like these. Not wholly blind to the hysteria of the Newgate regime nor incapable of some measured forms of sympathy, he needed some comfort. If so, he kept his balance chiefly through compulsive attendance to his charges' penitence, his beliefs serving him better than they did those to whom he ministered.

As a form of defence, this ministration worked well. Observing every nuance in the condemneds' attitudes, he kept the horrible business of execution from consciousness. He might be indignant at the imprisonment of boys and girls, but no indignation is to be heard at the randomized hanging of men and women. Rather, his diary repeats, week by week, this kind of thing:

> found two of the men with very great appearance of penitence, one of them exclaimed that his heart was so full, that it was ready to burst and complained that tears did not come to his relief.

> found them very humble, one of them very much depressed, from just having had an interview with his wife.

> [another] has been blessed with a greater accession of light, but from his being very incommunicative I cannot easily judge the state of his views.

> [the condemned women appeared] in a more proper frame of mind and considerably impressed with their situation.

And when a report was delivered 'for the execution of four men on Wednesday next out of the 43 capital convicts':

> These four received my announcement of their fate with becoming seriousness: not so some of the others who were respited—one said he would rather have been hanged and then he should have been happy—another that he had nothing to be thankful for—some of them received it with indifference, and some few exhibited the appearance of thankful surprise.

Given these recurrent performances, it is not surprising that Cotton preferred to consort with gentleman felons like Henry Fauntleroy, the doomed banker whose wealth, social standing, and spectacular forgeries made him the most celebrated metropolitan criminal of the 1820s. The repentance of people who at least spoke the same language as Cotton came to carry the chief burden of his anxiety. He became quite obsessed with Fauntleroy, visiting him daily in the five weeks before his execution, all but ignoring the poorer felons in the meanwhile. When the King in Council ordered Fauntleroy for execution, he noted that the forger 'received the communication with great humility and resignation', but also that he veered from contrition to 'agitation' and 'depression'. 'This man's firmness surprizes me, I confess; I trust however it is well founded and that he . . . is really penitent.' On the day before his hanging, Fauntleroy's 'firmness and at the same time apparent humility are wonderful, such as I do not recollect to have witnessed in any criminal before'. On the fatal morning Cotton administered the sacrament and led Fauntleroy out to the scaffold. The ordinary had stood on the scaffold hundreds of times before; he had helped women like Eliza Fenning hang, and one of the traitors of Cato Street, and any number of footpads and housebreakers. Sheriffs were often squeamish about their similar duties, but not Cotton, his mind on higher things. Fauntleroy's 'firmness' now 'exceeded any calculation I could have formed', he noted: 'I trust and hope he was sincerely penitent; firm I am sure he was to the last. Surely this firmness has arisen from the blessed assurance of pardoned sin and acceptance with God.'

But the truth was that of this much-needed assurance Cotton was chronically uncertain. How badly he needed it he had revealed at Fauntleroy's last service. There was something a little mad about the blood-curdling sermon with which he tried to terrify Fauntleroy into repentance in these last hours. Listening from the prisoners' pews, Cotton's free-thinking enemy Haley wrote that Cotton 'preached personalities by the half hour, at an unhappy man who was sufficiently tortured by his own reflections: can anything by possibility be more unfeeling than the manner in which you deport yourself towards condemned prisoners?' Cotton preached from Corinthians: 'Let him that thinketh that he standeth take heed lest he fall,' which made Haley wonder if 'the Doctor was about to favour his audience with a professional description of the new drop'. Hitherto Haley had thought Cotton 'a little soft-headed or so'. He now decided he was 'a perfect ninny'. The aldermanic gaol committee agreed. A few days after the execution it announced that Cotton had 'unnecessarily harassed [Fauntleroy's] feelings, and that the object of such sermons was solely to console the prisoner, and that from the time of his conviction nothing but what is consolatory should be addressed to a criminal'. Henceforth the public would not be admitted to the condemned service. Cotton was vexed by the rebuke: 'One, and perhaps the only good, of an execution, i.e. the solemn admonition to the public, will thereby be lost.'[40] But we may

[40] *Newgate Monthly Magazine*, 1: 226–8; Newgate Visitors' Book, 22 Oct.–23 Dec. 1824. The prohibi-

discern self-protection in this response. What Cotton was also lamenting was that the consolation which his urgent hell-fire sermonizing had offered to himself was now to be withdrawn, a needed sublimation denied.

4. *Piety and sensibility*

Few evangelicals did much to protest at the scaffold or the laws which sent people there. Some would intensify the condemneds' anguish if that increased their chances of penitence. Gibbon Wakefield said it was the chaplain's very job to break his targets' spirits. It is a bleak picture altogether. But the time for relief comes at last! Some pious people did protest, or at least they got close to the sinner's physical terror in this world without obliterating it in obsessive references to the world to come. Rare birds though they were, they incorporated a here-and-now sympathy *within* the effort to save souls; they did not simply sublimate or deny the horror. To a modern and secular sensibility, it is their capacity to do this that makes them both bearable and accessible.

Elizabeth Fry, as we have seen, was not one to rock political boats. But in 1817 she attended a Newgate woman about to be hanged for infanticide and confided to her diary that she was 'deeply affected'. The vocabulary to which she resorted at this point is important for our argument. 'The whole affair has been truly afflicting,' she wrote: 'This event has brought me into much feeling, attended by some distressingly nervous sensations in the night, so that this has been a time of deep humiliation to me; thus witnessing the effect and consequences of sin. The poor creature murdered her baby; and how inexpressibly awful to have her life taken away!'[41] We are inclined to believe her distress. Although consciousness of sin was part of this response, an imaginative projection was at its centre, passing from one woman to another. But, as much to the point, her vocabulary was less pious than sensationalist, and in that sense secular: *feeling*, *nervous sensations*, *inexpressibly awful*. The same content is discerned in one of Fry's Newgate helpers who in a similar situation simply allowed herself to be lost for words (she also used the term *inexpressible*) when faced by a condemned woman whose execution was delayed until she gave birth to her baby. The baby born, the woman was to be hanged within the week:

tion on public attendance at the condemned sermon was soon relaxed: see Radzinowicz, iv. 347–9, for ongoing concerns about the unseemliness of the occasion in the 1840s.

[41] *Memoirs of Fry*, i. 256.

I found poor Woodman lying-in, in the common ward, where she had been suddenly taken ill; herself and the little girl were each doing very well. She was awaiting her execution, at the end of the month. *What can be said of such sights as these*. . . . I read to Woodman, who is not in the state of mind we could wish for, indeed so unnatural is her situation, that *one can hardly tell how or in what manner to meet her case*. She seems afraid to love her baby.[42]

Nor should the empathetic capacities of some male sermonizers be denied either. Their pleas that the condemned turn to the Lord sometimes betrayed their own distress. They did not challenge the law which inflicted the agony, but their texts trembled with more than affected feeling. Fine delivery might make simple words electrifying as they too struggled with the pity of it all:

And now, my poor and miserable fellow-sinners, what more shall I say to you? How shall I bid you farewell to all eternity! I feel as a fellow-creature for the natural agitations of your mind, and could join your tears.—You are surrounded by nothing but gloominess—laden with heavy irons! and have this moment fixed before your eyes—awful sight!—the pattern [the coffin in the condemned pew] of what will soon contain your youthful bodies.

All this is alarming beyond description! but be not too much cast down; it will soon be over! cast your eyes beyond the gloomy confines of the grave!—look to the Lord, he hears your sighs. . . . What have you to fear, if Christ be yours? Seek him, O seek him then, while he may be found, call upon him while he is near!—And now, blessed Saviour! we commend these thy servants to thy tender mercies! wash them in thy precious blood, and present them spotless to thy Father! guide them through the valley and shadow of death—send thy holy spirit to conduct them to Abraham's bosom—Lord Jesus receive their souls![43]

What was heard in a declaration like this would depend on the hearer. A free-thinker would dismiss it as contemptuously as Haley dismissed Cotton's sermons. Still, in the simplicity of this language we seem to meet the open responses to the pain of others which we have been looking for in bourgeois culture, striving to cut across barriers of class. The real questions are where this capacity was conceived, and how and why a few pious people achieved it.

'What springs act upon the Quakers, which do not equally act upon other people?' Thomas Clarkson asked, as he pondered the Quakers' peculiar 'good-will' and 'tender feeling' for sufferers. He answered that their benevolence expressed their remoteness from worldly tempta-

[42] *Memoirs of Fry*, 275, 279 (my italics).

[43] Revd Edward Barry, 'A sermon preached to the convicts under sentence of death in Newgate on Sunday morning, April 20th, 1788, previous to their execution on the following Wednesday', in *Twelve sermons preached on particular occasions* (1789), 105.

tions, their democratic and disciplined organization, and their optimistic faith.[44] He might have added to his list the impulse in 'old dissent' to bear witness to moral standing in a world which denied their rights, their virtue scoring more than a point or two off the corruptions of ruling power. But even that would not quite close the case. What might also be added is the relationship which the religious expressions of sympathy bore to their secular equivalents.

Although Christians were advised to love, Quaker benevolence was as much generated within those processes of secular learning which were addressed in Chapter 7, entailing a capacity for projective identification which derived from the sensibility that they shared with all their class, as it was from within Christian belief. What belief did then was important, but facilitatively so. The passionate intensity of evangelical language and the collectivization of its practices, the solidarity of the sects and societies, the mutual reinforcement of commitment through the group's scrutiny of its own behaviour—all sustained the strongest in their encounters with distressing scenes, and enabled them to release sympathies they had learnt elsewhere. Even if the goal was the soul's salvation rather than bodily relief, piety enabled believers to face rather than to shrink from the earthly distress confronting them. The causative sequence suggested here—its fount a sensibility learnt in secular culture, then supported by disciplined belief—is not always difficult to unravel.

A starting-point lies in the recognition that the responses to others' sufferings achieved by the more sympathetic figures in the philanthropic narrative entailed the projection on to those others of their own emotional vulnerability or fragility. Some admitted that their vulnerability had infantile roots, and that their piety was developed to defend them against their own passionate natures or the remembered terrors of childhood. In the young Elizabeth Fry, notably, emotional sensitivity was both a posture learnt and an affliction suffered. If the adult was gifted in her capacity to feel others' misery, it can only have been because, until her conversion to the Quaker ministry in 1798, in her eighteenth year, misery had been her inheritance; she knew it as a familiar. Her revulsion at hanging, her capacity to feel for a condemned woman, and later the needful force of her faith, drew on the remembered fears of childhood and the despair of the irreligious young woman. In later life she recalled chronic childhood fears of violence: of

[44] T. Clarkson, *A portraiture of Quakerism* (3 vols., 1806), ii. 161–71.

guns, the dark, the sea. Abraham's sacrifice of Isaac induced the 'awful fear' that her parents might sacrifice her likewise. 'Fear was so strong a principle in my mind, as greatly to mar the natural pleasure of childhood' and 'I believe my nervous system was injured in consequence of it.' Her dread that her mother might die was such that 'my childlike wish was, that two large walls might crush us all together, that we might die at once, and thus avoid the misery of each other's death'. 'My religious impressions, such as I then had, were accompanied by gloom.' With only 'sceptical or deistical principles', she confessed to her diary the 'desolation' of her spirit. It was 'a great comfort to me that life is short and soon passes away': 'I am a bubble, without reason, without beauty of mind or person; I am a fool. I daily fall lower in my own estimation. What an infinite advantage it would be to me to occupy my time and thoughts well. I am now seventeen, and if some kind and great circumstance does not happen to me, I shall have my talents devoured by moth and rust.' Beneath the vague pinings for extinction, there was fear of death too, and the effusing need to find meanings for her life:

> *I must die! I shall die!* wonderful death is beyond comprehension. To leave life, and all its interests, and be almost forgotten by those we love. What a comfort must a real faith in religion be, in the hour of death. To have a firm belief of entering into everlasting joy. I have a notion of such a thing, but I am sorry to say, I have no real faith in any sort of religion; it must be a comfort and support in affliction, and I know enough of life to see, how great stimulus is wanted, to support through the evils that are inflicted, and to keep in the path of virtue. If religion is a support, why not get it? . . . I think it almost impossible to keep strictly in principle without religion.[45]

These early diary entries are uniquely revealing texts because they are so unguarded. Not yet encased in the controlled pieties of her maturer years, they speak for the compensatory meanings of her soon-to-be-found faith. (*If religion is a support why not get it?* Shortly after this she was converted, much to her relief, by the strict Quaker William Savary.) Above all they speak for the sources of her sympathy for the miseries of people she later encountered in Newgate. Many children experience the fears of the young Fry, and many teenagers the languid yearnings of the diarist. But her adult remembering and cataloguing of these things was a conscious admission that those experiences shaped her adult life.

Where was 'sensibility' in this? *I must die! I shall die!* The young Fry was raised in provincial Norfolk, the daughter of a wealthy banking family lax in Quaker practice and belief, and of a father with an un-

[45] Cresswell, *Fry*, 6–9, 19.

Quakerly taste for music. She was not protected from the pleasures of the world, the season in London, the Norwich balls and concerts—at one of which she was entranced by the prospect of meeting the duke of Gloucester. And there were always those novels again: 'We fly to novels and scandal . . . for entertainment,' she admitted (and was not alone in this in Quaker circles).[46] At this age she knew how to gush as well-bred girls of feeling must gush. On one occasion in her diary, aged 17, she gushed wonderfully, to an effect censored in later years—submitting herself to the cultured feeling from which polite girls could hardly escape, revealing impulses behind the career to come: 'there is a sort of luxury in giving way to the feelings! I love to feel for the sorrows of others, to pour wine and oil into the wounds of the afflicted; there is a luxury in feeling the heart glow, whether it be with joy or sorrow. . . . I love to feel good—I do what I can to be kind to everybody.' No sentimental heroine could have put it better.[47]

In ensuing years Fry's faith disciplined her feelings, as she meant it to. She needed 'the influence of principle, that would lead me to overcome these natural feelings, so far as they tend to my misery'.[48] The superego could not banish feelings so deeply planted and conscientiously cultivated, but it could empower her to subordinate them to religious practice and good works in conditions that were otherwise unbearable. With this 'influence of principle' achieved she could meet anything. She even met the death of her 4-year-old daughter in 1814. Her first desperately unreal response to this tragedy was that the child's death 'was so much like partaking with her in joy and glory, that I could not mourn if I would; only rejoice, almost with joy unspeakable and full of glory'.[49] The disciplined denials of belief indeed! But it was the same capacity to contain feeling within belief that sustained and held her true to her witnessing in Newgate. In choosing work of this kind, Fry had models enough to draw on in the secular literature to which she had been youthfully exposed, advising how true sensibility might make its own discriminating moral choices: Elizabeth Hamilton's widely read novel, *Memoirs of modern philosophers* (1800), for example, presented just such a heroine as Fry became.[50] But in that dreadful place it was faith that

[46] 'In those days (I am almost ashamed to say) I read many novels,' one Quaker diarist apologized in the 1820s for his youthful tastes of a half-century before: J. W. Frost (ed.), *The records and recollections of James Jenkins* (New York, 1984), 69.

[47] BL Add. MS 47456, 'The diary of Elizabeth Fry' (transcribed by her daughter Katharine Fry), i. 4, 9, 12–13, 16; *Memoirs of Fry*, i. 8–19, 23–4.

[48] Cresswell, *Fry*, 46.

[49] *Memoirs of Fry*, i. 238–9.

[50] S. Cox, 'Sensibility as argument', in S. M. Conger (ed.), *Sensibility in transformation: Creative resistance to sentiment from the Augustans to the Romantics* (1990), 77–8.

upheld Fry. She learnt not to flinch: 'What I have felt about the hanging them [*sic*] is also distressing; having had to visit another poor woman previous to her death; which again tried me a good deal; but I was permitted to be much more upheld, and not so much distressed as the time before.'[51]

Of these dispositions there are echoes in other evangelicals' biographies. The luxuriantly self-referential quality in Fry's instinct to good works was echoed in William Allen's feeling that 'there is so great a pleasure in every attempt to do good, that the attendant feelings afford a species of sublime enjoyment'.[52] And the notion that evangelicals' empathetic impulses originated in the secular culture, and that faith only disciplined that emotional charge, is confirmed in the case of the young and rising barrister William Page Wood, later lord chancellor.

Wood was no penal reformer in the 1820s, but he was earnestly evangelical, his diary recording his sense of his sinfulness and the need for an atonement. It happened that his diary also recorded horror at 'the dreadful state of our sanguinary code of criminal law'. The *dreadfulness* of execution hung heavily upon his *heart*, he noted—reaching for the sensationalist language of sensibility, of course. He recorded a revealing moment when the horror found its home in that sympathetic organ. One day in 1828 he visited the death-bed of a young woman relative and left her deeply moved. Her last words echoed in his mind: 'it seems hard to be called away at five-and-twenty,' she had said. Immediately after this

> I was obliged to pass by the Old Bailey, and the mob which surrounded me warned me that an execution was about to take place. I . . . learned to my horror that six victims of our accursed system of penal law were about to suffer death, of whom the eldest had not reached the age of five-and-twenty, and not one of them had been guilty of blood. Yet here was a crowd uttering coarse jokes, and rushing with savage ferocity to witness the perpetration of a legal enormity which was to desolate six families, whilst I had felt overwhelmed with grief at the anticipation of the possible loss of a young relation, dying in undisturbed tranquillity of mind, and surrounded by kind friends who would find and bestow consolation in reflecting upon the promises held out to us by our blessed Redeemer! . . . I trust that I shall never forget the scene, nor the feelings which it inspired.[53]

Here untrammelled was a projective pity for the death of felons quite unrelated to a preoccupation with the felon's salvation (or crime). Here

[51] BL Add. MS 47456, fo. 293.

[52] *The life of William Allen, with selections from his correspondence* (3 vols., 1846), i. 104.

[53] Stephens, *Memoir of W. P. Wood*, i. 145, 151–2. As a liberal MP in the 1840s, Wood was an abolitionist for all crimes except murder.

was an unexpected witnessing, a grief projected across social boundaries, an equation imaginatively achieved between unknown, socially remote, and suffering men on the one hand, and a known and suffering woman of similar age on the other—together with an acknowledgement of the men's exacerbated isolation and anguish, all combined within what for Wood was a repellant visual tableau. The Redeemer had a place in this witnessing but it was neither an initiating nor a primary place. If the Redeemer achieved anything here, it was to hold Wood true to the feelings that flowed from his witnessing. But he experienced the pain in the first place through his own projective imagining, a *cultured* response not given, he noticed, to the watching crowd.

In this spirit many evangelical reformers and their associates acknowledged 'the pain of witnessing a public execution', as Buxton put it. George Sinclair, passing an execution, had to avert his eyes from 'a sight so painful to humanity'.[54] Basil Montagu, too, told how once in 1801 he had rushed to Huntingdon with a last-minute reprieve for two men about to hang for sheep-stealing, breaking the news to them minutes before their execution. One of them collapsed and clung to Montagu as to his saviour. 'The pity which I felt for these poor men, and the horror at the supposed or real disappointment of the crowd, have made an impression upon me which no time can efface,' he wrote in 1812. Eighteen years later he wrote that 'there has not been an execution during the last thirty years without my remembering the two men at Huntingdon.'[55] In sudden shocks like this, others' pain could become such men's own.

Thus we put in place a range of available repertoires in early nineteenth-century piety, from the mean to the generous, and note their place in the secular culture, not exclusively the religious. The question which presses now is how far the repertoires were disseminated in that 'public opinion' which is commonly alleged to have derived mounting energy from the evangelical example.

[54] R. McGowen, 'A powerful sympathy: Terror, the prison, and humanitarian reform in early nineteenth-century Britain', *Journal of British Studies*, 25 (July 1986), 319–20.

[55] B. Montagu, *Account of the origin and object of the Society for the Diffusion of Knowledge upon the Punishment of Death, and the Improvement of Prison Discipline* (1812), 5; also his *Thoughts on the punishment of death* (1830), p. vii.

CHAPTER 15

FABRICATING OPINION

1. *Diffusing knowledge on the punishment of death*

IN FEBRUARY 1825 THE *EXAMINER* GAVE ITS USUAL COUPLE OF inches to the latest sentences at the Old Bailey. The paper found it less extraordinary that thirty men and women had been condemned to death than that the recorder of London had exhorted them 'to prepare for that awful fate to which a jury of their countrymen, influenced by no other motives but those of a just regard for the laws of God and civilized society, had doomed them'. Whereat the paper exploded: 'The laws of God!—where, Mr Recorder, do you find the punishment of death is ordered in the Bible for theft and such offences? We had thought that the sacred law runs thus: "Whosoever sheddeth man's blood, by man shall his blood be shed." '[1] Although the paper then passed on smoothly to other matters, what impresses us here is that it was now able to affect a *reflexive* contempt for the capital law and its agents which was unthinkable a couple of decades before. What had also changed is indicated in its appropriation of a biblical reference to endorse the contempt. A couple of decades earlier those who deplored the capital code would have claimed the high moral ground by appealing to efficiency and reason. Now most would claim it through the language of piety. A dozen and more years of evangelical agitation had shifted the terms of debate substantially.

Evangelical mobilization on the death penalty needs its due, then, even if we do cut it down to size later. Very few people were actively involved. Great swathes of opinion still went with Paley or with Peel on this question. Not all opinion was clear-headed or consistent. The middle classes still accommodated majority attitudes which evangelical appeals barely touched, and 'evangelical' may itself be a misleading attribution

[1] 27 Feb. 1825.

when the vast majority of evangelicals lent their names to causes far more congenial to the governing classes—and hence, as one of their historians put it, to causes more 'useful' to national moralization—than anything like the abolition of capital statutes. Given ruling-class obduracy, capital law reform was by and large a cause which evangelicals 'could not afford to take part in'.[2]

Still, 'public opinion' is defined by the noisiest voices, not the most numerous. It is a construct in collective fantasy, a working fiction in terms of which clamorous groups assert their interests and standing by suggesting that their high ground is more thickly populated than it is. Where it does not exist or exists thinly, a supportive opinion must be invented or exaggerated. Then (we have said), regardless of its real purchase, the very naming of an opinion may call it into being. In time its reality comes to be taken for granted if social and material contexts lend the naming credibility.

All this held true here. By the 1820s 'opinion' was taken seriously. In liberal circles much optimism began to attach to it. 'The form of government becomes liberal in the exact proportions as the power of public opinion increases,' it was claimed in 1828.[3] 'That with which we have to contend is public opinion,' Gibbon Wakefield agreed (ironically, in the voice of his fictitious hangman in 1833): 'Unfortunately, the judges have little power over those things which turn public opinion against hanging: the beastly education, as dear Mr Cobbett says, that is going on all over the country, the hankering after new-fangled laws, the growing inquisitiveness of the people, the obstinancy of the Quakers and the increase of newspapers and books.'[4] In this list the obstinacy of the Quakers and their friends needs attention. That they spoke from the narrowest of bases may itself measure their propagandist achievement.

Until the 1820s most evangelicals were niggardly supporters of capital statute reform, mostly endorsing it only so far as government led the way. Even Quakers came late to the question, and had not led the vanguard, as we have seen. Howard and Hanway had put prisons on the philanthropic agenda in the 1780s, but they ignored capital punishment. George Fox had spoken against execution a century before, and there was always a strong Quaker disposition against it; but most were as blind to it as the rest of their class. It only slowly acquired programmatic

[2] F. K. Brown, *Fathers of the Victorians: The age of Wilberforce* (Cambridge, 1961), 347–51.

[3] W. McKinnon, *On the rise, progress, and present state of public opinion in Great Britain and other parts of the world* (1828), 6–7.

[4] E. G. Wakefield, *The hangman and the judge, or A letter from Jack Ketch to Mr Justice Alderson* (1833).

significance as it became linked to questions of prison reform. It was rationalist circles that led the debate on the death penalty from the 1770s. Only with Romilly's reopening of the parliamentary question in 1808 did Quakers and their supporters climb on to the bandwagon. But climb on they did, and to important effect.

Romilly brought in his Bill to remove the death penalty from pocket-picking in May 1808. In July, with Romilly's approval, the Quakers William Allen, William Forster, and Luke Howard, along with the London sheriff Richard Phillips and the Benthamite barrister Basil Montagu, met 'to converse on the subject of our little society'. This was the Society for the Diffusion of Knowledge upon the Punishment of Death and the Improvement of Prison Discipline. Across the next twenty years the Society waxed, waned, and waxed again, modifying its name and focus as current concerns dictated, re-forming itself in 1828 when parliamentary debate on capital punishment for forgery made that expedient.[5]

Thomas Haskell identifies four preconditions for the mobilization of humanitarian action in these decades, and the emergence of the Society assuredly expressed their fulfilment. The actors must acknowledge ethical principles which endorse the alleviation of suffering; they must see themselves as responsible for the evils to be acted against (if only indirectly, should they abstain from acting); they must believe that they possess a sufficient technique for intervening; and the technique must be familiar from use in other contexts. These conditions were characteristically fulfilled among individuals, Haskell claims, whose expectations and personalities were honed in their experience of the market: businessmen mainly, scrupulous in the fulfilment of ethical maxims, accustomed to intervene in events, to achieve effects, and to anticipate future advantage, and possessed of a missionary zeal to disseminate their own habits and values among dependants and rising generations alike.[6] In a culture in which many other philosophical currents were already pointing towards human perfectibility, and in which the business and dissenting communities were relatively marginalized by aristocratic and gentry élites, these people were unavoidably confident about their own virtue and the prospects for national remoralization. Haskell talks mainly

[5] B. Montagu, *Account of the origin and object of the Society for the Diffusion of Knowledge upon the Punishment of Death, and the Improvement of Prison Discipline* (1812), 6, 11; W. Allen, *The life of William Allen* (3 vols., 1846), i. 104; Radzinowicz, i. 348–50.

[6] T. L. Haskell, 'Capitalism and the origins of the humanitarian sensibility', *American Historical Review*, 90/2 and 3 (Apr. and June 1985), 339–61 and 547–66.

about the rise of anti-slavery in the later eighteenth century. But his preconditions attach to engagement in penal reform as well.

To be sure, people were never as concerned with punishment as they were with anti-slavery in these decades. For one article on the penal code in Allen's *Philanthropist* (1811–18), there was a superfluity on slavery. The movement for penal reform drew on the anti-slavery model, and the suffering slave had an iconic standing which was transferred easily to the suffering felon.[7] When Montagu, Allen, and friends met in 1808, they exhorted each other to 'remember what by perseverance has been done on the Slave Trade', and most of those supporting their Punishment Society were already involved in the earlier agitation; membership overlapped. By 1817 the Punishment Society included Wilberforce and Clarkson, along with that tightly intermarried Quaker clan of Frys, Buxtons, Hoares, Gurneys, and Fosters which dominated the extra-parliamentary penal reform movement between 1808 and 1830 and was active in anti-slavery as well. And as Allen's journal became the *de facto* organ of anti-slavery and of the Punishment Society alike, it invited its readers to feel through their sympathies for both kinds of wretches.

The other model which the Punishment Society had in mind was that of the Philadelphia Quakers. They had already achieved a wonder of the modern world by discarding the punishment of death and 'by annexing as mild penalties as possible to the transgression' of laws. As Turnbull had shown in his *Visit to the Philadelphia prison* (1796), this policy combined 'the refined principles of reason and morality' with 'the true interests and wishes' of the community. Hence it was not surprising that Montagu should cite the Philadelphia prescription in his account of the Punishment Society's origins:

> Nothing can be more true, than [that] Christianity commands us to be tender-hearted one to another, to have a tender forbearance one with another; and to regard one another as brethren. . . . But where are our forbearance and our love; where is our regard for the temporal and eternal interests of man; where is our respect for the principles of the Gospel,—if we make the reformation of a criminal a less object than his punishment; or if we consign him to death in the midst of his sins, without having tried all the means in our power for his recovery?[8]

[7] See *Third Report of the Society for Diffusing Information respecting the Punishment of Death and the Improvement of Prison Discipline* (1816), facing p. 6, for depiction of prison conditions which directly harnessed familiar contemporary images of conditions on slave ships.

[8] R. J. Turnbull, *A visit to the Philadelphia prison . . . with observations on the impolicy and injustice of capital punishment* (Philadelphia, 1796), 1. Home Secretary Pelham took Turnbull's text seriously enough to have it précised for Home Office perusal in 1803: Pelham papers, BL Add. MS 33122, fos. 159 ff.

If Christian love thus inspired penal reform, the techniques of reform were organization, funding, and publicity. Here again the reformers were well equipped. Quaker hierarchy (the local monthly and quarterly meeting reporting to the national annual meeting) provided an efficient network of propagation and funding, much of it from banking. The list of 287 subscribers to the Punishment Society for the years 1810 and 1811 included five Barclays, five Gurneys, five of the Coalbrookdale Darbys, two of the Tukes of York, and Joseph Fry; a handful of evangelicals of other denominations contributed too. Between 1808 and 1814 the Society spent £1,149 on its publications, information-gathering, and advertisements,[9] and by 1818 it listed supporters as scattered as Yorkshire and Tunbridge Wells.

To all this should be added the caucus's axiomatic belief in progress and its skill at public relations. 'What may we not hope to effect by industry and zeal in a good cause?' Edward Harbord, later Lord Suffield, asked brightly in 1818: 'The fact is, gentlemen, we are now arrived at a period when from the extent of civilization and a general diffusion of knowledge it requires little more than to draw the public attention to a just and righteous cause to ensure its success in this land of freedom and benevolence.'[10] As with anti-slavery, the wooing of opinion was calculated, the force of example called into play. Elizabeth Fry herself knew how to play to the gallery. 'The prison and myself are become quite a show,' she noted of her efforts in Newgate in 1818; 'It does much good to the cause, in spreading amongst all ranks of society a considerable interest in the subject, also a knowledge of Friends and their principles.'[11] Information was gathered too. In 1817 the Punishment Society got its provincial supporters to enquire into 'interesting cases' of capital punishment in their vicinities, taking its cue from Romilly's belief that proof of prosecutors' and juries' alleged reluctance to prosecute or to convict on the full capital charge would best prompt parliamentary action. Accordingly it instructed its members to enquire into:

> the conduct of prosecutors, especially where reluctance to prosecute capitally is shown, or a recommendation to mercy is made.—The conduct of juries in recommending to mercy; and also, in cases of indictment for stealing, in the valuations they make of the stolen property so as to alter the legal character of the offence.—As to executions, the general feeling produced by them among the spectators and the inhabitants of the neighbourhood . . . .—As to the criminals themselves, what had been their previous character and habits, and

[9] Montagu, *Account*, 16–23; 'Second Report of the Society', *Philanthropist*, 5 (1815), 223.

[10] R. M. Bacon, *A memoir of the life of Edward, third Baron Suffield* (Norwich, 1838), 69.

[11] 'The diary of Elizabeth Fry', BL Add. MS 47456, fo. 299.

whether they had received the advantage of any education or religious instruction.[12]

This effort paid off when parliament appointed the select committee on the criminal law in 1819. Under Mackintosh's chairmanship and Buxton's guidance, the committee took evidence from the Society's supporters—Montagu, City radicals like Wood and Phillips, and the Quaker banking firms Fry, Gurney, and Hoare. So convincingly did it cull their prepackaged evidence about prosecutors' and juries' reluctance to bring felons to execution for non-violent theft in shops and dwellings and for forgery that the committee's evidence is still cited as proof of how public opinion was moving.

Cited to the same effect are the public petitions against the capital code, or parts of it. Nearly all the petitions between 1811 and 1830 seem to have been achieved through the Society's provincial networks or through the meeting-house which produced anti-slavery petitions in these decades.[13] In 1811 two petitions from 150 proprietors of Irish bleaching-grounds and 24 London calico-printers claimed that because capital punishments were 'by the common opinion of mankind considered as disproportioned to the offence', they deterred potential prosecutors from prosecuting and juries from convicting. Romilly took care to point out that the petitions expressed 'the sincere belief and conviction of experienced men' and not of 'speculative and theoretical reasoners'. This persuaded even die-hards like Ellenborough to vote for the repeal of the statute imposing death for theft from bleaching-grounds. Ellenborough might have paused had he known that both petitions were in fact 'set on foot and promoted by Quakers'.[14]

The same initiative determined the petitioning wave of 1819. The most widely publicized petition was adopted by the City's court of common council in December 1818. Quakers and their friends were behind this too. The meeting was attended by Montagu, Buxton, and 'many gentlemen of the Society of Friends'; and only they could have dictated its terms. Rehearsing the standard argument about prosecutorial and jury reluctance, the petition declared that the capital code was at odds with that 'tenderness for life . . . which, originating in the mild precepts of our religion, is advancing and will continue to advance as these doctrines become more deeply inculcated into the minds of the community'. The new tenderness ensued from 'the certain and general

[12] Tract 3 of the Society, *On the effects of capital punishment, as applied to forgery and theft* (1818), 26–8, listing the metropolitan committee's resolutions of May 1817.

[13] D. Turley, *The culture of English antislavery, 1780–1860* (1991), 65–7.

[14] Romilly, ii. 367, 361–2.

principles of our nature . . . the advanced state of civilization in the country . . . and the diffusion of Christianity, by which we are daily taught to love each other as Brethren, and to desire not the death of a sinner, but rather that he should turn from wickedness and live.'[15] This London petition opened the way to similar pleas using the same arguments and vocabulary. Between February and June 1819 over sixty petitions for criminal law reform were submitted by boroughs, public meetings, small groups of gentlemen, freeholders, burgesses, or simply 'inhabitants' of this place or that. They came from all parts of the country (though more from the south than the north), from Leeds, Bristol, and Norwich as well as from small rural parishes like Finchingfield in Essex.[16] That local meeting-houses were behind most of them is certain. Petitioning then lapsed until Peel's Forgery Bill in 1830 unleashed the biggest petition wave of all: nearly two hundred petitions. The most famous was signed by 735 country bankers and merchants from 214 towns, stressing as ever that 'where the practice condemns the law, the law ought to be altered'.[17] Again its origins are no mystery. 'It happened that one Sunday morning during this period, [Buxton] was visited at breakfast by Mr John Barry, who suggested the extreme importance of getting this feeling formally expressed; whereupon Mr Buxton dictated the . . . petition.'[18]

The impact of these campaigns is not denied. The 1819 petitions nudged the Commons into setting up the select committee. Those of 1830 at least put Peel on the defensive as he consolidated the forgery statutes (though he still contrived to leave the death sentence within them). But their main achievement was to intrude a pietistical language into the critique of capital law hitherto only tentatively expressed. 'The characters of Christian mercy, righteousness, and love, [are] the firmest bulwarks of society and government,' the petition resolved upon at the 1819 assembly of the Friends declared: 'were these principles received and acted upon to their full extent, were the genuine spirit and precepts of the gospel of our Lord Jesus Christ implicitly obeyed, way would ultimately be made for the abolition of this practice [of hanging] in all cases.' Admittedly, appeals to love were not always productive. With their 'thee's' and 'thou's' and funny clothes, Quakers were already deemed comical enough. MPs laughed when Wilberforce read out their petition in the Commons, and Mackintosh conceded that it speculated

[15] *Morning Chronicle*, 11 Dec. 1818; PD1 xxxix (1819), 81–6.

[16] *Journal of the House of Commons*, 74 (1818–19): page references fully listed in Radzinowicz, i. 527 n.

[17] Ibid. 348–50, 512–13, 526–8, 590–6.

[18] *Memoirs of Sir T. F. Buxton*, ed. C. Buxton (n.d.), 113.

extravagantly 'as to the future existence of some happier condition of society, in which mutual good-will may render severe punishments unnecessary'.[19] The 1819 committee—and Mackintosh thereafter—rested their case for mitigation of the capital code on pragmatic rather than ethical or utopian grounds. None the less, the moralized appeal against the capital code would now echo more widely than it had hitherto, in petitions for mercy, in newspaper reporting, and in parliamentary debate. The Punishment Society had put the terms in which the high-minded contemplated the hanging law securely in place, and it was an achievement.

2. *Propaganda and reality*

Lifted out of context like this, the Quaker effort against the capital code can be made to look impressive. But it was meant to look this way. Realists could easily have called its bluff. 'What shall we say of a code', Sinclair asked the Commons in 1819, 'which, if strictly or even moderately enforced during a single year, would deluge every part of the country with such an effusion of human blood as would appal the most unfeeling, and startle the most indifferent?'[20] *Not much*, realists might have replied. The code was not meant to be strictly enforced and it did not spill blood anyway. But it was becoming strangely difficult to say something as obvious as this, and when such things were said they began to be ignored. When the Common Council's petition in 1818 spoke about the 'public outcry' against the capital code, one alderman pointed out that 'he had never heard of this outcry unless from Sir Samuel Romilly and his adherents'. He had heard no outcry from parliament, he said, or from the judges, or from the recorder or common serjeant of London. The code was necessary because it was difficult to prevent crime without it; only a few were executed anyway.[21] This protest is as forgotten now as it was ignored then.

None the less it made a fair point. Not only was the outcry a good deal short of universal, but the commitment even of the penal reformers' natural constituencies also fell far short of their claims. There were only some 20,000 English Quakers in 1800 (4,000 fewer than that by 1840), and most were passive.[22] In the most comprehensive of Quaker diaries for the years 1761 to 1821, James Jenkins reflected on capital punishment only once, in a passing lament for the execution of 'no less than

[19] PD1 xxxix (1819), 799–800; the petition at 396–400.

[20] Ibid. 903–6.

[21] *Morning Chronicle*, 11 Dec. 1818.

[22] E. Isichei, *Victorian Quakers* (Oxford, 1970), 111.

twenty poor wretches' in 1785—'a sad proof of the profligacy of the times, but also, of the existence of those sanguinary laws, which have long disgraced our country': but not a word on the subject thereafter.[23] Those most active in the cause of penal reform were so very few that most of them—the Frys, Gurneys, Buxtons—had intermarried. Nor when they entered the fray was it as into a consuming preoccupation. William Allen spent more time on anti-slavery, Lancastrian or Bible schools, Spitalfields poor relief, or savings banks than ever he did on punishment. Apart from a meeting with Lord Sidmouth in 1813 to plead for a housebreaker and another meeting in 1818 to plead against the hanging of two boys for uttering forged notes ('we cleared our consciences, and, I think, made a little impression'), capital punishment seems to have cost him little more lost sleep than did the immolation of Hindu widows.[24] Even the good Elizabeth Fry had eleven children to worry about and did not spend more than a moiety of her time in Newgate, and that chiefly in 1817–18. How marginal capital law reform was to her daily preoccupations is conveyed in her chatty reference, in a letter to her brother, to the imminent execution of Joseph Hunton for forgery in 1828 (on whom more below): 'I am much obliged to Catherine for the gown which is arrived. I expect on 4th day [Thursday] to visit poor J Hunton with his wife. What a sad case—surely we ought to make a greater stir about capital punishment. I hope to dine at Upton today . . .'[25] Add to that the fact that penal issues were low in the priorities of the high political classes (Romilly's apart, but not excluding Mackintosh's, the political memoirs of the period never give penal debates a high profile), and the further fact that many reformers pursued ends not always acknowledged in public, and the heroic story begins to look shaky. To be sure, 'love' was part of the Quaker purpose, and the Punishment Society's manifesto did stress both 'the temporal' as well as the 'eternal interests of man'. A Quaker like J. J. Gurney had a direct and uncomplicated dislike of executions: 'such sudden and violent transfers from life to eternity are inexplicably awful, and are that which I feel fully convinced that man, under the gospel dispensation, ought never to inflict on his fellows.'[26] But we have already seen that many regarded capital law as an evil chiefly because it blocked the way

[23] J. W. Frost (ed.), *The records and recollections of James Jenkins* (New York, 1984), 180.

[24] *Life of William Allen*, i. 104, 341–2.

[25] E. Fry to her brother J. J. Gurney, 3 Nov. 1828 (Gurney papers, i, 216: Library of the Society of Friends, Temp. MS 434).

[26] J. J. Gurney to J. Hutchinson, 21 Aug. 1822 (Gurney papers, iii, 370: Library of the Society of Friends, Temp. MS 434).

to repentance, or because its uncertain and random mitigations diminished the pressure on felons to contemplate their immortal souls and allegedly increased crime. Though they never liked executions' effects on watching crowds, and some hated execution itself, it was chiefly the uncertainty of the law's implementation that was held to corrupt.[27]

Bald figures give the game away too. Three years after its formation the Punishment Society boasted only 287 subscribers in the nation. The £1,149 subscribed between 1808 and 1814 looks niggardly when compared with—say—the £5,500 which constituted the funds of the British and Foreign Schools Society in 1816 alone. Between 1808 and 1814 Joseph Fry donated only 4 guineas to the Punishment Society, the five Barclays gave 2 guineas each, the five Darbys of Coalbrookdale 7 guineas, the five Gurneys 13. Josiah Wedgwood and William Strutt gave 100 guineas apiece for the Schools Society in 1816 alone, and neither subscribed to the Punishment Society.[28] Elizabeth Fry's ladies' committees for prison visiting, with seventeen committees in the whole country in 1827, also remained weak in comparison with other charitable movements: 'the paucity of references to it in Quaker records suggests that very few were actively involved,' the historian of Victorian Quakers observes.[29] As for the petitions to parliament, the sixty or so of 1819 contained 12,000 signatures altogether, a piffling score compared with anti-slavery petitions.[30] Anti-slavery raised 519 petitions in 1792, 224 in 1824, and 5,020 in 1833. Anyway, this era saw a massive increase in parliamentary petitioning on all manner of subjects—nearly 5,000 annually in the late 1820s. Given the more urgent clamours on Catholic emancipation and parliamentary reform, MPs would not think the 200 petitions for penal reform in 1830 unduly impressive.[31]

Peel at the Home Office got some measure of all this. He had telling points to score against those who claimed that the movement of opinion against the capital code was ineluctable. When the provincial bankers petitioned against the capital statutes on forgery in 1830, he reminded the Commons that they did not speak for metropolitan

[27] Cf. Wilberforce presenting the Quaker petition in 1819, and Alderman Wood presenting the petition from the City: PD1 xxxix (1819), 85 and 398. 'Capital punishment and penal reform never really captured the imagination of Victorian Friends in general, or received the degree of enthusiastic support which they gave, for instance, to the temperance movement': Isichei, *Victorian Quakers*, 251–2.

[28] *Philanthropist*, 6 (1816), 72 ff.

[29] Isichei, *Victorian Quakers*, 249.

[30] Mackintosh: PD1 xxxix (1819), 903.

[31] Turley, *English antislavery*, 65–7. The number of public petitions to parliament increased from an average of 176 p.a. in the late 1780s and 205 in 1801–5, to 899 in 1811–15 and 4,656 in the politically expectant period 1828–32. Between 1833 and 1852 parliament received 251,488 petitions on public affairs, with 65 million signatures between them (PP 1852–3, lxxxiii. 104–5).

banking opinion. Thanks to the abolition of small notes, and because they hardly ever used cheques not drawn by known clients, country bankers were less imperilled by forgers than London bankers were. In London the danger of forgery 'was tenfold greater', and London bankers were significantly silent on the question of the death penalty, he said. A handful of Quaker bankers had told the 1819 committee that they would not prosecute lest those they prosecuted be hanged; but thirty-six 'of the most eminent London bankers' had formed an active association to prosecute forgers under the capital statutes as recently as 1825, and forgeries against them had declined. Nor was reluctance to prosecute and convict for capital forgery apparent in the fact that the proportions of forgery prosecutions abandoned or resulting in acquittal were no higher than they were for murder or other crimes.[32] Peel might have added that London juries were much happier to convict forgers than provincial juries were.[33] He might also have cited the mercy petition signed by forty-seven merchants and citizens of London (sixteen addressed from Lloyd's) on behalf of the forger Joseph Hunton in 1828. It accepted the verdict which finally hanged the man—'it being essentially requisite, indeed of the very last importance, in a great commercial country like England, that forgery should be deemed a capital offence, and punished with dreadful severity'.[34]

Peel also questioned the notion that public alienation from the capital code was proved by individuals' reluctance to prosecute or to give full evidence against those who had committed crimes, and by juries' reluctance to deliver full verdicts in capital cases. The argument to the contrary, that these variables were increasing, was at the heart of the reformers' case. Although reluctance to prosecute was sometimes 'conscientious', Peel admitted, it was usually mixed in with other motives like 'the just fear of trouble and expense' in bringing a prosecution.[35] Furthermore, he believed that juries' practice of undervaluing the goods stolen in cases of larceny from dwelling-houses reflected the difficulty of proving that a whole theft in excess of forty shillings had been committed at one time.[36] Peel was pushing it here: technical difficulties of proving a whole theft could not account for all the 555 par-

[32] *The Speeches of . . . Sir Robert Peel delivered in the House of Commons* (4 vols., 1853), ii. 164–6.

[33] Provincial assizes dealt with about three-quarters of all forgery prosecutions 1812–29, but provincial grand juries threw out proportionately more bills of indictment than London grand juries did, and, once convicted, London forgers were more likely to hang. At assizes, the proportion executed ranged from 24% in 1812–15 to 27% in 1816–18, down to 5% in 1824–9; at the Old Bailey the comparable proportions were 69%, 44%, and 36% respectively. Two-thirds of forgers hanged in the last 6 years of the death penalty were Londoners: PP 1819, viii, app., and PP 1830–1, xii. 495 ff.

[34] HO17.88/I (Hunton).

[35] Peel, *Speeches*, ii. 164–5, i. 406.

[36] Ibid. 247.

tial verdicts on stealing from dwellings which were delivered at the Old Bailey in the fifteen years 1814–29, for example,[37] and jury reluctance was often real enough. Yet Peel might have been closer to the mark than the reformers were. There was nothing new either in prosecutors' reluctance to prosecute or in juries' practice of pious perjury. Neither evinced a provably *growing* distaste for capital law. Like the 'plentiful tears' of the crowd which sometimes watered even Tudor executions, these practices betrayed a populist sympathy with the scaffold victim which antedated its naming within polite cultures. Indeed, thanks to the better management of prosecutions, the practices might well have been in *decline* when the 1819 committee adduced them as proof of public antipathy to capital punishment.[38] Moreover, judges had always connived at the jury's use of the partial verdict, because it was one of the standard ways in which élites wished the capital code to be mitigated. Similarly it had been as early as 1596 that Edward Hext noted the opinion of 'the simple country man and woman . . . that they would not procure a man's death for all the goods in the world' and that they would 'give faint evidence' in court if a restitution of their goods was promised. Dr Johnson similarly cited reluctance to prosecute as proof of disaffection with the severity of punishments in 1751.[39] Had this reluctance increased by the early nineteenth century? The mounting tide of prosecutions suggests otherwise. Thanks to deepening anxieties about 'crime', along with the extended reimbursement of prosecutors' and witnesses' costs, prosecutions at assizes and quarter sessions increased from 3,267 in 1805 to 18,107 in 1830, the rise between 1815 and 1820 being

[37] PD3 (Lords, 1833), c. 279.

[38] In late 16th-cent. London and in Sussex in the 1620s some 5–6% of property offenders benefited from juries' resort to partial verdicts (I. Archer, *The pursuit of stability: Social relations in Elizabethan London* (Cambridge, 1991), 245; C. B. Herrup, *The common peace: Participation and the criminal law in seventeenth-century England* (Cambridge, 1987), 157–8). At the Kent assizes in the 1660s some 31% of all convictions for property crimes entailed partial verdicts (J. S. Cockburn, 'Twelve silly men? The trial jury at assizes, 1560–1670', in J. S. Cockburn and T. A. Green (eds.), *Twelve good men and true: The criminal trial jury in England, 1200–1800* (Princeton, NJ, 1988), 172–3). At the Surrey assizes and quarter sessions 1660–99, 16.8% of trials of male property offenders resulted in partial verdicts, rising to 24.9% in 1700–39, then falling to 12.7% in 1740–79 and 7.5% in 1780–1802. Some of the decline might have resulted from magistrates' elimination of weaker cases at the committal stage and the replacement of the prisoner's implicit obligation to prove his innocence by the prosecution's obligation to prove the case: in these conditions, the jury's refusal to convict on the full count would have become a less necessary safeguard than hitherto (Beattie, p. 419 n. 32). Jury undervaluation, however, did become more important in the 18th cent., because of the multiplication of capital statutes (T. A. Green, *Verdict according to conscience: Perspectives of the English criminal trial jury, 1200–1800* (Chicago, 1985), 276).

[39] K. Wrightson, *English society, 1580–1680* (1982), 157; S. Johnson, 'The necessity of proportioning punishments to crimes' (1751), in W. J. Bate and A. B. Strauss (eds.), *Samuel Johnson:* The Rambler (3 vols., New Haven, Conn., 1969), ii. 246.

the steepest of the whole century.[40] By these tokens, noose or no noose, Peel could fairly have said that majority opinion still unambiguously supported the capital code.

3. *Quakers, forgers, and others*

To be secure, an opinion must be sustained as much in feeling, interest, and experience as in principle. Although those who hold an opinion will assert it *as* a principle, principle alone is unlikely to hold them true to it. Attention may always slip, other interests intrude, and emotional commitment fall short of good intentions. The fact that Quaker and evangelical concern about capital punishment showed all these human characteristics explains its inconsistencies and silences.

Active Quakers were concerned about a narrow range of capital statutes when all is said and done. They petitioned parliament for a wide-ranging revision of the criminal code in 1819; and many did deplore the whole code in principle. But class filters would keep getting in the way, so that in practice they were not active in protesting against the statutes which hanged poor men and women—for shoplifting, burglary, horse-stealing, and the like. Nor were they as prominent or noisy in petitioning for mercy in such cases as other people were. Even their response to forgery hangings was less than wholehearted. This is odd, since it was the forgery statutes that they made most noise about.

From Mackintosh's attack on the forgery statutes in 1816 on to the 1819 committee, on through the debates on Peel's consolidations in 1830, and on again to the statutes' repeal in 1832, it was understandable that more Quaker words flowed against these laws than against any others. First, as commercial men themselves, this area of capital law touched on their experiences most closely; it was imaginatively accessible to them. Secondly, it was tactically realistic to focus the attack in this way. Removing the death penalty from forgery and sheep- and horse-stealing alone, Buxton wrote in 1819, would save thirty lives annually—'which is something'.[41] Of recent provenance, the forgery statutes were a soft target, and had looked that way ever since Dr Johnson told Dr Dodd in 1777 that the forgery for which he hanged had 'no very deep dye of turpitude': 'It corrupted no man's principles: it attacked no man's life. It involved only a temporary and reparable injury.'[42] There was also an

[40] V. A. C. Gatrell and T. B. Hadden, 'Criminal statistics and their interpretation', in E. A. Wrigley (ed.), *Nineteenth-century society: Essays in the use of quantitative methods for the study of social data* (Cambridge, 1972), 392.

[41] *Memoirs of Sir T. F. Buxton*, 45.

[42] Radzinowicz, i. 468–9.

easily read absurdity in the application of the forgery laws which softened the target further. The fact that only fourteen people were hanged for forgery in 1817 looked ridiculous when 31,280 forged notes were presented to the Bank of England in that year.[43] By the 1820s, moreover, with the percentages of reprieves increasing apace and the will to hang for forgery weakening outside London, the discrepancy between frequent executions for forgery in London and infrequent executions in the provinces had become disconcertingly obvious.[44]

Yet even in respect of forgery laws Quaker concern was highly unstable. The greatest forgery case of the 1820s was Henry Fauntleroy's, the wealthy banker hanged in 1824. For nine years he had systematically diddled his partners and his bank's creditors to the tune of some half-million pounds, spending a fair number of them on a tantalizingly dissolute life with mistresses in his Brighton villa. That Fauntleroy's energetic labours for his own life got a good deal of public support is not denied. The petitions on his behalf were so bulky (a Home Office minute recorded) 'that they have been stored in the cupboard downstairs'. Alas, because they were never reunited with the original dossier,[45] it is now difficult to pin down what Quakers thought of his case; but it looks as if sympathy for him became audible only when it was sure that he would die, and even then it was muted. The main engine behind his appeals was his defence attorney James Harmer, always ready to score a point against the scaffold if he could. Reformers kept quiet on the whole. 'If a poll had been taken of the members of the Common Council—although most were radicals and reformers . . . nearly all of them would have decreed that the Berners Street banker must go to the scaffold. . . . Fauntleroy was hanged because the City of London thought that he ought to be hanged.'[46]

When Quakers got really worked up about a forgery case, it was when the culprit was most like them—Quaker and wealthy. Joseph Hunton's was the second biggest forgery trial of the 1820s. A merchant who had moved his firm to the City of London from Yarmouth and Bury St Edmunds as his fortune expanded, Hunton counted himself a Quaker even though his shady practices had caused his ejection from

[43] Ibid. 537–8.

[44] The national percentages of those reprieved after sentence of death for forgery rose from 62% in 1810–17, to 78% in 1818–24, to 92% in 1825–31 (PP 1831–2, xxxiii. 137). Of the 28 men prosecuted for forgery at the Old Bailey in the last 2 years of the death penalty for the crime, 1828–9, 9 were acquitted or discharged and 19 were sentenced to death: of these, 13 were pardoned and transported for life, and 6 hanged (PP 1835, xlv).

[45] HO17.87/I (Qk42).

[46] H. Bleackley, *The trial of Henry Fauntleroy and other famous trials for forgery* (1924), 47–8.

the Society five years before his downfall. But his family were still devout Friends, and Friends continued to sustain them energetically in their crisis, even if for some, as we noted earlier, the engagement was still marginal to their daily concerns.[47]

None of those who wrote or petitioned on Hunton's behalf excused his frauds. They supported him chiefly because his juries had recommended mercy 'on account of his large family and former respectability', and because before his attempted flight to New York Hunton had left securities to cover his forged bills. The appeals also chose to doubt Hunton's fraudulent *intentions*. But Joseph Geldart of Norwich gave another game away when he implied in a long letter to the Home Office that the peculiar 'calibre and twist' of Hunton's mind, and his apparent unconsciousness of 'moral or wilful guilt', were transparent to those who knew him. Even at his worst Hunton was intelligible to his supporters, for at issue was a moral turpitude to which all businessmen were vulnerable:

> He has fallen from innate deeply rooted avarice, the pure love of money for its own sake; for he had no idea of expense or personal enjoyment, but he had an ungovernable determination to be *rich*, and could he have accumulated ten large fortunes for his ten children, he would have been happy according to his own views, without enjoying any of it himself.

As one commercial man put it menacingly to the home secretary, 'the act for which it seems [Hunton] is doomed to suffer is one which is performed . . . by hundreds daily. . . . And the only difference between him and others is that he was discovered and they are not. . . . Should . . . the law take its course I truly believe an impression will be evinced in [commercial quarters] that a familiarity with scenes of blood has disqualified your august presence [from] rightly appreciating the life of man.'

This identification with Hunton permitted those who deplored his crime to plead that the whole man be seen—especially his general good conduct and the respectability of his connections, not presented in the trial. Geldart laid this out fully:

> [Hunton] has always been amiable and respectable in private life, as all his connections are, and some of them highly so. . . . He might have called the most respectable persons from Yarmouth and from London to attest his general character and mode of life, and who although they never could admire such a restless selfish man as he was, nor feel that warmth of friendship which they

[47] For the trial, see OBSP 1828: 993, 996; for the appeal documents upon which the following is based, see HO17.88/I. An engraved portrait of Hunton and 2 broadsides on his execution are in Hodgkin papers, BL Add. MS 38855, fos. 246–50: cf. also the Catnach broadside in BL 1888. c. 3., fo. 54.

> might for others, yet they must acknowledge that he lived highly respectably in private life, with his three sisters when they were young, two of whom married well, one of them to my worthy friend John Ransome of Manchester, the surgeon whose evidence was thought so important by the magistrates at the time of the Riots. His own marriage was into the family of the Sewells of Yarmouth, and his wife is an excellent woman, and the mother of ten children, the elder ones already appear of great promise. It is indescribable the horror into which the various branches of this family are thrown. There are four distinct families besides his own now in this situation, and their feelings are probably the sharper from the tenderness of their [Quaker] views on the subject of criminal punishment.

As Hunton's partner pleaded, the man had 'always discharged the duties of society, as a husband, a father, a master and the other relative obligations in life in an exemplary manner'.

When thus moved, pious men knew how to pull out the stops. Out came the well-rehearsed charges. Hudson Gurney told Peel that the death sentence for forgery only deterred prosecutors. Samuel Squire petitioned the king to say that when so many were charged with forgery but not prosecuted, 'the state of the law is . . . an impediment to the due administration of justice'. Hunton should be pardoned because of 'the too great severity of the law, on the score of humanity, and the avowed public feeling of horror and dismay at the frequent capital punishment inflicted on culprits for forgery whilst offences of greater magnitude are only punished by transportation'. Quakers mustered MPs' and peers' support to save Hunton. One enlisted Lord Faversham's help; Gurney enlisted Thomas Amyot's; J. A. Ransome of Manchester turned to Sir Astley Cooper and the Suffolk MP Sir Thomas Gooch; a Leicester associate used Robert Otway Cave; Abraham Sewell, Hunton's Quaker brother-in-law, enlisted George Anson. The Ipswich Quakers James Ransome and Jeremiah Head, along with Hunton's wife, sought personal interviews with Peel. Petitions came in from the Society of Friends, from seven Suffolk MPs, from inhabitants of Cirencester, from forty-four Manchester merchants, from the lord mayor of London and the inhabitants of Bishopsgate Street, and from the inhabitants of Yarmouth, Hunton's home town. This last petition is made of parchment, sewn end-to-end and seven feet long, and contains 680 names. Nearly all these papers testified (against the evidence one would think) to Hunton's 'amiable and most exemplary life', or to the fact that Hunton had always 'evinced that high moral feeling so generally characteristic of Quakers'.

How far such representations indicated the successful intrusion of evangelical language and values into the appeals of the 1820s may be measured in the confidence with which Hunton's sister-in-law Maria Sewell asked Peel for an interview, her piety not precluding a touch of menace when she prayed

that the sighs and tears of his afflicted and amiable wife and innocent children with their numerous relatives and friends may not appear in awful array against thyself and others equally responsible for the judgements pronounced upon earth. I can with heartfelt sincerity say Father forgive them they know not what they do forgive them their trespasses although they may not have forgiven.

Moreover, big cases like this tapped into the fantasies or offended feelings of numberless people remote from the drama, unstable, resentful, or principled as they might be. Into the Home Office came letters signed or anonymous, giving gratuitous opinions on the case, and in most of them appeals to the Deity loomed larger than they would have done several decades before. Fauntleroy's hanging in 1824 had produced its fair share of these: 'Merciful King. Spare Henry Fauntleroy's life: direct the false cause of it, of his wife's, of his children's [grief?], to the Glory of God, to the sanctification, consolation and salvation of their Souls, through Jesus Christ, our Lord and Saviour and most merciful intercessor'; or again more menacingly:

To His Majesty: of great importance: Sire, Your Majesty seldom hears the truth: *you* need mercy, at the tribunal of the Sovereign of the Universe, as much as Mr Fauntleroy needs mercy from you, as his Sovereign. Seek, by repentance towards God, and faith in the Saviour, that mercy without which, Your Majesty, like the meanest of your subjects, must perish. . . . Your Majesty can, in opposition to your council, exercise this Godlike attribute. Let Mr Fauntleroy be its first object. With what satisfaction will your Majesty, every night, resign yourself to sleep!

So too in Hunton's case; one Thomas Parkin of 14 Poultry sent the Home Office a personal text, emphatically entitled '*Solemn protest*':

as a member of the body politic, I enter my most solemn protest against the shedding of the blood of Hunton and Fenn [another forger contemporaneously in Newgate], and all others [guilty] of crimes not punishable, according to the Scriptures of Truth, with death; and if such men, after this protest, suffer on the scaffold, I am, as a member of the community, guiltless of their death. . . . This protest will be a witness before him who is King of Kings . . . against all who were accessory to the murder, or who had the means of preventing it.

These examples demonstrate how the language of piety had established the rhetorical grounds for dissent from the law. They would also offer the most impressive evidence of evangelical mobilization and influence were it not for one thing. The critical vision released in the Fauntleroy and Hunton campaigns was highly selective. 'Not a tongue appears to stir' for a young man sentenced at the December sessions, while in Fauntleroy's case 'petitions flow in from every side'.[48] Of the last eight men hanged for forgery in England (in 1827–9), Hunton was the only one whom Quakers strongly supported. Was this because nearly all other condemned forgers were humble but shady customers—clerks etc.—hovering on the edge of the commercial world, some of them quite on their beam-ends? So it would seem. The last forgery condemnations at the Old Bailey (checked in Old Bailey Sessions Papers and the Home Office appeal papers) all came from the shabbier classes. The very last to be hanged in England for forgery was Thomas Maynard, on 30 December 1829. His appeal papers do refer to 'some City Quakers' working for him, but the same papers of Henry Conway, Edward Martelli, and Richard Gifford, hanged earlier that year, contain no references to Quaker support. Gifford's appeal was supported by Lord Colchester, Stephen Lushington, Lord Melborne, and Arthur Onslow, chiefly because Gifford's father was a messenger in the House of Commons.[49] To point this out is not to belittle Quakers' humanity. It only recognizes that their sympathetic imagination had familiar and significant limits, and that 'opinion' met its limits at that point too.

It so happened that three other men hanged alongside Hunton in 1828—though you would not have thought so if you listened only to Quaker concern. It is not clear that any of the Quakers upset by Hunton's hanging gave these men's fates a second thought. The first, John James, was a 19-year-old tailor who hanged for housebreaking. The only, and vain, support that he got in his appeal was a bleak petition from his widowed mother, her 'mind bordering on distraction'. Joseph Mahoney, aged 26, was also hanged for breaking and entering, and the only support he got was from his father, who acknowledged 'the justice and impartiality of the learned judge' and who, in addressing the home secretary, could find nothing much to say other than that he 'implored the mercy of God to bless you'. The third to hang with Hunton was James Abbott, 28, a journeyman glove-maker who had tried to murder

[48] Sir Robert Deverell to Peel, 24 Nov. 1824, Peel papers, BL Add. MS 40730, fos. 194 ff.

[49] The last cases are listed in PP 1835, xlv; see also HO17.68 (M02) (Maynard); HO17.35/I (E04) (Conway and Martelli); HO17.54 (I010) (Gifford).

his wife. There was much more going for Abbott than for the other two, but this was no thanks to reformers, or Quakers, or evangelicals.[50] A single Worcester Quaker wrote a brief character reference on Abbott's behalf because he had once employed the man. No other pious voices are heard in the appeal dossier. Such middle-class support as Abbott got came from his defence attorney Harmer (again) and from the sheriff Mr Booth. They were the ones who mustered affidavits and mercy petitions for Abbott. And it signifies that they turned for their support to common people chiefly, who had no illusions about their place in the world—people who would affirm a sense of justice without piety, access to interest, or an eye to public effect.

Abbott hanged for maiming and wounding his unfaithful wife in a fit of jealous rage. The evidence brought in his support by neighbours, midwives, and publicans was about the murky sexual transactions of poor people, set in lodging-houses, on grimy staircases, among people who signed affidavits with crosses. An appeal like this takes us close to all those sexual economies in plebeian life which are difficult to access otherwise—to plebeian views of female obligation, for example, when husbands and wives were at loggerheads. But Quakers would hold aloof from such murky doings, not finding them as moving or as interesting as we might.

Abbott's wife, the affidavits said, defied her husband by working in a tavern near their lodgings in Fetter Lane because she could meet men that way. She neglected her three children. She told a neighbour that her husband was not man enough to get a child. She told the illiterate shoemaker's wife who acted as her midwife that she had sex with a gentleman for half a crown. A bricklayer's wife of Smithfield and a vendor of patent medicines testified that she had sex in a Fleet Street brothel, caught VD, and communicated it to her husband. Their three children were probably not Abbott's own. These were things which everyone knew.

One night after the husband had dragged the wife back from the tavern, both quarrelled violently in their lodging's privy and stairwell. When a widow lodging on the first floor intervened, the wife shouted that she would not submit to her husband in bed and 'was determined he should never have any connection with her in future'. The widow told her: 'it is a pity you should go on in this manner, you ought to yield to him. . . . For God's sake do not expose yourselves in this manner, why

50 HO17.82/I (Pn26) (Abbott), (Qn70) (James), (Qn15) (Mahoney).

not go upstairs and make yourselves comfortable?' 'That is all I want her to do, Mrs Sage,' Abbott replied: 'to stay at home and mind her family.' Abbott later mercilessly beat up his wife, allegedly provoked beyond endurance.

Sheriff Booth had all this material sworn in the Guildhall and sent it to Peel at the Home Office. He got a dozen master glove-makers and tradesmen shakily to sign petitions to say that Abbott was a quiet and inoffensive man and a reliable workman; they were 'moved by the strongest feelings of Humanity' in testifying. The jury signed a petition for mercy too, acknowledging that Abbott had been moved by 'a species of jealous insanity'. Several anonymous letters arrived at the Home Office saying that 'the number of executions have not the desired effect': there should be no hangings except for murder. There was much excitement in the neighbourhood of 1 Fetter Lane. But all this interest was class-specific. Middle-class people rarely stooped to have opinions about matters as grimy as these. If Booth was interested, it was for the reason that all sheriffs were interested. He had to face the condemned in their cell, to conduct them to the scaffold's foot, and, officially, to watch them choke.

The appeal failed, as did Hunton's.[51] So on 8 December 1828 the old pageant was enacted; 100,000 people attended, the papers said—as many as had attended Fauntleroy's hanging, and more than had attended Holloway's and Haggerty's in 1807 (which confirms motifs in Chapter 4, that reference points were as remote and memories as long as that). *The Times* said that there was a general belief that Abbott would be respited: 'hundreds of persons would not be convinced that execution would be done upon the culprit, till they actually saw him with the rope round his neck.'

Sheriff Booth was so upset at the failure of Abbott's appeal (he had travelled to Windsor to make a vain plea for a respite to the king) that he left his duties to the under-sheriffs. Hunton was attended in the prison by several Quakers, including ladies: a special apartment was set aside for them. They prayed with him for hours on the Sunday before his death, Elizabeth Fry and her brother among them; and two Quaker elders sat with him all Sunday night. But Quakers would not attend him to the scaffold; a minister of Leadenhall Street did that instead. In the last half-hour *The Times* reporter was transfixed by Hunton's every movement, ignoring the others. There was pity for Abbott, he reported;

[51] For the King in Council's decision on Hunton, see Ch. 20.

but in the pressroom what he watched was Hunton's praying, reading, murmuring, and the religious consolations he offered to the young and terrified James. 'Oh dear, is there any necessity to tie the cord so fast?' Hunton asked; and when he got no answer he murmured: 'Well, well, thou knowest best.' He asked if he could wear his gloves as he died, and was allowed to do so. On the scaffold he asked for his eyes to be bound with a blue handkerchief to which 'he was fondly attached'. Women in the crowd shrieked as usual as the four men dropped.[52]

Was 'opinion' formed by these newspaper detailings? Newspaper reports by the 1820s were a mine of images and information. For those who did not attend executions, little else could so shape collective awareness—even though their influences are unverifiable and we can only follow hints. But how else to explain the arrival after so many executions of so many anonymous letters to the king? 'Thou art a Murderer, Wretch that thou art,' one said after Hunton's death. And in the same hand, letters signed 'a friend of poor and murdered Hunton' were sent to Judge Park who had condemned him, threatening to cut the judge's throat. 'You are a bloody rascal a bloody thief and a bloody murderer of Hunton,' they said.[53]

It was not, then, that the reformers 'invented' opinion or spoke into a vacuum. They tapped into anxieties about harsh law which worked on many psychic and social levels by the 1820s. Queasiness was real enough among some prosecutors, juries, or pleaders for mercy on others' behalf. And pious reformers gave those who were at whatever level uneasy about executions new banners to name themselves by, and this naming of opinion helped to bring it into being. Mental deposits were laid which abolitionists would build upon much later. Quaker hectoring played its part in the extended process of social learning, converting diffuse feeling into opinion on Quaker terms. And yet there is no doubt that the reformers exaggerated the clarity, ubiquity, and intensity of the public concern for which they spoke. Much of the energy they themselves focused on the cause was spasmodic and prejudicially selective in its targets. Also, none of these good people was much heard protesting on the issue of *justice*. They hardly ever used the word. To find a concern about justice it is to other kinds of people entirely that we must turn.

[52] *The Times*, 9 Dec. 1828. [53] HO17.88/1.

CHAPTER 16

APPEALING FOR JUSTICE

1. *Appeals and opinion*

THE PEOPLE WHO PETITIONED FOR MERCY ON OTHERS' BEHALF in the 1820s were of a kind obscure to history, who wrote no books, contributed to no newspapers, and signed no *public* petitions. But they spoke as revealingly about punishment and justice as public propagandists did—more so indeed, since their opinions were not programmatic or for public effect. The documents they leave give significantly different meanings to public opinion from those acknowledged so far. Many petitioned out of obligation to dependants, and a few out of expedience. But most acted disinterestedly against the injustice perceived in particular verdicts or sentences, moved by commonsensical ideas of fair trial and proportioned punishment which were remote from the discretionary principles of the judges' law. 'We only want *fair play*,' as Sydney Smith exclaimed in 1826 when he inveighed against counsel's inability to sum up for defendants. Most petitioners could have said the same.[1]

Although the scaffold looms over the appeal archive, appeals did not address the hanging code exclusively. They flowed in for non-capital convictions also. They aimed at prosecutors' or constables' malice, judges' bias, the haste of trials and the failure to hear key witnesses, the absence of defence counsel or the veto on its summing-up, the absence of a regular appeal system, and the caprices of mercy. Sometimes, as we shall see in Chapter 17, a conflict between legal and communal values was the issue, the community claiming a privileged knowledge of protagonists in a case which the law was assumed not to possess. Many appeals did no more than home in on specific failings in trial procedure without generalizing from them. But in all cases the precision with

[1] S. Smith, *Works* (2 vols., 1856), i. 363. Ch. 6 above discusses the contexts and contents of appeals; App. 1 comments on the sample and the appeals' location and arrangement.

which failings were pin-pointed spoke for a well-anchored knowledge that the administration of justice should and could be fairer than it was—and, since the law was an arena of negotiation, that it was one's duty to say so. Sometimes petitioners were moved by indignation and incredulity that such things could be done. Indignation is an optimistic emotion on the whole. It implies belief that the world might be changed.

The petitions themselves gave notice of changing times. In the eighteenth-century provinces (less so in London) those who used their influence for convicted felons were usually led by gentry and occasionally by peers. A few of these early appeals were as large and well orchestrated as any subsequently presented. The appeal in favour of a burglar, Matthew Barker, condemned at Norwich in 1784, for example, leaves a huge dossier. Filled by the representations of Lord Leicester, the bishop of Norwich, Sir Thomas Beever, and scores of Norfolk gentry and JPs, it was forceful enough to oblige the judge to agree to a conditional pardon.[2] Gentry at this date might use the language of sympathy as freely as any of the middle classes. 'Humanity must ever be considered as a common cause,' the brother of Lord Vernon wrote axiomatically on behalf of a Derbyshire horse-thief in the same year.[3] But these early and high-status appellants were not on the whole men who would express that quality of *indignation* we have our eyes on. The system they worked with was their own. The most extraordinary cases could pass with no public outrage whatever. If great men of the locality let them pass, minor felons would hang.[4]

By the 1820s changes were marked. Grandees and gentry were still

[2] The judge, Loughborough, saw no legal grounds for changing his mind about hanging Barker, but privately admitted that 'the youth of the prisoner, the penitence he has expressed, and now the precedent of his escape from almost certain death [thanks to the sheriff's unauthorized respite] seem to me to have excited a zeal in his behalf in Norwich and the neighbourhood, which would counteract the effect of punishment in point of example: if the man should now suffer, his fate would seem hard, the execution would appear an act of cruelty, and the compassion for the sufferer would absorb the respect for the law by which he suffered. My idea is that example is the sole end of punishment, and that no case where a man is thought to suffer hardly, and dies an object of compassion in ordinary offences, is of much public utility.' The bishop's letter in favour of mercy gave great weight to this opinion, he added: HO47.1 (Barker's case, Sept. 1784).

[3] HO47.1 (Rose's case, Mar. 1784).

[4] Cf. John Carey, condemned for a highway robbery with violence by Gould J. at Essex assizes, 1784. Carey had to mount his own appeal. He told a tale which, 40 years later, would certainly have attracted local interventions. He had no counsel or attorney to represent him. He had entered the sea service 2 years before, had lost the sight of an eye, and had his skull fractured, which left him often delirious. In want, he had robbed a man of 10½*d*. He had returned the money 'on finding the person from whom he took it to be in equal distress with himself'. The judge recommended 'a speedy execution . . . in order to strike the greater terror' into the hearts of daring thieves. Such a sentence would have caused a local outcry in the 1820s, and would have been unthinkable. HO47.1 (Carey's case, Mar. 1784).

prominent petitioners. But more and more people of middle means were participating in this appellant procedure, as well as in the prosecutory part of the legal process, and they became the majority voice in the petition archive. Doctors, agents, businessmen, tradesmen, clergymen, farmers, and attorneys petitioned with growing confidence and familiarity with the process in which they were engaged. They altered the terms of mercy appeals significantly. The plea for mercy fused into a quest, overtly, for *justice*.

The bloody code did not collapse because of the mounting force of these voices, but home secretaries and judges knew they must be monitored. Those who protested too vehemently could be dismissed as sentimental, mad, or radical, and, if need be, checked up on by magistrates or informers. Most petitioners at the other extreme observed the most deferential of forms. It was between these poles that the telling voices were heard. Collectively they sound like the voices of the 'people'—seldom otherwise heard. At this level the appeal archive was becoming a monument to those 'milder sort of people' whose tenderness Burke had warned the government to respect lest excess in punishments make them 'consider government in a harsh and odious light'. It gave voice to those 'enlightened and respectable classes of the community' of whom Mackintosh spoke, whose support for the laws was 'essential to their efficiency'. Here if anywhere was that 'general opinion' which Paley had advised governments to treat 'with deference' and manage 'with delicacy and circumspection'.[5]

Not all middle-class petitioners were critical of the law's punishments or moved by sympathy for those they tried to save. Their petitions could just as well be conceived in obligation and hatched in interest as gentry petitions were. What was going on in 1822, for example, when Somerset churchwardens, chaplains, solicitors, and farmers asked for mercy on behalf of an old man whom Burrough J. had sentenced to death for bestiality with a sow? Forty years earlier a local constituency like this might have let him hang. Now, more practised in registering the signifying textures of an old man's life, perhaps they felt obliged to feel sorry for him. They pointed out that in a long life, twenty-one of them in the navy, he had supported his family honestly as a chimney-sweep. He had received

[5] Edmund Burke, 'Letters and reflections on the executions of the rioters in 1780', in *The works and correspondence of the Right Honorable Edmund Burke* (8 vols., 1852), v. 580–1; Sir J. Mackintosh, 'Speech on . . . committee to inquire into . . . criminal law . . . on the 2nd March, 1819', in *Miscellaneous works of the Right Honourable Sir James Mackintosh* (1851 edn.), 724; W. Paley, *The principles of moral and political philosophy* (1785), book VI, ch. 2.

severe wounds at Trafalgar—one to his head which was trepanned and left him often 'mentally deranged'. The evidence of the witness against him was unreliable, they added. The judge was unmoved, however. The detailed evidence was 'too gross to be unnecessarily repeated', he reported to the Home Office; he had no doubt whatever about the witness's veracity. So one day in a market square the old man was brought out in shame to his neighbours' view, and there he was hanged, all 71 years of him.[6] We cannot say whether the petitioners felt indignation at this outcome. Perhaps all they feared was that in tense times cruel punishment might inflame local feeling.

This was shown a year later when at Shrewsbury assizes Best J. condemned to death for arson a half-witted man called Richard Howells. Despite the man's pitiful demeanour in the dock, the judge told the Home Office that this 'miserable looking wretch' was 'more of an imposter than an idiot'. He stuck by this opinion although he agreed that the prisoner would 'probably arouse commiseration in those who are not acquainted with the case. He appears to be paralytick and . . . half witted but [to] have . . . sense enough to impose on many.' The local magistrates pressed for conditional pardon. Howells was deformed and half-witted, they wrote. But what mattered most was 'the effect of such a spectacle upon the scaffold, which in the present impression of the people's mind would rather excite sympathy for the sufferer than abhorrence for his crime'. This persuaded Best more than the poor Somerset bugger's plight had swayed Burrough. Peel respited the sentence and Howells was transported for life.[7]

In cases like these, sympathetic feeling cannot be assumed. But other appeals—scores of them every year in the 1820s—leave no doubt about petitioners' indignation at slap-happy justice or distaste for hanging. In a few specialized contexts some express distaste for the values of the law *in toto*, and this with such vigour and in terms so remote from our notions of what it was possible even to think in those days, that assumptions about past people have to be quite adjusted. Who would imagine, for example, that in 1828 someone would rhetorically ask the home secretary whether b–gg—y (thus both the Old Bailey Sessions Papers and the Home Office's annotations on the letter) was 'that kind of offence wherein man injures his fellow man, either in person, or property? How does it happen that it is visited in this country *alone* with death? What just right can any legislators, or legislature, have to deprive man of the

[6] HO47.63/7 (Chilcot's case). [7] HO47.64/8.

free use of any parts of his own body?' The anonymous writer was responding to a case in which two men had been condemned to death for buggering each other in the area of their lodging-house (on the evidence of a disreputable woman moved by malice, the men's desperate mercy petition claimed). Disclaiming connection with the case, the author asserted interestingly:

All, whether male or female, have certain inclinations or propensities which must be and are gratified, and for aught I see should be so: they are implanted in us by some unknown power . . . [or] superior agency, and over [them] we have little control. . . . I would ask you whether youth is not initiated at the most tender age into such practices, aye, long before they leave School. . . . It may be, and very likely was, your own case, for the practice pervades all classes of society, from the peer to the peasants, though unquestionably the former in far greater proportion to the latter. For God's sake therefore bestir yourself, and do not suffer these unfortunate wretches to die, when thousands besides, are as guilty and perhaps much more so (if guilt it be) than they are.

The King in Council commuted the men's death sentence to transportation for life, though not before they had waited an excruciating three months in Newgate's condemned cells.[8]

But it was opinions on routine hangings that are most commonly heard—like that of the Islington householders responsible for prosecuting two housebreakers. They take pity when at the Old Bailey in 1826 the latter are condemned to death. At first the petitioners plead with rotund deference: 'We entreat your Majesty to permit us to say that by granting our prayer you will impart peace to our minds, and give fresh proof to ourselves and to the prisoners of that clemency and benignity which has so eminently distinguished your gracious Majesty's dispensation to your subjects.' But this is what they say next: 'We further humbly beg leave to add that our repugnance to take away life, where no attempt to destroy life is made, is so insuperable, that we would rather endure the repetition of our misfortune, than by seeking a remedy thus sacrifice our best feelings by compromising altogether our consciences.'[9] Nor were attitudes like this dependent on a merely fashionable sympathy. They were evinced by people remote from

[8] Martin Millett's and James Farthing's cases: OBSP, Sept. 1828; HO17.88/I (Qn16); PP 1835, xlv.

[9] HO17.76 (dossier O120). About 60 people from Binfield's home village in Norfolk signed a separate petition for mercy, and the Norwich Quaker Samuel Gurney forwarded documents on the case to Peel via Thomas Buxton. Savage was also supported by his East End ex-employers, a tailor, a market man, and a greengrocer and cow-keeper. Peel's pencil scribble ran thus: 'the papers were fully considered by the King in Council and were not thought to afford sufficient ground for a commutation of the capital sentence of the law.' Both Binfield and Savage hanged. It was Savage who, on the scaffold, treated the ordinary of Newgate with the contempt noted above in Ch. 1: *Ann. Reg.* 1826: 91.

metropolitan culture who were moved by a sense of justice not very different from the compassion of those sixteenth-century country people who 'would not procure a man's death for all the goods in the world'. The best examples of this antique generosity come from those parts of the country where hanging was rare and therefore shocking.

At the Carnarvon assizes in 1822 one Lewis Owen was sentenced to death for shooting an excise officer in a robbery. The petitioning inhabitants—tradesmen mainly—strained to find extenuating circumstances. Owen had been wounded in the king's service, they pointed out; he also laboured under 'temporary insanity' when committing the crime, thanks to a recent blow on the head by a plank. But, most to the point, there had been no execution in Carnarvon for the past twenty years. 'In a county such as this, not used to crime . . . the people revolt at the idea.' The county's high sheriff felt the same. He reminded Home Secretary Sidmouth that 'the great end of justice is public example rather than personal punishment'. The trial had already 'created a sensation, equal in its effects, to that generally intended and supposed to be produced by execution. . . . I cannot endure the thought of having a human being executed during the time that I am in office.' Alas, Carnarvon had to endure it. The judge acknowledged that 'the feelings of the people revolt at the idea' and that 'happily North Wales has little need of such terrible examples'; but he did not yield and the execution proceeded.[10]

The memory of Owen's hanging fuelled an even more energetic campaign for mercy a year later, however. In 1823 one Henry Jones was condemned for stealing £38 from the house of a female relative on the day after the woman had been murdered by an unknown person. There was no evidence to connect the two crimes, so the murder was not tried. But local opinion was that Owen was condemned because the judge took the connection between the crimes for granted. Two petitions to the king from the high sheriff, the coroner, and the magistrates, clergy, and gentry of Carnarvon, 'deeply lamenting the sentence of death', begged for pardon on the grounds that the man was honest, inoffensive, and of weak understanding. There was not the slightest evidence that he had committed the murder, and if he had taken the money it was probably because he believed it was due to him on his

[10] HO47.63/6. In 1829 the attorney-general of the north Wales circuit claimed that only 4 people had been hanged in Anglesey, Carnarvonshire, and Merioneth in the preceding 40 years; 26, however, were hanged in Monmouthshire between 1800 and 1830, almost half the total for Wales: David J. V. Jones, *Crime in nineteenth-century Wales* (Cardiff, 1992), 225.

relative's death. Other letters from local MPs and the county's lord lieutenant confirmed that Jones was 'a decided idiot' but an honest one: 'his wages were always very low because another man must always be employed with him in order to show him what to do.' Then came a scathing letter to the judge from the prosecuting counsel at the trial—one of many lawyerly voices we shall soon meet in a similar context:

> I know that the laws of England doom a man, convicted of larceny in a dwelling house to the amount of 40 shillings to death, but public humanity does not permit that judgement to be executed, except in outrageous cases. . . . Allow me to call your recollections to the excellent character the man received from the most respectable witnesses? Allow me to remind you that the jury recommended the man to mercy? Allow me to state, that I, who prosecuted for the crown, ask also for mercy? . . . I cannot but anticipate the long, painful and never to be forgotten remorse, which must be felt for his execution.

For our purposes the most interesting petition was signed by twenty neighbouring farmers—humble men, they said. They apologized that they were only 'Country People at a loss for want of Education of Eloquency of Address'. They declared that had there been opportunity, 'hundreds of persons' would have wished to sign their petition too. They were full of remorse that they had not supported Jones with character evidence at the trial: 'We have through Ignorance and disheartendness [*sic*] not knowing what steps to take been too indolent—yet God cause us to succeed and prosper.' Had the neighbourhood had notice of the trial, the whole of it 'would have gone at their own expense although 25 miles'. They reminded the king that 'the poor idiot's life is valuable to *him*'. They begged the king's mercy. The judge did not capitulate to this storm either: his duty was 'painful', but he saw no reason to revise the sentence. For reasons the dossier does not record, Peel overruled him. On the day before the execution Jones was respited for transportation. The judge, weasel-worded, wrote to Peel afterwards that this made him 'very happy'.[11]

2. *Indignation*

The quality of justice and how it was perceived is best assessed as 'ordinary people' have always and rightly assessed it: that is, particularistically—not in terms of the rightness of abstract principle or the credibility of procedures, but in terms of the fairness of outcomes in

[11] HO47.64/11.

practice. At this point individual cases signify. They signify for the historian too. They have vividness in them, the vividness and even paradoxical freshness experienced when the dustiest documents are opened to reveal, as if from yesterday, long-dead judges' pomposity and self-love, common people's sad entanglements, and the fresh-penned protests of people never again or otherwise to be heard. The two appeals to be summarized now were not spectacular or important. But they are representative of a clash between old values and new, as the judge betrayed his merciless fortitude, and disinterested parties stood incredulous at it. In close-focused instances like these, we see how a quiet but indignant middle-class opinion was shaped piecemeal. When people got involved in or heard about appeals like these, they acquired a slightly surer sense of what was wrong with the world.

At the Chelmsford assizes in March 1824 Best J. recorded the sentence of death against one Benjamin Ellis for highway robbery, namely for attacking a keeper of Waltham forest and stealing his gun.[12] Dice were loaded against Ellis from the start. The son of a market gardener, he was a rough man, a drinker and a fighter, fond of displaying his pugilistic skills and barred from some public houses as a result. He signed his mercy petition with a cross. In a community with a long history of conflict between forest owners and customary users conviction was likely—not least because Ellis's prosecutor was the forest's lord warden.

After the trial Ellis's family had gumption and means enough to employ an attorney to draw up his petition, reiterating the alibi Ellis had sought vainly to establish in the trial. He could not have committed the crime, it said, because he was at home asleep at the time, and his three brothers could prove it. With them he had travelled that night from the village of Loughton where they had visited their sick sister. After drinking at the Crooked Billet in Waltham forest, they had all gone to bed at home in Barking. 'Such a defence,' the petition acknowledged (an attorney's words), 'when it fails of proof, is generally considered to be, and would, if untrue, be adding subordination of Perjury to the crime'. But the alibi had been rejected by the jury only because one of the brothers involved had been flustered and confused by fierce cross-examination in court. Since he was 'quite illiterate, and had never been in a Court of Justice before', he had contradicted the other brothers' testimony about the food they had eaten in the tavern. This had been the only firm basis of Ellis's conviction. The petition added character evi-

[12] HO47.65, Mar. 1825.

dence from the minister of his parish (not heard in court) and from a former employer in Covent Garden who would employ Ellis again if he were pardoned. A gentleman farmer and a custom-house clerk who had bought wood from the family and believed them to be honest likewise testified to Ellis's integrity and industriousness.

Armed with petition and letters, Ellis's brothers walked or hitched lifts the fifteen miles to Mr Justice Best's house in Bedford Square. Neither this long travelling nor this direct approach was exceptional. The latter expressed widely understood reciprocities, and visits to a judge's house or even to the Home Office were tolerated.[13] At Best's they swore an affidavit (signing by marks) to support the alibi, and were peremptorily dismissed to make their way tediously home again. When the judge forwarded the affidavit to Peel, the home secretary followed usual practice; he instructed the Chelmsford assize clerk to enquire into Ellis's character more closely.

Impartial opinions were not sought. The first to be consulted was Ellis's prosecutor. He opined unsurprisingly that the brothers were bad characters. He was sure that two of them had committed the robbery, even though the gamekeeper who had been attacked had only been able to swear to Benjamin. He gave no evidence for this belief. The local magistrates were also consulted. As unsurprisingly, they too were sure that the conviction was right. The publican's wife at the Crooked Billet was consulted. Alas for Ellis, she denied that the brothers had gone to their sister's that night. That looked like the end of it. When it came down to it (the clerk reported back to Peel), Ellis was only 'a laborer often employed near the Forest where I know he has been ascertained to be *too busy* with the deer and wood'.

Then comes the intervention crucial to this argument. The dossier next discloses a beautifully scripted and lengthy letter to the home secretary from a Mr C. H. Lichfield. We know nothing about him other than that he lived at the gentlemanly address of Dolphin Place, London, and that he must have been wealthy. His letter politely informed Peel that although the Ellises were strangers to him, he had been 'asked' to take up their case. He had employed his personal agent, a Mr Orme who was 'now conducting my business', to set in train a personal investigation.

This impartial investigation, so unlike the Home Office's own, brought dividends. In Waltham, Orme found two men ready to depose

[13] Cf. Harwood's mother in her approach to the lord mayor of London on her son's behalf. Washerwoman though she was, she swore her affidavit in the Mansion House: see Ch. 1, above.

that they had done the deed. One, dressed like Ellis and of his height, could have been mistaken for Ellis. On the night of the supposed crime (their depositions read) they had accosted the forest keeper in Waltham forest, believing him to be a poacher. The keeper had tried to fire at them, but the powder had only flashed in the pan. They had wrested the gun from him and thrown it into the undergrowth. There (crucially) it was discovered by a farmer a few days later, hidden by bushes. In short, there had been no theft at all: it was the theft of the gun which had been the basis of Ellis's conviction for robbery.

With this information and the two men in tow, Mr Lichfield's agent called on the judge in Bedford Square. He need not have bothered. Best was high and mighty. He warned Orme to 'beware of the parties imposing on him as they were bad characters altogether', as if that answered the case. When Orme pointed out that the two men were waiting downstairs and could give Best a personal account of themselves, the great man replied that if he set eyes on them, he would commit them to trial as Ellis's accomplices. The deputation retreated.

Orme pursued his enquiries in Ilford next. He discovered that although Ellis 'was apt to be merry when he got a glass too much, and [was] fond of showing his pugilistic knowledge', his character was otherwise intact. What had blemished his reputation was a false rumour put about by a parish constable against whom Ellis had an action pending for an assault, the imminence of which action gave the constable every interest in speeding Ellis's departure from the country or from life. His malicious report had prejudiced the committing magistrate against Ellis in the first place. Orme got hold of the constable's letter to the magistrate to prove the point. Back in London Mr Lichfield was convinced that this clinched the case. He wrote to Peel in high indignation. Thus 'the life or liberty of an innocent man must become forfeited through the wicked and evil insinuations of others. Thus, Sir, may the magistrates have been deceived . . . the character of the unfortunate man has been whispered away, his evidence discredited.'

Peel had no choice but to consult the committing magistrate about the case, though with predictable results. The magistrate dismissed the veracity of the two men who had confessed to the attack on the forest keeper. It was true, he surprisingly admitted, that the gun had been found in the bushes by a third person, so that no theft could be proved. But he had no doubt that it had been Benjamin Ellis who had wrested it from the forest keeper. Again he gave no grounds for this belief beyond the fact that the Ellises were not widely esteemed and that their claims to respectability were false: 'the Father being described as a market farmer, is to carry the appearance of respectability, whereas he is only a jobber'. True, there was no criminal record against Benjamin—'yet he is one of the low pugilistic amateurs and has been forbidden entry to several of the public houses at Ilford and its environs; and . . . the minister and churchwardens of Barking feel that they would be ill advised in signing a recommendation in favour of the prisoner.' When these views were forwarded to Best, the judge contentedly advised Peel that the alibi which the parties were now seeking to prove had obviously led to perjury. The conviction and sentence should stand. Peel acquiesced. On the back of Ellis's appeal he pencilled the familiar word 'Refused'.

We cannot now know the full truth of the story. The brothers' alibi had been rejected by a jury and had not been supported in subsequent enquiry. It was also a firm rule that an alibi sworn but not proved in court could not subsequently be accepted merely by dint of its energetic repetition. The brothers might well have been over-active in the forest by night. The forest keeper had claimed to identify Benjamin as one of his assailants (although it was dark). Two other men might have been bribed to accept responsibility for the misdemeanour of assault (with its lighter prison sentence). But the case still stank. The gun was not stolen; a conviction for highway robbery should not have stood. The conviction was an exemplary attack on local roughs, as the judge admitted. Nothing contradicted Orme's claim that the conviction rested on the constable's malicious character evidence against Ellis.

How and why the benevolent Lichfield got involved in the case is not easily answered. His manner of writing shows that he was not a Quaker. He is not to be found on the Punishment Society's subscription lists. Perhaps he had some personal connection with the locality. Our best assumption is that the intervention of a commercial or professional gentleman in a case like this expressed a combination of disinterested benevolence and a dislike of the landed or forest interests which loomed behind the prosecution—or a mistrust of justice *tout court* as its malfunctionings were reported to him.

Ellis was not hanged. He was sent to the prison hulks instead. Two years later his illiterate mother employed a Barking freeholder to ask the Home Office to confirm the rumour that her son was now on the ship *Earl St Vincent* bound for Botany Bay. Nobody had told her. A clerk replied curtly that Ellis had volunteered to go to Australia, feeling 'it would be better for him'. His file was closed. Thus Ellis disappeared from his mother's knowing and from English history alike.

A second representative tale follows, more briefly. In Richmond a maidservant Sarah Wharmby had loyally attended her wealthy mistress for several years.[14] When the mistress died in 1824, the servant was prosecuted by the lady's family for stealing a quilt and other articles. After a speedy hearing at the Surrey assizes, Best J. (him again) recorded the death sentence against her (mitigating it, like Ellis's, to transportation). Then the appeals began, as they invariably did in such cases. The mistress's lifelong female friend protested to Peel that it would have been in character for the dying woman to have given some of her possessions to her servant and not to have told the rest of the family about the gift; anyway, even if this had not happened, the sentence was excessive. A gentleman living in St James supported this view, and so did the Revd Brooke Boothby, rector of the parish in which Sarah Wharmby's father lived. He wrote to the home secretary on intimate terms ('Dear Peel'), pressing him actually to *read* the petitions he enclosed, implying that he might otherwise not. Boothby also wrote to the king's favourite, Sir William Knighton, for more immediate access to the king, to say that he had himself once employed Sarah Wharmby and found her full of 'zeal and affectionate solicitude'. The evidence against her was only circumstantial. Moreover, 'it is a severe punishment for a first offence; the most hardened villains get off for much less'. He next wrote a petition for Sarah's sister to sign. This was full of 'heartfelt grief', 'tears and sighs'. 'The law must convict,' it acknowledged, but was not 'endless banishment for the first crime she was ever accused of . . . a severe visitation upon a defenceless woman . . . a punishment which, in Your Majesty's Merciful Reign, is scarcely ever inflicted but upon the most hardened and dissolute of Your Majesty's subjects?' Boothby petitioned further. The father was dying of grief, he wrote, and the family was 'wretched at the hopeless doom'. Sarah Wharmby herself was 'naturally of a weak constitution; . . . she can have little hope but that the awful change of scene she must now contemplate will speedily close her natural life'; should the king heed these fervent prayers, 'her future life will be devoted in prayers for thankfulness to the Almighty, and gratitude to your Majesty'.

It was quite a barrage. The letters and petitions conveyed energy, concern, compassion, indignation; they appealed to justice. When Peel consulted the judge, however, Best replied with his curt charm. He had 'not the least doubt of her guilt', he wrote. It was 'a very bad offence'

[14] HO47.65/13.

for a servant to steal from her mistress. The jury had agreed. 'His Majesty in sparing her life has shown her as much mercy as she is entitled to.' Then his master-stroke: 'You will observe that only one of the witnesses to her character knew her before she went to Mrs Bayley's. I suspected that she had not such a good character before.' On this suspicion, Peel refused the plea for mitigation. When a year later Boothby asked Peel if the term of transportation could be shortened, Sarah Wharmby's father having died of grief in the meanwhile, Peel was advised that 'she sailed for Vandiemansland in September 1824, and nothing has since been heard of her'. 'Have this copied and forwarded to Mr Boothby,' Peel scribbled on the memo, closing the case. So Sarah Wharmby disappeared from English history too.

3. *Mercy, justice, and the lawyers*

The opinion embedded in mercy campaigns like these was expressed by all manner of people in the 1820s. But a good part of it was concentrated within well-informed constituencies. The most visible were lawyers, or at least men with legal training. It was the rise of the lawyers that underpinned the acuity of the more ambitious appeals in the early nineteenth century, just as it had underpinned improvements in evidential rules in court and the most informed criticisms of legal process since the 1770s. In these decades, lawyers' influence, professionalism, and ambition were increasing apace. While the first edition of Peake's *Compendium of the law of evidence* in 1801 had numbered 190 pages, the fifth edition of 1822 had 426, plus appendices. 'When the author first turned his attention to the law of evidence, he was treading on almost new and unbroken ground,' Peake wrote in 1822. The multiplication of law books thereafter was such that by 1840 a barrister observed that if things went on like this, 'he should be pointed out as the only man in Lincoln's Inn, who had not written a book'.[15]

Lawyers were at least as important as radicals and Quakers in undermining the credibility of the *ancien régime*'s judicial processes and penalties—more so, probably, though the constituencies overlapped. They deserve larger credit as agents of change than that accorded them so far.[16] They deserve it in their role as appellants too, whether in their

[15] [A. Polson], *Law and lawyers, or Sketches and illustrations of legal history and biography* (2 vols., 1840), ii. 53.

[16] We still await a solid study of lawyers' critiques of the law across these crucial decades which does not concentrate narrowly on the great names of Bentham etc. Only Hay and Snyder have noted 'the creation of a public, *professional* critique, by young barristers writing in the new professional journals . . .

clients' names or in their own confrontations with judges and home secretaries. We cannot prove that they were more successful in winning reprieves or mitigations than other people were. But more than any other single group they challenged the law in its own language. Appeals with lawyers behind them were the most difficult for executive and judicial élites to shrug off.

The half-century before 1820 had witnessed an explosion in the numbers of barristers and attorneys.[17] Prospering from the growth of commerce, they multiplied most on the northern, Oxford, and western circuits, where textiles, mining, pottery, ironwork, and shipping created profitable business. (In the 1820s the most ambitious petitions seem to have come from the same circuits, though this is only an impression.) But not all barristers trained in the later eighteenth or early nineteenth century intended to practise for long or at all. Many entered the Inns in pursuit of a wide education, often self-administered, since legal training was only slowly systematized. From among such people attacks on the irrationality of the old legal regime came energetically in the 1770s and 1780s, sustained subsequently by Romilly and Mackintosh in parliament and Basil Montagu outside it (lawyers all), then sustained again by those who wrote in the professional journals of the 1820s and testified to the parliamentary committees of 1819 and the 1830s.

Most practising barristers were as self-serving then as they are now. Advancement depended on keeping on good terms with judges, and habituation to the law in general and to gallows business in particular did serious damage to both the personality and the imagination. 'Death makes as little impression upon the minds of those who are occupied in the profession of the law as it does in an army,' Boswell noticed.[18] After one capital conviction the young journalist William Jerdan was astonished when the barristers spent a convivial evening drinking together, while he could think only of 'the miserable being in the condemned cell'; Serjeant Ballantine recorded the same scene and had similar feelings about it.[19] Macaulay cared little about the fate of the 'stout hand-

interested in the professional logic of replacing genteel laymen by qualified lawyers and often allied with reform attacks on aristocratic and gentry government': D. Hay and F. Snyder (eds.), *Policing and prosecution in Britain, 1750–1850* (Oxford, 1989), 301–40, 384–8.

[17] For occupational information, see D. Duman, 'The English Bar in the Georgian era', in W. Prest (ed.), *Lawyers in early modern Europe and America* (1981); D. Duman, 'Pathway to professionalism: The English Bar in the 18th and 19th centuries', *Journal of Social History*, 13 (1980), 619–21; D. Duman, *The judicial bench in England, 1727–1875: The reshaping of a professional élite* (1982); B. Abel-Smith and R. Stevens, *Lawyers and the courts: A sociological study of the English legal system, 1750–1965* (1967), 55–6.

[18] W. K. Wimsatt and F. A. Pottle (eds.), *Boswell for the defence, 1769–1774* (1960), 266.

[19] W. Jerdan, *Autobiography* (4 vols., 1852–3), i. 99–100; W. Ballantine, *Some experiences of a barrister's life* (5th edn., 1882), 65.

some swaggering' footpad at whose trial he served in 1827. Ladies liked him, he wrote breezily to his sister, 'but in spite of their prettiness and of their pity I would not give two pence for the life of this Adonis if he should be convicted. Love to every body at home. Ever your affectionate Brother, T B M.'[20]

But there is always a bright side. There were some barristers in the 1820s who would be moved to indignation against casual judges, the absence of defence counsel, or the irrationalities of the appeal process. Forged in long practice, theirs were pragmatic rather than theoretical critiques of the law, but all the better for that. A case tried at the 1827 winter gaol delivery in Essex gives astonishing evidence of what it was possible for barristers to *know* by now, and to say privately to their friends. It also exposes a source of the emotional energy which fuelled their pressure for procedural reform in these decades. Day after day on circuit they had to live with bullying judges in their sight-lines, and sometimes even barristers were appalled by the view.

One William Leggatt, a village butcher, was sentenced by Park J. to seven years' transportation for the theft of some pigs.[21] Immediately after the trial Leggatt's defence counsel, John Jessop, sprang into action, high dudgeon his name. He collected affidavits testifying to Leggatt's probity as a tradesman, husband, and parent; then he got some fifty farmers and shopkeepers to petition that they had known Leggatt for years and had done business with him to the tune of hundreds of pounds. But his most startling step was to send the home secretary a list of charges against Park's conduct of the case, and they were of the gravest kind. The conviction, he wrote, was 'produced by the partial summing up of the presiding judge, who throughout the trial appeared to me to have been impressed with the most unfavourable, erroneous and prejudiced opinion of the prisoner's conduct and character, as well as of that of his principle witness, Sarah Cooper'. The prosecution was based on the dubious identification of the prisoner by a local horse-doctor who had seen a man driving the stolen pigs at night and whose evidence in court had been contradictory. 'Mr Justice Park treated these contradictions as very unimportant . . . [even though] the whole strength of the case rested upon the consistency and accuracy of a solitary witness.' Park had further ignored the evidence of the key alibi. Sarah Cooper had spent the night at Leggatt's house visiting his wife and had sworn that Leggatt went to bed there at nightfall:

[20] T. Pinney (ed.), *Selected letters of Thomas Babington Macaulay* (Cambridge, 1982), 35.

[21] HO47.73/14.

Her evidence [had] . . . all the characteristics of truth, and her manner had nothing of embarrassment beyond what might have been expected from a woman in humble station attacked (and that in no very measured phrase and manner) by a practised advocate and a judge, whose sneering and sarcastic mode of questioning her proved sufficiently how little he believed the truth of her story.

Taking their cue from Park, Jessop went on, the 'jury did not hesitate, nor deliberate, *for one single minute*, but immediately found the prisoner guilty'. 'I would humbly ask your Lordship', Jessop addressed the home secretary (Lansdowne) dangerously, 'whether any human being, Mr Justice Park excepted, *could* consider this more than a doubtful case':

My object is to assist an unfortunate person whom in my conscience I believe to be innocent, and to have been most unrighteously convicted and sentenced. . . . I shall therefore say nothing further than this, that having been at the bar upwards of 26 years, and extensively employed in the courts of criminal justice, *I have never before witnessed a trial*, a conviction or a sentence, which caused me so much pain, or which I reflect on with such entire dissatisfaction.

Jessop was not done yet. Invoking for comparison the rules of civil procedure, in which 'the rights and property of individuals are guarded with the most cautious anxiety; and any error in law or any misdirection on the part of the judge, as well as any verdict of the jury, contrary to the evidence, are grounds for applying for a new trial,' he went on to denounce the current mercy system itself:

There is no court in which the mistakes in misapprehending facts or statements; or the prejudices, the partialities, or the erroneous impressions, of a judge; or the finding without, or against evidence, of a stupid, an obstinate, or an unintelligent jury can be corrected or revised. . . . [A] prisoner once convicted loses liberty, life, fortune and character without appeal, except to the official quarters to which I am now respectfully addressing myself; and even then, though perhaps the two former may be restored to him (even against the fearful opposition of the convicting judge's notes), yet *then* he is *pardoned* only, not *acquitted of guilt*, and released after a long association with felons and outcasts, degraded in his own estimation and that of those who once esteemed him . . . and leaving behind him the recollection of his disgrace and shame as the inheritance of his wife and offspring. In this respect, my Lord, I believe the criminal jurisprudence of England differs from that of every other civilised state in Europe; and it is a defect in its administration to remedy which would entitle a British statesman to the lasting gratitude of his country.

Jessop was venting a critique of the ways in which the prerogative was administered which became increasingly audible over the next decade, to which it took the state three-quarters of a century to respond.

Park's angry response to this attack is a rare document in the appeal archive, the defence of a judge put under judgement, it too a marvel of its kind. So little prejudiced was he against the prisoner, he fumed in reply to the home secretary, that he actually put off the trial for a day to give the defence witnesses time to arrive. As for the impartiality of his summing-up—it was clear in his verbatim notes which he enclosed: 'this is my usual mode of summing up, and I am not conscious of having varied from it.' An alibi was always strictly examined, and defence counsel did not object at the time. The case was fully left as a *fact* to the jury. As for his sneering at the witnesses, that was 'wickedly stated': 'If I did so, it is what I never did before during the twelve years I have been an English judge; and it is a thing so abhorrent to my nature, that I am astonished how such an assertion could be made by a person who, I suppose, claims for himself the character of a man of veracity.' He inveighed against 'the unbecoming, indecent and unwarrantable manner in which a barrister has presumed to calumniate my character upon this trifling case' (the last adjective revealing). Then the disdain deepened wonderfully: 'The absurdity of supposing that a judge could have adopted a prejudice against a *pig stealer* and *his witnesses*, whom he never saw or heard of before, is too gross to be conceived by any but a madman.' The imputation was so insulting 'that nothing but my good nature, and the consideration I feel to be due to my own character . . . prevented me from laying [it] before my brethren in the next term'.

Yet for all this blustering, a hit had been scored, as the home secretary's trimming response to it revealed. Lansdowne scribbled on the back of the dossier: 'Not to be transported but sent to the penitentiary.' This outcome confirmed the force of Jessop's larger charge, needless to say. But it was the best that could be hoped for. The appeal could be counted a success.

Even if Jessop was a rare bird among barristers, his indignation was of its time, symptomatic of the profession's rising confidence and expectations, as well as of mounting unease.

4. *Attorneys ascendant*

The role of attorneys in mobilizing opinion on harsh punishments and rough justice was greater than that of barristers. Attorneys had closer dealings with the accused, their families, or their patrons than barristers did. Their growing importance in shaping appeals followed naturally on their increasing importance by the later eighteenth century in committal hearings as well.[22] They also had less to lose in the way of judicial favour. Indeed by sticking their necks out they had much to gain. When you were as low as an attorney, the only possible movement was up.

'The caterpillars of the nation', attorneys were called in the seventeenth century. Although with the growth of business and agricultural wealth they were putting more fingers in great men's affairs, their repute only slowly improved. Tom Clarke in Smollett's *Sir Launcelot Greaves* 'never owned himself an attorney without blushing'. Barristers excluded attorneys from the Inns and fined themselves in their assize messes if ever they made the mistake of 'shaking hands with an attorney, drinking tea with his wife, dancing with his daughter, calling him in open court "a highly respectable and worthy individual", &c.' Garrow J. hated attorneys. Sir Vicary Gibbs 'punished one of the tribe by boxing his ears in open court'. Abbott CJ's 'dislike to an attorney amounted almost to an aversion'.[23] In 1813 counsel queried a witness's evidence in court simply because he *was* an attorney. In 1822 it counted against a condemned rapist that his attorney and chief supporter was 'a very shabby fellow'.[24] In his *Treatise on the law of attorneys, solicitors and agents* (1825) Robert Maugham admitted that 'pettifoggers in the law' were as ubiquitous as quacks in medicine, charged 'not merely with derelictions of moral duty, but with a want of knowledge and erudition'. There were too many attorneys too. Maugham estimated that there were 8,000, with 2,400 in London, excluding numberless uncertificated practitioners.

Like medical men at this time, then, attorneys had every interest in displaying expertise and claiming status. There had been earlier efforts at this,[25] but it was in the 1820s that they entered that uneasy territory known to all ascendant professions in which standing was systematically claimed. A recent Act entitled Oxford, Cambridge, and Dublin graduates to be admitted to service after three years, and in London a professional institution to control access was set up. Maugham's *Treatise* was itself a turning-point in professionalization. Whereas Blackstone had given little more than a page to the attorney's duties, Maugham's was the first to invite them into 'that station in the community . . . to be removed from which would be both degradation and disgrace'.[26] Many

22 Beattie, pp. 276–8.

23 [Polson], *Law and lawyers*, i. 137, 358–9, ii. 305. For barristers' fines when over-nice to attorneys, see J. Arnould, *Memoir of Thomas, first Lord Denman* (2 vols., 1873), i. 65.

24 HO47.64/15.

25 R. Robson, *The attorney in eighteenth-century England* (Cambridge, 1959), ch. 3.

26 R. Maugham, *A treatise on the law of attorneys, solicitors, and agents* (1825), pp. iii, xii–xvii, 23.

pettifoggers continued to extort their miserable shillings from poor petitioners, debtors, bankrupts, and litigants. But by the 1820s better-established attorneys were probably giving more acute advice to petitioners who could pay than would have been available half a century before. Some were powerful voices in their own right as they campaigned against rough justice, even if self-publicity was seldom remote from their minds.

The prime example is inescapably that of the London attorney James Harmer, prototype of Jaggers in *Great Expectations*. Born the son of a Spitalfields weaver in 1777, Harmer contrived to set up as an attorney on his own account by 1799; when he retired, his practice was said to be worth £4,000 a year, his estate much more. That his 'appearance indicated good living and good nature' was not unrelated to the profitability of clients like Lord Hertford, the libertine upon whom the unspeakable Lord Steyn was modelled in *Vanity Fair*.[27] It also came from his management of famous criminal defences with money behind them: Fauntleroy's for example, and that of Hunt, the murderer Thurtell's co-conspirator.

He made enemies in this work. An ex-sheriff complained to Peel that Harmer would work more happily for a man guilty of forging £170,000 than for any of those whom he, the sheriff, had 'attended to the gallows, for passing a one pound note . . . to purchase the necessaries of life for a starving family'. All Harmer wanted was 'to make the Bank of England smart'.[28] But this was unfair. While working for Fauntleroy, Harmer also worked for the wife-batterer Abbott, a poor man, as we have seen. In fact he 'possessed amongst the classes whose natural destination was the Old Bailey, an immense reputation'.[29] He was alone among attorneys in having published three exposures of wrongful conviction and condemnation to death. He helped to break open the thief-taking scandal in 1816, and to mount the campaign against the corrupt effects of offering rewards for arrests and prosecutions. This drew him into contact with William Hone and City reformers like Matthew Wood or Grey Bennet.[30] He also acted as the Peterloo radical Samuel Bamford's attorney in 1819, and for the Cato Street conspirators in 1820. He gave evidence against the capital code to both the 1819 select

[27] Hertford papers, BL Eg. 3261, fos. 153–60, 170, 174, 219.

[28] J. W. Parkins to Peel, HO17.87/I (Fauntleroy dossier, Qk42): Parkins disliked Harmer because he, Parkins, was a victim of Fauntleroy's frauds: H. Bleackley, *Trial of Henry Fauntleroy* (1924), 17.

[29] Ballantine, *Experiences*, 83.

[30] J. R. Dinwiddy, 'Robert Waithman and the revival of radicalism in the City of London, 1795–1818', in his *Radicalism and reform in Britain, 1780–1850* (1992).

committee and the 1836 royal commission on the criminal law. The 1819 committee pointed out that Harmer had dealt with more than 2,000 capital prisoners and that his arguments against the death penalty were authoritative. An alderman in 1833 and a sheriff in 1834, his radical instincts were attested in his lengthy proprietorship of the *Weekly Despatch*, its populism so mistrusted in the City that in 1840 it cost him the mayoralty.[31] In all these commitments Harmer shared kinship with men like Hone or Place as one of those succeeding *menu peuple* widely encountered in these decades, men hatched in giddy radical days who never quite betrayed their origins or political training as they rose.

In tense years campaigns against injustice could always cause their authors trouble. Harmer usually guarded his language more carefully than oddballs like Lofft or propagandists like Hone did, but the 89-page book he published at his own expense to attack the sentencing and hangings of Holloway and Haggerty in 1807 pulled no punches. The criminal law 'affords much less security to life than the rest of our laws do to the most insignificant property', he wrote. He later claimed that the book earned threats of expulsion from the attorneys' roll, and it is obvious why it did. The book was the first to appeal against a dubious trial and sentence by speaking directly to a general readership about the 'fairness' of legal process rather than deferentially to the authorities of state. It antedated Hone's book on Eliza Fenning in this.[32] Hone's success in 1815 doubtless spurred Harmer to mount further campaigns, two of which resulted in books in 1819 and 1825 respectively. But he continued to pay a price for his temerity. In 1825 his defence of another condemned man, Harris, was (he said) attributed to his 'passion for notoriety', his 'spirit of opposition', and his habit of 'obtruding a speculative opinion on the world'. He proclaimed loftier motives, however: 'Humble as my situation is in my profession, I am bound to use my utmost exertions, however feeble they may be, to see the law fairly and justly administered.'[33] It would be churlish to doubt him. He saved at least one life and probably many more.

[31] For Harmer's career, see entry in *DNB*; S. Bamford, *Passages in the life of a radical* (orig. pub. 1884; Oxford, 1984), 208, 212, 243, 250; A. Somerville, *Public and personal affairs . . . an exposure of treacherous patriots and drunken lawyers . . . connected with Alderman Harmer and the* Weekly Despatch (1839); *Mr Alderman Harmer and Mr Walter of* The Times: *From the* Morning Advertiser *of 25 September 1840*.

[32] J. Harmer, *The murder of Mr Steele: Documents and observations, tending to shew a probability of the innocence of John Holloway and Owen Haggerty, who were executed on Monday the 23rd of February 1807, as the murderers of the above gentleman* (1807), 1. Harmer's next book on miscarriages of justice was his *Account of the . . . case of George Mathews, who was capitally convicted . . . and afterwards . . . pardoned . . . containing Mr Harmer's letter, etc.* (1819).

[33] J. Harmer, *The case of Edward Harris, who was executed at Newgate for robbing and ill-treating Sarah Drew, investigated, and facts and arguments adduced, to prove his innocence* (1825), 82, 86.

Harmer's efforts in 1824–5 to save Edward Harris from execution for robbing a woman in London Fields was his best-publicized campaign, and it is worth dwelling on for a moment before we assess the impact of such things on 'opinion'. As Harmer aired the injustices of the Harris case in the *Weekly Despatch*, he pulled in other benevolent activists. The appeal dossier reveals a Mr Bayley of the 'Society for mitigating punishments' who interviewed the committing magistrate on Harris's behalf. It also reveals a Mr Brogden of the City's Common Council, an evangelical to whom it mattered that in his last hours Harris 'prayed to the Almighty for forgiveness of all his sins as he is innocent of this particular one laid to his charge'. In begging an interview with Peel, Brogden announced that his only object was 'to forward the ends of justice', which was probably right.[34] Brogden was Harmer's chief helper in unravelling motive and counter-motive, collecting affidavits, and establishing highly telling grounds for doubt. The case is one of the most solidly documented of the 1820s. Harmer's book was 112 pages long; it is supplemented by a huge dossier which is still in the appeal archive. The size of the dossier itself measures the campaign's impact. It also measures the Home Office's increasing need to allow time for investigation whenever a fuss like this was made.

Harris's was a grimy tale too convoluted to be detailed here. A known thief, Harris was convicted of capital assault and theft on the basis of an uncertain identification by the attacked woman. The grounds of the appeal were that the identification was false and Harris's alibis true. As he got to know Harris in the condemned cell, Harmer's belief in his innocence grew. First, Harris was a thief but never a violent one. 'I will take care never to go above the shoulders,' he boasted—meaning that he would never commit crimes which might hang him. Secondly, Harmer was certain that Harris was the victim of a conspiracy on the part of the officers of the Worship Street police office which arranged the prosecution: 'when a man of loose character has become obnoxious to the police, and it is desired to excite a strong feeling against him, convenient persons are always at hand, in what are called "penny-a-line" men', to invent a case against him. Most of Harmer's public campaigns concerned malicious prosecutions and police corruption (as well they might, recent research has shown[35]). In this case he

[34] HO17.107/II (dossier Vk1).

[35] 'People from all social classes turned to the criminal law . . . to further personal conflicts, some of them only once, some of them habitually'—so often as to 'make it virtually impossible to be sure that even the plainest of cases reported in trial reports, where the evidence appeared clear and the

marshalled telling evidence of conspiratorial intent in police and magisterial investigation, naming names freely. He bombarded the king and Peel not only with petitions in Harris's name and with maps depicting the scene of the crime, but also with a small mountain of affidavits to prove the Worship Street officers' malignity.

It was all in vain, however. The barrage forced Peel to allow some time before Harris was hanged; but that was all. Four months after Harris's conviction the King in Council turned the appeal down. Harmer went to Peel's house. Peel was 'evidently displeased by my visit; and, on my entering, put the question to me "whether I could be justified in importuning him again on this subject, and distressing his feelings unnecessarily". I answered that the urgency of the case must be my apology.' Peel heard him out in ill grace and then was adamant that 'twelve men had decided on the case in the first instance; that great pains had subsequently been taken to investigate it; and that, as all the facts had been laid before His Majesty and the Privy Council, when the Recorder made his report, he could not think of interfering with their decision'. With Peel's feelings in no further danger of being distressed, Harris hanged on 22 February 1825.

Again, at this distance we can never be certain how justified appeals like this were. Benefactors who had axes to grind were sometimes taken in by the guiltiest of felons.[36] But Harmer had made out a case to be answered. A letter in the appeal dossier from Vansittart to Peel two days before the execution shows the narrow grounds on which the Council allowed Harris to hang. Vansittart felt that Harris was guilty because the young woman whose evidence convicted him was of good character; it was not '*likely*', therefore, that she 'should deliberately swear away the life of an innocent man without any necessity'. How likely it was that Harris had been framed might as easily be inferred from his last days in Newgate as chronicled by the ordinary:

> *16 February 1825*. Harris . . . said that he was innocent and that he should die declaring this, and that he was murdered. Wood exclaimed that he was innocent and should tell the people so, when he was brought out to suffer, and would die firm. Durham received the fatal intelligence in silence and with seeming resignation.

defendants at a complete loss for a defence, are not in fact successful conspiracies against innocent men and women': Hay and Snyder (eds.), *Policing and prosecution*, 47–9, 344, 380–4.

[36] Basil Montagu had a book printed in 1810 in which he pleaded the case of one Dennis Shiel on the grounds that key evidence had been unheard, only to stop publication at the last minute when he found that Shiel had suppressed a material fact and was rightly convicted: *The case of Dennis Shiel condemned to die and now sentenced to be transported for life to Botany Bay* (1810). The BL copy contains an MS note explaining the book's withdrawal.

18 Feb. [Harris] very determined in denying his guilt, nor could I fix his attention to the concerns of eternity, but he was constantly reverting to the hardship of his fate.

19 Feb. Harris at chapel this morning turned to the prisoners and protested his innocence most vehemently. . . .

20 Feb. [In the evening Harris was] more violent than ever.

21 Feb. [Harris] almost frantic upon the subject of his innocence. When I expressed a doubt whether I should administer the Sacrament to him or not, he showed great anxiety to partake—however he became calm in the course of this interview, and exhibited some degree of penitence for his sins generally.

22 Feb. [On execution day] Harris became much disturbed and very violent—from the time the sheriffs arrived to the very moment of his execution, he kept asservating his innocence and shrieking murder. Durham was very humble and sustained with religious hope through this shocking scene. When Harris came before the public, he was unable to make any speech as he had intended, and could only vociferate innocent, murder! The crowd was immense.[37]

Actually Harris said more than the Revd Cotton reported. On the scaffold he said to the sheriff: 'Indeed, sir, you are going to hang a man that is entirely innocent; what bad laws ours are, to hang an innocent man. . . . Is it not a shame to keep a man five months in gaol, and then bring him out, and hang him?'[38] The last letter in Harris's Home Office dossier is from the keeper of Newgate. Written to the Home Office on the day of the execution, it reported bleakly that Harris did not acknowledge his crime. He 'was in a very frantic state and kept crying out "murder" from the cells (and indeed on the platform)'. The authorities had to be informed on such things, though what they did with their knowledge we cannot imagine.

5. *Influences and effects?*

There is no way of proving that the growing number and ambition of mercy appeals by the 1820s had the slightest influence on the retreat from harsh punishment. Petitions did not express a popular rebellion against the law, and anyway those who challenged verdicts or sentences were an eccentric sample of the nation at large. The most numerous and interested appellants—felons, their families, and neighbours—counted for nothing in public debates, while petitions by professional or commercial men only rarely made a national splash. Policy-makers

[37] Newgate Visitors' Books, London Corporation RO, 209C.

[38] Harmer, *Edward Harris*, 101–2.

might also console themselves that appeals addressed the failures rather than the successes of English justice, and since most verdicts and sentences were accepted passively, they could believe that they were largely just. In any case, shielded by Olympian privileges and power, élites were often blind to the scripts which historians, with hindsight, sometimes wish to read on the walls.

And yet something was shifting. There were symptoms of élite anxiety, even if little more. Individual members of the king's Council were becoming a little queasy about the ways in which they conducted their death-dealing business (Chapter 20), and Peel's holding action against root-and-branch reform had some defensive meanings (Chapter 21). Whig attacks added to a sense of absurdity as these agents pardoned increasing proportions of capital convicts while hanging ever more arbitrarily chosen handfuls. The petitions and the thickening dossiers accompanying them could only feed this unease. Even if only two or three innocent people were convicted or condemned each year, 'it is quite a sufficient reason why the law should be altered', people like Sydney Smith were saying. Writing in 1826, Smith was convinced—indeed, knew—that many more people were wrongly judged than that.[39] We are after all in quest of those hazy conjunctions where attitudes were subliminally as well as consciously shaped and new ways of relating to the social world were developed, and in the swelling tide of petitions there were scripts enough for those who would read them. By the 1820s a home secretary must sense that opinion would have to be heeded if the images of justice were to be kept well repaired.

Campaigns against harsh justice had obvious impact on the 'public' too, but how lasting and of what kind? It is not always easy to reply. Eliza Fenning's case in 1815 raised a famous ruckus, and in Chapter 13 we have affirmed its residues. But most cases passed without such notice. Wronged people carried resentments back to their home communities and they might fester there; stories like Ellis's would cause local sensations. But even in plebeian London residues are not easily proven. Harris's misfortunes excited much 'contrariety of opinion, and considerable alehouse discussion': but this sounds like gossip rather than a raising of consciousness.[40] Middle-class attitudes were distorted by party allegiances. Metropolitan campaigns could seldom be dissociated from City politics when magistrates, the recorder of London, or Common Councilmen were involved. Press comment divided on party

[39] *Works* (2 vols., 1856), ii. 117.

[40] Harmer, *Edward Harris*, 6–7, 87, 98, 105.

lines. In Harris's case *The Times* attacked Harmer for becoming a friend to 'this supposed martyr' of the law. The occasional Quaker might lend some support, though not overwhelmingly when all is said and done.

But if few appeal campaigns had direct impact on policy-makers or wider opinion, we may still feel that effects were delivered. That English criminal law was capable not only of cruelty but also of injustice had entered the realm of the thinkable, the sayable, and even the *known*. These things, hardly said at all in the 1770s, were axiomatic in some circles after 1815, and even Peel's friend Sydney Smith could say them a decade later. The age of the steam-intellect society was dawning. There was that 'beastly education' to which Wakefield referred—that 'hankering after new-fangled laws, the growing inquisitiveness of the people, the obstinancy of the quakers and the increase of newspapers and books': these were beginning to bite.[41] And so old mentalities shifted a little, at least among those ready to engage in others' plights.

Our best confirmation of residues comes from later debates. What was said in the whig-governed 1830s could not have been said without reference to earlier mercy campaigns. One example will do. Attorney-General Pollock told the royal commission on criminal law in 1836 how he remembered the exertions to save men from execution which were made in 1828 by the City attorney E. A. Wilde, then sheriff of London. He found it a 'very appalling fact' that Wilde saved so many lives in a single year. It convinced Pollock that 'some legal constitutional mode ought to be adopted by which errors and mistakes, from whatever source arising, should be corrected in criminal trials'.

Impressed, the commissioners got Wilde to tell them his stories. In 1828, he said, he won reprieves for men condemned for forgery and buggery respectively (the last probably the case which this chapter has already noted). A third case concerned two illiterate labourers, Anderson and Morris, condemned at the Old Bailey in 1828 for robbery with violence. To Wilde's satisfaction both were innocent. Here was another convoluted tale of perjured evidence and malicious prosecution, eloquent on magistrates' readiness to commit rough men to trial on the prosecutor's merest say-so and on Recorder Knowlys's conduct of the trial in slapdash spirit. The case survives copiously in the appeal papers. These show that Wilde's labours were as impressive as Harmer's for Harris. They paid off too; but only just: Peel was never one to avoid cliff-hangers if he could. 'After several communications

[41] E. G. Wakefield, *The hangman and the judge, or A letter from Jack Ketch to Mr Justice Alderson* (1833).

with Sir Robert Peel,' Wilde said, 'but not until half-past eleven o'clock on the night before they were to have been hanged . . . was I able to procure a reprieve.' Nor did Peel free them from prison, however innocent he must tacitly admit them to be. They stayed in prison for two more years, since no home secretary could allow egg to stick visibly on the law's dignified visage.[42]

Experiences like this did leave residues, and the royal commissioners admitted their meaning. They recorded Wilde's protest about the chancy ways in which miscarriages of justice came to light:

> From the state of destitution and ignorance in which prisoners generally are, and from want of a court with proper officers, the prisoners have not the means of bringing their cases in a proper state for consideration. . . . Practically speaking, unless sheriffs, magistrates, and governors of prisons and their officers, or charitable individuals, exert themselves in collecting and authenticating the grounds upon which a prisoner seeks for the relief, and subsequently take the trouble of communicating with the Secretary of State, the prisoners have little or no chance of a successful result to their application.

So long as petitions were 'rather an appeal to mercy, than the assertion of a right', Wilde added, he was 'perfectly satisfied that many persons have suffered punishments where they have been positively innocent of the crime'. A more certain procedure was 'indispensably called for': 'From the want of it the sheriffs and the officers of the prison, who, of course, are the only persons in constant communication with the prisoners, are often placed in a most painful situation in having to judge how far, consistently with the discharge of their own duties, they ought to interfere.' Wilde was not prepared to recommend a court of appeal, but he was certain that time should be allowed for serious investigation. 'As the law at present stands, a man may be apprehended in London for a capital offence, tried, convicted and executed in less than a week,' and although this celerity might induce greater terror in the criminal classes, 'you have no right to make the prisoner a victim to effect a collateral object. . . . The prisoner is entitled to every doubt, and all the mercy that can be extended to the particular case.' It confirms the growing influence of these City attorneys that James Harmer was interrogated by the commission to similar effect. He deserves the last word:

[42] *RC on criminal law, 2nd report* (PP 1836, xxxvi), 79, 99–103; HO17.45 (dossiers Gn1, Gn21, and Gn28). Peel's scribbles on these dossiers ran thus: 'Do not let them be transported and give me this two months hence, RP: Sep. 12 1828'; 'Nov. 26: Let this case be brought under the consideration of the Secretary of State when the Prisoners have been in confinement two years from the date of their sentence, R Peel.' Only with the whigs in office does this note appear: 'Free pardon prepd. 1 June 1830.'

Have you had experience of any cases where persons you believed innocent were convicted, and where you had reason to suppose that had they been defended by Counsel, with power to address the Jury, a different result would have taken place?—*I think so. I have known certainly instances in which I had not a particle of doubt the men were innocent, who have been convicted and executed. . . .* Were those persons capable of defending themselves?—*Certainly not. In one of the cases to which I alluded, I was quite satisfied the offence was never committed; and if there had been any body to remark upon the evidence, it was impossible that he could have been convicted.*[43]

Did much change? Apparently not. In 1866 Sir Fitzroy Kelly, recent attorney-general and soon lord chief baron, said that it was not 'reasonably to be doubted that in many instances innocent men have been capitally convicted, and in certain numbers of instances, few of course, but yet formidable numbers, have been actually executed'.[44] And in 1868 the abolitionist Charles Gilpin reminded parliament of two Italians recently awaiting execution who had been proved innocent in the nick of time.[45] It was scant consolation that the number of innocents who in Paley's words might 'be considered as falling for their country' were now confined to those charged with murder. But although the appeal court did not come in the nineteenth century, Ewart's Act of 1836 did give defence counsels the right to sum up before juries.

We are still not done with 'opinion'. The microhistory in the next part of the book shows how attitudes to particular cases were embedded in local understandings and communal relationships that have not been attended to so far. But enough has been said to reveal that the changes in perception and feeling of concern to us had many roots: the growth of a reading public, cheaper print, more competitive newspapers; the energy of post-war radicalism, fuelled by its famous victories over gagging laws; the linking of radical campaigning to exposures of injustice; the exposures of corrupt policing; the heightening of awareness achieved by Quakers and radical whigs; the increasing assertiveness of lawyers. Admittedly until these legacies were carried into parliament by more confidently independent MPs after 1832, we are not dealing with

[43] *RC on criminal law, 2nd report*, 4.

[44] RC 1866: 1054–5, 1063–4. Kelly referred to an investigation which 'established satisfactorily to those who investigated the matter, and in most of the cases to the satisfaction of the advisers of the Crown', that 22 capital convictions could be identified between 1802 and 1840, 7 of which were executed, in which the innocence of the convicts was established. 'There is presumptive ground for believing', he added, 'that [other] innocent persons have suffered death for want of having influential or wealthy friends to procure an investigation of their case.'

[45] PD3 cxci (1868), c. 1037.

the opening of floodgates. Still, these several kinds of campaigning contributed to that larger mobilization of feeling and indignation to which the whig repeal of the capital statutes in the 1830s gave some satisfaction. Later abolitionists built on these foundations.

PART V

MERCY, JUSTICE, AND COMMUNITY

CHAPTER 17

THE RAPE OF ELIZABETH CURETON

A Microhistory

1. *Microhistories*

THIS CHAPTER ADDRESSES A DIFFERENT KIND OF OPINION ABOUT capital justice from those surveyed so far. It also uses a different method. It offers a microhistory, rooted in neighbourhood and community, playing to a sense of place, and of incident—to the imagination, in short.

On a June evening in 1829 Elizabeth Cureton of Coalbrookdale in Shropshire let John Noden into her parents' cottage after her parents had gone to bed. The couple then had sex on the parlour floor. A couple of days later she told the constable that he had raped her. This at once enmeshed both of them in a legal process which stripped her of privacy and reputation and threatened him with death. She prosecuted him; and at the Shrewsbury assizes in August Noden was tried for rape, convicted, and sentenced to hang. Their neighbours and betters became excited commentators upon the drama. Her sexual character was stripped bare in accusations and gossip as the community appealed to save Noden's life at her reputation's expense.[1]

I have already said in the Preface that in terms of its bulky documentation this was one of the biggest appeal campaigns in the 1820s. But it was not of importance otherwise. Coalbrookdale was no Montaillou and Noden no Martin Guerre. No new law was made in the case. And it concerned only one among the 18,674 other felonies tried in 1829. So

[1] The story is based on the trial transcript, petitions, affidavits, and correspondence in HO47.75; other unacknowledged detail from *Shrewsbury Chronicle and North Wales Advertiser*, 21 Aug., *Salopian Journal*, 19 Aug., *The Times*, 19 Aug. 1829.

what might stories like this achieve? One story 'is all very well, bristling, jumping with novelty', a sceptic insists; but ten or twelve stories later, 'recurrent factors begin to emerge': it is these that are the stuff of serious history. They should be organized into series, 'to avoid redundancies, repetition, the *déjà vu*'. The rest is merest anecdote.[2]

One reply is that all stories are *sui generis*, and that although microhistories must relate to the preoccupations of mainstream historiography if they are to bite, the best assert a creative independence of academic ratiocinations. They may seek to recapture Cobb's 'sense of place', for example, or they may be read as novels are read, their appeal depending on a resonance and energy within them which may take the historian by surprise.[3] They put us in touch with the lived textures of past times, and helpfully too. The narrow universes in which most people experienced the exactions of power may often be better apprised in the microcosm than through aggregative analyses, the microcosm illuminating the universal.[4] It was no accident that the broader visions which sustain this book were hatched in my discovery of Cureton's story, as the Preface explains.

Most such studies have been based on court records, and it is obvious why. The trial of one or more of a community's members is one of the rare moments in its history likely to be copiously recorded. It entails a quest for hidden truths, when obscure people have to articulate motives, interests, and buried values and assumptions. This reliance is not without its dangers, however. Trial records are official and public texts. They construct accounts of behaviour in terms of legal criteria of guilt or innocence, or in terms intelligible within dominant value systems, or in terms that newspapers can report. They tell how the law deals with an event, and how spectators are meant to receive it as the state extends its definitions to social and geographical peripheries. They conceal as well as reveal.

The appeal papers from which the present story is taken reach further than this. They extend to the negotiations and attitudes which

[2] M. Vovelle, *Ideologies and mentalities*, trans. E. O'Flaherty (1990), 241–2.

[3] Cf. (among others) R. Cobb, *A sense of place* (London, 1975); E. L. Ladurie, *Montaillou: Cathars and Catholics in a French village, 1294–1324* (1978); C. Ginzburg, *The cheese and the worms: The cosmos of a sixteenth-century miller* (1980); R. Darnton, *The great cat massacre and other episodes in French cultural history* (1984); N. Z. Davis, *The return of Martin Guerre* (1985); G. Weber, *Giovanni and Lussanna: Love and marriage in Renaissance Florence* (Berkeley, Calif., 1986); G. Levi, *Inheriting power: The story of an exorcist* (Chicago and London, 1988); E. Muir and G. Ruggiero (eds.), *Microhistory and the lost peoples of Europe* (Baltimore, 1991).

[4] E. Muir, 'Observing trifles', in Muir and Ruggiero (eds.), *Microhistory*, 1–10; L. Stone, 'The revival of narrative', in his *The past and the present* (1981), 86.

determined protagonists' fates in their home communities. They reveal quite other understandings than those heard in court, rooted in customs and material lives to which judges were blind. They can take us 'behind the formal words of standard legal phraseology, to penetrate a language rich in hidden assumptions, and hinting at the elements of a popular collective morality, of an alternative system of justice . . . parallel to, and not necessarily hostile to, official morality, official religion, and the carefully defined rights and wrongs of the law'.[5] The stories told in this mode are of varying ambition and most are out-of-the-way. None is 'representative'. But then the most representative stance that people have always adopted towards the criminal law (when not ensnared in it) has been one of ignorance or indifference, and any story which gives the lie to this must break the rule. Such stories may still be emblematic. They expose fractured moments when people were in exceptional crisis, or observers were moved to exceptional passion. The English appeal archive is a major source for quests of this kind. What follows is one story among the thousands buried there.

Most simply, this is about one community's attitudes to a woman who challenged male assumptions about her body's accessibility and who paid a familiar not to say predictable price. It is a cruel story which feeds into the long history of collective connivance in male violence against women. Coalbrookdale disagreed with the court's verdict and the judge's condemnation of Noden, because it suspected that Elizabeth Cureton prosecuted him from malice. The law was more generous than this. Despite its enduring and notorious suspicion of the prosecutrix's motives in a rape trial, legal rulings insisted that 'particular evidence' of the woman's sexual reputation should not be given in court. Since the community thought otherwise, and was to that extent the more misogynist and patriarchal in its assumptions, it was the law, paradoxically, that proved to be Elizabeth Cureton's chief friend.

Secondly, the story says much about how communities negotiated with legal values which in contexts like these were at odds with their own understandings and knowledge. Although none of Noden's supporters imagined that they were part of a national debate, they exhibited a confident engagement with the law which was widespread by the 1820s, as we have seen. It was an integrative relationship that ensued by and large, of value to the state. Although central definitions of justice in England had been assimilated long before the nineteenth century, the

[5] 'A view on the street: Seduction and pregnancy in revolutionary Lyon', in Cobb, *Sense of place*, 124–5.

assimilation was still far from complete. Some communities kept the law at bay for years yet: remote coastal communities living off plundered shipwrecks, Celtic outposts where 'barbarous custom' still ruled, parishes where gleaning rights were defended, communities which refused to collaborate with magistrates.[6] Beyond these instances was the quieter dissent of more settled communities at odds with a legal system which failed to live up to its own promises and values. And then there were communities like this one, whose dissent drew on a local knowledge of protagonists in trials, sustained in webs of custom, neighbourliness, and kinship as understood by people who might never have travelled far from their parishes in their lives. In such places and contexts alternative conceptions of justice took shape beyond the justice of the courts and delicately subversive of it. 'Two conceptions of order'[7] still did gentle battle in many communities, the central one at odds with the local and the customary; so that appeals were negotiations between legal knowledge and communal knowledge, between judges and judged.

The 'opinion' we encounter here differed from that addressed hitherto. Noden was to hang, and of course this was an issue. But in this campaign for mercy, as in many others, there was no audible critique of punishment. That Coalbrookdale was a Quaker stronghold dominated by the iron-founding Darbys seems to have been by the way. Five Darbys had subscribed a meagre seven guineas to the Society for the Diffusion of Information about the Punishment of Death in 1810–11. And Madeley parish, of which Coalbrookdale was part, had been one of the sixty or so communities that had petitioned the Commons against the capital code in 1819—'sincerely regretting the frequency of capital punishment, believing that instead of impressing with salutary awe, it too often renders callous the hearts of surviving offenders'.[8] Yet whatever was thought privately, nobody spoke of the capital code in this case; and the Coalbrookdale Quakers were marginal players in the campaign. The audible issue was the right relations between men and women, to which the law failed to attend. Not all 'opinion' bore the progressive messages that we might most like to read. But enough now of commentary. Henceforth let the story speak for itself.

[6] The 1839 *SC on the rural police* (PP 1839, xix) is full of examples; see also P. King, 'Gleaners, farmers, and the failure of legal sanctions in England 1750–1850', *Past and Present*, 125 (1989), 146.

[7] K. Wrightson, 'Two conceptions of order: Justices, constables and jurymen in seventeenth-century England', in J. Brewer and J. Styles (eds.), *An ungovernable people: The English and their law in the seventeenth and eighteenth centuries* (1980).

[8] Ch. 15, above; *Journal of the House of Commons*, 74 (1819), 221.

2. *The crime*

When Elizabeth Cureton told the parish constable that John Noden had sexually assaulted her on the floor of her parents' parlour in Teakettle Row, she was aged 24, unmarried, and living with her parents. Noden was an odd-job man and wheelwright aged 27 who lodged across the road. On and off, he had been harassing her for five years. At the Shrewsbury summer assizes, in broad Shropshire accents, she gave this account of a courtship gone spectacularly wrong.

Noden had first flirted with her when she was 17, she said. She had rejected him when he 'acted imprudent'. Although the relationship never overcame these beginnings, he had persisted; and this past January he had gained her parents' permission to visit her 'on the footing of courting her in the way of marriage'. On the Sunday before the denouement the couple quarrelled. When Elizabeth refused him access to the cottage, Noden threw stones through the kitchen and parlour windows, making such 'a great rumpus' that her mother forbade him to call again. None the less, on the night of Wednesday 17 June she capitulated, and everything followed from that:

> My father and mother went to bed between 9 and 10. . . . When I got upstairs I heard a knock at the door—about half an hour after father and mother went upstairs. I opened the window of my bed room and asked who was there. The prisoner answered it is me. . . . I went downstairs and dressed. I opened the door. He asked me to go walking with him. He had a smock frock on in which he never came before. I refused to walk with him. He asked me how long father and mother had been gone to bed. I told him not long.
>
> He then shut the stairs passage door and forced me into the parlour. He caught hold of me and threw me down immediately. . . . I received a blow on the back part of my head which stunned me. He said he would have it dead or alive and I told him I would die first and I screamed out. He put his arm across my mouth.
>
> He then forcibly and against my will heaved up my cloaths and entered my body with his private parts. His trowsers were undone before he came into the house because he had no trouble. There was no time or opportunity for him to undo his dress; it must have been undone when he came into the room. I cannot tell how long he was upon me. I think a quarter of an hour or 20 minutes.
>
> I screamed out and made as much noise as I possibly could. . . . He told me he would have it dead or alive. He entered my body, got up and went out. My mother heard me scream and came down stairs. My father is rather hard of hearing. . . . She came down stairs as the prisoner ran out at the door. I told my mother all that had happened.

Sarah Cureton confirmed that she found her daughter on the parlour floor. 'Her cloaths were up and she was very ill. . . . She said Noden had used her very ill and she was undone for ever.' Mrs Cureton went to where Noden lived and asked him what he had been doing. 'He said "nothing", that was all.'

Noden's trial took six or seven hours in a three-day assize. At the end of it the jury found Noden guilty of rape and Sir John Vaughan sentenced him to hang. Vaughan was an undistinguished judge, ignorant of large tracts of the law and famous chiefly for his jocularity in court.[9] But he rose to this occasion with well-practised solemnity. Donning the black coif, he reminded Noden that he had been convicted of a foul and atrocious offence against the laws of God and man. If he were not subjected to the highest penalty, 'every woman and woman's child would be left a prey to men of such unbridled passion and lusts as you have proved yourself to be'. It was his duty to give Noden the heart-rending intelligence that he must expiate his offence with his life. He noted that the jury had recommended mercy, and he would forward that to the proper quarter; but he could hold out no hope of mercy whatever. Noden must prepare for the awful change he was about to undergo:

> That pardon you are denied on earth may be secured in heaven by sincere repentance and contrition; and if earthly judges are deaf to your prayers for mercy, there is a tribunal in heaven where it may yet be granted. Prepare therefore for death; repent, and your prayers will be accepted; for even at the eleventh hour there is mercy with your Saviour, to the worst of sinners.

He then told Noden that he was to be taken away to a place of execution and hanged by the neck until he was dead.

We know surprisingly little about Noden. We do not even know if he was literate. We do know that he had lived in the parish of Madeley all his life.[10] Of medium height and 'by no means . . . ill-looking', witnesses gave him the character of 'a sober, modest, well-behaved, orderly man', a hard worker, and kind to his parents. It counted a little in his favour that he had surrendered his bail voluntarily. It also counted in his favour that during his trial (the press reported) he watched Elizabeth with composure, apparently convinced that he had done no wrong. Beyond that his character remains hidden. We hear his voice once and loudly,

[9] E. Foss, *The judges of England* (1864), ix. 289–90.

[10] The son of Robert and Ann Noden, he was baptized at Madeley on 7 Feb. 1802 (Madeley registers). Variants of Noden's name (Roden, Nodin) were and are common around Madeley, but the marriage of Robert and Ann is not recorded in any Shropshire marriage register, and they were probably immigrants to the area. (For this and subsequent information on baptisms and births, I am indebted to Mr M. Lewis and Mr R. Gwynne, who are collating the registers.)

however, when Vaughan delivered his sentence. Weeping bitterly, he fell to his knees. 'Spare my life, my lord,' he exclaimed, 'oh spare my life! It's a wrong concern I've done, but spare my life.' As the judge placed the cap on his head, 'the cries and sobs of the prisoner were redoubled, and he continued to entreat mercy throughout the whole of his lordship's address'. He was led from the bar in a 'paroxysm of grief' to the condemned cell in Shrewsbury gaol, outside whose gates he must expect to hang.

He would have seen others hang there. Coalbrookdale people often walked the twelve miles to Shrewsbury to watch executions; the ironmasters let them have the day off.[11] Only last April he might have watched John Evans die for robbery and attempted murder. As Evans had ascended the scaffold, the newspaper reported, he had raised his eyes to the castle behind him, to the sun, and the sky above. He had trembled when he looked at the scaffold, passing his handkerchief over his face. The executioner could not get the noose over his ears, and had to try again. The chain had to be lowered; Evans was a short man. The crowd was large, and the boys of Shrewsbury School fought at their classroom windows opposite to watch him swing into space, kicking.[12]

How Elizabeth Cureton felt about Noden's sentence we cannot say. She recommended him to mercy, but had good cause to do so. Strong sentiments against her could be relied upon. Tormenting weeks of rough music and hangings in effigy of women who accused men of rape were known elsewhere. She would have been told that Coalbrookdale would mount an appeal. Its only grounds could be that she had lied in court or that she had asked for what she got.

The woman in such a case drew more attention than the man, so we know more about her than about Noden. Newspapers described her as a tall, robust, decent, clean young woman. She had 'nothing particularly marked about her countenance' but she was 'becomingly' and 'dashingly' dressed. In the all-male courtroom she gave intimate evidence against Noden 'with considerable excitement but without much embarrassment'. Very much a local girl, she was also well known. Her parents had married outside Shropshire but had baptized their first child Mary a dozen miles distant at Lilleshall in 1799. Their next four children were baptized in Madeley; two had died as infants in 1804. Elizabeth was the next born, in 1805, and Ann followed in 1808. Ann married a Dawley man when she was 18; Mary married the shoemaker Francis Franklyn at

[11] B. Trinder, *The industrial revolution in Shropshire* (1973), 366–7.

[12] *Shrewsbury Chronicles*, 10 Apr. 1829.

Wrockwardine when she was 17, but lived with him in Coalbrookdale by the time of this story.[13] In 1829 Elizabeth was the only daughter still unmarried and living with her parents.

Richard Cureton earned a fair wage as a moulder at the Coalbrookdale Company's ironworks, a few hundred yards from the cottage he rented from the Company in Teakettle Row. The family could afford to live respectably. The in-law Franklyns were respectable too. The community was a small and a close one. For Elizabeth none of this boded well. At home she was not going to escape the secondary trial inflicted on all women who challenged communal values as she did, and her harshest judges were to be those among whom she had grown up.

3. *The community*

Noden's and Elizabeth Cureton's story unfolded in a landscape which still survives nearly intact from their times. 'The most extraordinary place in the world', one observer called it in 1837. The Severn gorge and the dales feeding it from the north were anciently an area of miners and ironworkers, rich in coal, iron ore, clay, and limestone. In 1709 Abraham Darby I was the first in the world to smelt iron with coke (not two hundred yards from the Curetons' parlour), giving Coalbrookdale its claim to be a birthplace of the industrial revolution. In 1779 the world's first iron bridge was thrown across the Severn a mile away. By 1800 there were more blast furnaces on a half-dozen-mile axis with Coalbrookdale at its centre than in any similar space in Britain.[14]

Many visitors came to wonder at the Iron Bridge and usually made their way up from the Severn into the smoky horrors of Coalbrookdale itself. De Loutherbourg's painting *Coalbrookdale by night* (1801)[15] depicts a Stygian landscape centred on the ghastly crimsons, oranges, and yellows of an open furnace, smoke billowing up against a lowering night sky, the foreground cluttered by wagon teams and discarded castings, the buildings and chimneys bathed in the reflected glow. In 1775 Arthur Young found the dale 'very romantic'—'a winding glen between two immense hills which break into various forms, and all thickly covered with wood, forming the most beautiful sheets of hanging wood'. But he found the beauty of the heights out of keeping with 'that variety of horrors art has spread at the bottom: the noise of the forges, mills, &c. with all their vast machinery, the flames bursting from the furnaces with the

[13] Cf. n. 10, above.

[14] J. Plymley, *General view of the agriculture of Salop* (1803), 340.

[15] Science Museum, London.

burning of the coal and the smoak of the lime kilns, are altogether sublime'. Coalbrookdale, a visitor wrote in 1801–2, 'wants nothing but Cerberus to give you an idea of the heathen hell'. He found it impossible to stay there, because 'the prodigious piles of coal burning to coke, the furnaces, the forges, and the other tremendous objects emitting fire and smoke to an immense extent, together with the intolerable stench of the sulphur, approached very nearly to an idea of being placed in an air pump.'[16]

By 1829 Coalbrookdale's heroic days were fading. Obsolescent techniques, tardy management, the competition of Welsh and Staffordshire ironmasters, and post-war depression took their toll. The Company furnaces were blown out around 1818, and when it directed its energies to its modern blast furnace at Horsehay, the Darbys' old yards became simply those of another large foundry and engineering shop. Even so, in 1829 the Curetons lived on the edge of a tortured and still lively landscape.

The dale's main axis wound up from the Severn for a mile, past the lower forge and boring-pools, moulding-rooms, and workshops, some of them converted to tenements as forge work contracted. Scattered tenements, terraces, and allotment gardens accompanied the route to the original upper furnace site below the furnace pool at the dale's northern end. Here there was a congestion of moulding-rooms, joiners', carpenters', and smiths' shops, and warehouses, with Company tenements and terraces higgledy-piggledy on the perimeters, and the corn-mill to the east. We must imagine a muddle of wagons, rails, horses, and dark-faced busy men, heaps of coal, coke, and discarded ironwork, much noise, and, when orders were high, the sulphurous pall by which visitors were incommoded.[17]

The dale's population was small none the less, and still in close relationship with the surrounding countryside. Ironmaking is not labour-intensive; the Darbys were ungenerous in their housing provision, too.[18] So even by 1841 only 1,367 people lived there in some 280

[16] B. Trinder (ed.), *'The most extraordinary place in the world': Ironbridge and Coalbrookdale* (1977), 65–6, 33.

[17] A. Raistrick, *Dynasty of ironfounders: The Darbys and Coalbrookdale* (1970), 248–9, 253–7. The lease map of 1827 depicts Coalbrookdale at this time (ibid. 250–1). The fullest contemporary itinerary of the dale's works is the anonymous account in B. Trinder (ed.), *Coalbrookdale in 1801* (Ironbridge, 1979): buildings had changed little by 1829.

[18] The ironmasters controlled all housing in the dale, having built it to attract labour. Supply lagged behind demand. In 1782 Madeley's 440 houses each accommodated an average of 6.1 people; by 1841 the average in the parish's 1,476 dwellings had shrunk to 4.9. Well into the 19th cent. many of the dale work-force still had to come in daily on foot; overcrowding remained chronic: 1841 census (PP 1841, ii); W. G. Muter, *The buildings of an industrial community: Coalbrookdale and Ironbridge* (1979), 45; Trinder, *Industrial revolution*, 321–6.

households.[19] The result was that as you ascended the mile to the upper furnace, threading your way between the wagons and iron-railed slipways which took castings to the Severn, you never lost sight of the hillside coppices into which, as one observer said, the terraced cottages were tucked as if by accident; allotments and vegetable gardens were ubiquitous, and pigs rooted. Then from the upper furnace pool you could escape the clamour of the works in a five-minute walk. One of the oldest roads into the upper dale, doubling back behind the furnace yard, would return you to the Severn, high up through the secluded woods of Captain's Coppice above the bustling valley below. You could safely make merry with your sweetheart here, as we know Elizabeth Cureton once did with Noden. If alternatively you bore north-westwards, all that held the countryside at bay were the few hundred yards of climbing track which led to the pastoral uplands of Little Wenlock. Four miles away, the volcanic cone of the Wrekin summoned the Company's workforce annually to the May festivities of that pastoral England which still stretches westward to the Welsh hills.[20]

The bottom of this road, at the upper edge of the dale, is where the Curetons lived; so did John Noden. Others to be encountered in this story also lived there: Edwards the surgeon, Elizabeth's chief maligner, and in the grandest houses the captains of the community's fortunes, the Darbys. Walk there now and little has changed. The road is manicured and tarred, and you have to obliterate the curving railway viaduct of 1864 and some of the early Victorian cottages to the west of the yard. The upper furnace pool was prettier then than the stagnant pond it is today. The Darbys had landscaped its upper reaches and once put a pleasure-boat on it. By the 1830s a miniature Iron Bridge crossed it to give access to Green Bank Farm, the property of Dr Peter Wright, for whom Noden had done odd jobs for fifteen years.[21] Its banks were

[19] MS census schedules (PRO). About 450 lived there and 500 worked there in 1756. Around 1760 the population of Madeley amounted to 900 families, perhaps 4,500 people. By 1801 this had increased only to 4,758; by 1831 to 5,822. There had been little migration: only 10% had been born outside the county. Welsh names abounded from much earlier movements. Trinder (ed.), *'The most extraordinary place'*, 19; J. Randall, *A history of Madeley* (1880), 166; Trinder, *Industrial revolution*, 313–15.

[20] To the east, a few fields were scarred by the opencast mining and squatters' huts of Coalmoor, and these became more numerous towards Dawley; but cows grazed mainly, and farm tracks replaced cobbles (*A history of Shropshire*, ed. G. C. Baugh (*The Victoria history of the counties of England*, vol. xi (1985))). Perry's engraving of the Coalbrookdale works (1758) suggests that down the Little Wenlock road came the 2-mile wooden railway track of 1749 from Coalmoor, on which 4-horse-teams drew wagons of clod coal from mines in Little Wenlock to the dale furnaces and the Severn wharves: reproduced in Trinder, *Industrial revolution*, 122, 153 n. 47; cf. Shropshire RO 1224.259: *Plan of a waggon way design'd to convey Mr Forrester's clod coals from his coalworks to CBD.*

[21] Trinder (ed.), *Coalbrookdale in 1801*, 12.

planted with conifers to provide the master families with a romantic view from their front windows[22]—a view they could not prevent the Curetons from sharing. But not much else has changed; no new houses have been built.

So it is still possible to trace the sequence of fine Darby houses climbing past the pool which Elizabeth's father passed daily to and from the moulding-sheds, and to see what he saw. Dale House first, begun by Abraham Darby I and completed after his death in 1717, was where Alfred Darby and his family lived (in 1841, at least). Rosehill comes next, with pedimented door-case, arched window-light, and stone quoins enlivening the brick facing. Richard Darby had moved into this house in 1811, and his family still lived there in 1851. Past its coach-house you come to three eighteenth-century villas flanking either side of the road—more plebeian dwellings. In one of them in 1829, or in one demolished since then,[23] a Mrs Scotton boarded John Noden and his mother. Next comes the early nineteenth-century house flanking Teakettle Row in which the surgeon Edward Edwards lived with his mother, sister, and younger brother: we shall hear much of him. Further up you pass the Chestnuts, built by Abraham Darby III's youngest sister. The Quaker meeting-house followed, now demolished, its graveyard still full of dead Darbys. On the crest of the hill, before the open countryside began, stood Sunnyside, built by Abraham II in 1750 and demolished in 1852; and then the White House where Francis Darby lived.

It was under the surveillance of this Darby ghetto that the Curetons lived in a terrace of skilled working families like them. Just before the surgeon Edwards's house you turn left and climb a dozen yards to Teakettle Row.[24] Today the Row's elevation is tidied up but little changed. Built in 1735–46, it is one of the earliest industrial workers' terraces in Britain, and the earliest of several such in Coalbrookdale. Six cramped one-and-a-half storey dwellings, with the upper attic bedrooms extending into the roof-space and lit by dormer windows, stretch in a terrace some thirty-three yards long. Each had a separate pig-store to the rear, and each shared a communal brewhouse, not only to brew in,

[22] Engraving by William Westwood, *c.*1835.

[23] Perry's 1758 engraving shows small cottages north of Rosehill which no longer stand.

[24] Teakettle Row had been built by Richard Ford, Abraham Darby I's successor as controller of the Coalbrookdale works, and was subsequently owned by Abraham and Francis Darby. In 1801 Quaker meetings were held in one of the Teakettle Row cottages. The Darby family owned all the houses in the neighbourhood, with the exception (in 1847) of surgeon Edwards's house, which was owned by Barnard Dickinson, till 1838 the Company's general manager and joint-leaseholder (Randall, *History of Madeley*, 289; Madeley tithe map, 1847 (Ironbridge Museum Library); Raistrick, *Dynasty*, 249, 253, 257).

but also to wash clothes in its cast-iron boiler and to hang slaughtered pigs. Behind and up the hill each cottage had its own allotment backed by rising woodlands and open country.[25]

Evidence at the trial makes it clear that the Curetons lived in the first-built cottage, at the northern end nearest to surgeon Edwards's house (so he would know Elizabeth only too well). This and its neighbour were the smallest and shallowest cottages. Its ground floor consisted of a small living-room, parlour, and pantry. Upstairs were two bedrooms. Cooking was done in the living-room on a cast-iron range. The floor was tile or brick. These were not conventional two-up and two-downs, however. The cottages interlocked, in a design derived from a rural squatter vernacular.[26] Bedrooms extended over neighbours' downstairs pantries or parlours; some parlours and living-rooms occupied the frontage jointly, their neighbours' pantries occupying the space to the rear. This plan saved on access space. It also meant that you could hear what your neighbours were up to, as Elizabeth found to her cost. Thus her garrulous neighbour, Sarah Williams, swearing in her affidavit against her:

> from the nearness of our house to Cureton's our doors being no more than four or five feet from each other and a thin brick partition only between our houses if there had been any cries or noise that night I must have heard it as I have often and many times heard chairs grating against the floors in Cureton's house and have plainly heard the conversation even the plain words made use of while they have been talking in Cureton's house when I have been sitting down at night on my own and when my door has been shut close.

There were plenty of people to do the listening. At the 1841 census the six cottages were occupied by eleven families averaging 6.4 members. Where families were not doubled up, lodgers were taken in to make ends meet. Altogether there were seventy people, forty aged under 16. The congestion was grim. It was an unusual luxury for the three Curetons to live alone. A dozen years later their one-time neighbours Sarah and David Williams lived in four rooms with six children. Another witness at the trial, William Hazleden, aged 70 by 1841, then had only two children still living with him and his 45-year-old wife, but had taken in a married couple and their 10-week-old baby to make ends meet. Michael Fletcher and his wife, 40 in 1841, had a 15-year-old son who worked as a moulder; but there were six further children aged between 13 and 2. Tenants changed cottages as families grew, and turnover was

[25] Muter, *Buildings of an industrial community*, 37, 27, and fig. 28.

[26] J. Alfrey, *Coalbrookdale* (1986), 120–6.

high. In 1829 the Williamses, with only a 3-year-old and a baby, had lived in one of the smallest cottages next to the Curetons; but twelve years later, with their family of eight, they had moved into one of the middle and larger cottages. Only five of the households in the 1841 census were still resident in 1847.[27] The Curetons had left by 1841, and are to be found nowhere else in the Coalbrookdale schedules of that date.

So this was intimate and crowded human territory, a far cry from the neat Row of today. It was above all a neighbourhood, in which personal recognition and reputation, mutual indebtedness and emotional support, curiosity and long-harboured enmity, bonded people dependently.[28] It was also a world which must have been peppered with now-forgotten sights, noises, smells. On a patch of hillside some thirty yards wide, we must envisage a score of shrieking children playing at doorsteps or in the woods behind, and half a dozen babies crying. In these half-dozen dwellings were a dozen mothers to scold them, and through the walls and at doorsteps and over washing and in child-minding to know each other's business as well. The odours of coal smoke, dampness, privies, pigsties, and Rosehill's stables flavoured the air, trapped in the hazy pall which on windless days drifted from the forges and darkened the sky. The track in front of the Row was muddy in winter, and much frequented too. Passing human figures were as frequent on the Wenlock road, as the straining horses and the groaning coal-wagons on the rails. Before the Row's pump was installed, women passed back and forth to the works yard to bring back water in buckets balanced on cloth pads on their heads. They also met there to mangle their washing. Coughing miners from Dawley lurked on the wall opposite Mr Edwards's house, reluctant to ring his bell; farmers from Little Wenlock made their way to Ironbridge. The master class and their visitors came and went aloofly, in gigs. Horse dung was trodden deep into the mire of the trackway, and children collected what lumps of coal they could, fallen on to the rails.

4. *Men and women*

One of the nastier accusations Edwards the surgeon levelled at Elizabeth Cureton after the trial was that she and her mother 'had made up the case against [Noden] . . . for the purpose of getting him to marry' her. Noden also thought that he had a chance of evading trial if

[27] Madeley tithe map, 1847.

[28] Cf. D. Levine and K. Wrightson, *The making of an industrial society: Whickham, 1560–1765* (Oxford, 1991), 279–81.

he offered to marry Elizabeth; he toyed with this solution several times after his arrest. Defence counsel put questions implying that Elizabeth had asked Noden to marry her by way of settlement two days after the attack. So the idea that this was a malicious prosecution to force Noden into marriage adhered to the case.

The charge that Elizabeth Cureton was overripe for a husband looks reasonable at first. Her sisters had married when they were 16 and 17. She, by contrast, had lived alone with her parents for some five years, and her continuing dependence on them might be unwelcome. Enemies could also have said that her needfulness was obvious because none of her past suitors came from the skilled work-force and two came from outside the dale. One was a warehouseman; another a Horsehays shoemaker; Thomas Small was a brick-maker; and John Nicholas was a bricklayer from nearby Buildwas. Noden himself was no great catch for a skilled man's daughter. For the daughter of a skilled moulder, there were plenty of more eligible bachelors to look to among the skilled work-force: why was she looking so far down-market if not in panic?[29] Add to this the fact that her reputation was tarnished, and her plight looks grimmer. She had had a half-dozen years' vigorous sexual life and at least five lovers, Noden included. She had miscarried a child and had been venereally infected, the doctor's affidavits later revealed. Teakettle Row must have known this. If unseemly behaviour escaped the attention of the Quaker ironmasters, it would not escape that of Methodist neighbours.[30] The Curetons' immediate neighbour, the Waterloo veteran David Williams, was a pious Methodist.[31] He and his wife were to testify against Elizabeth enthusiastically.

[29] In her 20–24-year-old age-group in 1841 there were 46 unmarried women in the dale, but 17 were in domestic service, weak competitors in the skilled marriage market; 44 of the men in the same age-group in the dale were unmarried (only 10 being married), 32 of whom were skilled moulders, fitters, turners, or pattern-makers earning between 18*s*. and 30*s*. a week. This was the occupational and age range in which the daughter of a moulder would look for her suitors. In the older 25–29-year-old age-group, nearly all the women and two-thirds of the men were married. The men over 25 who had not married were poorer labourers or craftsmen, struggling for security and earnings enough for marriage: like Noden himself—still hunting a wife at 27.

[30] Madeley had been a centre of Wesleyan revivalism since John Fletcher, once Wesley's designated successor, served the parish as vicar 1760–85; he was often host to Wesley. Fletcher set no precedent for opposing the capital code. Once he refused to petition to save a local youth from the gallows, but wrote a prayer for him instead, acknowledging the justice of his condemnation and hoping that he would escape hell: H. Potter, *Hanging in judgment: Religion and the death penalty in England* (1993), 15. By 1816, with 161 members worshipping locally, the Methodists dominated Coalbrookdale's religious life: R. F. Skinner, *Nonconformity in Shropshire, 1662–1816* (1964); B. Trinder, 'The Methodist new connexion in Dawley and Madeley', *Wesley Historical Society Occasional papers*, 1 (1967).

[31] C. S. Peskin, 'Memories of Old Coalbrookdale', *Transactions of the Caradoc and Severn Valley Field Club*, 11 (1941); C. S. Peskin, *Local notes, 1850–1901* (typescript, n.d., Ironbridge Museum Library, CBD.A.404).

But men have always been able to construct plausible fantasies about female needfulness, and this is one of them. Few of these suspicions survive closer examination; it is not here that we shall find Elizabeth Cureton's motives. Neither her dependency on her parents, nor her spinsterhood, nor her sexual adventurism was unusual. Parental pressure on the child to leave home was lower than it was in more impoverished communities, because dale families were skilled and relatively affluent.[32] Most unmarried men and women of Elizabeth's age-group lived with their parents in Coalbrookdale,[33] and most could earn their keep too. Nor is it likely that she lived idly. A few women, highly skilled, walked daily to the porcelain factories of Coalport. Others worked seasonally as gleaners or harvesters or as milkwomen or washerwomen.[34] (We know from an allusion in the trial that Mrs Cureton worked, though we do not know at what.) Elizabeth's spinsterhood at 24 was not abnormal either. She was still a year or so away from the average national age of female marriage.[35] In a community in which eligible men were abundant she had no pressing need as yet to worry about her prospects.

Also these Coalbrookdale girls enjoyed some independence. They were not like the forge women who 'often enter beer shops, call for their pints and smoke their pipes like men', or the pit women picking ore at a shilling a day, or the colliers' women who migrated every spring to work the market gardens of London. But sexual expectations in the parish were extended by these more vigorous and vulgar sisters, and pre-marital sex was commonplace in Elizabeth's class. Coalbrookdale was sexually more conformist than rural Shropshire (illegitimate births in the county were 50 per cent higher than in the dale),[36] but the local

[32] In Teakettle Row in 1841 unskilled men were few. Francis Rushton supported his wife and family of 6 by gardening. The only other unskilled men were a nightwatchman and William Edwards aged 41, who had to support his wife and 4 children as a labourer. The rest were skilled ironworkers: 1 engine-smith, 1 pattern-maker, 4 blacksmiths, 2 moulders, 1 packer. Of the 6 sons of employable age, 1 was a labourer, but 5 were moulders, the youngest of them aged 12. On attitudes to continuing child-residence in such communities, see J. R. Gillis, *For better, for worse: British marriages, 1600 to the present* (1985), 114.

[33] 59% of the dale's 20–24-year-old spinsters did so in 1841.

[34] I. Pinchbeck, *Women workers and the industrial revolution 1750–1850* (orig. pub. 1930; 1969), 273; Trinder, *Industrial revolution*, 353–4.

[35] In 1841 the average man married at 27.3 years old and the average woman at 25.35 years.

[36] Shropshire illegitimacy rates were high by national standards. Bastardy rates in England peaked in the mid-1840s; some two-fifths of all first conceptions were extra-marital and a high proportion of all spouses had sex before marriage. Illegitimate births per 100 births in 1842 in Wellington and Madeley were 6.8; in Shropshire, 9.3; in England, 6.7. 'Locality persistence' in rank ordering is a marked feature of these indices; Shropshire remained notorious for high illegitimacies throughout the 19th cent.: *Registrar-General's sixth report*, PP 1844, xix. 24 ff.; P. Laslett, 'Introduction', in P. Laslett, K. Oosterveen, and R. M. Smith (eds.), *Bastardy and its comparative history* (1980), 54–5, 30, 63, table 1.5.

sexual economy was vigorous enough to ensure that women like Elizabeth had their own sexual power.

The Severn gorge was a transitional society located on the frontier between rural and industrial England. Old custom and belief retained their meanings in such a place, and although Dissent was strong, piety was not pervasive. In Elizabeth's day it was still believed that Puck played tricks in kitchens and fields. One Madeley inhabitant saw spirits dancing by moonlight. A witch nearby changed herself into a hare. As late as the 1870s old women read omens of disaster in comets and earth tremors.[37] Custom ruled sexual relations too, and not all were simply exploitative of women. In 1830 'Francis Yates did penance in Madeley church for calling Bartlam's the Mason's wife a Whore in a show at Iron Bridge the owner Yates the Saltman's Son.'[38] And rough music delivered its old penalties upon those who abused women:

> when a man was known to have . . .'thrashed' his wife an effigy was made of cloth and straw and an old hat and coat put on it. It was then soaked with paraffin and tar and with the usual tin can and tin whistle band a procession was formed to the culprit's home and set on fire. The inscription chalked on the door 'Straw for sale' indicating that thrashing had been done.[39]

It was in the area's wakes and fairs that Elizabeth Cureton would have found spaces to display her charms. Here men displayed prowess in animal-baiting and rough play while young women flaunted such finery as they possessed and bartered full or partial sexual favours for male attention.[40] 'Bowling, drinking and Rebbelry Cursing Swearing and fighting on the Knowle and a dead corpse [in the river] going by at the same time,' as Thomas Beard of Jackfield nicely described a scene of this kind in his diary in 1832.[41] At Ironbridge wake they baited badgers, at Broseley they baited bulls (until the curate of Madeley suppressed the sport around 1825). And Elizabeth was frisky in these age-old courtship arenas. We learn from the trial that it was at 'a show of wild beasts at the Ironbridge' in 1824 that the warehouseman William Edwards neatly relieved Noden of her company and commenced his seduction of her. It

[37] Randall, *History of Madeley*, 120–1; Peskin, *Local notes*.

[38] *The diary of Thomas Beard, trowman of Jackfield* (Alderman Jones's notebooks, Shropshire RO 1649.1; typescript in Ironbridge Museum Library).

[39] Peskin, *Local notes*; Gillis, *For better, for worse*, 131.

[40] Randall, *History of Madeley*, 348; Trinder, *Industrial revolution*, 366; D. A. Reid, 'Interpreting the festival calendar: Wakes and fairs as carnivals', and J. K. Walton and R. Poole, 'The Lancashire wakes in the nineteenth century', in R. D. Storch (ed.), *Popular culture and custom in nineteenth-century England* (1982).

[41] *Diary of Thomas Beard*.

was at Madeley wake in Michaelmas 1826 that the bricklayer John Nicholas likewise 'commenced a more intimate relationship with her'.

If wakes were arenas of female display, night-visiting was a practice that girls like Elizabeth controlled. By regulating nocturnal access to the parental house, the custom pleasantly combined the semi-illicit with the functional as girls played off suitors against each other, giving tokens to some, turning away others, the parents conniving by going to bed. In places where live-in agricultural labour was declining or economic expectations were rising 'more direct, spontaneous forms of intimacy' were taking over.[42] But in Coalbrookdale, with its home-dwelling sons and daughters, the ritual still flourished. The girl's favour was apparently symbolized by the giving of wine, as Elizabeth showed in court: 'I . . . asked [Thomas Small] to take a little wine. My father and mother were not present when I gave him a little wine. They were just gone to bed.'

Local élites took a dim view of night-visiting, so Elizabeth Cureton was defensive when cross-examined about it in court, and her neighbours made much of it when they saw how it might save Noden's neck. But before the trial she was probably able to see her affairs through to their several consummations without much embarrassment. Courtship was still semi-public, and her neighbours and equals would not have thought twice about her visitors. She had not been secretive about them. David Williams next door could name the dates and times of Noden's nocturnal callings.

It was not ostracism that was the problem but gossip. Gossip flowed among Elizabeth's five-score neighbours, and between the women gathering at the communal mangle. It flowed back down the dale, through Ironbridge eastwards along the Severn, south from Dawley and across the bridge to Broseley—as carters, barge-men, market people, artisans, miners, and drinkers at public houses went about their business. At Jackfield, three miles from Teakettle Row, the barge-owner Thomas Beard used gossip to keep compulsive track of the parish's sexual misadventures: '*1830. 3m. 24.* John Transom was Catchd in Bed with Another Man's Wife at Worcester by the whomans Husband and the Man stuck a Pickel in his Backside Which Caused him to rannaway without his Clothes & after that offered the Man 15*s.* to make it up.' Read his commonplace-book and the illicit passions of plebeian Madeley come literally to life:

[42] Gillis, *For better, for worse*, 33–4, 109–10, 121.

1827. 4m. 9. Young Nancy Round had a Bastard Born by Young Jack Amphlett son of George.

1828. 9m. 16. Jack Amphlett had a Bastard Swore on him by A gards daughter and Jack Roden was Bound per him.

1829. 9m. 18. Will Greenwood had a Bastard served on him by Sam Rodens servant.

1830. 2m. 2. Tom Boden had a Child Born the Next Day after he was Maried to Cullys.

1838. 10m. 15. John Lloyds Sweethear was up here after him from the Black Horse at Arley to know if he would Marie here Being Pregnant by him her Name is Newland she slept at William Cullis's.

1839. 7m. 29. Hereford Asises where John Lloyd, Thos Davis and Benjamin Dodd went and had a trial about John Lloyd's Bastard which was sworn upon by the Landlord of the Black Horse at Arley's daughter Jeny Newnham. Robert the Horse Driver went to Hereford to swear that the Gardener belonging to Earl M.—— was catchd in Bed with Jesy Newnham's daughter at the Black Horse, Upper Arley the 22n of November 1838. This case was tried at Hereford 29th July 1839 and lost by John Loyd.

Passages like these reveal attitudes to women which bonded plebeian and professional males alike. For all their independence, women were fair game, even commodified. 'Wm Bennett of Coalport had is wife Cried Down at Madeley Market,' Beard noted in 1836—a denunciation for adultery or a wife-sale. One John Aston declared as an old man in the 1870s that he got the best of all his four wives by buying her for 1*s.* 6*d.* at Wenlock fair, clinching the bargain by bringing her back to Coalbrookdale with a halter rope around her neck.[43] And then there was always the violence:

1829. 8m. 4. Saml. Roden Was Commited to Prison for Beating Black Molley's Daughter.

1832. 11m. 3. Tom Beard beat his wife very bad.

1833. 3m. 17. Jack Lowe beat his sweetheart and gave her 2 black eyes at the Calcutts.

1836. 6m. 9. George Ball beat his wife and beat 3 women that stood by at the same time.

1836. 6m. 9. Thos. Doughty Beat his Wife and Beat Jno. Reynolds for taking her Part.

1839. 5m. 21. Owner Adam Yates horsewhipt his wife Lidia untill she cried Murder for Getting Drunk at Jackfield.

1839. 7m. 19. The Constable took Jemmy Latham for Beating his wife at Loyd Head.

[43] Peskin, *Local notes*; S. P. Menefee, *Wives for sale* (1981), 244, 253, 254.

Beyond these records of assault and beatings were the rapes that Beard knew of, and sure enough Elizabeth Cureton's was among them:

1829. 8m. 4. Daniel Adams and yong Thomas the Welchman Where Fined 10*s*. each for Breaking in the House of John Parker Benthall with intent to Committ a Rape upon his wife and daughter.

1829. 8m. 12. a yong man In the Name of Robert Noden was Tryed for a Rape upon Miss Cureton of the Dale Sister to Franklin the Shoemaker's Wife and condemned.

1829. 9m. 25. Noden was sent from Salop Gaol to Woolwich for Transportation.

1829. 9m. 16. Jack Parker was sent to Shrewsbury Gaol for Fighting and a Rape.

1838. 6m. 13. Hugh Culliss was Commited to Shrewsbury Gaol for a Rape on a Wench 13 year of age and he was the age of 65.

1838. 7m. 25. Hugh Cullis from the Rape of Ellen Oakley was Quitt at Shrewsbury Assises.

This recording brings us to the heart of Elizabeth Cureton's dilemmas. By burying her and Noden's misfortune in a catalogue of like disasters, Beard's diary put it into the perspective it assumed for most people. Beard did not know Noden; he got his name wrong. And he signalled his remoteness from Elizabeth's station in life as a skilled man's daughter by giving her the rare title of 'Miss'. But his detachment speaks not only for that malicious voyeurism which pervaded all small communities. It also speaks for the fact that in male eyes Noden's alleged attack on Elizabeth was no deviation from the norms of heterosexual life. It expressed something more common.

The borderline between an aggressive seduction and wooing on the one hand, and an act which could be defined as a rape on the other, was always hazy to the male view. If English law has had notorious difficulty in determining its location, so, historically and self-servingly, have most men. Most rapes have been unrecorded not only because women feared going to law, but also because violence was accommodated within patterns of seduction and courtship in which women were forced to acquiesce. This was how things between the sexes had always been, and Beard could not have been mealy-mouthed about it if he had tried. Elizabeth Cureton lived in a society in which sexual harassment was ubiquitous and gossiped about, but otherwise unremarked. Her neighbours would have seen little difference between a rape and 'a sweetheart business' (as surgeon Edwards termed it) which happened to have gone a little sour. In such a community, to prosecute a man for rape, especially if he was a past or would-be suitor, was a dangerous thing to do. If Elizabeth was not anxious to trap Noden into marriage, we have a long way to go before we understand why she took so rash a step.

5. *Noden's trial*

It was unusual for common people to prosecute at assizes. Rape prosecutions were especially rare. The difficulties of proof were great, the secondary trial of the woman often more vicious than that of the man. Most rape prosecutions were thrown out by the grand juries which vetted indictments beforehand, or were converted into trials for attempted rape; and few prosecutions which went their full course succeeded. In the 1820s an average of forty-three men a year were prosecuted for rape in England and Wales (only thirty-one in the 1810s). More than eight in ten were acquitted; only one of the remaining two would hang. In Shropshire in 1829 there had not been a rape trial for a quarter-century, and the last execution for rape had been in 1789.[44]

For these reasons alone Elizabeth Cureton was assured maximum publicity when she brought Noden to book. The theatricality of the assizes guaranteed more. The 20,000 citizens of Shrewsbury, together with country people, extracted what excitements they could from their assizes. People would walk vast distances to attend them,[45] and there was plenty to gaze on. Assizes paraded the dignity and splendour of the county élites; for élites too they were focal moments in the social calendar.

On Wednesday 9 August the county's high sheriff entered Shrewsbury in court dress, riding a caparisoned horse and accompanied by javelin-men and a cavalcade of friends and tenants. Having met Sir John Vaughan in his carriage, he escorted the judge to the guildhall to open the commission (the second judge, for the civil cases, arrived later). On Thursday morning both judges attended divine service at St Chad's, where an appropriate sermon was preached by the sheriff's brother-in-law.[46] Amid more coming and going of carriages, the court reassembled at noon. Vaughan met the grand jury, eighteen county notables in their finery arrived to fulfil their biennial and not uninteresting duties. Waiting for the judge to read the king's proclamation, the courtroom would be tainted by the jubilee mood, and by much confusion too. We must imagine a degree of unruliness difficult to envisage

[44] Annual judicial statistics are published in PP. Execution for rape was abandoned in the 1830s; national rape prosecutions for that decade rose to an annual average of 57, but the conviction rate dropped. The retreat from the death penalty for rape might have encouraged this rise in prosecution, but probably no more than the abolition of the proof of emission: see below.

[45] John Parker, labourer, persuaded 13-year-old Susan Ripper to walk with him from their home village to Thetford assizes and back in the same day to watch the sentencing of a murderer in 1822—a distance of 20 miles. On the way back, he raped her: HO47.63.

[46] *Shrewsbury Chronicle*, 14 Aug. 1829; *Salopian Journal*, 19 Aug. 1829.

today: much chewing and spitting and jostling and muttering, and the odour of unwashed bodies and clothes in the thickening air.

Noden's trial on the last day of the assizes, Saturday 12 August, was given heightened attention because the preceding cases had been unspectacular. The 29 who had been tried were a humble crew: 19 were aged 26 or under, 3 in their late teens; 6 were women; 12 were labourers, and the rest included a tin-man, a baker, a weaver, a nailer, a pig-jobber, a fishmonger, a chimney-sweep, a collier, a glass-cutter, and 2 shoemakers; 12 were illiterate and 11 could read only a little.[47] Vaughan skipped through their cases with well-practised celerity. On the first day thirteen cases, involving twenty people, were tried by six o'clock. A horse-stealer had a judgement of death recorded against him. A labourer was sentenced to seven years' transportation for breaking and entering. A sheep-stealer was found insane and kept in custody without sentence. A man found guilty of manslaughter was fined a shilling because of mitigating circumstances. Most of the rest were petty larcenists and none got more than six months' hard labour in the house of correction. On the second day, Friday, twelve were tried. Eight were found guilty and five had the death sentence recorded against them.[48]

Noden's trial also raised excitement because Noden had been released on bail after his committal, so his crime was not on the calendar: a rape trial had not been anticipated. That is why on Thursday Vaughan had congratulated the grand jury that the calendar was 'not stained with charges of murder, or the violation of female chastity, or arson, or those more filthy offences [sodomy] which disgust the human ear'—an achievement he attributed to the gentry's 'scrupulous observation of [their] duty as magistrates'. Only on the Friday did the grand jury bring its true bill against Noden, requiring him to present himself next day.

So now the packed court buzzed with expectation. The girl was young, striking, poised, and dashingly dressed. The man, surrendering himself from bail and now sitting in the dock with impressive composure—therefore innocent?—was fair-looking. The charge was rape. If things went well for her, a hanging was possible. If things went well for him, it could only be because proof was offered that she had asked for

[47] ASSI 5.149/I (PRO).

[48] Those with a sentence of death recorded against them before Noden's trial were Edward Backe, an illiterate labourer aged 42, and Richard Sims, a 26-year-old shoemaker, for horse-stealing; Jonathan Toomey, 21, a glass-cutter and literate, for breaking and entering; Thomas Plimmer, 24, an illiterate nailer, for counselling Toomey to commit his crime; Samuel Jones, 26, a weaver, for confessing to breaking and entering.

what she duly got: and what she asked for and what she got, the assembly expected to hear in detail.

The lawyers rose to the occasion zealously. Two attorneys conducted the prosecution, and two more led the 'acute and severe' cross-examinations for the defence. Vaughan himself acknowledged the status of the case by allowing the trial to run from nine in the morning to nearly four in the afternoon. The crowded chamber enjoyed an anticipatory *frisson* when it was cleared of women, as was customary in sexual cases. This meant that Elizabeth took the witness-box before a sea of expectant male faces. She also faced an avid press. The *Shrewsbury Chronicle* reported the case in forty-eight column inches, omitting only the clinical details of the rape and the medical evidence called. The *Salopian Journal* found the evidence 'not of a nature to bear publicity', but gave as many inches to it none the less. Even in London *The Times* reprinted the *Chronicle*'s report in abbreviated form.

The case rested not only on Elizabeth's account of Noden's attack, noted already.[49] It also hinged on whether Noden's and the Curetons' behaviour in the ensuing fortnight suggested his guilt or their malice. Cross-examination unravelled events after the assault in rare detail. This was because the defence case partly rested on the Curetons' failure to prosecute immediately.

The sequence unravelled in the trial was as follows. After Noden's attack on the evening of Wednesday 17 June Elizabeth had taken to her bed, so shaken that she could not leave it for ten days. On Thursday evening Edwards the surgeon was called to examine her. He found her 'feverish and a good deal agitated'—from 'mental excitement', he said. She was bruised on her left breast and lower back, and her neck was stiff. But as yet there was no suggestion of rape. He did not examine Elizabeth's genitals; her mother dismissed the idea. Richard Cureton next called for the Madeley parish constable Walters. It was he who urged the family to permit a full medical examination. So surgeon Edwards came again on Friday morning. Finding 'two discolorations between the lips next to the pubis, a slight laceration at the posterior [part] of the labia', he agreed that Elizabeth had been 'recently penetrated', but he did not believe in the rape. The hymen was not recently ruptured, though he admitted that 'women destroy it themselves'.

[49] Sarah Cureton and the neighbouring surgeon Edward Edwards were the most important prosecution witnesses. Richard Cureton, Elizabeth's brother-in-law Francis Franklyn, and the parish constable Walters gave support.

Noden also visited Elizabeth's bedside that Friday morning. According to Elizabeth, he made an incriminating admission: 'He asked me how I did and whether my father knew of it. I said yes and he asked me if I had had Mr Edwards the surgeon. I said I had. He said "Betsy if you say anything about me you'll surely hang me."' Yet still, for ten days, no complaint was laid against him. He continued to live across the road.

Then somehow things got moving. The family must have pondered its choices—to let things be, to compromise with Noden, or to press a charge. Elizabeth opted for the last. Rising from her bed ten days after the attack, she and her mother went to the magistrate five miles away at his Benthall manor-house and lodged her complaint on oath. The constable arrested Noden at a Coalbrookdale public house that evening, left him overnight in custody at the Plough inn, and on Sunday walked him to the magistrate.

Noden was scared. He would marry Elizabeth, he told Walters anxiously. Or he would pay compensation. 'What was the charge against him—was it a capital charge?' he asked. When the constable replied that 'it was merely for an attempt . . . he then said I am ruined. He then said if she proves all she can against me I know I shall be hanged.' This was the second self-incriminating comment which clinched the case against him. Had he only known it, attempted rape was only a misdemeanour, never harshly punished.[50]

The Benthall magistrate sent Noden on to the Wenlock sessions. So another ten miles' walking, now in the charge of Elizabeth's shoemaker brother-in-law Francis Franklyn. No handcuffs, no chains: there was no escaping in this countryside. But as Noden's panic mounted, self-incriminations multiplied. 'I hope you are not going to hang or transport me,' Franklyn reported him as saying, and again he offered to settle with the Curetons by marrying and paying money. When Edwards the surgeon overtook them in a gig on his way to the sessions, Noden exclaimed 'I am done': 'He said he understood he was charged with an assault with intent to commit a rape, that he feared she was going on too far. I said to Noden it appears to me that she is not accusing you wrongfully. He said, no for if she said the worst of him all the men on earth could not save him.' Again he asked Franklyn to arrange a settlement with

[50] Punishment for attempted rape could not exceed a fine or local imprisonment and finding sureties for good behaviour for life. This charge 'should not be adopted where there is any probability that the higher offence will be proved': W. O. Russell, *A treatise on crimes and indictable misdemeanours* (2 vols., 2nd edn., 1826), i. 563.

Elizabeth. And again, before the magistrate, 'he said he would have her or pay a sum of money provided they would make it up with him'. But it was too late. The Curetons were dining with an attorney, while Noden was without support.

In most petty sessions in this period attorney's advice for the prisoner was 'more a matter of favour than of right'.[51] But Noden did not do badly in what ensued. The magistrate took medical testimony on oath from surgeon Edwards and, convinced by Edwards's scepticism, questioned Elizabeth only cursorily; her mother was not examined at all.[52] A rape was not in question. He bound Noden to appear before the next quarter sessions to answer only for the attempt. Even better, the scantiness of the evidence induced him to admit Noden to bail.[53] Several weeks later, perhaps prompted by reports of Noden's self-incriminations, the magistrates announced that the charge was 'too serious' for them to investigate, and they transferred the prosecution to the assizes. Even so, Noden had cause for hope.

6. *The law on rape*

By the 1820s case-law on rape was becoming a little more generous to the prosecuting woman than hitherto. It was certainly more generous than communal opinion and custom was inclined to be, as we shall see. None the less, like all women in her plight Elizabeth Cureton was confronted by a judicial procedure which was required to treat her testimony with suspicion and to subject her personal character to trial.[54]

The ancient justification for this bias, cited in Noden's day as till

[51] G. Bolton, *The practice of the criminal courts, including the proceedings before magistrates, in petty and quarter sessions, and at the assizes* (1835), ch. 3. At the assize trial Noden had 2 defence barristers, at a cost of 1–2 guineas each. His Coalbrookdale patrons might have paid.

[52] The magistrate should have sworn Cureton to her deposition in the presence and hearing of the accused; he did not even commit Elizabeth's answers to paper. The defence at the trial dwelt on these omissions to show that the proper forms were neglected. Bolton (*Criminal courts*, ch. 2) observed caustically that country magistrates were often guilty of 'a culpable neglect of the requisite ceremonies on investigations of this kind, in often signing their names to the depositions without an attestation, or before the witness is sworn, and without knowing whether it is necessary or not'.

[53] Under the recent Act of 1826, bail might apply in felonies where the evidence 'shall . . . not be such as to raise a strong presumption of the guilt of the person charged' (Russell, *Treatise*, i, pp. xxxv ff.).

[54] For sexual crime in this period, see A. E. Simpson, 'Rape and the malicious prosecutrix: The blackmail myth and its origins in 18th-century legal and popular traditions', *Journal of Criminal Law and Criminology* (1986); A. E. Simpson, 'Vulnerability and the age of female consent: Legal innovation and its effect on prosecutions for rape in 18th-century London', in G. S. Rousseau and R. Porter (eds.), *Sexual underworlds of the Enlightenment* (Manchester, 1987); C. A. Conley, 'Rape and justice in Victorian England', *Victorian Studies*, 29/4 (1986), 519–56; Anna Clark, *Women's silence men's violence: Sexual assault in England 1770–1845* (1987); R. Porter, 'Rape: Does it have a historical meaning?', in S. Tomaselli and R. Porter (eds.), *Rape* (1986).

recently in ours, lay in Sir Matthew Hale's cautionary dictum that although rape was 'a most detestable crime', 'it must be remembered, that it is an accusation easily to be made and hard to be proved, and harder to be defended by the party accused, though never so innocent'.[55] The conditions which had to be fulfilled in a successful rape prosecution were stringent. The victim was to 'make fresh discovery and pursuit of the offence and offender, otherwise it carries a presumption that her suit is but malicious and feigned'. The prosecution had to be speedy for the same reason, for 'long delay of prosecution in such case of rape always carries a presumption of a malicious prosecution'. Beyond that:

> If the witness be of good fame, if she presently [speedily] discovered the offence and made pursuit after the offender, shewd circumstances and signes of the injury, whereof many are of that nature, that only women are the most proper examiners and inspectors, if the place, wherein the fact was done, was remote from people, inhabitants or passengers, if the offender fled for it; these and the like are concurring evidences to give greater probability to her testimony, when proved by others as well as herself.
>
> But on the other side, if she conceald the injury for any considerable time after she had opportunity to complain, if the place, where the fact was supposed to be committed, were near to inhabitants or common recourse or passage of passengers, and she made no outcry when the fact was supposed to be done, when and where it is probable she might be heard by others; these and the like circumstances carry a strong presumption, that her testimony is false or feigned.

Underpinning these principles was Hale's insistence that the character and reputation of the woman was an essential object of enquiry: 'the excellence of the trial by jury is in that they are the triers of the credit of the witnesses as well as the truth of the fact.'[56]

Traditionally, two further conditions had to be met if a rape charge was to be proved: penetration and ejaculation. Penetration was the basis of the offence. 'To make a rape,' Hale wrote, 'there must be an actual penetration.' 'A very slight penetration is sufficient,' Russell observed. And Burn: 'Any penetration, however trifling, though it do not break the hymen, is sufficient for this purpose.' About ejaculation there was more dispute. Hale had rejected emission as a prerequisite of rape. Emission by itself made neither rape nor buggery, only an attempt at either; it might be referred to only as 'an evidence of penetration'. It

[55] M. Hale, *The history of the pleas of the crown* (2 vols., facsimile reprint of 1736 edn., 1971), i. 635–6.

[56] Ibid. 632–3. Russell repeated these principles verbatim in *Treatise*, 562, and so did R. Burn in *The justice of the peace and the parish officer* (25th edn., 1830), v ('Rape').

could not be a condition of rape, since ('as physicians tell') some may rape who are incapable of emission.[57] Despite Hale's authority, however, this was a question upon which 'very different opinions have been holden', and proof of emission remained necessary in many trials until the 1820s.[58] It was only in *R.* v. *Burrow* (1823) that Holroyd J. gave a clear ruling. The prosecutrix's admission that she was not conscious of her assailant's ejaculation need not render a conviction for rape impossible, because 'although the woman may not perceive the emission, the crime may nevertheless be completed; as where the time is fully sufficient, and there is not interruption, or other circumstance to raise a contrary presumption. Emission may in fact be presumed unless where the probability is to the contrary.'[59] Shortly before Noden's trial, parliament had resolved the difficulty in 9 Geo. IV, c. 31, s. 18. Noting that 'offenders frequently escape by reason of the difficulty of the proof' of emission, the statute ruled that henceforth 'the carnal knowledge shall be deemed complete upon proof of penetration only'.[60] This ruling could only make successful rape prosecution easier: it might help to explain why there were more rape prosecutions in the 1820s than in the 1810s. This was the first small mercy for which Elizabeth Cureton might thank the law.

Was there any comparable movement with regard to the woman's moral character? On the face of it, yes. *R.* v. *Hodgson* (1811–12) ruled that the woman need not answer whether she had had connections with other men before the rape; such evidence was inadmissible.[61] This is still cited as a binding and 'progressive' ruling, expressing a newly chivalrous attitude to women.[62] The truth is that in the next quarter-century the Hodgson ruling was much eroded.[63] In *R.* v. *James Barker*

[57] Hale, *Pleas*, 628.

[58] Russell, *Treatise*, ch. 6.

[59] G. A. Lewin, *Cases determined on the crown side of the northern circuit* (2 vols., 1834–9), i. 91, 288.

[60] Burn, *Justice of the peace*, 'Rape'.

[61] Harriet Halliday, prosecutrix, had been asked by defence counsel if she had previously had intercourse with other persons or another person named. The judge ruled that she was not bound to answer questions which incriminated and disgraced herself. Defence called a witness to prove that she had been caught in bed with a young man a year earlier, and offered the man to prove that he had had intercourse with her. The judge ruled that this was inadmissible because the prosecution was unprepared to answer the charge. The prisoner was found guilty, but judgement was respited for the judges' collective consideration. At their meeting on 30 Jan. 1812 the judges agreed that, in cases of rape, the character of the prosecutrix as to general chastity might be impeached, but only by general evidence: *York Herald and County Advertiser*, 10 Aug. 1811; W. O. Russell and E. Ryan, *Cross cases reserved . . . from the year 1799 to the year 1824* (1825), 211–12.

[62] Clark, *Women's silence*, ch. 7.

[63] In 1836 Sir Frederick Pollock recalled his dissatisfaction with the Hodgson ruling: 'if [Hodgson's] counsel could have addressed the jury [on the girl's moral character] he must have been acquitted, and it was a conviction that gave dissatisfaction to the whole Bar, and I believe to the public. . . . There can be no doubt of his innocence of the crime as charged upon him, though probably he had been in a situ-

(1829), Park J. held that questions might be put to the woman to show her *general* 'light character'.[64] In *R.* v. *Moses Martin and Aaron Martin* (1834) Williams J. added that the prosecutrix might be asked the *specific* question of whether the prisoner had had intercourse with her, with her consent, before the offence. As he explained:

> I was one of the counsel in the case of Rex v. Hodgson. The question in the present case is as to previous intercourse with the prisoner and the question there was as to intercourse with *other* men. . . . I must say that I never could understand the case of Rex v. Hodgson. The doctrine, that you may go into general evidence of bad character in the prosecutrix, and yet not cross-examine as to specific facts, I confess does appear to me to be not quite in strict accordance with the general rules of evidence.[65]

Had this ruling been available in 1829, the judge might have permitted a far more searching examination of Elizabeth Cureton's sexual history than he did.

Suspicion of female deviousness was not the only basis of the protections which the law extended to the defendant at the woman's expense. Other biases in the defendant's favour were generalized out of evidential rules applicable to all offences. No witness (or prosecutor) was obliged to answer questions designed to expose him or her to any punishment or criminal charge, even though 'it should seem that such questions may be legally asked'. At the same time, however, authorities were divided as to whether witnesses or prosecutors, to establish their reliability, could be compelled to answer questions designed to discredit, disgrace, or degrade them. *R.* v. *Hodgson* had said not; but that ruling was contradicted by many others (not on rape).[66] As Best J. had said in *Cundell* v. *Pratt*: 'Until I am told by the House of Lords that I am wrong, the rule I shall always act on is, to protect witnesses from questions, the answers to which may expose them to punishment; if they are protected beyond this, from questions that tend to degrade them, many an innocent man would unjustly suffer.'[67] In other words, evidence was

ation to create considerable suspicion. . . . I have no doubt she was at first a consenting party, but that she was forcibly detained a longer time than was convenient to the domestic arrangements of the family in which she was a servant, and having been detained, probably by some degree of force, during the latter part of the interview, she converted the whole matter into a charge of rape': *RC on the criminal law, 2nd report* (PP 1836, xxxvi), 4–5.

[64] F. A. Carrington and J. Payne, *Reports of cases argued and ruled at* nisi prius *in the Court of King's Bench and Common Pleas and on the circuit*, iii (1829), 589–91.

[65] Ibid. vi. 562–3.

[66] Russell, *Treatise*, ii, bk. VI, 'Of evidence'; T. Peake, *Compendium of the law of evidence* (5th edn., 1822), 6.

[67] Russell, *Treatise*, ii. 625–30.

admissible that from the witness's 'general bad character' he or she ought not to be believed on oath.

What was distinctive in rape cases was that evidence was admissible about the woman's 'character as to general *chastity*'. At this point patriarchal values are not difficult to discern, and their prejudicial nature has been widely noted nowadays. Early nineteenth-century judges made no effort to hide them. In rape, Park J. said in 1830, a woman who has consented will sometimes deny that she consented, because it was in women's nature so to behave:

> In transactions of this kind, the injured party is not always altogether innocent. She has perhaps made some advances, and, it sometimes happens, that a person, for the purpose of protecting her character, [is] ready to deny it, and support it at any expense. There is some degree of resistance generally [even] when the desire of the woman goes along with that of the man.[68]

Although in this climate Elizabeth Cureton must expect hard questioning about her character and credibility, Vaughan allowed her an easy time of it on the whole. He allowed no particular evidence to be advanced. Medical evidence was interrupted when surgeon Edwards wanted to state that she was no virgin. And the questions put to her and her mother were restrained compared with the evidence put to Peel after the trial. Her cruellest trial lay ahead, outside the courtroom, not in it.

Still, Noden's defence was not undamaging to her. Four farmers, an innkeeper, and surgeon Wright, Noden's one-time employer, testified briefly to his good character. Then six defence witnesses were called to prove first that the sexual assault was not completed, secondly that Elizabeth had consented to Noden's advances, and thirdly (since she and her parents had not disclosed the attack speedily and had been prepared to settle with Noden out of court) that the prosecution was malicious. One-sixth of the cross-examining sought to undermine her credit by exposing inconsistencies in her account of relations with other men.

First, had the rape been completed? Doubts were easily cast. Mrs Cureton had not examined Elizabeth closely on the night of the attack and could not swear that intercourse had occurred. The Curetons failed to call surgeon Edwards for twenty-four hours and refused to let him examine Elizabeth's genitals on his first visit. Then, most damagingly, Edwards told the court that he had had good reason not to conduct a

[68] Lewin, *Cases on the northern circuit*, i. 293.

genital examination that first evening, because Elizabeth had very clearly told him that Noden had not 'effected his purpose':

> I am sure the girl gave me that answer. I am as sure of her giving me that answer as I am of my existence. If she has sworn today that she never told me that he had not effected his purpose, she has sworn falsely. . . . I was determined to have had an examination of the parts if they had not denied it. That was the only reason I did not examine her that evening.

Secondly, there had been no audible proof that night that Elizabeth had withheld her consent. Though she said she 'made as much noise as I possibly could', it could not help her credibility that by her own admission Noden was upon her for fifteen or twenty minutes. Nor did it help that although her mother heard Noden enter the cottage, she then fell asleep. Elizabeth's cry awoke her and she descended only as Noden was leaving. The next-door neighbour David Williams testified similarly that he went to bed and fell asleep at 8.45 p.m. His wife woke him when she came to bed at ten, but he 'went to sleep again very soon and heard no disturbance all night': 'I heard no screams or noise from Cureton's house.' (To be sure, the Curetons did have explanations to offer for all this. Elizabeth's parents slept over the kitchen not the parlour, their door was shut, and the father was deaf. The Williamses' bedroom was on the furthest side from the Curetons' parlour, ten to twenty yards off, said Mrs Cureton, and there were 'double bricks, seemingly, between us'.)

Thirdly, the defence tried to establish that the Curetons had maliciously invented the rape charge to get Noden to settle by payment. Noden had certainly believed that a settlement was possible. And the Curetons probably did put out feelers. Surgeon Edwards testified that two evenings after the incident Mr Cureton was anxious only that Noden should pay the surgeon's bill. Noden's one-time employer Wright confirmed this and added that Richard Cureton 'desired me to interfere and settle this business between the parties'. This he had been willing to do, though in the event he advised Noden to pay no money.

It was further suggested by the defence that when Noden first visited Elizabeth's bedside, she had asked him to marry her. One Valentine Leighton testified that when Noden had sent him with a placatory gift of peas and strawberries to the Cureton cottage, Mrs Cureton had told him that 'there would have been nothing said about the matter' if only Noden had returned with her on the night of the attack to agree to some settlement. And that same night the son of Noden's landlady had heard Mrs Cureton urge Noden to come back to talk the matter over. At this stage a rape prosecution had not been in the Curetons' minds.

In court the Curetons had no choice but to deny each of these charges outright. 'I never authorised anyone to compromise this charge,' Richard Cureton declared. 'I never desired anyone to endeavour to settle it. I never said to anyone that all I wanted was that Noden should pay the expenses.' Pressed harder, however, he conceded that he had asked Edwards to take Noden as paymaster for his medical bill. Pressed harder still, he admitted that 'I asked Peter Wright to talk to Noden concerning the matter but I never mentioned settling it.'

Finally there was the delicate question of Elizabeth's moral character. Defence had to tread carefully here. Since particular evidence was inadmissible, not much in her sexual life was uncovered at this stage that was disreputable—other than the pattern of night-visiting. But Elizabeth and her mother equivocated defensively in the face of questions whose relevance they must have wondered at and for which they were unprepared.

The aim was to establish how far Elizabeth had made herself accessible to past suitors as well as to Noden himself. To the first such query she made the mistake of replying that she 'never went out to any man at night nor let any one into my father's house but the prisoner Noden'. Under pressure she retracted this, conceding first that over a year ago she had treated one Thomas Small to wine in the course of his nocturnal visits, and secondly that her relationship with Small had continued and had precipitated the quarrel with Noden on the Sunday before the rape. She next admitted letting two other men visit her by night: William Edwards the warehouseman and John Nicholas. We must imagine the probing and insistent questions of defence counsel, not recorded here, as she admitted that

> I have been acquainted with others besides Small and the prisoner. William Edwards and John Nicholas—I cannot exactly tell. I have let Nicholas into the house at night. I cannot recollect how often. He has come once or twice late and I have let him in. I did not give Nicholas any wine. I cannot say how often I have let Nicholas in at night. Not ten times. I do not know that I ever have let him in after my father and mother have been gone to bed. I cannot swear I may not have let him in after father and mother have gone to bed four times. I dare say they have gone to bed and left me with Nicholas below. I did not permit liberties to be taken with me.

Sarah Cureton also had to admit that Noden 'had been at the house on the evenings before. I have heard some one in the house before after I have been in my bed and I have supposed it to be the prisoner.' Cross-examined, her confusion mounted as her daughter's had:

I cannot say whether suitors may have come to my house after my daughter without my knowing it. Noden has come when I have known nothing about it—my daughter has told me so. There have [been] suitors. One in particular, Thomas Small, and William Edwards—I am certain there were no more to the best of my knowledge. Oh! yes there is another, Nicholas. I cannot say that Nicholas was not visiting six months as a suitor to my daughter to my knowledge. I had forgotten the name of Nicholas. I do not know in particular how long he visited. I never knew of any man visiting at night my daughter after my husband and I went to bed. Nicholas did but I cannot say whether any other man did. I cannot say whether Nicholas did not come 20 times after my husband went to bed. I did not know of any other man visiting my daughter after my husband went to bed. I cannot recollect or say whether Small visited after my husband went to bed. I do not recollect any thing of Small being closeted with my daughter in the parlour at night when I have been in the kitchen. I knew nothing at all of my daughter treating Small with wine and I do not think it is so. If my daughter has sworn it I would believe her.

Two defence witnesses developed the theme. The Curetons' neighbour David Williams reported at least one late visit from Noden. He had known the couple walk out twice together about seven o'clock in the evening in summer, but 'I once heard him let in between 12 and 1 at night on 21st of March. . . . I had been out on business. . . . I was not tired that night. I thought somebody was rapping at our door. I would not swear he was let in but I swear I spoke to him.' And the hapless Thomas Small was brought in too:

I have known Cureton and the daughter 12 or 14 years. I used to visit at the house and at nights. Not very often. Sometimes. It was mostly at nights. When the father and mother have been in bed I have been in the house at night several times. I have staid sometimes until 2 or 3 o'clock in the morning. I have sometimes drank fresh beer and sometimes ale and have taken wine with her. Besides drinking it I have taken wine away. Sometimes one bottle and sometimes two. I generally went about 9 or 10 o'clock and she let me in. . . . I visited her as a lover. I generally got there when she was up. I never took any indecent liberties with her. Never did more than kiss her. The wine was taken away twice, elder wine.

After some half-dozen hours the case was over. Defence counsel had established a few points and conveyed more by innuendo. Noden's conduct before and after his arrest was not that of a guilty man. Despite 'every opportunity of going away', he had lived across the way from Elizabeth from the time of his bail until he presented himself at Shrewsbury assizes on 12 August. If he was angry on the night of the attack, he was justifiably so: Elizabeth seems to have been playing fast and loose. Elizabeth had apparently resisted the attack neither audibly nor strenuously. Her mother intervened late; neighbours had heard nothing: and this despite Elizabeth's claim that the rape lasted fifteen or twenty minutes and that she had screamed as loud as she could. It was twenty-four hours before the Curetons had called in the surgeon and the constable, and then, as surgeon Edwards testified, Elizabeth told him that Noden had not penetrated her and he had been refused permission to examine her genitals. The alleged efforts of the family to arrange a financial settlement with Noden out of court, and their readiness to leave him at large for ten days after the attack, threw doubt on the purposes of the prosecution. Finally, it was implied that Elizabeth's conduct with previous suitors and with Noden himself had not been decorous, nor had her mother's in permitting their visits. The women's equivocations created an unfavourable impression. Elizabeth, too, it had been established through a deftly inserted reference on Edwards's part, was no virgin. Her general character had been impeached. What was in question, in short, was the degree of her consent as well as her credibility as a witness; and Noden's apparent equanimity suggested that he believed that some degree of consent had been forthcoming.

Against this, Noden had been making a nuisance of himself before the attack; Elizabeth had rebuffed him several times; she had certainly been 'penetrated' that night; and (most damagingly) Noden after his arrest thrice spoke of his peril if she told all.

To analyse the defence case in this way is to give a misleading impression of its coherence in the courtroom. Noden himself made no speech in his own defence, and defence counsel in this era was not allowed to do so at the end of the trial. So the points which the defence scored emerged confusingly in cross-examinations interlarded with examinations by the prosecution. It was not defence counsel, therefore, but Vaughan who summed up the case. Having done so at length (in what terms is not reported), he left it to the jury to decide

> whether the contradictions which had been given to several statements made by the prosecutrix and her family, were any thing more than those accidental slips which the most circumspect witness, when subjected to a severe cross-examination, is likely to make; or whether those contradictions were part of a mass of perjury, invented to take away the prisoner's life.

His own opinion was that the contradictions were 'chiefly upon minor points, as to the conduct of the prosecutrix and her family, while all the evidence tending to the prisoner's conviction remained unshaken'. This was not far short of a direction to the jury, qualified only by his drawing attention to Noden's good character and to the favourable fact that he had surrendered himself to trial when he had an opportunity to escape.

For the one and only time in those assizes, the jury—five farmers, two shopkeepers, two victuallers, a maltster, a tailor, and a shoemaker—retired bodily from the court to consider their verdict. A bailiff was sworn to keep them without meat, drink, fire, or candle till they had agreed on one. They took thirty minutes. The foreman delivered the verdict. Noden was guilty, he said, but 'we recommend him to mercy with all the power we possess'.

MR BARON VAUGHAN. Gentlemen, have you weighed well your recommendation? Have you any doubt of the prisoner's guilt, that you so earnestly recommend him to mercy?

FOREMAN. My Lord, we feel no doubt whatever that the real act was perpetrated.

At this point, the prosecution attorney announced that Elizabeth Cureton added her strongest recommendation for mercy to that of the jury.

Amidst Noden's loud lamentations, which Vaughan ordered to be silenced, the judge then uttered his terrible words. Noden was led away in chains to the cart which conveyed him to the condemned cell in Shrewsbury Castle. The jury was dismissed and a new one sworn, and in the couple of hours remaining that day, three people were briskly acquitted of larceny, one was convicted of keeping a disorderly house, and one more of stealing a watch. Vaughan dined with the high sheriff that evening, and on Sunday with his retinue left Shrewsbury for Hereford, where his next assizes opened on Monday.

At the conclusion of Noden's trial, Sir John Vaughan gave every indication that he meant Noden to hang. He could have let the court know otherwise had he wished. In 1823 parliament had permitted judges merely to 'record' a judgement of death against capitally convicted felons who were 'under the particular circumstances of any case . . . a fit and proper subject to be recommended for the Royal Mercy'.[69] Vaughan had already resorted to this device to record the death sentence against the six other capitally convicted felons whose fates he decided at Shrewsbury, leaving them in no doubt that their lives would be saved.[70] His refusal to do so in Noden's case was pointed enough for

[69] This was sanctioned by 4 Geo. IV, c. 48, s. 1: Bolton, *Criminal courts*, ch. 7.

[70] Thus he told Charles Rose, convicted of horse-stealing, that as this was the only offence with which he had been charged, 'he would save his life . . . on condition that he left this country for ever'. Another convicted horse-stealer, Edward Backe, with excellent character references, was told that

Noden's lamentations in the dock to be appropriate. The hapless man would be told as he was led shackled to the wagon conveying him to Shrewsbury gaol that he had a week to live.

7. *The first appeal*

Yet after the court adjourned, Vaughan inexplicably ordered a temporary stay of execution. He might have meant to do this all along. Judges were not averse to playing games with the psyches of terrified felons. But a better reason presents itself in the intervention of surgeon Edwards after the trial. Edwards played a key role in what ensued.

Vaughan was well disposed towards Edwards. He was the only educated witness in the trial.[71] The judge praised the manner in which he had given evidence. But as Edwards himself later complained, he had been roughly treated in the witness-box. Both the judge and the defence had checked him when he had offered evidence about Elizabeth's injuries and about a revealing conversation he had had with Walters the constable, even though this evidence, he believed, would have 'tended to unravel the case' in Noden's favour.

Vaughan noticed his indignation and after the trial he called Edwards for 'a long private conference', ostensibly about the discolorations on Elizabeth's body. Whatever he learnt impressed him. Vaughan's postponement of execution could only have encouraged Edwards to press the case further. Over the next weeks persons of influence in Coalbrookdale were to mount an efficient campaign to get Noden pardoned, and Edwards was probably its instigator.

Edwards had personal and professional investments in the case. He had the most privileged knowledge of the parties and the strongest doubts about the Curetons' credibility, Elizabeth's in particular. He lived next door to the family and had treated their ailments. He had a professional and possibly a prurient interest in the uses to which Elizabeth put the body whose development he had observed over half a dozen years. And his pique at his treatment in court was that of a professional man whose self-esteem and expertise had been slighted.

A 31-year-old bachelor, living with his mother, brother, and sisters in the big house adjacent to Teakettle Row, Edwards was prospering. Like

horse-stealing was now unhappily so prevalent 'that no man could trust his horses in the field at night with anything like certainty that he should see them safe in the morning'; he would, none the less, 'once more try what lenity would effect'. The housebreakers Jones, Toomey, and Plimmer were told at once that their necks would be saved.

[71] *Shrewsbury Chronicle*, 21 Aug. 1829.

those of most country surgeons, his patients were mainly modest or plebeian people; his treatments were modest too. He was remembered for his faith in scuffing, bleeding, leeching, and Epsom salts, which last remedy coughing miners from Dawley threw into the bushes when they emerged from his surgery.[72] But he shared these shortcomings with most of his profession. ('Met Mr Forbes the surgeon going to kill a few patients,' as a Somerset parson nicely noted in his diary at this time.[73]) Since there were no neighbourhood physicians to take away the smarter custom, he must have profited from the ironmasters' clientage as well. He ran a large house, employed two living-in domestics, was a protégé as well as tenant of the Darbys, and gave modestly to charity.[74]

Edwards would also have had a high opinion of his professional standing. This was an era when no social group was taking a keener interest in establishing its credentials than the recently despised surgeon-apothecary. 'All pride, and business, bustle and conceit', Crabbe said of the profession.[75] Edwards was of the generation which had begun to benefit from recent improvements in medical education. Hospital training was displacing the old systems of apprenticeship to local surgeons through which practitioners used to acquire their skills. A proud member of the London College of Surgeons, Edwards had probably spent a year as a pupil in a local hospital and another year in London attending lectures and walking hospital wards. These were costly investments on which he expected returns.[76] He had another reason for wanting to be taken seriously in court. Medical men were claiming expertise in territories hitherto closed to them. It was their claim to *forensic* expertise—on the indications of poisoning, infanticide, as well as of rape—which had the most particular professional utility, for even lawyers, in the greatest of all professions, must defer to it. The first chair in 'medical jurisprudence' was established in Edinburgh in 1807, while lecture courses in forensic science were multiplying in London, along with textbooks.[77] If he had taken his London training seriously,

[72] C. S. Peskin, *Quaint Sayings* (n.d., Ironbridge Museum Library), p. 2.

[73] J. Ayres (ed.), *Paupers and pig killers: The diary of William Holland, a Somerset Parson, 1799–1818* (1986), 24.

[74] He lived in Rosehill in 1841 (census schedules). For charitable work, see Randall, *History of Madeley*, 260, 353–4.

[75] A. G. Hill (ed.), *Letters of William Wordsworth: A new selection* (Oxford, 1984), 115.

[76] I. Loudon, *Medical care and the general practitioner, 1750–1850* (Oxford, 1986), 35, 48–51, 132.

[77] See J. G. Smith, *Principles of forensic medicine* (1821), 2; T. R. and J. B. Beck, *Elements of medical jurisprudence* (5th edn., 1836), p. xiii; G. E. Male, *An epitome of juridical or forensic medicine: For the use of medical men, coroners, and barristers* (1816); J. G. Smith, *Hints for the examination of medical witnesses* (1829); J. A. Paris and J. S. M. Fonblanque, *Medical jurisprudence* (1823). Cf. M. A. Crowther and B. White, *On soul and conscience: The medical expert and crime* (Aberdeen, 1988), ch. 1; C. Crawford, 'A scientific

Edwards might have absorbed two messages pertinent to the Noden case. The first was that doctors had a professional entitlement to be heard as forensic experts in court. The second was that some medical experts doubted the very possibility of rape.

For all its scientific pretensions, medical knowledge was (and is) socially constructed, at no point more obviously than in doctors' belief in the alleged rape victim's capacity for deception. 'To give the appearance of violence having been committed,' one expert warned, some women 'have used acrid and stimulating substances to produce inflammation.' Hence doctors must always sensitize themselves to the woman's character by checking up on 'the enlarged state of the vagina'. Rape was usually impossible anyway, for 'no man can effect a felonious purpose on a woman in possession of her senses without her consent.'[78] This last notion was to inform the medical literature for a century yet, and by a ruling of 1845 (*R.* v. *Camplin*) the law accommodated it.[79] As Paris and Fonblanque opined: 'It is at all times difficult to believe that in a mere conflict of strength, any woman of moderate power of body and mind, could suffer violation, so long as she retained her self-possession: . . . all accusation therefore must be viewed with suspicion, if there be not a great disparity of strength in favour of the assailant.' As another text put it in 1815, rape was only possible if 'some very extraordinary circumstances occur'. Except when rape was perpetrated while the woman was stupefied, the violence necessary to achieve it must be such as to wound the body, for every decent woman would resist even to death; but even those signs of violence might be treated with suspicion, since consenting women sometimes struggle for the pleasure of it.[80]

In these and like contexts medicine's authority was exiguous, notwithstanding its extended Latinized speculations about the existence of the hymen and proofs of virginity or otherwise. But, in country prac-

profession: Medical reform and forensic medicine in British periodicals of the early nineteenth century', in R. French and A. Wear (eds.), *British medicine in an age of reform* (1991), 203–30.

[78] Male, *Epitome*, 127–33.

[79] S. Edwards, *Female sexuality and the law* (Oxford, 1981), 126, quoting R. L. Tait, *Diseases of women and abdominal surgery* (1889).

[80] Paris and Fonblanque, *Medical jurisprudence*, 417, 423; S. Farr, *Elements of medical jurisprudence* (1815), 44–5. It had been an ancient notion that conception could be achieved only if the woman took pleasure in the sexual act and thus released her own seed to fuse with the man's. Rape law had accordingly long maintained that if pregnancy followed an alleged rape, there could have been no rape: the woman must have consented. E. H. East in 1803 was the first legal writer to accommodate embryological truths and acknowledge that consequent pregnancy did not disprove rape (*Treatise of the pleas of the crown*); Male and Smith followed him. But Farr in 1815 (*Elements*, 45) and Paris and Fonblanque as late as 1823 still held that rape was disproved by subsequent pregnancy: see A. McLaren, *Reproductive rituals: The perception of fertility in England from the sixteenth century to the nineteenth century* (1984), ch. 1.

titioners like Edwards, it induced a sense that their opinions were worth listening to. Metropolitan society might debate for a long time yet whether doctors were gentlemen, but Edwards would have had no truck with this. Distancing himself from the contaminating status of apothecary, and working in a district where higher-status physicians were few and the dominant ironmasters spoke in Quaker 'thee's' and 'thou's', Edwards was not a man to tolerate doubts about his expertise. He had every incentive to lead the campaign to save Noden's life.

Events in Coalbrookdale and its neighbourhood now moved swiftly. The organization of the campaign on Noden's behalf was efficient and unscrupulous, its main target being Elizabeth Cureton's reputation and credibility. In the week following the trial, eight men were cajoled or bullied into swearing affidavits before the magistrates about her sexual experience. Surgeon Edwards swore one too. He was joined by a juryman at Noden's trial, two of the Curetons' female neighbours, and an ex-neighbour. The most startling affidavits were extracted from Elizabeth's ex-lovers. We can only imagine the pressures brought to bear on these men. Since Thomas Small's affidavit contradicted his evidence in court, he exposed himself as a perjurer. At least two of the other men, Nicholas and Owen, seem to have been recently married.[81]

On 22 August the organizers circulated a finely inscribed parchment petition for signature in and around Madeley parish. The affidavits accompanied them. Addressed to Vaughan, the petition declared that the signatories knew Noden and were convinced 'that he has always conducted himself as an inoffensive, moral, sober honest and industrious man, kind and dutiful to his parents and in every respect hath sustained an unimpeachable character'. It also stated that those signing had perused the affidavits. They firmly believed 'that the said unfortunate John Noden was not guilty of the offence with which he was charged'. They implored Vaughan's merciful intervention on his behalf.

The petition was signed by 192 people. They were headed by the Darby family: Abraham, Alfred, Francis, and Lucy. Every local surgeon signed—not only Edwards but also Wright of Coalbrookdale, two from Madeley, one from Ironbridge, and two from Dawley Green. The parish curates signed, and so did iron-founders like Boycott and Thomson, and a charter-master of Dawley. A few gentlemen signed. The innkeeper Robert Chilton of Dawley Bank inscribed next to his signature the information that he had been a juryman at Noden's trial. Then there were

[81] A John Nicholas was married in Madeley in Dec. 1827, and a Robert Owen in Broseley in the same month: I am grateful to Mr R. W. Gwynne for this information.

farmers, innkeepers, accountants, shoemakers, grocers, butchers, schoolmasters, and nine women. In haste the petition and the affidavits were posted to Vaughan. On the 26th he and his fellow judges would be in Gloucester for the last assizes of the circuit. There they would sign the formal circuit letter to recommend to the king's mercy the names of all those convicts against whom death had been recorded.

The petition worked. Vaughan's mercy list comprised the three horse-thieves and the three housebreakers he had dealt with in Shrewsbury. To it he now added the name of John Noden.[82] Peel signed the Home Office's standard reply on 5 September, informing the judges that the king had accepted their recommendations; the men named were to be 'transported to New South Wales, or Van Diemen's Land, or some one or other of the islands adjacent thereto'. On 18 September the Home Office dispatched its instructions to Shropshire's high sheriff to remove the seven men to the hulk *Dolphin* at Chatham, ready for embarkation.[83]

And so a month after conviction Noden and his fellows in misfortune left the rural landscapes in which they had lived all their lives. Manacled and in a horse-drawn wagon, they began their bumpy ride towards London, the Thames, and for all they knew Australia. Noden might rejoice that he still lived. But he must have contemplated the new scenes which revealed themselves to him with bewilderment. Confined on the hulk with some 900 felons, and quickly set to hard labour, he now had to wait for his destination to be decided upon.

8. *The second appeal*

It was not exceptional to win a commutation of a death sentence to transportation; 95 per cent of English death sentences were commuted by the late 1820s. Four of the seven rapists sentenced to death in 1829 were reprieved too.[84] Most were reprieved on weaker community support than Noden was given. What distinguished the Coalbrookdale campaign from most such endeavours was that Noden's reprieve did not end the matter. Exceptionally, this campaign continued, aiming at a full royal pardon and his release from servitude. Coalbrookdale people never once said that they were activated by hostility to capital punishment. This extension of their campaign proves that what mattered was their conviction that Cureton had accused Noden wrongfully. It was not mercy that Coalbrookdale thought it was seeking, but justice.

[82] HO6.14. [83] HO13.54: 67, 95. [84] See App. 2, below.

So surgeon Edwards must have argued. When news came that Vaughan was recommending the reprieve, the zealous doctor next day prepared a more complete medical statement than his earlier one. It was probably he who now circulated a second petition on parchment, this time to the king. It repeated the first petition's view of Noden's character but it also overtly impugned the Curetons' credibility. Acquainted as they were with the character of and the evidence given by the prosecution witnesses and with the affidavits subsequently sworn, the petitioners were 'quite convinced and satisfied in their own minds of the entire innocence of the said John Noden of the charge upon which he has been . . . convicted': they sought a full pardon.

Nearly twice as many people signed the second petition as had signed the first—358 in all. They were drawn from a wider area and carried more social weight too. The Wenlock bailiff, magistrates, and town clerk signed, and (again) the Darby family, male and female alike, along with ironmasters Boycott, Thomson, and Onions of Broseley, the ministers and curates of Madeley, Dawley, and Buildwas, and churchwardens and parish clerks. The surgeons who signed the first petition were joined in this one by surgeons from Much Wenlock, Broseley, Ironbridge, and Shrewsbury, and by two of the county's physicians. Three local gentry signed, and so did eight clerks, accountants, and excise officers, several farmers, and nearly every shopkeeper in the district. Elizabeth Cureton had no chance of privacy or reputation. The import of the affidavits read by this lot was damning and her destruction complete.

Of all the statements, the two from surgeon Edwards were the cruellest and most telling. Betraying his patient's confidentiality without compunction, they offered a graphic account of Elizabeth's sexual history which would not have been admitted in court.[85] They told how in April 1824 he had been called in to examine Elizabeth for 'disease on the private parts'. He 'discovered that the hymen was gone' and that 'the disease was occasioned by sexual intercourse'. Again, when Elizabeth was consorting with Noden, her mother called him in because she feared that she had the disease again. In a gratuitous vignette he described how he found Elizabeth 'with her clothes on lying on her back upon the bed and her person ready exposed for examination'. Her

[85] In court doctors could be compelled to give confidential evidence 'when the ends of justice absolutely require the disclosure' (Paris and Fonblanque, *Medical jurisprudence*, 160, and cases cited there; cf. also Peake, *Compendium of the law of evidence*, 175–6). But the ethics of voluntarily publishing confidential information out of court were something else.

thighs were covered with blood. Since he believed this resulted from a miscarriage, he gave no treatment.

Next he rehearsed the details about the Curetons' equivocations. They led to the conclusion that this was just a 'sweetheart business' gone a little wrong. Elizabeth and her mother 'made up the case against the said John Noden for the purpose of getting him to marry' her; they 'took every means to exaggerate the case'. Mother and daughter had twice denied that Noden had effected his purposes. None the less, Sarah Cureton 'thought he ought to be punished for doing what he had done, using her daughter so ill, and asked me what course she should pursue': 'Considering it then as I still do that it was nothing more than a sweetheart quarrel, I . . . told her as she had called in Walters he was the most proper person to advise her, not doubting but that as a peace officer he would endeavour to compromise the quarrel.' Walters, however, had foolishly advised the Curetons to prosecute. It was only thereafter that 'the mother said she did not tell me the whole of it,—that he did effect his purpose, that he did what he liked. The daughter replied yes, he did what he liked with me, yes he did.'

Then came Edwards's opinions on rape, lasciviously detailed. Examining her, he found 'two small discolorations between the labia against the pubis about the size of a pea each and a slight breach of continuity at the "fourchette" . . . which was occasioned (to use her own expression that "he was a very large man") by a large penis. He being in stature a middle-sized man.' Sexual intercourse had taken place, but 'no great violence was used . . . it was impossible such a thing could have been effected without much greater violence upon so robust a subject, as she is, without a tacit consent'. It was only in proving the penetration of a child, 'where the subject is not of sufficient strength to make resistance', that the absence of major injuries might be discounted.

Finally Edwards vented himself of his irritation that he had not been properly examined in court. 'I was not, to my great surprise, asked a single question by the court as to the extent of the violence (except as to the discolorations on the person by the judge after the trial), and [this] half examination made an unfavourable impression upon the jury.' Nor was he allowed to give his firm opinion that Elizabeth's ensuing confinement to bed was 'only a cloak to absent herself from publick notice; . . . had she been disposed, there was neither illness or injury which would have prevented her going about as usual during one day'. He felt that these points, if sworn in court, would have 'unravelled the case'.

The affidavits of Elizabeth's two female neighbours might have unravelled the case too. The first came from the Curetons' immediate neighbour, Sarah Williams, wife of the Methodist blacksmith who had testified at Noden's trial. The only illiterate to testify,[86] her breathless garrulity exudes satisfaction at being asked her opinion at all. Relations with the Curetons could not have been amiable at the best of times. Relations would have become less amiable as she now ventured the devastating information that at about ten o'clock on the night of 17 June:

> I was going out of my door and on looking towards the end of the row of houses and at the distance of ten or fifteen yards from where I stood I saw Elizabeth Cureton going from . . . her own house towards Mrs. Sorton's [or Scotton's, where Noden lodged]. She went along the road out of my sight and in about three minutes I saw Elizabeth Cureton and John Noden coming back along the road towards Cureton's house. They came to the door of Cureton's house and Elizabeth Cureton went in first and John Noden followed her and then the door was shut.

We can imagine the straining of Mrs Williams's ears thereafter:

> I had occasion to go out at the front of my house shortly after to shut our shutters and then I went into our house again and then went out at the back door to fetch a raking coal to rake the fire with. I then raked up the fire and swept up the hearth and shortly I went up stairs to bed. . . . On going up stairs to bed I did not go to sleep for some time for I had a child to suckle and I gave it the breast before I went to sleep and from the time I went to bed till I went to sleep I am sure it was ten minutes or more. . . . During the whole time after seeing Elizabeth Cureton and Noden go into Cureton's house till the time I went to sleep I never heard any the least noise or cries of any kind in Cureton's house.

Even with the doors shut she could normally hear talking in the Curetons' house, and the grating of chairs. Noden, she added, had been a frequent nocturnal visitor at the Curetons'; through the wall she had come to know his voice well. Moreover:

> It was only Friday that I heard that Elizabeth Cureton had said Noden had ill used her and on that Friday evening I was asked by several if I had heard any noise or cries in Curetons house about ten o'clock on the Wednesday night before and I said no. . . . I remember all that I had been doing and all that I had seen on that Wednesday night the talk then about it makes me remember all about it and I know that all I have said is as true as God is true.

The plot thickened when Mary Hazleden, the 33-year-old wife of another smith in the Row, swore that she passed the Curetons' house

[86] Her affidavit was marked with a cross. Aged 21 in 1829, she had a 3-year-old and a baby to care for; another 4 children were to be born by 1841 (1841 census schedules).

between ten and eleven o'clock on the night of 17 June and noticed fire- or candle-light within. She stood at her own door for a time, went inside, and remained up until midnight. 'She heard no screaming or noise whatever in the house of the said Elizabeth Cureton and she is quite certain if there had been any screaming or noise she . . . must have heard it.' Neighbourhood tensions must have been at breaking-point as these women betrayed Elizabeth Cureton. Mary Hazleden had known Elizabeth for eleven years or so, she claimed. But their evidence was consistent.

The evidence against Noden had already worried a juryman at Noden's trial, Robert Chilton, a victualler of Dawley. He now swore that in the jury chamber he had remained 'fully and perfectly satisfied in his own mind from the evidence given upon the . . . trial that . . . Noden had not committed such offence forcibly or against the will of Elizabeth Cureton'. The jury foreman had told him that the jury could do without his consent. Under pressure he had agreed to the guilty verdict 'under a promise from the foreman and several others of the jury that they would strongly recommend the said John Noden to mercy'. It was rare in those days for jurymen to break ranks in the jury-room. To break ranks after the trial was rarer still. The home secretary noted on his affidavit that Chilton's testimony was 'very singular'.

And finally Elizabeth Cureton's ex-lovers added their extraordinary contributions to the dossier: systematic and detailed revelations of her dalliances with them. These were in fact the sexual exchanges commonplace in her world. When sworn on paper, however, circulated to local dignitaries in their parlours, and sent from there to the Home Office, they looked shocking.

William Edwards first, warehouseman to the Coalbrookdale Company, swore that he had known Elizabeth since her infancy. Five years ago he had seen her and Noden together in Captain's Coppice. Noden's 'hand was up her petticoats'. He took it for granted that 'she was a person of unchaste habits'. So a week later, when he had seen her with Noden at a show of wild beasts at Ironbridge, he had asked her to leave Noden and go with him. She had done so willingly. In the ensuing fortnight he visited her frequently. On one visit, her parents having gone to Wellington market while Elizabeth stayed to do the ironing, she shut the door and fastened it. 'He conceived that such circumstance was meant to encourage his advances and he therefore proceeded to put her to lie on the table where he had connection with her and . . . she made no resistance whatever but seemed quite as willing as this deponent.'

There was, he added, no mention of marriage between them. He continued to visit her for about eight months, having sex with her as often as he pleased, sometimes in her father's house, sometimes in his own father's house, sometimes in the fields. She gave him presents. Several times 'she has of her own accord put her hand into his breeches and handled his private parts'. Frequently she asked for sex when her monthly courses were upon her [to avoid conception], but he refused. Her mother sometimes gave him gin to drink, and allowed the couple to stay downstairs when she went to bed. Elizabeth frequently drank wine with him. She put paper in the keyhole of the parlour door to prevent anyone spying on them. Then she informed him that she was with child (the child presumably that surgeon Edwards said miscarried). He ran away prudently to Staffordshire. Now living again within ten yards of the Curetons, and at home during the critical night of 17 June, he too swore that he had heard no screaming or any other noise whatever that night.

Next Robert Owen, shoemaker of Dawley, swore that he had known Elizabeth for ten years and more. About five or six years ago she had visited his mother's house, come into his bedroom where he was lying on the bed, and kissed him. 'He pulled her on the bed to him and put his hand up her petticoats to which she made no resistance.' Once when she sat on his knee 'he put her hand into his small clothes where she continued it of her own accord for some time'. Disturbed by his mother, they agreed to meet in a coppice, 'but he did not go there fearing that she may be pregnant and charge him with being the father'.

John Nickless (or Nicholas: the spellings of names were unstable), bricklayer of Buildwas, swore that he had known Elizabeth for over ten years and became sexually intimate with her in 1826 at Madeley wake. Visiting her thereafter at her father's house, he had sex with her 'whenever he pleased, sometimes in the parlour . . . and sometimes in a large arm chair which stood by the fire place in the kitchen'. Often he remained in the house for most of the night, she treating him with as much wine as he chose to drink and she drinking with him. 'His intimacy continued for about six months during the whole of which time he had connection with her whenever and wherever he pleased and without any objection on her part.'

Thomas Small, brick-maker of Madeley, who also had known Elizabeth for ten years, swore that for a considerable time he visited her after her parents had gone to bed. 'On most of these occasions and whenever he pleased he had connection with her and . . . on several of such occasions she made the first advances by unbuttoning his small clothes and putting her hand within the same.' He had first been intimate with her about a year ago and remained so for about three months.

Finally, one Francis Rushton or Rustleton swore that he had known Elizabeth for fifteen years and more. In June 1824, when he lived about fifteen yards from the Curetons' house, he had seen her sitting on a garden bench one evening with William Edwards. Edwards's breeches were unbuttoned and Elizabeth 'had hold of his private parts and was playing with them for some time'. Then again, passing the garden one summer Sunday evening in 1827, he 'saw her lying upon her back and some man upon her. . . . Her petticoats were up as he this examinant saw her bare knees and at this time it was quite light.'

At this distance the interpretation of Elizabeth Cureton's motives in prosecuting Noden against her own best interest may be filtered through many anachronistic as well as ideological prisms. Who can say where the truth lay? But in this catalogue of dalliance and desertion we may at last find some credible explanation for her rash decision.

Many pressures were at work on her. The constable was one of them: for him there was money in prosecution—fees for executing warrants and for Noden's detention.[87] He might have told her that her expenses would be reimbursed.[88] Her parents were another. But it is her personal interest in pressing the case that is the issue, for a pattern seems apparent here. She was a vigorous, unabashed, and confident woman; she knew her sexual power and might once have enjoyed it. But she also found that she was fair game. Passing her from one to the other, men pestered, betrayed, infected, and at last impregnated her, deserting her at once lest the child's maintenance be laid against them. Watched by disapproving neighbours, recurrently humiliated, she grew angry and

[87] Randall (*History of Madeley*, 232) gives an undated story about a Madeley constable Samuel Walters, 'a broken-down tradesman' and son of the incumbent of Madeley, who once tried to enlist his own father by giving him the king's shilling in the dark. He also made use of blank warrants (filled in as occasion required) to arrest 3 innocent girls whom he drove to Shrewsbury gaol to his own profit. This injustice came before parliament and resulted in the abolition of the system of granting blank warrants throughout the kingdom. I have no proof that this was the Constable Walters of 1829.

[88] Bennet's Act of 1818 and Peel's Criminal Justice Act of 1826 help to explain why the perceived crime rate escalated in the 1820s and 1830s. The sense that, now that they were paid for, many more prosecutions than hitherto were likely to be malicious, vexatious, and ungrounded worried lawyers for many years yet, until the police became the main instigators of prosecutions later in the century: D. Philips, *Crime and authority in Victorian England: The Black Country, 1835–1860* (1977), 104–5, 112–14, 119; Beattie, 48; Radzinowicz, ii. 74–82; D. Hay, 'Prosecution and power: Malicious prosecution in the English courts, 1750–1850', in D. Hay and F. Snyder (eds.), *Policing and prosecution in Britain 1750–1850* (Oxford, 1989).

ashamed. Noden's jealous clamour, exacting sexual compliance through overfamiliar pressures that left little physical mark, was the last straw. He was rejected utterly. She took to her bed, and in prosecuting him sought a kind of blind revenge upon them all.

Oddly enough, far away in the Home Office, Robert Peel got this balance roughly right. After he had read the dossier and the judge's opinion on it, he scribbled on the back that Noden had 'received some previous encouragement'. But then he added that he thought that Noden was guilty and ought to be punished. This was not a verdict which law books could sanction: if she had encouraged Noden there could be no rape. But there was rough justice in his decision none the less.

9. *Outcomes*

In 1829 some two thousand appeals for pardon landed on Peel's desk.[89] Most required little attention, sent in as a matter of form with the thinnest of supporting evidence.[90] In principle Noden's case was more demanding. In the 1820s only one other appeal against a rape conviction had resulted in so substantial a dossier, and even that had not elicited as sustained a campaign as this one had; that man had hanged.[91] Even so, Peel would have skimmed the Noden dossier cursorily. The judge must advise. On 24 September, therefore, he sent the Coalbrookdale petition and affidavits to Sir John Vaughan at his home in Langham Place, asking for his report. Vaughan took three weeks before replying on 17 October, in a letter which hedged its bets with painful caution.

Vaughan was not dissatisfied with the verdict, he said, on the evidence presented in court at least. Admittedly there had been 'some discrepancies . . . in the testimony of the witnesses for the prosecutrix as it regarded some minor circumstantial material to the main enquiry, and

[89] Petition Registers, HO19.4.

[90] e.g. from the 1829 Shrewsbury assizes, the Shropshire MP Robert Slaney forwarded a petition for Thomas Plimmer, one of the housebreakers whose death sentence Vaughan had commuted to 14 years' transportation. Slaney observed that this was Plimmer's first offence, and that 'his father, a very aged man, has implored me to write for him'. He had no knowledge otherwise of Plimmer or of the 14 Shrewsbury citizens who supported him. Their signatures were blotted and clumsy. Their petition could claim only that up to June 1829 Plimmer was 'a quiet, industrious, honest young man'. It took only a minute or two for the Home Office to deal with this. Peel's secretary replied curtly to Slaney that Mr Peel saw no ground for extending the mercy already granted. Plimmer (along with his fellow Shrewsbury convicts) was shipped to New South Wales on 25 Nov. 1829 (HO17.45/I (bundle Go19), HO13.54, HO8.22).

[91] The case of John Pattern, convicted in 1825 at the Warwickshire assizes of a rape in Birmingham: HO47.68.

the story of the prosecutrix was calculated to excite doubts and much suspicion'. It was this, he said, that induced him to spare the prisoner's life. None the less he had felt that the conviction was right, and 'in some degree' still did so, on two grounds. First, 'the intercourse was effected against the consent of the complainant', even though 'the accused might not have anticipated an earnest and *bona fide* resistance to his wishes'. Secondly, to the constable Walters and to his brother-in-law Franklyn Noden had made incriminating 'declarations inconsistent with the presumption of his innocence'. The credit of these two witnesses seemed unimpeached.

What Vaughan should have added was that the new evidence on Elizabeth's sexual experience was improper: the prosecutrix's character should not be impeached in detail. It was Peel or his secretary who scribbled on the back of each affidavit that 'this would not have been admissible evidence in the trial'. What the judge did say, however, was conclusive: 'If I were now called upon to declare what would have been the impression of my mind if the facts disclosed in the various documents [now] presented to me had been proved regularly in court upon the trial I should have thrown my weight into the scales of justice to have insured the acquittal of the prisoner.' If he had any objection to the post-trial affidavits, it rested only on the formal grounds that they threatened the dignity of the court and were not to be relied upon, because their authors could not be cross-examined. He could not, he wrote,

> overlook the danger to be apprehended from giving too easy credence to depositions made by persons *in favorem vitae* in contradiction to what they solemnly swore upon the trial, or to the voluntary depositions of persons not examined in court and consequently not subjected to the test of my cross-examination, and above all to the voluntary affidavits of jurymen made after they are released from the solemn obligations under which their verdict was pronounced.

Then, with a rotund flourish, he put the ball back in Peel's court: 'How far under all the circumstances of this singular case, a pardon or any and what further mitigation of the punishment of the prisoner beyond a dispensation with the rigorous execution of the law, would best answer the great ends of criminal justice, is a question upon which I do not presume to express any opinion.' He returned the affidavits to the Home Office along with a transcript of his notes on the trial.

Noden was the victim of his times. However unsound the verdict, there was no question in this era of release or retrial. So after reading

Vaughan's letter, it is not likely that Peel agonized for long over the decision which he now pencilled in his own hand on its reverse. He too found it politic to trim his judgement as he closed the case finally: 'Let the prisoner be sent to the Hulks but *not* transported. I believe him to have been guilty but to have received some previous encouragement.'

In October 1829 John Noden's name and those of the six convicted with him were among the 920 convicts under sentence of transportation on board the hulk *Dolphin* at Chatham. On 22 October they were transferred to the hulk *Retribution*. Five were sent from there to New South Wales on 25 November. Noden stayed behind. All we know of him so far is that his behaviour and his 'bodily state' were good. He was, the hulk records made clear, there for life. In the event he was shipped to a fate worse than transportation, some said to a fate worse than death. On 15 August 1831 Noden was sent to the British hulk establishment's most notorious outpost, in Bermuda, established by Peel some years before. It was a terrible place. Yellow fever ended the lives of most who were sent there, if they did not die of hard labour on the fortifications beforehand. In 1829 the men attempted to kill three keepers. They were 'daring, mutinous, violent and evil minded', the governor wrote, and the Home Office agreed: it was 'very requisite that convicts should have as little relaxation as possible . . . as they are exceedingly prone to idleness'. Not surprisingly, convicts were in short supply. 'Casualties and deaths' reduced the quota, the Ordnance Office complained: on average ninety out of seven hundred were ill at any one time.[92] Somehow Noden survived this hell-hole for over twenty years. Then he returned to Coalbrookdale on ticket-of-leave, a hardened man or a broken one. We find a John Noden occupying a croft and pastures in the dale in 1854, and a local diary notes that John Noden 'the returned convict' died there in 1873.[93]

And Elizabeth Cureton? After the trial she disappears from local records. Curetons lived in Coalbrookdale in 1841, but neither her name nor those of her parents appear in the census. If she managed to find a husband, it was not in Madeley. But an Elizabeth Cureton was married in the not distant parish of Wembridge in 1831, and another Elizabeth Cureton was married in the parish of Wellington in 1832. Perhaps she found peace after all.

[92] PC1. 77/I, 28 May, 23 Sept., 23 Dec. 1829.

[93] The earl of Craven's estate map (1854) (Ironbridge Museum Library, maps 231, 233, 237–8); Peskin, *Local notes*.

PART VI

THE OLD ORDER RESISTS

CHAPTER 18

FURRED HOMICIDES, SABLE BIGOTS

The Judges

1. *Hanging kinds of men*

WE MAY AS CHEERFULLY AS WE LIKE (IN PART IV) SURVEY THE penal critiques of the 1770s and 1780s, their rearticulation in parliament in and after 1808, the mounting vociferousness of radicals, prison reformers, and Quakers, and the gentler pressure of those who used the mercy process on others' behalf. Yet the outcome of none of these pressures was assured; for decades yet there was little for reformers to celebrate. With conditions in prisons still remote from those envisaged by advocates of separation and solitude, men and women hanged as numerously as ever. It bears repeating that in the post-war quinquennium 1816–20 over three times as many people were condemned to death in England and Wales as in 1806–10, and in 1826–30 three and a half times as many. Moreover, as many were hanged in the 1820s as in the pre-1815 decade: 65 hanged in London between 1816 and 1830 as against 79 in the *eighty* years 1701–80.[1] Also, Home Secretary Peel still defended the main bastions of the bloody code, convinced that his holding operations would last a generation or two. In England and Wales in 1829, the last full year of his office, 74 people were hanged. With hindsight we know it could not last. As the Introduction suggested, too many more people were being prosecuted after the peace, too many more had to be capitally convicted under the statutes—and too many more were therefore having to be pardoned—for the hanging law's credibility to be long sustained. Yet decision-makers at all levels were determined to perpetuate it so far as possible. So on the part of

[1] See App. 2.

the judges, home secretaries, and Cabinets who are the focus of this last part of the book, what mentalities informed this long resistance to penal change before 1830? How could they not see the writing on walls?

A first answer is that they could rationalize their obduracy, advancing reasons from Paley, for example. An apparently savage penal code was the small 'price we pay for our liberties', an MP told parliament in 1811: 'we love freedom and happiness . . . we are jealous of previous restraint and control of our actions . . . we wish to avoid the teasing vigilance of the perpetual superintendence of the law.' Better a harsh penal code, affecting relatively few lawbreakers, than a Frenchified and centralized police system inimical to a free people which would affect countless more. A great commercial nation needed the protection of capital law; and anyway, given the English distaste for torture, alternatives could not be easily conceived.[2] Economic and status self-interest underpinned these animadversions also: in so unequal a society, how could they not? Historians get aerated when they debate this exhausting and exhausted question; we shall not join them yet. We recall merely that it was Adam Smith, not Marx, who said that 'laws and governments may be considered in this and in every case as a combination of the rich to oppress the poor, and preserve to themselves the inequality of goods which would otherwise be soon destroyed by the attacks of the poor'.[3]

We pursue less conscious resistances to change than those: the kinds rooted in half-conscious assumptions or in the cramped imaginations élites brought to the contemplation of people weaker and smellier than themselves. At these levels the darker stories that can be told about the law are not about the law really. They are about the law's agents and their mental map of the world. However many felons were justly convicted or mercifully reprieved in our period, large numbers were hanged or transported simply because élites did not find their plights and possible innocence worth thinking about closely.[4] To explore this

[2] *Hansard*, xix, 29 Mar. 1811, c. 647.

[3] Adam Smith, cited by D. Hay, 'Prosecution and power: Malicious prosecution in the English courts, 1750–1850', in D. Hay and F. Snyder (eds.), *Policing and prosecution in Britain, 1750–1850* (Oxford, 1989), 344. Sir Thomas More had said the same many centuries before, in *Utopia* (1516): 'I can perceive nothing but a certain conspiracy of rich men procuring their own commodities under the name and title of the commonwealth. They invent . . . all means and crafts, first how to keep safely without fear of losing, that [which] they have unjustly gathered together, and next how to hire and abuse the work and labour of the poor for as little money as may be. These devices, which the rich men have decreed to be kept and observed under colour of the communalty, that is to say, also [by] the poor people, then they be made laws.' Cited by P. Jenkins, *Crime and justice: Issues and ideas* (Monterey, Calif., 1984), 64.

[4] The point here, nicely, is not a 'left-wing' one. It is Sir Ian Gilmore's, who describes 18th-cent. criminal law as 'avowedly terrorist': *Riot, risings and revolution: Governance and violence in 18th-century England* (1992), 156 ff.

awesome possibility it is time to put the judges in our sight-lines. They were the front-line agents behind most of the horrors recounted in this book.

There are several ways of contemplating these dislikable beings. One way *not* recommended is to play into the illusion encouraged in some kinds of history-writing that they were accessibly modern kinds of people—friends, as it were, who went to the same colleges. Admittedly they used a Latinized and abstract language like that of historians. Also their profession survives, and continues to evince similar dispositions and power; and a natural servility will always induce some to accord such men the *post mortem* respect which they demanded in their own lifetimes. But we should follow a different course and dissociate ourselves from them utterly. See the judges remotely, as anthropologists might—note their peculiar and isolating privileges, the narrowing socializations of their childhood histories and career structures, and above all the peculiar emotional structures which ensued—and the subjects may be quite transformed, their oddness and unpleasantness freely registered. Crippled as much as elevated by their power and standing in the world, the judges confront us with a peculiar and diminished species of being in whom benevolence, sympathy, love, and the imaginative faculties were denied. If they had no access to the postures of sensibility, it was because they had no social need of them. Representatives of a cultural *ancien régime*, they spoke for long continuities in élite cultures against which the bourgeois man of feeling must oppose himself to establish a standing in the world.

Contemporaries shared this view of the judges, so what may sound like anachronistic judgement here merely rephrases that of the time. It was the whig Dr Parr in 1808 who called the judges 'sable bigots' and 'furred homicides'. It was Sydney Smith in 1822 who saw in the judge 'a cold, slow, parchment and precedent man without passions or *praecordia*'. William Hone likewise knew that 'a hard head, a cold unfeeling heart' were the judge's chief attributes: 'some of the finest faculties of the human constitution, the imagination and sentimental affections' were unknown to him, for his mind was 'obliged to labour in the trammels of dismal formalities, like the horse in harness, dragging a heavy vehicle in the wheel-ruts made by those who have gone before'.[5] Graphic satire endorsed these verdicts, pushing them, indeed, to

[5] Romilly, ii. 253 n.; Sydney Smith, *Works* (2 vols., 1856), i. 363, ii. 118; W. Hone, *The spirit of despotism: Dedicated to Lord Castlereagh* (5th edn., 1821), 77.

38. Joseph Meadows, *Vignette*, 1836.
Meadows's image of a black-capped and bewigged deathshead, surrounded by emblems of judicial office, would have had real bite in the 1810s and 1820s. But by 1836 bloodthirsty judges were easier targets than hitherto, since the bloody code was collapsing and their time was assumed to be past.

wondrous extremes (Plate 38). The verdicts got the judges' number well enough. We shall hold on to them in this chapter (leaving more balanced views to the next one). All we need add to the verdicts here is the reminder that it was to the last ditch, and vindictively, that these men defended their power to hang people.

The jurist Blackstone praised the common-law judges. Their contributions to the liberties of England were not to be sneezed at, he said (in effect). He lauded the wisdom of the common law which they dispensed, and deplored the failures of statutes to keep the law tidy. Also, mercy came chiefly from the judges. Had reference been made to the judges, the cutting down of a cherry tree in an orchard could never have been made capital, he argued.[6] The trouble with this cheering construction is that not one judicial protest can be found against adding new capital offences to the statute-book, hundreds against their removal. It was the judges' generous interpretation of the Black Act which allowed it to encompass so many capital crimes. And when the debates about repealing capital statutes began in earnest in 1808, the judges' taste for hanging came into its own.

Romilly's Bill to abolish death for shoplifting is a case in point. It was defeated in the Lords six times between 1810 and 1820. Lord Chief Justice Ellenborough led the opposition by announcing in 1810 that if the Bill were passed, there was no knowing where the rush to benevolence would stop: 'we shall not know where to stand; we shall not know whether we are upon our heads or our feet.' The way things were going there would soon be a Bill even to repeal the death sentence on stealing

[6] D. Lieberman, *The province of legislation determined: Legal theory in eighteenth-century Britain* (Cambridge, 1989), 215.

from a dwelling-house; and he waxed eloquent on the plight of the poor cottager which would result. It was not as if transportation could take the place of hanging, he added. Felons considered transportation only 'a summer's excursion, in an easy migration, to a happier and better climate'. And had not the abolition of death for picking pockets two years before merely increased the crime? Perhaps, in fact, judges recently had not hanged enough. Too much mercy made the law look soft, and opened the door to needless reform. As Best J. chimed in in the 1813 debates, 'we do not wish the laws of England to be changed': the Shoplifting Act had been passed 'in the best period of our history, and there was no reason for hazarding an experiment'.[7]

Ellenborough's was a menacing intervention and Romilly was right to be nonplussed by it. To claim that pickpocketing prosecutions had increased because milder punishment encouraged the crime rather than the prosecution was disingenuous to a degree. A month later he suspected that judges were acting on their supremo's heavy hint. When Heath J. sent three men to the gallows in Kent, that judge countered Romilly's plea for mercy on behalf of one of them by announcing that nobody had appeared for his character. 'It surely ought to be made generally known, that the not producing witnesses to the character of a prisoner leads . . . to such important consequences,' Romilly snorted. His fears deepened when Ellenborough boasted to Lauderdale that he had left for execution a man convicted of shoplifting because 'when he came to the bar, [he] lolled out his tongue and acted the part of an idiot'. Lauderdale asked Ellenborough to identify the law that stipulated execution for counterfeiting idiocy in court; the reply is not recorded. National and London death sentences and executions did go up a bit in 1810.[8]

This judicial obduracy hardly melted in the period of this book. The petition archive gives whiffs of it in plenty, right to the end. One example will do. In 1837, with the capital code collapsing, two rapists were sentenced to death by Coleridge J. at Chester assizes. They fainted in court when Coleridge told them that their crime 'still remained and he supposed always would remain, punishable with death'. The judge would have hanged if he could; had he succeeded, these would have been the last men hanged for rape in England. They were saved by public petitions. Some fifty people pleaded that the men were hitherto of good character, that the evidence against them was murky, and above

[7] PD1 (Lords, 1810), c. 324.
[8] Romilly, ii. 325–31.

all that the punishment was 'far disproportionate to the offence'. 'It would be shocking to hurry two fellow beings into Eternity when their being sent out of the country may have as sufficient effects for the future prevention of such cases.' The Quaker John Barry reminded the home secretary that Coleridge was 'a severe judge', not forgotten for ordering the deaths of two men for robbery in 1835 when the House of Commons had already removed the death penalty from the crime in the preceding session. Barry added that Coleridge entertained 'some antiquated opinions regarding crime and punishment which are fast disappearing at the present day': it was not for nothing that he had edited the tory *Quarterly Review*.[9] Easily multiplied, cases like this offer few surprises. Judges' hanging tastes are part of folk memory as well as of record all the way down to the unspeakable Godard CJ. The more interesting question concerns what they say about judges' temperaments and views of the world. How do we explain the hard heads and cold hearts that Hone commented upon?

2. *Mediocrity*

One way to begin is by discarding any notion we may have that the twelve judges of this era were men of intellectual or legal distinction. The contrary point is endorsed time and again in the delicate irony of contemporary legal biographers like Edward Foss, however polite they had to be ('Lord Wynford's countenance, though not handsome, was very attractive').[10] Charles Abbott, lord chief justice after Ellenborough from 1818 (Baron Tenterden from 1827), stands out as a rare exception in Foss's catalogue of mediocrities. The son of a wig-maker and hence made much of by those who lauded the supposed openings that a legal career offered humble men of talent, Foss thought Abbott 'a complete master of every branch of the law'. Erudition was also claimed for Hullock and Holroyd. But most judges owed advancement to competence rather than brilliance, and more often to the prince of Wales's quest for minions to launder his financial and marital problems, or to Lord Chancellor Eldon's determination to reserve legal honours and offices for tory supporters.

Many judges when barristers had wangled themselves government seats in the Commons with advancement in mind: Abbott was not the

[9] HO17.56 (dossier Ix17), Apr. 1837.

[10] E. Foss, *The judges of England: With sketches of their lives* (9 vols., 1864), ix. 9–54, 68–98, surveys the judges alphabetically across 1820–37.

only one famous for a 'leaning towards those who sat in high places'.[11] Where political interest did not point, professional socialization did the job just as well. 'In a community in which lawyers are allowed to occupy without opposition the high station which naturally belongs to them,' de Tocqueville wrote, 'their general spirit will be eminently conservative and anti-democratic'—which put it mildly.[12] Whence men like Sir James Burrough, master of bankruptcy but not of criminal law, sitting in the Commons on the government side till he was put on the bench in 1813—deemed not so much a lawyer fit enough, as an upholder of right principles staunch enough, to send the Derbyshire Luddites and the Cato Street conspirators to the executioner's axe in 1817 and 1820.

Then there were judges like the Lord Wynford of the attractive countenance, earlier Sir William Draper Best, suppressing his once liberal leanings and turning tory to feast off the royal patronage which got him his knighthood. Best was a judge 'superficial in legal knowledge,' Foss noted, 'apt to form hasty and questionable opinions'. More, he was 'considered by the profession as an indifferent judge, [who] brought himself into bad odour, as well by the political bias he displayed, as by his occasional irritability and intemperance on the bench'—Foss's delicacy not concealing that 'sometimes in the ardour of his exertions he would disturb the dignity of his court'. 'He had no imagination,' Grant recalled bluntly: 'his manner was cold and inanimate, and his speeches were monotonous and tiresome.'[13] Not for Best the effusions of the 'many persons who entertain false notions of humanity', as he confessed in his one published work: 'the number of [prisoners] whose repentance has produced lasting habits of industry and integrity . . . is too small for legislators to trust much to the conversion of criminals.' If not hanging, then 'transportation for life is the only way in which an intolerable nuisance can be abated,' he knew. 'The increase of crime is in a great degree to be attributed to the letting loose of old offenders after short confinements in hulks and prisons,' he declared.[14]

Then there was Robert Graham, who also owed his elevation to toadying to the prince. His 'reputation as a lawyer was not very high' either. But he had one distinction: in his long service he 'sentenced more unfortunate human beings to death than any other judge who ever presided at a county assizes'. A man of antiquated politeness, his

[11] 59% of judges between 1790 and 1820 had been MPs: see D. Duman, *The judicial bench in England, 1727–1875: The reshaping of a professional élite* (1982), 78; [A. Polson], *Law and lawyers, or Sketches and illustrations of legal history and biography* (2 vols., 1840), i. 358–9.

[12] Duman, *Judicial bench*, 102.

[13] J. Grant, *The bench and the Bar* (2 vols., 1837), i. 82.

[14] *The substance of a charge delivered to the grand jury of Wiltshire* (1827), 19–23.

clothing proclaimed his adherence to the old ways—a three-cornered hat, black coat, knee-breeches, and buckled shoes so old-fashioned that people stared as he and his wife took evening walks along Oxford Street in the 1820s.[15]

William Alexander, next, was an equity lawyer and friend of Eldon, also 'aware of his limited acquaintance with criminal law and the practice of the common law courts'. William Garrow was a schoolmaster's son, who owed 'his ascendancy to his natural talent and sagacity more than to any deep knowledge of law'. Nor had J. A. Park any 'particular eminence as a lawyer', Foss said, confirming Grant's view that his attainments 'did not surpass mediocrity'. Known for his 'irritability about trifles', he had tory 'prejudices of the oldest', Ballantine recalled. Although he might not have worn a pigtail in the 1820s, he 'ought to have if he did not'. We met him at work in Chapter 16. Then there was Sir John Vaughan, he who dealt with Noden—'celebrated for possessing not the slightest notion of law, and a bold, rough and ready style of address'. 'In his glory . . . in any case in which plain country people were the parties,' Vaughan browbeat and badgered them with a 'low, vulgar sort of wit'.[16] His brother was the king's physician, whence the much chortled-over witticism that he became a judge only *by prescription*.[17] These were the quality of men, owing their place to favour, ignorant of the law they dispensed, who hanged people so that order would be sustained.

Historians of patronage in this era will find none of the above surprising. Society operated through connection, and this was how business was done. At least things were better than they had been. Judges never escaped their time-honoured role in radical satire as bloodthirsty enemies of the poor, nor the charge that there was 'too great a vicinity between Westminster Hall and St James', as Wilkesites had put it. As they enjoyed pensions and places at ministerial pleasure and played partisan roles in sedition and treason trials, the charge of an excessive political intimacy still echoed in radical attacks as they had a century before. None the less, a relative independence of political pressure had been

[15] Grant, *Bench and Bar*, i. 79.

[16] Ibid. 299.

[17] J. Arnould, *Memoir of Thomas, first Lord Denman* (2 vols., 1873), i. 58–9; W. Ballantine, *Some experiences of a barrister's life* (5th edn., 1882), 125. The ponderousness of judicial and barristerial wit has always measured the profession's isolation and self-love: it should be attended to one day. Foss ho-hoed characteristically: 'The squibs that circulate in Westminster Hall are so numerous, and often so good, that it is to be regretted that a legal album is not kept, and a recorder appointed . . . to delight the rising generation . . . [with the] fun and frolic' of lawyers. An example? When Garrow tried in court to 'extract from an old spinster the proof of a tender of the money in dispute, Jekyll tossed him these lines:—"Garrow, forbear; that tough old jade / Will never prove a *tender maid*"': Foss, *Judges*, viii. 224–5. The strong-minded will find many gems in W. H. Grimmer, *Anecdotes of the bench and Bar* (1852).

achieved. Judicial tenure during good behaviour regardless of the sovereign's pleasure was guaranteed by the Acts of Settlement of 1701 and 1714. This was one of several changes which ended the days of the 'covetous, venal, cruel, abandoned' judicial 'timeservers and sycophants' of earlier times. By the mid-eighteenth century 'allegations and denunciations of corrupt malpractice by individual judges, or bribery strictly defined, dwindled away to nothing'.[18] With the nice bump-up in salaries and pensions in 1824 which gave the lord chief justice £10,000 a year and puisne judges and barons a not bad £5,500 (up from £3,000 in 1799), the sale of offices from which judges had hitherto made some honest pennies was abolished too.[19] And, as we shall see in the next chapter, the judicial bench had acquired authority and independence enough by the 1820s to resist most of the more intrusive demands of political and social grandees for favours. At this formal level these were not corrupt men, and legal rules were followed.

Still, mediocrity was never rectified. The criminal courts were not places to attract the keenest minds or those hunting the largest fortunes. Some circuit judges practised habitually on the civil side, since it was through judgements there that fame might come. Criminal business was marginal to their interests and reputation. Few judges were chosen from the most successful barristers in practice. These last often refused the bench because they earned more at the Bar or anticipated higher rewards later.[20] After a long but unprofitable spell on circuit, a gifted barrister like Romilly preferred to stay in London coining £8,000 or so a year through *nisi prius* and Chancery business. Abbott made the same money in practice, refusing the bench 'from prudential motives' until his energies and eyesight began to flag, when he began to 'yearn for the comparative tranquillity of the bench' and went there aged 44.[21] So it was not surprising that the puisne judges of the common-law courts were the lowest of the judicial hierarchy. The circuit called for hard and bumpy travelling, and the assize day was long and tedious, while sending horse-thieves and footpads to the gallows was not deeply interesting, once you got used to it.

The high status and cosseting on circuit gave some compensation, with fanfares, dinners, and all that 'My Lording' in court. 'We have been made much of,' Denman wrote on circuit in 1833: 'Lord Feversham

[18] W. Prest, 'Judicial corruption in early modern England', *Past and Present*, 133 (Nov. 1991), 82–5, 89; T. A. Green, *Verdict according to conscience: The English criminal trial jury, 1200–1800* (Chicago, 1985), 334–6.

[19] Foss, *Judges*, viii. 199, ix. 1.

[20] Duman, *Judicial bench*, 78–82.

[21] Romilly, i. 45, 92–4; W. C. Townsend, *The lives of twelve eminent judges of the last and of the present century* (2 vols., 1846), ii. 246–7.

sends us a buck; Lord Fitzwilliam a whole paradise of pine-apples, melons, and grapes.'[22] Anyway, what else was there for them to do as they drifted into senility? 'Some [judges] resemble ancient females—some have the gout—some are eighty years old—some are blind, deaf, and have lost the power of smelling,' Sydney Smith noted in 1826.[23] The average age of the twelve judges on circuit in 1825 was 65, advanced, given life expectancies of that time; and then as now judges who were past it could only be gently nudged towards resignation by friends, however crazed their sentencing. The gout-smitten Ellenborough clung to his job 'with adhesive grasp', even when 'he could scarcely totter to his seat'.[24] The 79-year-old Burrough was 'wholly incompetent', Under-Secretary Hobhouse wrote to Peel in 1829 after Burrough was smitten by a seizure at Bridgwater assizes: 'he has neither eyes or ears or voice.' 'No opportunity should be lost of accepting or even of pressing the resignation of Burrough,' Peel agreed: 'From all quarters I hear complaints which it is quite evident are but too well founded.'[25]

'Lawyers belong to the people by birth and interest,' de Tocqueville wrote, 'to the aristocracy by habit and taste.' The people to whom they belonged had capital or property. Despite rare cases like Abbott, the costs of the Inns and the later difficulties of getting briefs meant that only landed families or (increasingly) merchants, manufacturers, attorneys, or surgeons normally sent their scions to the Bar. Both Bar and judiciary lived behind 'an almost impenetrable barrier' across which men even of middling means could seldom pass. Once a barrister, moreover, as the future Lord Campbell wrote in 1810, you lived 'on a footing of perfect equality with men of high birth, of the best education, and the most elegant manners'. On circuit you were not even permitted to travel by public transport or lodge at public inns (to stop unseemly association with attorneys).[26] Isolation took its toll. The pursuit of briefs, place, and favour, devotion to precedent, and the business of always defending what *was*, cramped imagination and feeling alike, as it still may in these bizarre callings.

3. *Bad tempers*

Even allowing for the robust machismo of that John Bullish age, judges' collective *unhappiness*—their bad tempers—is the next thing to notice.

[22] Arnould, *Denman*, i. 432. [23] *Works*, ii. 118. [24] Townsend, *Lives*, i. 389–90.

[25] Peel papers, BL Add. MS 40399, fos. 309–11. Burrough did resign that year.

[26] D. Duman, 'The English Bar in the Georgian era', in W. Prest (ed.), *Lawyers in early modern Europe and America* (1981), 99; Duman, *Judicial bench*, 27, 70.

Their harsher reflexes were laid upon them in childhood; this was an age in which a schoolmaster could write to the *Gentleman's Magazine* that his young pupils were properly lashed when they misbehaved, since 'when all subordination is destroyed, anarchy and confusion will necessarily ensue'.[27] The lash was here accommodated as a political philosophy. Etonians regarded a dozen or so daily floggings 'as a natural incident of the day'. Many hated the regime and projected their resentment on to others, bullies efficiently made. But some loved it. As that later-life flagellant Lord Melbourne said of his Eton tutor: 'I don't think he flogged me enough, it would have been better if he had flogged me more.' The regent himself apparently bared his bottom to the disciplinary attentions of the young ladies at Mrs Collet's Covent Garden establishment.[28]

In their homes it was no wonder that such men thought of themselves as 'created in the image of a punitive, flagellant God', as the disciplines of school spilt over into those applied to children and servants, and, though it was a more secret practice, to their wives. It was Buller J. who had ruled in 1782 that husbands might beat their wives with a stick provided that the stick were no thicker than a thumb.[29] Thanks to these and like habituations, judges and statesmen accepted disciplinary violence in the public realm with diminished difficulty. Their projection of this familiarity into penal values was direct. Most, for example, thought the long campaigns against military flogging ridiculously tenderhearted. While army and navy floggings had long been abolished in France, the duke of York recommended in 1812 that the number of lashes inflicted on wrongdoers be restricted to a maximum of three hundred. In the parliamentary debate on penal whipping in 1823 Peel joked that if Grey Bennet's motion for its abolition were accepted—why, then even 'the punishment of flogging in public schools ought to be abolished by act of Parliament'. MPs would have chortled at the absurdity of the thought, as Peel for once meant them to.[30]

In such a culture a judge's self-image was patriarchal, and sternly so. 'In the moral government of the world . . . the actions of men must be

[27] *Gentleman's Magazine*, 50 (1780), 618.

[28] I. Gibson, *The English vice: Beating, sex and shame in Victorian England and after* (1979), 18, 52, 99, 183; J. Chandos, *Boys together: English public schools 1800–1864* (Oxford, 1984), 230, 237.

[29] L. Stone, *The road to divorce: England 1530–1987* (Oxford, 1990), 198–206, and his *Family, sex, and marriage in England, 1500–1800* (1977), pl. 29; J. A. Sharpe, *Crime and the law in English satirical prints, 1600–1832* (Cambridge, 1986), 176 (Gillray cartoon on the subject).

[30] J. Dinwiddy, 'The early nineteenth-century campaigns against flogging in the army', in his *Radicalism and reform in England, 1780–1850* (1992), 125–48; *The speeches of . . . Sir Robert Peel delivered in the House of Commons* (4 vols., 1853), i. 239–40.

regulated by those rules, which the wise and the good in all ages have declared,' Judge Christian intoned in 1819; and it went without saying that the wise and the good were male. Women were unfitted for action and decision. Sensibility assailed them; they felt too much. On questions of mercy or hanging, Christian continued, women would always vote weakly for the softer options:

> I know how to appreciate their judgments in the general *civil* transactions of life: it is often superior to that of their husbands and brothers . . . but such are the timidity and sensibility of the sex, their greatest ornaments, that I certainly should never employ them, or consult them, as generals, judges, or legislators, in any case where their judgment could be influenced by fear, pity, or *love*.[31]

The judges' professional need to eschew *love*, indeed their inability to accommodate it, will be central to our understanding of them.

It was not admittedly the judge's business to feel deeply. But there was no *social* pressure on him to cultivate feeling anyway. In so far as sensibility was functional to the middle classes—to purchase social standing and repute and to affirm a moral status which they liked to think élites lacked—judges, with status enough already, could do without it. One or two like Dudley Ryder wept fashionably as they sent young women to the gallows and subsequent dissection. After nudging a reluctant jury in 1754 to convict a possibly insane young woman of infanticide and then, 'very well satisfied', condemning her to the noose and the anatomist, Ryder made a speech which 'so affected' him, he told his diary, 'that the tears were gushing out several times against my will. . . . A lady gave me her handkerchief dipped in lavender water to help me.'[32] For young women sent to the noose for infanticide this affecting fashion continued: as late as 1813 Grose J. 'was so much affected in passing sentence that he could not refrain from tears'.[33] But you would not otherwise catch many judges in tears by the 1800s.

Temper is the word which rings through even the most deferential biographies of judges in these years. Thus: Ellenborough's 'severity of demeanour . . . intolerant manner, and . . . frequent petulance, naturally

[31] E. Christian, *Charges delivered to grand juries in the Isle of Ely* (1819), 284, 285 n. Christian was professor of law at Cambridge and chief justice of the Isle of Ely (regarded as a county palatine until 1837). He was in effect a magistrate at £150 p.a., though called to the Bar. A man of famous mediocrity, his cousin Ellenborough thought him fit only to rule a copy-book. He died in 1823, the quip went, 'in the full vigour of his incapacities' (*DNB*).

[32] Gilmore, *Riot, risings and revolution*, 150; J. Langbein, 'Shaping the eighteenth-century criminal trial', *University of Chicago Law Review*, 50 (1983), 118–19.

[33] Anon., *Murder! The trial at large of Ann Arnold, for the wilful murder of her infant child . . . at the assizes . . . Bury, March 26, 1813* (1813), 15.

produced more fear than love'; or: 'The defects which marked and shaded the character of the Chief Justice are those which peculiarly belong to Englishmen, and chiefly resolve themselves into faults of temper.'[34] Abbott likewise was 'a sour old man, with the manners of a pedagogue', 'more overbearing than one expects to find in a man of superior understanding', 'exceedingly harsh and morose'. His 'temper was naturally bad; it was hasty and it was violent'. He hated the modern poets with their feeling hearts; the fact that he 'had read more law than almost any man of his day' helps to explain why. The irascible Gaselee J.—swelling like a balloon (Denman noted) when knighted by the king, but so short in stature that when he stood up the king thought that he was still kneeling; his faculties 'considerably impaired' before he retired—was the model for Dickens's blustering Mr Justice Stareleigh in *Bardell* v. *Pickwick*: not, it seems, a portrait overdrawn.[35] There was no judge who was not, by the standards of sensibility, an emotionally crippled being.

Then there is their culpable inefficiency to be noted. At the bottom of the pile the recorder of London was the worst of the lot. He presided at the Old Bailey, with relays of judges or serjeants assisting him as the court dealt with its several hundred prisoners per session. The Court of Aldermen, themselves deemed judges within their City by ancient privilege, elected him as their deputy from among the common serjeants; the serjeants in turn were elected by the Common Council, 'a parcel of by no means the highest class of tradesmen, who were quite incompetent to form a judgement' (so the lawyers said).[36] None of this guaranteed learning or competence in the recorder. The two who held the office in the early nineteenth century, Sir John Silvester (1803–22) and Newman Knowlys (1822–33), were dismal specimens, both more or less corrupt. The former was notorious for touching up female supplicants when he had them at his mercy. He died aged 76 after carousing a night away with the duke of York.[37] On Silvester's death, Knowlys owed his job to aldermanic opposition to Sir James Mackintosh's candidacy for

[34] Townsend, *Lives*, i. 396, ii. 248.

[35] Foss, *Judges*, viii. 323; Ballantine, *Experiences*, 172; Polson, *Law and lawyers*, ii. 358; Townsend, *Lives*, ii. 246–7; Arnould, *Denman*, i. 265, 409; E. Henderson, *Recollections of the public career and private life of the late John Adolphus* (1871), 187.

[36] Ballantine, *Experiences*, i. 79; Arnould, *Denman*, 198.

[37] Of Silvester, *c.*1811–14, it was said that, to plead for her husband's life, a 'Mrs Hammon was shown up stairs, whilst her friend and I waited in the room below. What was our astonishment, in a short while, to see her rush in, aghast and panting with terror—the Angelo of that infamous hour, had proposed to her to go with him to Richmond from the Saturday to Monday, as the condition of his recommending the remission of her husband's life!!! This fact may appear incredible: it is literally true!' W. Jerdan, *The autobiography of William Jerdan* (1852), i. 130–1.

the office. *The Times* thought Mackintosh would have made a recorder 'philosophically conversant with the principles of law', while Knowlys was 'a mere practitioner in the courts', 'a mere technical lawyer', lacking in personal dignity and legal ability. 'An old twaddle', the younger Ellenborough wrote more bluntly in 1830 after observing Knowlys's prattlings in court.[38]

Under Knowlys's direction the bad name of Old Bailey justice got worse, so that 'the mode in which business was conducted in that tribunal made it a term of opprobrium to be called an Old Bailey barrister'. Appealing on behalf of two men whom Knowlys had sentenced to transportation for stealing from a ship, the rector of Stepney reported that the recorder's 'sharp altercations' with counsel had caused such 'noise and confusion . . . as to make it quite impossible for the jury to pay that attention to the matter' which justice required.[39] Knowlys turned the Old Bailey 'into a bear-garden', a complainant wrote to the Home Office on another occasion. When Knowlys was forced to resign in 1833, it was because he had absent-mindedly ordered the hanging of one Cox whom the King in Council had just reprieved on Knowlys's own advice a couple of days before—Knowlys having been at the Council. At a rowdily exultant meeting of the City's Common Council Knowlys was called an imbecile old man to his face, and points were scored off the aldermen whose creature he was thought to be. The Council put the 'mildest and most charitable construction' on the error, namely that it resulted from 'some mental infirmity incident to his advanced age'. But it was not to be forgotten that Cox was saved from the noose by the slimmest of chances. Sir Thomas Denman, who had attended the Council at which Cox was reprieved, in astonishment read the announcement of Cox's imminent execution in the newspaper, just in time.[40]

Judges' political prejudices are familiar from many accounts of the uses of the law against radicals, libellers, blasphemers, combiners, rick-burners, and rioters in these years. Sounding out abundantly in their charges to grand juries, their texts rang changes on standard motifs: 'The great and important principles of this constitution are the best, the

[38] *The Times*, 1 and 12 Apr. 1824; Lord Ellenborough, *A political diary*, ed. R. C. E. Abbot (2 vols., 1881), ii. 229.

[39] The recorder convicted on the basis of the polluted evidence of 2 other felons. Knowlys conceded that 'a fair case' could be made for mercy, although 'the floating property of merchants is so exposed that nothing but the terrors of the law can prevent plunder even to the extent of ruin and bankruptcy': HO47.73/4 (Rudolph's and Wilson's case), Aug. 1827.

[40] *Morning Chronicle*, 22, 25, 29 June 1833.

wisest, and the freest, that the sun ever yet saw,' as a Scottish judge declared.[41] They opposed legal innovation and lamented the passing of pillory and noose alike, convinced that the 'clamour now existing against the laws of this country is founded in a misrepresentation of facts, in a misapplied humanity, and a misconception and ignorance of the laws of England'.[42] In the 1820s Peel pushed his criminal law reforms past the bench and the Lords by massaging judicial egos. But well into the 1830s die-hards like Coleridge could always be found to contrive a last hanging or two in protest against changing times.

No less chilling were judges' airy views of exemplary justice. Instances from the 1820s fill the appeal archive by the score. Thus Vaughan condemning three burglars at Gloucester—character references saved two, but not the third: Vaughan thought 'one sacrifice necessary to the due administation of criminal justice, and [I] therefore left Kay for execution. Upon this part of the subject my mind is at ease.'[43] A horse-thief sentenced to life transportation in Wiltshire elicited Burrough's observation that 'in the present state of the country I think much relaxation would be unsafe. . . . In consequence of the great extent of this crime I have thought it absolutely necessary that punishments should be inflicted that are likely to deter.'[44] Advising Peel to reject the petition of the parents of Thomas Lewis, sentenced at the Old Bailey to life transportation for the robbery of a watch, the recorder commented self-sufficiently that 'the offence of street theft and robbery have risen to a very alarming height; the prisoner's case was one of eight that I had tried on the London side that evening and of these eight six were for street theft and robberies'.[45] In 1818 Park sentenced a labourer to eighteen months in prison for stealing a sack of oats, and then raised the sentence to seven years' transportation because the man asked from whom he might recover the wages due to him.[46]

Then there was the luckless John Crane, sentenced by Abbott at Huntingdon to life transportation for burglary, while his accomplices were sentenced to three years in prison. Crane was younger than they (18), had not entered the house or shared the booty equally, and had secured their conviction by informing. A peer's younger son applied to Lord John Russell, who had served on the grand jury: 'The only reason I can make out for the difference of sentence is that some of the

[41] The lord president at the Stirling treason trials in 1820: see C. J. Green, *Trials for high treason in Scotland, under a Special Commission held at Stirling, Glasgow etc. in 1820* (3 vols., Edinburgh, 1825), iii. 540.

[42] Christian, *Charges*, 219.

[43] HO47.73/1.

[44] HO47.73/3.

[45] HO47.63/2.

[46] HO42.180 (cited in J. L. and B. Hammond, *The town labourer* (1949 edn.), i. 83).

magistrates who were acquainted with the families of the two interfered with the judge in their behalf.' Russell forwarded a petition signed by the prosecutor and overseer, and by the parents with crosses; it was full of sad self-abasement. Abbott, consulted, could not find or remember his notes. This gave Abbott 'no reason to change my opinion about the sentences when the matter was still fresh' in his mind. He did not see fit to admit that his sentences reflected pre-trial influence; to the contrary, he knew, he said, that Crane was a bad sort.[47]

The judge was a creature of his times and of the systems which gave him standing. Up to a point, this was how justice and sentencing were supposed to work, as we shall see in the next chapter. Add to this the truth that with Eldon's departure as lord chancellor in 1827 fresh air was breathed through the law, a little air anyway, and the judicial trade begins to look human. Lyndhurst, asked as lord chancellor to state the principles upon which he selected a judge, could still say that he looked out 'for a gentleman, and if he knows a little law so much the better'.[48] But although pitiless monsters like Gurney or Campbell continued to abound on the bench in later decades, at least advancement was made possible for a few more genial men like Denman, chief justice from 1832, or to conscientious and pious men like William Page Wood, lord chancellor by 1868.

Still, there were ogres enough—like Baron Hotham in 1800, condemning a 10-year-old boy to death for stealing notes at the Chelmsford post office. He saw in the lad 'art and contrivance beyond his years', and told the court of 'the infinite danger of its going abroad into the world that a child might commit such a crime with impunity, when it was clear that he knew what he was doing'. Hotham meant the lad to be hanged all right, and he would have hanged him but for a problem in the way. In court the boy wore a pinafore and looked so much an 'absolute child', Hotham admitted to the Home Office, that 'the scene was dreadful, on [my] passing sentence'. The crowd in court 'expressed their horror of such a child being hanged, by their looks and manners'. Hotham was forced to hint that the case was open to the royal clemency; the boy was sent to labour in Grenada for fourteen years instead.[49]

Should we, in all these cases, recognize the otherness of élite emotional structures before new forms of affective expression took hold? Or, assuming a common human potential for generosity in all eras, should

[47] HO47.65/5. [48] Ballantine, *Experiences*, 120.
[49] HO42.49 (Hammond, *Town labourer*, i. 83).

THE CLERICAL MAGISTRATE.

"*The Bishop.* Will you be diligent in Prayers—laying aside the study of the world and the flesh ?——*The Priest.* I will.

The Bishop. Will you maintain and set forwards, as much as lieth in you, quietness, peace, and love, among all Christian People ?——*Priest.* I will.

¶ The Bishop laying his hand upon the head of him that receiveth the order of Priesthood, shall say, RECEIVE THE HOLY GHOST."

The Form of Ordination for a Priest.

39. George Cruikshank, *The clerical magistrate*, 1819.

This was one of Cruikshank's illustrations in Hone's *The political house that Jack built*, published after Peterloo. The huge success of Hone's little books (this one sold some 100,000 copies in several editions) indicates that the law's structured hypocrisies were well understood by the artisan and middling classes at whom the books were aimed.

we decide merely to damn these men as stunted villains who consulted their own dignity more than justice or humanity? It can be no anachronism to reply that there was a magnificence in their awfulness, because contemporaries saw it as well as we: 'What would a judge be, without his black cap? Who would respect his horse-hair wig, when it should not remind one of the terrible vengeance of the law? Of what use crimson robes, but as they give the idea of blood?'[50] This was an age when satire from Gillray to Dickens flourished at these monsters' expense, and there was meaning in this commentary. It is at the cultural interfaces where powerful people meet weaker, or old manners clash with new, that the most passionately oppositional expressions will be nurtured. This was an era when clashes on these frontiers were multiple. Two cultures within the polite classes—that of benevolence, and that of whip, pillory, and scaffold—stood in opposition to each other. The threshold at which they clashed was often the courtroom, a flashpoint indeed for the creatively satirical imagination. In the end it is a consolation that the justice handed out by the sable bigots and their henchmen on the magistrates' bench is something to which satirists did their own justice (Plate 39). It is also a consolation that the satirists' words have endured longer than the judges' own:

> Next came Fraud, and he had on,
> Like Eldon, an ermined gown;
> His big tears, for he wept well,
> Turned to mill-stones as they fell.
>
> And the little children, who
> Round his feet played to and fro,
> Thinking every tear a gem,
> Had their brains knocked out by them.[51]

[50] E. Gibbon Wakefield, *The hangman and the judge, or A letter from Jack Ketch to Mr Justice Alderson* (1833), 2.

[51] Shelley's *Masque of anarchy* (written *c.*1819, after Peterloo; published 1832).

CHAPTER 19

QUALITIES OF JUSTICE

1. *Rough justice?*

HOW, IN WRITING ABOUT INJUSTICE, DOES ONE DO JUSTICE TO the furred homicides who so often and cavalierly delivered it? Why not just a curse upon their names? Dislike is easily called into play in writing about justice's unjust agents. But we cannot go on like this without causing offence. There must be a point beyond which we should protect even these monsters from posterity's contempt. It may be impossible to locate them within what Carlo Ginzburg has called a 'really dead history', free from here-and-now prejudice; they deserved the last chapter, some may agree. But we might still strive to admit the otherness of the traditions they worked by and by which they were trapped.

Hanoverian judges worked within a criminal justice system which quite purposefully upheld propertied hierarchy first and delivered justice second, in which respectable patronage, perceived character, and local tensions were *meant* to affect sentences and appeals, in which judges and home secretaries *had* to placate great men even if not to capitulate to them, in which juries were often timid, mercy grudging, pardons rare, and compensations for wrongful punishment unthinkable. Everyone who mattered knew these things and generally accepted them. Very few mercy petitions disputed the influence of 'character' on sentencing; nearly all of them used it to validate appeals. Similarly, judges' prejudicial and limited views of the prisoner's rights had historic sources both in the ancient expediencies of monarchical rule and in providential views of sin; it was not, on their part, premeditated individually or maliciously. A larger principle is at stake in these concessions too. If we see *ancien régime* criminal justice as simply a sham, and forget both how much its own rules mattered to it and how those rules could be applied against its own failings, we deprive ourselves of one explanation for its capacity to change and to become more rule-bound with time.

Three ways forward are open. The first is to allow past judicial values some integrity by recognizing that to ruling élites 'justice' meant something subtly different from what it has come to mean to us. The second is to note the legal arenas in which rules, with whatever difficulty, did on the whole rule. The third is to acknowledge (*none the less*) that the customs of the courtroom all but ordained the shaky rulings which judges delivered. With that behind us, we might feel wryly impressed by such justice as was achieved, because it was achieved against the odds.

In 1751 Dr Johnson denounced the 'vindictive and coercive justice' which rested on 'so many disproportions between crimes and punishments, such capricious distinctions of guilt, and such confusions of remissness and severity, as can scarcely be believed to have been produced by public wisdom sincerely and calmly studious of public happiness'.[1] Tory though he was, he spoke for the advanced opinion of his times, even if in vain. It was being widely reaffirmed in the Enlightenment that just punishment resided in its being matched to the socially agreed heinousness of the offence, and that just administration of the law depended on judges' impartiality, on no man's judging in his own cause, and on judges repudiating questions unrelated to the crime. From Beccarian or Benthamite sources these ideas directed attacks on the penal code after 1770, along with the decay of old understandings about bodily punishment as the body was naturalized. In the long though not the short term, the outcome was a formulation of the essence of justice which we unreflectingly inherit. Mackintosh's definition of justice in 1830 now sounds axiomatic. Deploring 'that excess of punishment beyond the average feelings of good men which turns the indignation of the calm by-stander against the culprit into pity', he defined a right sense of justice as follows:

> When anger is duly moderated,—when it is proportioned to the wrong,—when it is detached from personal considerations,—when dispositions and actions are its ultimate object, it [anger] becomes a sense of justice . . . Whenever [criminal laws] carefully conform to the moral sentiments of the age and country,—when they are withheld from approaching the limits within which the disapprobation of good men would confine punishment, they contribute in the highest degree . . . to nourish and mature the sense of justice.[2]

[1] S. Johnson, 'The necessity of proportioning punishments to crimes' (1751), in W. J. Bate and A. Strauss (eds.), *Collected works of Samuel Johnson* (New Haven, Conn., 1958–), ii. 242.

[2] J. Mackintosh, 'The progress of ethical philosophy' (1830), in *The miscellaneous works of . . . Sir James Mackintosh* (1851 edn.), 122.

When Mackintosh offered it, however, this statement had only lately ceased to be self-consciously oppositional. It was not that those who defended the old legal regime failed to acknowledge the beauty of this conception of justice, proportion at its centre; but how they bypassed it was hugely significant. Paley in 1785 admitted that 'true justice' would indeed entail 'the retribution of so much pain for so much guilt'. This, he wrote, was 'the dispensation we expect at the hand of God' and was 'the order of things that perfect justice dictates and requires':

> From the justice of God we are taught to look for a gradation of punishment, exactly proportioned to the guilt of the offender. . . . A Being whose knowledge penetrates every concealment . . . may conduct the moral government of his creation, in the best and wisest manner, by pronouncing a law that every crime should receive a punishment proportioned to the guilt which it contains, abstracted from any foreign consideration whatever.

Paley none the less had no difficulty in exempting human law from so exacting an ambition, and he did so by asserting that true justice was unattainable in an imperfect human society, since men could not aspire to the full knowledge of innocence or guilt which was given only to God:

> when the care of the public safety is entrusted to men, whose authority over their fellow creatures is limited by defects of power and knowledge; from whose utmost vigilance and sagacity the greatest offenders often lie hid; whose wisest provisions and speediest pursuit may be eluded by artifice or concealment; [then] a different necessity, and new rule of proceeding results from the very imperfection of their faculties.

Law therefore had to go for lesser goals. Socially agreed notions of proportion *might* have to be taken into account to prevent public outrage, but otherwise 'the proper end of human punishment is, not the satisfaction of justice, but the prevention of crimes': 'The fear lest the escape of the criminal should encourage him, or others by his example, to repeat the same crime, or to commit different crimes, is the sole consideration which authorizes the infliction of punishment by human laws.' To this end, punishment need only be determined according to 'the difficulty and the necessity of preventing' crimes, while 'the uncertainty of punishment must be compensated by [its] severity'. It was 'necessary to the good order of society, and to the reputation and authority of government' that people *believe* that convictions depended upon the proof of guilt and that sentences were appropriate to the crime. But should mistakes happen and the innocent sometimes hang, 'he who falls by a mistaken sentence, may be considered as falling for his country'. The chief end of the criminal law was not justice but 'the welfare of the community':

> The security of civil life, which is essential to the value and the enjoyment of every blessing it contains, and the interruption of which is followed by universal misery and confusion, is protected chiefly by the dread of punishment. The misfortune of an individual, for such may the sufferings, or even the death of an innocent person be called, when they are occasioned by no evil intention, cannot be placed in competition with this object.[3]

With hindsight, one may believe that the increasing elaborateness of justificatory texts like Paley's reveals a mounting anxiety about hanging law. The simple certainties of punitive pamphlets like *Hanging not punishment enough* (1701) were gone. After all, Paley's was an era when (before French events silenced these claims) Fox could announce of the capital statutes that 'their inhumanity was manifest, their absurdity ridiculous', when Blackstone believed it 'a kind of quackery in government . . . to apply the same universal remedy . . . to every case of difficulty'; when Burke could declare the system 'radically defective' and the criminal law 'abominable'; and when Wilberforce pronounced the system 'barbarous'.[4] So it is tempting to discern in Paley's exposition a whiff of special and defensive pleading.

But the truth is a little bleaker than that. His argument held sway for nearly half a century after 1785, and there was little defensiveness at these levels. This was because his expediential and discretionary arguments resonated with (and sustained) mentalities which went beyond reason, past interest even, into the anciently internalized assumption (among middling people as well as élites) that the primary function of criminal law was not to deliver justice so much as to sustain a right moral and social order with property at its centre. You had only to reject the premiss that human justice could not hope to reveal *truth* to see Paley's edifice as an apologia for maintaining hierarchy by legal force when need be. Acquiescence in the argument also expressed that pessimistic view of plebeian incorrigibility which characterized tory and Anglican penal thought for half a century yet. At issue in much of the debate about capital punishment were two competing views of human nature. The reformers hoped to bind the classes through the diffusion of sympathy and civility, a process which they believed the death penalty subverted. Most tories, contrarily, thought that the principle of

[3] W. Paley, *Principles of moral and political philosophy* (1785), 527–8, 531, 535, 552–3.

[4] I. Gilmore, *Riot, risings and revolution: Governance and violence in 18th-century England* (1992), 158–9.

humanity was merely speculative: only terror would protect a public interest threatened by a populace beyond the reach of finer feeling.[5]

It can be argued that in no society has justice been the *primary* value to which law devotes itself. The primary legal value to which all law has been dedicated has been order. It has been understood since classical times that without order there can be no justice, or liberty either for that matter.[6] Paley did at least accept the desirability of a jury to check 'corrupt particularities' among judges; he did insist that the capital code was 'never meant to be carried into indiscriminate execution', and that the prerogative of mercy ensured that it never could be; and he did justify the system of discretionary justice, however harsh, chiefly because it was preferable to 'that inspection, scrutiny, and control which are exercised with success in arbitrary governments' and which were inimical to 'the liberties of a free people'.[7] Still, justice was the remotest of his concerns, and the criminal law's disciplinary purposes were candidly detached from questions of proportion. They were only approximately, subordinately, or even (as Paley himself hinted) cosmetically concerned with guilt and innocence. And discretionary punishment was mediated through an alleged communal interest, as élites defined it. Altogether, even though he was widely challenged and by some contemptuously dismissed, Paley provided a theoretical sanction for punishing lawbreakers as harshly as perceived or invented necessity determined. Had other things been equal, punitive discretion could have given off the high and unmitigated odour of tyranny. Other things were not equal, of course. And the fact that they were not equal is the point upon which any tribute to the rule of law in *ancien régime* England rests.

A free punitive discretion could not prevail in the eighteenth and early nineteenth centuries because countervailing principles and practices contained it. These were enshrined in the principles of equal access to and equality before the law (valuable to the gentry in their own defence against absolutism); in the rule-bound nature of judicial process (which meant that prosecutions failed even if there were small mistakes in forms of indictment); and in ever stricter rules of evidence and increasing emphasis on the prosecution's obligation to prove its case.[8] The

[5] R. McGowen, 'A powerful sympathy: Terror, the prison, and humanitarian reform in early nineteenth-century Britain', *Journal of British Studies*, 25 (July 1986), 316.

[6] P. Stein and J. Shand, *Legal values in western society* (Edinburgh, 1974), ch. 2.

[7] *Principles*, 504, 534, 541–2, 552–3.

[8] Both were facilitated by counsel's increasing participation in trials after the 1730s, and the former by the printing of reports: Beattie, pp. 352–76; D. Duman, 'The English Bar in the Georgian era', in W. Prest (ed.), *Lawyers in early modern Europe and America* (1981), 86–128.

largest constraints referred to Blackstone's 'grand bulwark' of English liberty, the jury system; and to the growing independence from the crown and from lay élites of both the judiciary and the Home Office in its judicial capacity. Some other constraints worked to similar effect, but accidentally. The high costs and inconvenience of private prosecution meant that most lawbreakers escaped the courts entirely (again, the addiction to private prosecution expressed a suspicion of state prosecutions which was fuelled by seventeenth-century memories).[9] Almost despite itself, moreover, the criminal law was structured to achieve something like justice, if only because it was chronically vulnerable to accusations that it failed to live up to its own principles. It was not merely tactical cynicism that against much evidence persuaded most judges that justice was what they delivered. They might not have been thinking deeply about what they did, but they were captured by their own rhetoric, just as gentry élites were whose ancestral memories were shaped by the battles against seventeenth-century absolutism.[10]

The royal prerogative of mercy is a good example. It can be interpreted merely as a legitimizing system, as Blackstone and others implied that it might be (see Chapter 6). But to serve this function it had also sometimes to be the agency of just restitution it purported to be. When called upon, the appeal system did sometimes restore an order in things which humble as well as great people would acknowledge as right. Home secretaries or members of the king's Council who would have deplored the intrusion of Beccarian or Benthamite certainties into sentencing could sometimes be nudged towards an adjustment of verdict, sentence, or punishment by appeals to fairness and proportion.

The prerogative could operate thus when it was forced to address the unfairness of small sentences as well as of capital ones. Two examples will do. The Revd Charles Ingham was one of those many petty tyrants of the magistrate's bench who were overactive in applying their 'narrow-minded opinions and prejudices', as Brougham put it, and loath to let an 'incorrigibly bad character' escape lightly once he got him in his clutches.[11] At the Oundle petty sessions in 1824 he fined a day-labourer

[9] D. Hay and F. Snyder, 'Using the criminal law, 1750–1850: Policing, private prosecution, and the state', in D. Hay and F. Snyder (eds.), *Policing and prosecution in Britain, 1750–1850* (Oxford, 1989), 25–6, 34, 43–4. For examples of high prosecution costs, see D. Philips, '"Good men to associate and bad men to conspire": Associations for the prosecution of felons in England, 1760–1860', in Hay and Snyder (eds.), *Policing and prosecution*, 115–17.

[10] E. P. Thompson, *Whigs and hunters: The origin of the Black Act* (1975), ch. 10.

[11] The tone of decision-making at this level is exemplified by the Revd Thomas Methold of Suffolk when, in 1827, he sentenced 3 men to 7 years' transportation for stealing poultry, because they had been 'a terror to the country for years'. When prisoners' counsel protested to the Home Office at the

James Baker £20 with £1 costs for cutting an ash-tree bough valued at a shilling. The tyranny here was perfect. Ingham knew very well that Baker could not afford such sums: so he had a neat excuse to put Baker into prison for a year for non-payment. This was desirable, he later explained to Peel when called to account, since Baker had already been imprisoned once for poaching, and short periods of imprisonment served no useful purpose whatever.

Ingham had not, however, counted on the vigilance of Samuel Wells, proprietor of the *Huntingdon and Cambridge Gazette*. For some weeks (Ingham complained) the newspaper had been giving 'unjustifiable accounts' of the proceedings of the Oundle bench, betraying a 'disrespect and contempt' which 'merit[ed] the animadversion of the highest legal authority'. Ingham huffed and puffed that Wells was out only for publicity; but Wells, undaunted, wrote to Peel demanding to know why the prosecution had not proceeded under the milder Summary Trespass Act instead of the 1766 Preservation of Timber Act[12]—a statute which 'God knows is most intolerably cruel' and which anyway punished the offence of damaging growing trees in a plantation: Baker's stolen ash bough had already blown down in the wind. It had been Baker's poaching conviction that earned him his sentence, Wells pointed out: and 'it ought never to be endured that a JP convict harshly with reference to suspicions unconnected with the trial. Had the poor fellow the proper means I have no doubt the Counsel of Kings Bench would read a useful lesson to the magistrate.' Peel agreed and gave Baker a pardon.[13]

Similarly, in 1824 three men were sentenced at Shrewsbury quarter sessions to seven years' transportation for poaching while armed at night. The men sent a petition signed by crosses to affirm that they had previously been of good character and now left destitute families. What

'flippancy' with which Methold had conducted the trial (the conviction had been obtained 'contrary to established practice' on an accomplice's evidence and turned on character, not proven guilt), Methold replied robustly that 'a more desperate set of men (I believe) never stood at the bar of a court of justice. . . . There cannot be three worse men.' Methold was also riled by counsel's presumption in complaining at all: counsel 'behaved in a very improper manner to the magistrates, but unfortunately the person who took down his expressions has lost the paper, so that we cannot proceed, if advised to do so in the King's Bench': HO49.73 (Hawes, Curtis, and Jordan's case), Nov. 1827. Not all clerical justices were like this. The Revd Mr Haggitt, committing magistrate of Cambridge, took up the question of his own sentencing of a thief to 7 years' transportation when he learnt that the man's 'connections are respectable' and that 'the robbery of the shawl was committed in revenge for the ill treatment of an artful dangerous woman on whom he had laid his heart and who fully encouraged him till interest opened other projects to her'. Peel's memo noted that 'Mr Haggitt's mind has been strangely changed', but two years later Haggitt was still badgering him for a pardon 'absolutely as a matter of justice', and even went to London to see Peel in person. A year later Peel conceded the case on the grounds of a report of good behaviour on the hulks. A free pardon was belatedly granted (HO47.64/1 (Peachy's case), 1822–6).

[12] Radzinowicz, i. 22, 61–6. [13] HO49.65.

tipped the balance in their favour was information about the magistrates' conduct of the case which was given by their employer Mr Foster, 'the largest ironmaster in Staffordshire'. It was supported by a huge parchment petition from most of the employers of Stourbridge and its neighbourhood. Foster reported that the men's attorney was refused admission to their examinations and access to the depositions afterwards; an impossibly hefty bail of £25–£50 was demanded; and, most damningly,

> the prosecutor sat upon the Bench with the magistrates during the trial, and was seen to converse with some of them. All the witnesses were the prosecutor's servants, and though they swore to having received very serious injuries in the assault, the surgeon who attended them was not produced. . . . After the summing up and while the jury were deliberating on the verdict, they were irregularly called back by the assistant chairman, who repeated the strongest evidence against the prisoners, and was evidently anxious for a conviction.

Peel pardoned these three men too.[14]

Readers of this book will not expect it to paint a rosy picture of the criminal law. But *if* shifts in perception and sensibility contributed to the repudiation of the capital code or to the more impartial administration of the law (not forgetting our introductory reminder that material shifts determined these things more forcefully), it was because they worked upon a system which was equipped to accommodate them, and which had every interest in doing so to keep the law's repute in good repair. None of the principles or practices enshrined in the rule of law before 1830 *required* a commitment to proportion in punishment. But they did allow space for the expression of that ideal—through jury mitigations and recommendations to mercy, for example, and in the petitioning process itself. Also the principle of impartiality, however often betrayed, could not but undermine faith in discretionary punishment once there was a will to extend it there. It also counted that many curbs on interested or casual justice were more firmly in place by the 1820s than they had been a century before—not least thanks to the increasing intervention of professional lawyers in the courtroom and the appeal process. The law could be an ass, and class interest ranged freely within it. But it is not our intention to deny that it was an arena of negotiation too.

[14] HO49.65.

2. *Containing illegitimate influence*

The two most applauded constraints on the abuse of law were the jury on the one hand and the independence of the judiciary and the Home Office in its judicial role on the other. The law might have to jump to the interests of statute-makers in parliament; but independent juries and judges ensured that the overweening interest of lay élites would be kept in check in the trial and sentencing process at least.

The independence of juries should not be overstated. Beattie talks of a general agreement between judges and juries about the purposes of the criminal law throughout the eighteenth century. This was ominous: juries owned property after all. Moreover, in most trials jury deliberations were casual to a high degree. This enabled judges to deploy 'crafty distinctions and ensnaring eloquence' to 'throw dust in the eyes, and confound the sense of a well-meaning jury', determining outcomes with little difficulty. Judges bullied juries with directions which 'were brief, but pointed and leading, if not coercive'.[15] They had no compunction about shaping verdicts through their dominance in court and, especially before the arrival of the lawyers, their control of trial evidence. They ignored jury recommendations to mercy when it suited them. And although most radicals lauded the jury, others knew better. The radical Samuel Bamford, his Peterloo experience behind him, thought the jury 'one of the most bungling pieces of judicial machinery which could have been put together'. It was composed of men 'who had just the brute instinct of beavers to scrape a little substance together, and to keep it; but who for all other purposes, were far behind their neighbours'.[16]

Despite these limitations, juries' self-confidence and independence seem to have increased steadily. Independence from judicial coercion, conceded in the judgement on Bushel's case in 1670, was in principle never in question thereafter except in trials for seditious libel; Fox's Libel Act of 1792 affirmed it there too. That juries were obstacles to those who sought expediential justice was proved by their critics. Paley paid lip-service to the jury, but he still found them 'inadequate', on grounds eloquently different from Bamford's. The social status of these farmers, tradesmen, publicans, and yeomen biased them, Paley

[15] Beattie, pp. 406–10; T. A. Green, *Verdict according to conscience: Perspectives on the English criminal jury trial, 1200–1800* (Chicago, 1985), 334, 271 (citing *Bingley's Journal*, 1770); D. Hay, 'The class composition of the palladium of liberty: Trial jurors in the eighteenth century', in J. Cockburn and T. Green (eds.), *Twelve good men and true: The criminal trial jury, 1216–1800* (Princeton, NJ, 1988), 305–57.

[16] S. Bamford, *Passages in the life of a radical* (orig. pub. 1884; Oxford, 1984), 295.

thought, against clergymen, revenue officers, bailiffs, 'and other low administrators of the law'. They annoyingly favoured tenants over landlords and estate-holders over lords of manors. And they were easily inflamed by 'political dissensions or religious hatred': 'these prejudices act most powerfully upon the common people, of which order juries are made up.'[17] Ensuing battles over sedition revealed real tensions between juries and judges in which London juries did stand up for themselves. Élites knew that it was better to live with the jury than not, but their mistrust suggests that it could upset apple-carts. Its popular standing after its victories over ministerial diktat in the 1780s deepened its self-esteem, while increasingly systematic presentation of trial evidence facilitated independent decision-making too. Peter King has found that by the later eighteenth century Essex juries were more experienced in law than hitherto, and more independent.[18] Denman, when common serjeant of London, found Old Bailey juries acute.[19] By the 1820s Beattie notes a 'mental shift' in the acknowledgement of jury independence, in Phillips's *Golden rules for jurymen* (1820), for example, which exhorted jurors to give benefit of doubt to the accused and to stand up against judges (though Phillips's was an oppositional voice).[20] Juries' undervaluing of stolen goods to avoid a capital sentence had a longer history than reformers knew or admitted in the 1810s; pious perjury might have been decreasing in the early nineteenth century rather than otherwise.[21] Still, the fact that reformers' claims on this score were taken seriously speaks not only for a sensitivity to the jury's independence but also a suspicion that with time it could only grow.

The appeal archive of the 1820s also reflects jury confidence. Vexed judges frequently blamed juries for delivering verdicts against their advice. Juries protested conversely when their mercy recommendations were ignored by judges. Their protests often turned on unease about the proof of guilt or distaste for capital sentences. What was new was not so much the distaste as the fact that the distaste was now overtly expressed. A juryman at Noden's rape trial in 1829, as we saw in Chapter 17, refused to accept the foreman's insistence that Noden was guilty and forced the jury to retire to argue the case, insisting on a recommenda-

[17] Paley added that the requirement of unanimity in their verdict meant 'less assurance that the conclusion is founded in reasons of apparent truth and justice, than if the decision were left to a plurality, or some certain majority of voices': *Principles*, 505, 522–3; Gilmore, *Riot, rising and revolution*, 149.

[18] P. King, '"Illiterate plebeians easily misled": Jury composition, experience, and behaviour in Essex, 1735–1815', in Cockburn and Green (eds.), *Twelve good men and true*, 254–304.

[19] J. Arnould, *Memoir of Thomas, first Lord Denman* (2 vols., 1873), i. 203.

[20] Beattie, p. 375.

[21] See p. 407, above.

tion to mercy. When despite this Noden was left to hang, the juryman broke ranks by signing the mercy petition and sent the Home Office an affidavit dissociating himself from the verdict. This behaviour, the Home Office under-secretary noted tellingly, was 'very singular'. When, similarly, Park J. sentenced William Astbury to death for coining at Warwick assizes in 1823, it was the jury foreman who mobilized the mercy petition and got the whole jury to sign it. He wrote passionately to Peel that 'the unsupportable guilt of having to answer for the life of a fellow creature' tormented him: 'Is there nothing that can influence you? Mighty God! turn I pray thee the heart of the honourable individual [Peel] to mercy! mercy! and mercy rest upon him, when he stands in need.' Judges, and doubtless Peel, took a dim view of such outbursts. But how little power they had to silence them Park J.'s report to Peel in Astbury's case makes clear. Park had told the jury that

> having once recommended [Astbury to mercy] through me, . . . they had no further right, as a Body, to interfere. [The foreman] said, however, that the jury had determined to petition His Majesty through you [Peel]: I told him I would have nothing to do with it. But I expressed my astonishment that they would take such an interest for an *actual wholesale* coiner, where the quantity of the money which he had *just* coined, and the quantity found ready for the press, were so enormous; and that I thought if ever an example was to be made, this was probably the case.

The jury did petition, but in the event the judge could not be denied his discretion. Peel let Astbury hang.[22]

Where did judges stand in this cheering depiction? As far as standard capital cases were concerned (lesser crimes were different), what is generally impressive in judges' responses to appeals was their crabbed integrity. They did rely on magistrates' advice about the prisoner's record or the state of crime in the locality, but they were seldom otherwise influenced by factors outside their personal views about social needs or the heinousness of the offence. By the 1820s they resisted lay influence, however highly placed. When Lord Nugent, Sir John King, and others on the Buckinghamshire grand jury appealed for the lives of three horse-thieves in 1823 (visiting Peel in person to press him to mercy), Garrow J. announced that while his 'personal feelings' would be 'gratified' if other grounds could be found for reprieve, no such grounds could be found in the evidence. The men were hanged.[23]

Élite expectations and pressures were chronic problems for the

[22] HO47.64. [23] Ibid.

Home Office too, but were similarly contained. Many gentry still thought of legal penalties rather as they thought of their own property. They plied judges or home secretaries with their opinions on this case or that as if applying for favours in a gentleman's club. But by the 1820s this habit was generally tamed; and anyway such pressures as continued were usually in favour of mercy and about lesser crimes. The days of casually multiplied capital statutes were past. *Noblesse* now chiefly obliged aristocrats, gentry, and gentlefolk to come to the aid of sad cases in their domain, with ladies playing this benevolent game as keenly as men: 'Judge Moore told me, the first Connaught circuit he went to two murderers were condemned; one of the Grand Jury said to him, "One of those men will be hanged." "*Both* will," said he. "Oh no," replied the Grand Juror, "*Lady Sarah* always *saves one*." '[24]

In the late 1830s whigs like Lord Normanby (just cited) looked back ironically on the 'good old tory times' when élite influence was common, and congratulated themselves that it was no more. In fact when Peel took over the Home Office in 1822 the larger despotisms of which powerful men were capable had already been curbed. Petitions not backed by new evidence, or pressures other than from magistrates cognizant of cases or offenders' reputation, usually met with short shrift. When in 1822 Sir Thomas Champneys accused Peel of not responding positively to a petition from him—never before had he been so rebuffed—Peel scribbled on the back of his letter: '*mere impudence*'.[25] When in 1829 the earl of Denbigh objected to Burrough J.'s lenient sentencing of 'a very numerous and desperate gang of poachers', one of whom had shot the earl's gamekeeper, Peel replied ingratiatingly but did not change the sentence. Pressure of this kind was enduring.[26] Unless other evidence weighed, it was resisted.

In the 1820s it was chiefly in respect of non-capital cases that responses were more positive, but again usually to mitigate punishment, not enhance it. One day in a fit of benevolence the earl of Montmorris

[24] Normanby continued, of his lord lieutenancy in Ireland: 'there has, I know, been no *personal* remission of sentence for any offence, however trivial, for the last two years and a half . . . and in consideration of all the more serious cases I am known to take more pains, and more invariably to consult the judges, and more generally to be guided by their opinions than any of my predecessors': Normanby to Melbourne, 29 Jan. 1839, in L. C. Sanders (ed.), *Lord Melbourne's papers* (1889), 392–3. See HO47.73 for a typical petition from Lady Sarah Bayley and local gentry in Cook's case, tried at Suffolk assizes, Aug. 1827.

[25] PC1.77.

[26] The earl's letter came just after Peel's brother passed on to him the opinions of 2 Tamworth gentlemen that a petition from a horse-thief should be ignored, since the man was a notorious poacher: 'they sincerely hope you will not sanction the slightest mitigation of sentence' (PC1.77/I).

visited Worcester gaol and found a plausible 'object of compassion' in Sarah Randall, sentenced by Garrow J. to two years for receiving stolen goods. When he applied to Peel for her discharge a year before her time, the judge lengthily justified his sentence in familiar terms:

> the offence of breaking into cottages and other places during the absence of the family and robbing them, very often of everything portable, to the great distress and some times the absolute ruin of poor families occurs with distressing frequency; I have therefore found it necessary to repress the offence by severity of punishment. At Worcester it has appeared in many instances that a number of young men lead a life of profligacy and plunder associated with unhappy women of the wretched class to which Sarah Randall belongs.

But the earl replied that far from being one of the 'unhappy women of the wretched class' of Garrow's imaginings, her parents were honest, she had behaved well in prison, and she was dying of consumption. The judge conceded the point and Peel let Sarah Randall free.[27]

When pressure was self-interested, positive responses were sometimes made at low levels of sentencing, provided they were in the direction of mercy. In 1823 Lord Ravenscroft prosecuted two of Sir Robert Shafto Hawkes's workmen at Durham for shooting at night in his woods. Hawkes, however, missed the men's labour, and after six months of their year's prison sentence had passed, Ravenscroft came under pressure from their employer. Hawkes 'said [his] manufactory was at no time so much in want of workmen of the same class' as these two. Ravenscroft wrote that the men were of good character after all, they supported wives and children, and might as well be released. The judge deferred meekly, adding that he fully understood 'the trifling degree of guilt that class of people attach to the crime for which they are suffering'.[28]

Aristocratic influence was brought to bear likewise for an Irishman sentenced at Middlesex sessions to two years for attempted rape. Since he was the son of a widow on his estate, the earl of Fingall felt that he should write on his behalf. The man had merely 'got into a scrape'; the girl had probably fabricated her story to get him to marry her; the sentence was 'very severe'. The sessions chairman reported defensively that 'the evidence of his brutality and violence was unquestionable, and it did not depend on the girl's account' alone; indeed, 'it was impossible to elicit anything favourable out of the evidence'. Perhaps, however (he

[27] HO47.73. [28] HO47.64.

meekly added), the sentence *was* severe and *part* of it might be remitted: and so, duly, it was.[29]

Then there were pleas for special treatment in punishment when prisoners were more gently born or educated than most desperadoes. These too were sympathetically received. Thomas Chilton was sentenced at Dover to seven years' transportation for embezzlement. Aged 50, the son of a respectable farmer, and a failed brewer but married to a respectable woman with property, he was of excellent character (his well-placed supporters petitioned); scrutiny of his employer's accounts now suggested that he was wrongly convicted too. Now on the hulk *Retribution*, he was in acute mental distress in the company of 'such a set of ruffians as are on board the hulks'. Chilton was pardoned.[30]

Likewise the plight of the embezzler John Fearnhead, sentenced to fourteen years' transportation. The local MP was 'pressed by some of my most respectable constituents' to plead for migitation in view of Fearnhead's education, youth, and good character, and to ask whether 'he may be separated both previously and during his voyage from the other convicts with whom his education and preceding habits make it most grievous both to himself and to his friends that he should be forced to an association'. The hope was that Fearnhead might avoid being 'sent out in an ignominious manner' and might 'be allowed to take out somewhat more than is usually permitted to persons like him, say two trunks'. 'No mitigation,' Peel scribbled on the file—but then also: 'Attend to the other parts of Mr Wallace's request.'[31]

The social interests implicit in such interventions do not speak for impartial justice. But since they worked for mercy and endorsed the purposes of the prerogative, let it pass. Lay pressure tended to work benevolently. I have found no cases in which it sought hangings or transportations which judges had not already ordered.

The only effective interventions in favour of *harsher* punishment at serious levels came from magistrates. This was thought proper. Judges acknowledged in many appeal papers that their sentences would have been harsher or lighter if before the trial magistrates had said this or that to them, or not said this or that, as might be. In 1822 the chairman of the Kent bench, Sir C. Knatchbull, was astonished that Park J. had commuted the capital sentence of a sheep-stealer to a mere year's imprisonment. He asked Peel to give the community 'the sentence it required and to which it was entitled by not allowing this prisoner to

[29] HO47.73. [30] Ibid. [31] PC1.70

return to the home of his long continued depradations'. For once the judge was softer than the local community. Park pointed out that nobody deplored sheep-stealing more than he. But he always distinguished between stealing sheep to sell and stealing a single sheep to eat (which was what Post did). 'If I had known him, at the time I was recommending him [to mercy], to be a man of notorious bad character, I should probably have been more severe; but certainly I should not have carried the sentence into execution.' Knatchbull persisted, however: 'Personally I have no wish to press for severity of punishment, but . . . sheepstealing is an offence of almost daily occurrence in this neighbourhood' and Post was a man of the very worst character. 'The magistrates expect to be supported, the people have a right to be protected—the administration of the law, as exemplified in Post's case . . . will defeat the hopes of both.' Peel gave way: 'I think a year's imprisonment much too slight a punishment in this case. . . . 14 years transportation and detain him' on the hulks.[32]

Peel did not always defer to magisterial pressure. When in 1822 the prosecutor and six of the jury petitioned on behalf of a man sentenced at Canterbury to seven years' transportation for stealing a shilling's worth of corn and a pennyworth of chaff ('the sentence . . . is very severe, and such as the nature of the case does not in our opinion call for'), the recorder of Canterbury admitted that the magistrates had told him beforehand that the man had been acquitted at the sessions some years before for passing bad money—'but under circumstances which left very little or rather no doubt of his guilt'. Peel reduced the sentence to a year's hard labour.[33]

3. *Justice in the courtroom*

It is cheering to know that some independence from lay influence was achieved in the criminal law's administration, and that interest at its most naked was curbed. These things would certainly cheer historians who deplore representations of *ancien régime* law as being chiefly supportive of property interests or ruling-class rule. They have striven hard to show how open it was, relatively, to poorer people, how little biased it was, relatively, in administration, how rule-bound, relatively, it was,

[32] PC1.79. Knatchbull made a habit of writing to the HO in these terms: see PC1.74. The PC files are full of communications from local magistrates 'humbly begging that you will, in your wisdom, not permit this sentence to be shortened' (PC1.74).

[33] HO47.63.

and so on. They have healthily corrected a one-dimensional view of the law as malign.

Sometimes they have gone to the other extreme, however: Panglossian histories of English criminal justice are not hard to come by. Much of the energy invested in celebrating themes like these has been fuelled by fantasies about what 'marxist' historians have said about the law (these last seldom as vulgar as it has been convenient for their enemies to represent them); or else they have been fuelled by tacit reference to a counterfactual consideration—namely, the celebrants' sense of the despotisms which *would* have prevailed in eighteenth-century to modern England in the absence of legal rules, and which in much of the world do prevail. Britain, relatively, had much to be thankful for, they insist: paint not the island story, therefore, in darkest hues, for when all is said and done justice was pretty adequately done, relatively speaking.

All this is true; it is not in the interests of this book to deny it. The only trouble is that if you argue this way, you silence further debate. Implied or actual references to a hypothetical worst scenario will inevitably make the better scenario seem the best. Neatly evaded then are those darker shadowings which still cry to be filled in. You don't have to be a marxist to see that what happened in the early nineteenth-century courtroom takes us to the heart of the shadows. These things should not be left wholly off our island record.

The mentalities embedded in courtroom practice were in their way quite of a piece with (say) the horrors of execution. Rule of law or no rule of law, procedural constraints could hardly check judicial harshness, since procedures were to a degree designed to accommodate it. The contexts in which evidence was heard, verdicts reached, and sentences decided upon not only expressed the rule-bound curbs on arbitrary justice which have normally been emphasized; they also expressed the working-out of the attitudes to the world and its hierarchical order which were discussed in the preceding chapter. To say exculpatorily that an undeveloped bureaucracy or time-bound values explained the cavalier processes to which we now refer merely reaffirms the point that the procedures of law had no life independent of those who operated them and thought them sufficient.

Slap-happy justice, then? It depends on one's expectations. By ideal standards, though also by those of contemporary critics, *ancien régime* courts did offend against justice when other considerations intruded. And other considerations extraneous to the crime often intruded because their legitimacy was taken for granted; nor was there any great incentive to think hard about what was done in their name. In a society of massively imbalanced wealth and power, élite support for established legal processes was watertight, and for decades critical voices outside the charmed circles could, within limits, be ignored. There was always some concern about excessive hanging, some careful treading around the questions of whom and how to prosecute for sedition and treason, and mounting tensions about the noose. Some deny that there was such a thing as Pitt's 'terror', so circumspect were great men in the 1790s in going not for blanket prosecutions but only for those in which seditious targets would be squarely hit.[34] Still, even at its most beleaguered the law got away with murder enough. In inflammatory times juries were packed, spies employed, show trials arranged, aged radicals pilloried, men and women easily hanged, traitors decapitated, Six Acts issued, yeomanry called out—liberties, rule of law, and public opinion notwithstanding.

All this is often ascribed to ruling-class fear in and after the revolutionary decades. But the Ellenboroughs of the bench did not quail before 'opinion' so much as ignore or wonderfully defy it. After Hone's acquittal for blasphemous libel in 1817, the lord chief justice rode into the jeering Westminster crowd and 'laughed at the hooting and tumultuous mob, who surrounded his carriage, remarking that their saliva was more dangerous than their bite'. Emerging from the crowd he stopped his coach so that his coachman might buy some kippers.[35]

Much followed from these secure casts of mind. The most important was a relaxed attitude to procedures inherited from past times which were defended reflexively as self-validated traditions, if they were thought about at all. These did defendants so few favours that the wonder is that any justice was done to them whatsoever. The catalogue which concludes this chapter draws on the evidence of the appeal papers, among other sources. As always, the dossiers address the most clumsily judged cases; but these assuredly count in assessing the qualities of justice. Each year hundreds of examples accumulated to show how unambiguously (and palpably) these attitudes damaged defendants' interests; and there would have been more had patrons been on hand to bring them to light. Nothing much was done about them, however. What follows comments on (i) the disorders of court ritual; (ii) the in-built biases against prisoners, including the restrictions on defence

[34] C. Emsley, 'Pitt's terror: Prosecutions for sedition in the 1790s', *Social History*, 6/2 (1981), 155–84.

[35] W. C. Townsend, *The lives of twelve eminent judges of the last and the present century* (2 vols., 1846), i. 388.

counsel and the implausible fiction that the judge was the prisoner's friend; (iii) the hastiness of trials, the absence of lawyers, and the inattention to witnesses; and (iv) the principles and practice of exemplary sentencing and the influences on sentencing of 'character'.

(i) *Mayhem in court*

The pomp of assizes has lately impressed historians as 'a formidable spectacle in a country town, the most visible and elaborate manifestation of state power to be seen in the countryside, apart from the presence of a regiment'. It was 'a meaningful ceremony in which gestures, images, and phrases were all full of significance', we are told elsewhere. John Wesley thought that assize rituals 'by means of the eye or ear . . . affect the heart: and when viewed in this light, trumpets, staves, apparel, are no longer trifling or insignificant, but subservient, in their kind and degree, to the most valuable ends of society'.[36] The trouble with pushing this argument too far is twofold. In the first place, courts had less need to attend to their own legitimation than we tend anachronistically to expect; and the opinion of the populace mattered less, arguably, than it implies. As at the scaffold, the authority expressed in court did not depend on the smooth orderliness which later generations required to persuade them of the reaching controls of majesty. The king's commission, the power of the noose, and the support of the great and middling men in the nation sufficed for the law to be able to afford disdain for effects if effects went awry.

Secondly, even discounting the likelihood that executions delivered far more significant effects of these kinds than assizes did (see Chapter 3), the question remains open as to whether people really were so 'mystified by the first man who puts on a wig'.[37] Later sophisticates knew that, with trumpets flourishing 'not necessarily in harmony', the high sheriff 'in a mysterious costume', the judge charging the grand jury of magistrates 'to perform the not very arduous duties of indorsing their own previous committals', it was all mildly comic as well;[38] Dickens's *Bardell* v. *Pickwick* reinforced the point. Moreover, assize procedures would go awry more often than not. Madan in 1785 wrote of the 'low rabble' who attended trials, and of the 'noise, crowd and confusion'

[36] D. Hay, 'Property, authority, and the criminal law', in D. Hay *et al.* (eds.), *Albion's fatal tree: Crime and society in eighteenth-century England*, 27–8; McGowen, citing Wesley (1758), in '"He beareth not the sword in vain": Religion and the criminal law in eighteenth-century England', *Eighteenth-Century Studies*, 21 (1987/8), 192.

[37] Thompson, *Whigs and hunters*, 262.

[38] W. Ballantine, *Some experiences of a barrister's life* (5th edn., 1882), 51–5.

when trials continued after dinner and judges, witnesses, and jury were 'muzzy' with drink. In Scotland some judges swigged port in court.[39] Old Bailey justice was especially well lubricated. In the course of each sessions day in the 1830s,

> two luxurious dinners were provided, one at three o'clock, the other at five. . . . The scenes in the evening may be imagined, the actors in them having generally dined at the first dinner. . . . One cannot but look back with a feeling of disgust to the mode in which eating and drinking, transporting and hanging, were shuffled together . . . [with] the City judges rushing from the table to take their seats upon the bench, the leading counsel scurrying after them, the jokes of the table scarcely out of their lips, and the amount of wine drunk, not rendered less apparent from having been drunk quickly.[40]

As the 1836 royal commission delicately put it, 'instances may be adduced, in which the learned judge, towards the conclusion of a long trial, has been so much affected by the fatigue he has undergone, as to be incapacitated not only from commenting upon, but even from summing up the evidence'.[41]

Nor had procedural order improved by then. An assize day was more like a market-day than a solemn moment of state:

> the witnesses, totally unacquainted with, and equally uncertain of the hour when, and the place where, they are required to be in attendance, are first dragged into court through a crowd of persons (who are as much at a loss as themselves where to go, some being present from idle curiosity, and others, from want of knowledge, wandering about the avenues and purlieus of the court), to be sworn to the evidence to be given by them upon the bill of indictment before the grand jury; from thence they are taken back, through the crowd, to the entrance of the grand jury chamber, to await their being called on the bill, after which they are required to be in readiness to give evidence in court on the trial; but at what hour, or even on what day, they know not; nor is the attorney, on whom they rely for information, able, with any degree of certainty, to assist them.

Idlers occupied the witnesses' waiting-rooms, converting them to the same uses as alehouses 'for drinking and smoking at pleasure'; no respectable person would willingly enter them.[42]

(ii) *Biases against 'prisoners'*

These images do not encourage the notion that courts were places in which even-handed decision-making was palpable either. There was

[39] Gilmore, *Riot, risings and revolution*, 150–1.

[40] Ballantine, *Experiences*, 65.

[41] *RC on criminal law, 2nd report* (PP 1836, xxxvi), 8 n.

[42] G. Bolton, *The practice of the criminal courts, including the proceedings before magistrates, in petty and quarter sessions, and at the assizes* (1835), pp. ix–xi.

little press supervision. Newspapers were sometimes critical by the 1820s, but randomly so and lightly. The grievances of the law's targets counted for little. Carried into prisons, on to hulks, to Australia, or to home communities, they went unheard. Hence a fair degree of laxness in hearing evidence and in sentencing, and a frequency of trials in which rules might rule but did so alongside quasi-institutionalized assumptions and procedures which quite subverted defendants' interests.

Even before they were proved guilty, accused people were denied any right of the free-born English except the right to rule-bound trial. Pending trial, they might be imprisoned for months in grim conditions as they awaited the next assizes or quarter sessions (until the 1826 Act bail was rare). They were not only treated as but were also called 'prisoners'. 'Acquittal does not shelter them from punishment,' Sydney Smith observed in 1824, 'for they have already been punished. It does not screen them from infamy, for they have already been treated as if they were infamous. . . . The principle is, because a man is very wretched, there is no harm in making him a little more so.'[43] Accused persons, in short, were at once identified as emblems of a corruption whose extent was discerned less in their guilt than in the fragility of local or national order. Some part of the criminal trial's casual and rowdy conduct, the exclusion of defence counsel, or the admission of circumstantial evidence, derived from this tacit negation of individuality after arrest. Although this was changing with the intrusion of the lawyers in courtrooms, innocence was still not fully assumed until proved otherwise. Indeed there was 'little apparent anxiety about guilt or innocence as abstract issues'.[44]

Until 1836 counsel's labours for defendants were restricted. Defence counsel could address the jury in cases of misdemeanour or high treason (or in civil cases), but never in cases of felony. Ancient monarchical interests survived residually here. The 1836 royal commission traced the restriction back to 'a remote and illiterate age' when the forfeitures consequent on felony were lucrative additions to the royal revenue, not lightly to be argued away; and also when the jury were assumed to have been eyewitnesses of the crime. Thus:

> Witnesses for a prisoner were allowed to be sworn in misdemeanour a long time before this was allowed in trials for felony. Indeed it was not till the reign of Queen Mary, that . . . prisoners indicted for felony were allowed to call witnesses for their defence. There was a time . . . when counsel for a prisoner were not allowed even to cross-examine; when it was deemed criminal in a

[43] *Works* (2 vols., 1859), ii. 40–1.

[44] Beattie, pp. 341, 420–3.

counsel to give any suggestion by word or writing to a prisoner as to the management of his case; and when written suggestions to a prisoner have been taken away from him by order of the court. Thus the prohibition of speeches by counsel appears to be only a relic of still more unfair prohibitions exercised against prisoners, and not a regulation advisedly contrived in a free and enlightened age.

This history had to be rehearsed, the commission added, because otherwise the exclusion of counsel and other biases against the prisoner 'would appear absurd'.[45]

Suitably rationalized, these mentalities endured. Attorney-General Pollock thought that most lawyers would defend the veto on counsel's summing-up by claiming that it saved time, and that it would anyway be shocking 'to make a man's life the subject of contention and advocacy in a way to which we are not accustomed'.[46] Some even said that the veto saved the defendants expense ('as if anything was so expensive as being hanged!' Sydney Smith snorted). Pollock knew that such arguments were bogus; but they were underpinned by another long-dying fiction—that counsel could be dispensed with because the judge acted as the prisoner's friend. Sydney Smith got to the point as usual. To say that the judge was the prisoner's friend was 'an unmeaning phrase, invented to defend a pernicious abuse': 'To force an ignorant man into a court of justice, and to tell him that the Judge is his counsel, appears to us quite as foolish as to set a hungry man down to his meals, and to tell him that the table was his dinner.'[47] 'It is remarkable', the 1836 commission said, 'that this position . . . is left to rest on mere gratuitous assumption.' It bore little relationship to how judges did behave in court. Where they did not simply bully defendants, as Vaughan or Park liked to do, judges often had pleasanter businesses in mind than being friends to prisoners. At assizes crowded with the county's prettiest women 'in the most elegant négligé', they were not immune to a little peering and prodding.[48] Or: 'I am this moment returned from a Judges' dinner with the Lord Mayor,' Graham J. scribbled to Hobhouse to explain why he could not find time to report on an 18-year-old who was to be hanged in three days' time.[49] Or as the 1836 commission again delicately put it: 'the attempt to finish the list of causes and deliver the gaol, often occasions late sittings, and is attended with a degree of

[45] *RC on criminal law*, 11–12 and n.

[46] Ibid. 74.

[47] *Works*, ii. 110.

[48] Cited by Hay, 'Property, authority and the criminal law', 31.

[49] HO47.64.

mental and bodily fatigue which must in some degree impair the energies of the most powerful mind.'[50]

(iii) *The speed and quality of trials*

Beattie's analysis of cases in Surrey in the later eighteenth century gives a trial duration of about half an hour per case, all formalities included. Langbein notes that a mid-eighteenth-century assize judge would see more felony trials in a day or two than a modern judge would expect to see in a year. He explains this rapid processing in terms of

> the scheduling of trials close to the happening of the crimes, sometimes within a few days; prompt pretrial evidence-gathering and sifting by the JPs; the virtual absence of lawyers for the prosecution or defence; the conversational informality of the trial; the constant resort to the accused as a testimonial resource; the recurrent use of jurors who were long experienced in jury work, men who needed comparatively little formal instruction on the essentials of criminal law and procedure; and the guidance that the jury received from the judge, who exercised an unrestricted power to comment on the merits of the case.[51]

Even in the 1820s and 1830s, with an increased presence of lawyers in court and more regular procedures, trials remained uncommonly hasty. The Shrewsbury assizes of 1829 at which Noden was tried got through a dozen cases in a couple of eight-hour days, with seven death sentences recorded by the end of them. Old Bailey trials were quicker still, with between three hundred and five hundred cases tried by relays of judges across several days, eight times a year. In 1833 Old Bailey trials were found to last an average of eight and a half minutes each (though capital trials took longer). When Denman was common serjeant at the Old Bailey, his 'method of trying prisoners' was to rush through the easy cases to leave more time for the difficult ones. He admitted that in this he had an eye on witnesses' and jurymen's 'impatience'. Denman convinced himself that there was no 'single conviction that appeared to me unjust', but it is doubtful whether defendants took that view. As the 1833 observer noted, 'the rapidity with which the trials are despatched throws the prisoners into the utmost confusion':

> seeing their fellow-prisoners return tried and found guilty in a minute or two after having been taken up, [they] become so alarmed and nervous, in consequence of losing all prospect of having a patient trial, that in their efforts . . . to re-arrange their ideas, plan of defence, and put the strongest feature of their

[50] *RC on criminal law*, 7–8.

[51] Beattie, p. 378; J. Langbein, 'Shaping the eighteenth-century criminal trial: A view from the Ryder sources', *University of Chicago Law Review*, 50 (1983), 115.

> cases before the court as speedily as possible, they lose all command over themselves, and are then, to use their own language, taken up to be knocked down like bullocks, unheard. Full two-thirds of the prisoners, on their return from their trials, cannot tell of any thing which has passed in the court, not even, very frequently, whether they have been tried.[52]

Juries likewise rarely retired from court, usually announcing their verdicts after a minute or two's huddling in public, guided by the instant opinions of their dominant members.[53] The fact that the jury *did* retire to consider their verdict on a highway robbery in 1822 was cited in one mercy petition as itself justifying an appeal.[54]

Few prisoners were in a condition to prepare defences when each came into court 'squalid and depressed from long confinement—utterly unable to tell his own story for want of words and want of confidence'.[55] Even had they been in better condition, defences could not be prepared when they were denied access to depositions or to the names of witnesses against them. Sometimes charges were changed at the shortest of notice. Noden arrived at Shrewsbury assizes in 1829 believing that he was to be tried for the misdemeanour of attempted rape, only to find that it was for the felony of rape that he was tried and capitally convicted.

Few could afford counsel; even those who could would not get good counsel unless there was fame in the case. When Denman defended Brandreth *et al.* in the Derby treason trials in 1817 it was with ill grace. He worried that 'the fees . . . will not make any man's fortune, as every man among them is as destitute of money as common sense'.[56] In lesser cases the counsel employed or assigned by the judge would be the least experienced or talented. Attorneys usually chose them at random on the day of trial from among the junior barristers who in their rickety post-chaises would traipse the circuit behind the judge and leaders, hungry for briefs and half-guineas and counting themselves lucky if they got some.[57] Defences would be slapdash in such cases. Even so, at the Old Bailey in 1800 between three-quarters to two-thirds of those tried for property crimes lacked counsel; the proportion was probably higher in the provinces.[58]

[52] Arnould, *Denman*, i. 202–3; anon., 'The schoolmaster's experience in Newgate, II: Hurried trials', *Fraser's Magazine*, 5 (July 1832), 736; Ballantine, *Experiences*, 64.

[53] Beattie, pp. 397–8.

[54] HO47.63.

[55] Sydney Smith, *Works*, ii. 108.

[56] Arnould, *Denman*, i. 115.

[57] For images, e.g. of the tiro counsel's stage fright in court, see Ballantine, *Experiences*, 33, 37–8, 41, 51–3. Ballantine earned 4½ guineas in his first year at the Bar in 1834, 30 in his second, and 75 in his third: Romilly, *Memoirs*, i. 91–4, and *RC on criminal law*, 76, on judges' assigning counsel.

[58] Beattie, pp. 360–1. Data on costs are scarce. Hay and Snyder cite a murder trial of 1753 in which the defendant had to pay 3 counsel 30 guineas to defend him: 'Using the criminal law, 1750–1850', 23.

The most casual efforts were made to ensure that the defendants' witnesses attended the trial, however life-threatening the charge. This was a source of endless complaint in petitions for mercy in the 1820s. Supported by a lengthy list of farmers, gentlemen, and tradesmen, a Gloucester housebreaker petitioned representatively on the grounds that he was 'tried at Nisi Prius Bar instead of the Crown Bar where your petitioner's friends were in waiting': character witnesses and key evidence as to an incriminating knife left at the scene of the crime were therefore not heard.[59] One of those rare men whose wrongful conviction on a mistaken identification was proved to the Home Office's satisfaction and who thus achieved a free pardon wrote that 'the trial came on unexpectedly to him and . . . the said witnesses [who would have proved the alibi] arrived in court whilst the jury were considering their verdict.'[60] A 'poor girl' enticed into her employment by an Abergavenny prostitute and prosecuted by the latter for absconding with stolen clothes was 'quite unprepared for her defence and had no one to speak to her character', the sheriff's chaplain wrote to Under-Secretary Hobhouse. He enclosed a petition to the effect that witnesses arrived at the trial too late to support her; and the mayor (acting for her) had not prepared the defence properly.[61] At the Old Bailey a man sentenced to life transportation for street robbery claimed vainly that his master and friends, attending to give character evidence, had been told that the trial would not be on that evening and so had left.[62] A woman imprisoned for infanticide had her 'respectable witnesses' sent home because her defence counsel did not expect the trial to come on when it did.[63] Character witnesses for a Nottingham poacher transported for seven years were not called because the trial had commenced at 7.30 p.m., after they had left.[64] Thomas Lea, sentenced at Stafford to seven years' transportation for stealing a fishing-net, appealed on the grounds that he was tried only twenty-four hours after the grand jury found a true bill against him: 'in consequence thereof and of the great distance of his residence from Stafford being 30 miles . . . he was unable to sufficiently prepare himself for his defence.'[65] And so on and on: such revelations multiply in scores.

[59] HO47.64 Best J. had no doubt of Ford's guilt, claiming that all necessary witnesses were called. So this appeal against a death sentence already commuted to life transportation was refused.

[60] HO47.64.

[61] Best reported that, had he known the prosecutrix's true character, he would have made sure that she was acquitted; she was given a free pardon: HO47.64.

[62] Ibid. [63] HO47.74. [64] HO47.63.

[65] HO47.64. His character was unimpeachable, the vicar, constable, churchwardens, and local farmers attested, and he had resided all his life upon a farm rented at over £250—very respectable, in other

Their implications were understood perfectly at the time. How could it be otherwise, Sydney Smith asked in 1826, than that 'not a year elapses in which many innocent persons are . . . found guilty'?

There are seventy or eighty persons to be tried for various offences at the Assizes, who have lain in prison for some months; and fifty of whom, perhaps, are of the lowest order of the people, without friends in any better condition than themselves, and without one single penny to employ in their defence. How are they to obtain witnesses? . . . The witnesses are fifty miles off, perhaps—totally uninstructed—living from hand to mouth—utterly unable to give up their daily occupation, to pay for their journey, or for their support when arrived at the town of trial—and, if they could get there, not knowing where to go, or what to do. It is impossible but that a human being, in such a helpless situation, must be found guilty; for as he cannot give evidence for himself, and has not a penny to fetch those who can give it for him, any story told against him must be taken for true (however false). . . . A case of life and death will rouse the poorest persons, every now and then, to extraordinary exertions, and they may tramp through mud and dirt to the Assize town to save a life: but imprisonment, hard labour, or transportation, appeal less forcibly than death . . . to the feeble and limited resources of extreme poverty.[66]

(iv) *Sentencing: Example and character*

Finally, the conditions which determined sentences extraneously, over and above culpability. These conditions were sanctioned in juristic thought, as we have seen: judges were not in principle betraying a trust. Example was all; and punishment was not meant to be proportionate to crime. But what of the judicial attitudes which took shelter under these sanctions? Judges surrendered *casually* to fantasies about the deterrent lessons which communities needed, straining little, as prisoners' friends, to hear the full accounts which might have saved many of those they sent to slaughter. This was where trust was betrayed.

Examples are legion. Two 17- and 18-year-old lads whom Garrow J. left to hang in 1824 for robbing a Frenchman in a Southwark brothel were of 'good character and industrious habits', Henry Drummond wrote to Peel. They had gone into the brothel only 'upon hearing the screams of the women whom the Frenchman had struck, and finding him and the women fighting they took the part of the ladies; the police officer came in and caught them in the act'. Well supported though this

words. He had, he swore, paid £1 for the net; his servant bore witness to this. Defending his sentence, the chairman of the Stafford quarter sessions stated that the news that 'his house is the Rendezvous of all the poachers in the country . . . produced an immediate agreement in my opinion and that of the magistrates'. Peel pardoned Lea outright.

[66] 'Counsel for prisoners', reprinted in *Works*, ii. 107.

story was, it had not been tested in court. The case attracted much attention, the petitions—sign of the times!—being supplemented by floods of letters from people who had read about the judge's prejudicial summing-up in the newspapers. Garrow justified his sentence rotundly:

> The state of morals of a large portion of the inhabitants of that part of the Borough of Southwark in which this transaction took place, and the frequency of violent outrage committed upon persons who are enticed into houses of ill fame, induced me to think it might be attended with benefit to the public to impress upon the minds of a large assembly of persons from that quarter that it might be expected that such offences would be visited by the most severe punishment.

Under pressure he agreed that he had erred in his zeal, so Peel respited the death sentence, and off to Australia these two possible innocents were sent, outright pardon being beneath the court's dignity, needless to say.[67]

'In the present state of the country I think much relaxation would be unsafe,' Burrough J. wrote, defending a seven-year transportation sentence in Wiltshire for the theft of a mare: 'in consequence of the great extent of this crime I have thought it absolutely necessary that punishments should be inflicted that are likely to deter.'[68] 'The offence of street theft has risen to a very alarming height,' the recorder of London reported similarly in 1822, justifying a sentence of life transportation.[69] An Old Bailey judge's aversion to the pickpockets who 'constantly infest the streets of London' prompted him to sentence a man to seven years' transportation, even though the appeal proved to Peel's satisfaction that he had been an innocent bystander: he was pardoned.[70] In a burglary case in 1823 involving the theft of one wellington boot and three pairs of shoes, Park J. admitted that he had had grave doubts about the propriety of a conviction, but he had left the verdict to the jury, who went for guilty, 'though I perhaps should have been better pleased if the verdict had been the other way, yet as the jury had thought otherwise, I did not feel myself at liberty to do more than save their lives, as the town of Nottingham was stated by the magistrates to me to be extremely beset by men of this description'.[71]

And finally the question of 'character' around which so many sentences flowed. Character remained as central to judges' decisions in the 1820s as ever in the previous century. It was the key variable in sentencing. It is referred to in nearly every appeal mentioned in this chapter and this book.

[67] HO47.74. [68] HO47.73. [69] HO47.63. [70] Ibid. [71] HO47.64.

Character was a highly negotiable variable, and thus a pivotal element in judicial discretion. What was good character to one judge was not good character to another; and character which was sworn to defendants' advantage in one kind of case could be turned against them in another. In 1803, for example, the jury recommended the traitor Despard to mercy on what would normally be exculpatory grounds: 'the very high character given of him by Lord Nelson and Sir A. Clarke and Sir Evan Nepean, for good conduct and services as a military officer'. Lord Ellenborough knew how to bend this to the sentence he wanted, however. He professed himself 'perfectly at a loss to understand upon what reasonable principle' this recommendation should work to Despard's advantage. A high character in a traitor, far from rendering guilt more doubtful, could only 'operate very materially in aggravation of it'.[72]

The negotiable quality of character evidence was apparent on less important cases too. The chairman of the Surrey sessions in 1822 had come to doubt the guilt of a man he had sentenced to fourteen years' transportation for stealing a shawl. Much depended, Peel's secretary replied, on the characters of the witnesses he listed; he noted how some appeared to be 'all persons of the lowest and worst character'; and the appeal was refused.[73] It saved an Exeter housebreaker's life that 'an eminent solicitor of Exeter and the respectable people' gave him 'a very excellent character for honesty and sobriety' (true, he also had a wife and family to support).[74] The witnesses against a man he had sentenced for perjury, Abbott CJ reported, 'were respectable in their appearance and demeanour, more so than those who were examined for the defence': he gave no other reasons for withholding mercy.[75] 'The prisoner had to my eye the appearance of a hypocrite and a liar,' the same judge wrote to Peel, leaving a rapist to hang in 1824 on those most elemental of grounds.[76] In 1823 a free pardon was won for one Humphrey Goodwin, sentenced by Best to seven years' transportation at Winchester assizes for receiving a pair of shoes as a prize in gambling,

[72] Pelham papers, BL Add. MS 33122, fos. 116 ff.

[73] HO47.63.

[74] HO47.64.

[75] HO47.73.

[76] HO47.65. Cf. a homely example from Edward Christian: 'I know one instance, where a prisoner, for stealing from the person, was sentenced by the Chairman [of quarter sessions] to be imprisoned for seven years. He took off his iron-heeled shoe, and was preparing to throw it at the Chairman, but was prevented. The Chairman and the Court then ordered him to be transported for life. This was perfectly correct. . . . The judgments ought to be proportional to the character and former conduct of the prisoner: and the Court were clearly in error in the first judgement, as the conduct of the prisoner proved that severer judgment was necessary to protect the people of England from his future crimes, and as a salutary example to others': *Charges delivered to grand juries in the Isle of Ely* (1819), 286–7.

knowing them to be stolen. Here the appeal was initiated by the prisoner's naval commander on whose ship he had served four years: his conduct at sea had been gallant, he had saved lives, he had lost a leg in active service; he now supported a family by selling oranges from a horse and cart. These poignant details had not been heard at the trial, and the under-sheriff forwarded them with the view that 'there was but slight evidence to support the prosecution', and certainly no character evidence had been called. Best duly reported to Peel that the case against Goodwin was very clear, and because 'no person appeared for his character . . . I thought him a fit person to be removed out of the country'. Now, 'in consideration of the service that he has rendered his country and his general good character before this transaction', and since he had already been seven months in custody on the hulk *York*, Best would relent and 'humbly recommend His Majesty to grant him a free pardon'.[77]

These are some of the things we hear in the appeal archive. None endorses the notion that English justice was pretty 'fair' all things considered. 'It worries rich and comfortable people', Sydney Smith observed, 'to hear the humanity of our penal laws called in question': but 'if Harrison or Johnson has been condemned, after regular trial by jury, to six months' tread-mill, because Harrison and Johnson were without a penny to procure evidence—who knows or cares about Harrison or Johnson?'[78] Although many of the failures expressed 'mentality' rather than 'law', although contemporaries could not help being creatures of their times any more than we can, and although there is little point in lamenting the absence of better structures and procedures whose time had not yet come and still has not come—still, the failures refracted in the appeal archive were as clearly perceived as culpable by contemporaries as they may be by us. Even yet, however, those minorities who cared about these matters hardly knew the half of them. In the next chapter we hide ourselves behind the arras in the king's Council. It is time to get an earful of the jovial attitudes which informed life-and-death decision-making among the *very* great and good, among people even more important than judges.

[77] HO47.64. [78] *Works*, ii. 108.

CHAPTER 20

THE KING IN HIS COUNCIL

1. *Mercy in London*

EVERYONE KNEW THAT THE QUALITY OF MERCY WAS SUPPOSED TO drop as the gentle rain from heaven upon the place beneath; that it was an attribute to awe, majesty, and God himself; and that royal power was most like God's when mercy seasoned justice. Peel the home secretary told the Commons that the exercise of mercy 'ought to be as prompt and as pure as the visitation of justice'. Alas for the sorrier truth of the matter. Eavesdroppers on the king's Council as it reviewed the Old Bailey's capital sentences might have given away their presence behind the arras by involuntary snorts. In no other context was the power of life and death wielded with such remote and capricious disdain. Enmeshed in courtly rituals, insulated from the cramped worlds of those whom they decided to kill, half-attentive to the death-dealing business in hand—most of the Council's members present themselves as aloof, arrogant, and merciless beings. As ever in this book, this judgement is endorsed by the fact that some contemporaries said so too—at least in so far as they knew what went on, which was not much.

A full measure of the Council's decision-making and of the reflexes informing it may best be gained by examining particular cases. So this chapter ends with a representative episode when Peel worked hard to persuade the Council, against the king's wishes, to send eight men to the noose in 1822. But there is other evidence to draw on first, much of it provided by those few Council members who were becoming uneasy about their roles in the later 1820s and who wrote things down indiscreetly. Some of them did become a little anxious as public disquiet mounted and as pardons increased to avoid an excess of hangings. Even Eldon acknowledged new realities: 'Times are gone by when so many persons can be executed at once as were so dealt with twenty years ago,' he told Peel in 1822.[1] None the less, the murmurings we are going to eavesdrop upon were never much more than murmurings, confined to diaries or letters to friends. What remains striking is how disinclined the murmurers were to make their disquiet more audible. This charmed circle closed its ranks against scrutiny, their own scrutiny included. The illusion of merciful justice must be upheld through secrecy and by not thinking too much.

It was the job of the sovereign in Cabinet Council to review the capital sentences of the Old Bailey. In London the king exercised his prerogative directly. This was a relic of the direct rule which the medieval monarch claimed in his capital.[2] There was plenty of work to do, because the Old Bailey was the busiest criminal court of the kingdom. Sitting eight times a year for up to a dozen days on end, the recorder, two or three judges, the common serjeant, and a serjeant-at-law tried up to five hundred cases between them in any sessions by the 1820s—an awesome pace, as we have seen. They tried some quarter of all felonies in England and Wales between 1805 and 1820, and about one in five in the 1820s (the contraction due to the increasingly energetic prosecution of provincial crime: see Appendix 2).

The Old Bailey was also the bloodiest court in the kingdom. At the peak of the crime wave of the 1780s its judges presided over what has aptly been termed a judicial carnage. 'A little army of fellow creatures,' as the *Gentleman's Magazine* put it, were 'hanged like dogs' a dozen at a time.[3] In 1785, 85 of the 153 Londoners capitally convicted were executed, a number and proportion unprecedented in the eighteenth century and never to be repeated. Year by year thereafter, however, more were *sentenced* to death in the Old Bailey than anywhere else, until it came to this: 1820: 211, 1821: 151, 1822: 138, 1823: 124, 1824: 149, 1825: 168, 1826: 204, 1827: 214, 1828: 175, 1829: 131, 1830: 133.

These figures alone would guarantee that the pardoning process assumed an increasing role in the legitimation of the judicial system. You might plausibly kill 56 per cent of the capitally convicted in 1785, but you could not sustain that proportion of deaths in the 1820s without outrage. Still, although the proportion of pardons had to increase, more people condemned at the Old Bailey were hanged in the 1820s than in most earlier decades for over a century. In fact the bloodiness of Old

[1] Eldon to Peel, n.d., 1822, Peel papers, BL Add. MS 40315, fo. 63.

[2] A. Aspinall, 'The Grand Cabinet, 1800–1837', *Politica*, 3 (1938), 339, 333.

[3] Beattie, p. 384. But judges on the Norfolk and the home circuits similarly hanged 60% of those capitally convicted in 1785, up to 1 in 10 of those tried: SC 1819, appendices.

Bailey justice increased in relation to that of other courts in the country. In 1806–10 a fifth of those hanged in England and Wales had been condemned there; by 1826–30 that proportion had risen to 29 per cent.[4] Overall, 463 of the 1,849 executions in England and Wales from 1806 to 1830 were ordered at the Old Bailey, the numbers of hangings in the 1820s running as follows: 1820: 46, 1821: 29, 1822: 28, 1823: 11, 1824: 12, 1825: 16, 1826: 20, 1827: 17, 1828: 21, 1829: 25, 1830, 6.

This tide of executions was the responsibility not so much of the Old Bailey judges as of the King in Council, for this august body had to sanction each of them when they met to hear the recorder's report on the capital convicts cooped up in Newgate. Before the Council met, the recorder and the home secretary prepared the business. After each Old Bailey sessions the recorder would draw up a neat little table of the condemned's names, with cursory notes on their crimes, on whether or not there were petitions supporting mercy, and by whom (if anyone) the prisoners were recommended. If petitions did come in, he would send the home secretary some notes on each case, and the latter would forward the petitions to the king, with his notes too.[5] Then the king and Council would be advised by the home secretary that the report was ready,[6] and at the king's convenience the officers of state in their most lavish court plumage would assemble at Carlton House (occasionally at Windsor after 1828, and under William IV at St James's).

This meeting was technically called the Grand or Nominal Cabinet (to distinguish it from the executive cabinet), although it was sometimes confused with the Privy Council because it often followed Privy Council meetings. But not all the Privy Council stayed on for the Grand Cabinet, and outsiders came in too. The lord chief justice invariably attended, and the recorder of London or his deputy had to attend, because his report determined the agenda. Judges might attend, and the archbishop of Canterbury sometimes did. All Cabinet members could attend as well, and apparently did so unless the Council's business appeared routine and the meeting was hard to get to (Windsor was two hours' coach-ride from Westminster).[7] The regent or king would be

[4] See App. 2, below.

[5] For examples, see HO42.2: 192–200 (Apr. 1783).

[6] Eldon had to tell Peel this when Peel took office: Eldon to Peel, 7 Nov. 1822, Peel papers, BL Add. MS 40315, fo. 68.

[7] Thus Liverpool apologized to Peel for his absence from the next report because he had to receive the prince and princess of Denmark—'unless there were any cases of a very special nature which required a full Cabinet and this is not likely': 20 June 1822, Peel papers, BL Add. MS 40304. On the other hand, on 20 Nov. 1824 Wellington travelled to London from the Arbuthnots' seat to attend a report, returning next day: he discussed Irish Catholics with the king. See F. Bamford and duke of Wellington (eds.), *The journal of Mrs Arbuthnot, 1820–1832* (2 vols., 1950), i. 356–7.

attended by flunkeys—the master of the horse, the groom of the stole, the lord chamberlain, and lord steward.[8]

Then the proceedings began. The king slumped before a fire in one of the state apartments and sometimes fell asleep. The Council gathered elegantly round; the recorder read the lists of the condemned, commenting on each one; the home secretary and sometimes the chief justice gave advice and drew attention to petitions if they were worth considering; and everyone who had opinions gave them. It was not always a serious debate which ensued. At the first Council he attended Eldon professed himself 'exceedingly shocked' at 'the careless manner' in which it was conducted: 'We were called upon to decide on sentences, affecting no less than the *lives* of men, and yet there was nothing laid before us to enable us to judge whether there had or had not been any extenuating circumstances; it was merely a recapitulation of the Judge's opinion, and the sentence.' On one occasion Eldon disagreed that a man who had robbed in Bedford Square should be hanged: the man had not used violence, he pointed out. He wryly recalled the king's amusement: 'Very well. Since the learned judge *who lives in* Bedford Square, does not think there is any great harm in robberies there, the poor fellow shall *not* be hanged.'[9]

The tenderness of the Council's decision-making was not helped by the judges. Abbott CJ was a famous hanger of men. The recorder of London, Knowlys, was worse. 'He is much more bloody minded than I am,' Eldon noted as he objected to the numbers of deaths the recorder recommended in 1822.[10] But the king would give his assent one way or the other, and the recorder's pencil would duly note 'law to take its course' against the names of those to be hanged, and 'respited' against those who could live. This was the list which he took to Newgate next day.

Most members of the Council did no preparation for their duties. Only in 1829 was the recorder required to provide printed trial reports beforehand, for example.[11] But perhaps things got better with Peel at the helm. As home secretary, Peel was always well briefed, often wearily familiar with cases' details from the petitions and letters concerning

[8] See HO42.100, for list for 7 Feb. 1810, reprinted in A. Aspinall and E. A. Smith (eds.), *English historical documents, 1783–1832* (1959), 86–7. For the later judicial functions of the Privy Council (Judicial Committee Act, 1833), see H. J. Hanham, *The nineteenth century constitution* (Cambridge, 1969), 406.

[9] Aspinall, 'Grand Cabinet', 336, quoting H. Twiss (ed.), *The public and private life of Lord Chancellor Eldon* (3 vols., 1844), i. 398–9, and Lord J. Campbell, *The lives of the lord chancellors and keepers of the great seal of England* (10 vols., 5th edn., 1808), ix. 262.

[10] Eldon to Peel, n.d., 1822, Peel papers, BL Add. MS 40315, fo. 63.

[11] Lord Ellenborough, *A political diary*, ed. R. C. E. Abbot (2 vols., 1881), i. 229 (8 Dec. 1829).

them. Sometimes he consulted the chancellor or chief justice before the Council. In unusual cases (like Fauntleroy's) he met privately to consult with the prime minister as well.[12] Eldon was also conscientious. He claimed to read through every case three times if he could before Council meetings, though in what source is not explained. If he was too busy to attend the Council, he was at pains to give Peel his opinion on the more difficult cases.[13] By contrast, Lord Liverpool was conscientious chiefly in apologizing for his absences.[14] Unexplained absences from Council were frequent. When it was announced in the press in 1824 that Peel had to preside at Council in place of the lord president during the latter's sojourn in Italy, one Sir Robert Deverell wrote indignantly to Peel that the lord president's absence rendered the proceedings constitutionally invalid. When lives were at stake, forms of conduct signified deeply: 'are the great offices of State thus convertible in their nature, by mere courtesy? Are they not, more properly . . . checks upon each other?'[15] He complained in vain.

The central issue for us is the quality of the Council's decision-making. Although the death sentence was supposed to be a discretionary instrument, it was also supposed to be a closely meditated instrument. Eldon had already registered that it was less than that. In the late 1820s a few other Council members began to worry in their diaries too. Why this rather specific dating (after silence hitherto)? With the old tory hegemony cracking, did they think they might one day be accountable, as corrupt politicians cover their traces now? Or, more likely, was it just that old meanings were fading? Was the system at last becoming as ridiculous to them as it was to external critics, when an absurd 90 per cent of the condemned in London had to be pardoned to keep capital law in good repair?

As lord privy seal, the younger Ellenborough was one who seemingly read writing on walls. In 1828 his doubts struck at the principle of discretionary justice itself. It did this with a disingenuousness which could

[12] Peel to Liverpool, 24 Nov. 1824, Peel papers, BL Add. MS 40304, fos. 293–4.

[13] Eldon to Peel, n.d., 1822, ibid. 40315, fo. 63. Characteristic of Eldon's interventions is a note to Peel in 1826 advising him to read the cases of 2 separate robberies 'at length' because 'Bishop is convicted upon M Tallis' oath as to the identity of his person, when three persons are acquitted, equally sworn to by him, as to whom (the whole three) the Jury thought him mistaken. Bishop . . . must I think have a respite to have the whole evidence read as to the three acquitted.' He hoped that there would be a close discussion of this case: Eldon to Peel, 3 Dec. 1826, ibid., fo. 276.

[14] In his busyness, 'I would spare myself as much as possible': Liverpool to Peel, 20 June 1822, 10 Mar. and 29 June 1824: ibid. 40304.

[15] Deverell to Peel, 24 Nov. 1824, ibid. 40730, fos. 194 ff.

only have been affected, for the doctrine was not newly coined and had always legitimized the law:

> Four people ordered for execution, one for forgery, one for burglary, two for beating and robbing a man in a house of ill-fame.[16] There was a woman engaged, who was spared on account of her sex, but she was the most guilty of all.
>
> I do not like the Recorder's reports. I am shocked by the inequality of punishment. At one time a man is hanged for a crime . . . because there are few to be hanged, and it is some time since an example has been made of capital punishment for his particular offence. At another time a man escapes for the same crime . . . because it is a heavy calendar, and there are many to be executed. The actual delinquency of the individual is comparatively little taken into consideration. Extraneous circumstances determine his fate.[17]

'The thing that distresses me in all these cases', Ellenborough noted a few months later,

> is that men are punished not with reference to the extent of their own crimes, unless they be very great, but with reference to the number and circumstances of similar crimes committed by others at the same time. Our laws are so framed that all cannot be executed who incur the penalty of death—and wisely and humanely so framed, I think; but still the consequence is that in every case it is not the law but individuals who decide whether a man shall suffer or not.

Ellenborough found all this 'very difficult and painful'—a 'duty, executed, I believe, most conscientiously; but I wish it did not fall to the king's Ministers to execute it'.[18] He seems to have kept quiet, however.

The Council's clerk, Greville, also shared the disquiet. 'What a curious supplementary trial this is,' he wrote in his diary at the end of 1830: 'how many accidents may determine the life or death of the culprit.' He noted that the outcome of a vote on whether or not to send a man to the scaffold frequently depended on the chance membership of the Council meeting in question. He also noted how the mood changed when the whigs took office. Although in December 1830 Tenterden CJ (Abbott) had insisted on sentencing a forger to hang, the new lord chancellor, Brougham, insisted on a vote and the man was saved:

> Little did the criminal know when there was a change of ministry that he owed his life to it, for if Lyndhurst had been Chancellor he would most assuredly have been hanged; not that Lyndhurst was particularly severe or cruel, but he would have concurred with the Chief Justice and have regarded the case solely

[16] Ellenborough was referring to the cases of James Anderson, George Morris, Mary Young, saved by the exertions of the attorney E. A. Wilde: see Ch. 16.

[17] Ellenborough, *Diary*, 154–5 (28 June 1828).

[18] Ibid. 267–8 (1 Dec. 1828).

in a judicial point of view, whereas the mind of . . . [Brougham] was probably biassed by some theory about the crime of forgery or by some fancy of his strange brain.[19]

Again, when Lady Burghersh championed a convicted forger (a good-looking forger, Greville wryly noted), it was at first in vain that she 'went to all the late Ministers and the Judges to beg him off'. But the new government were 'in better humour, or of softer natures, [and] suffered themselves to be persuaded, and the wretch was saved'. Even so, according to Greville, the new home secretary could himself be capricious. When a man was hanged in 1832 on thin circumstantial evidence, 'we attacked Melbourne about it, who defended it by saying that he never would interpose the royal prerogative between the law and its exercise, when in fact he does so every day. This case shocked us very much.'[20]

If well-worn routines were newly thought to be shocking, the Council gave no public sign of it. Secrecy kept the law's private parts decently veiled. Secrecy suited a government averse to radical law reform too; but the truth is that most probably cared little. Merciful business was not the main business in hand for many Council members. The report was invariably preceded or followed by Privy Council appointments or a royal reception, so meetings were often about place, favour, and jolly convivialities rather than justice. Ellenborough again:

7 May 1828. The report very heavy. There were at least ten cases in which the punishment of death ought to have been inflicted. We chose six. Cabinet dinner at Goulburn's.
15 January 1829. Lord Liverpool was at the Council to deliver the garter of the late lord, his brother. This new Lord Liverpool has no son and three girls, the prettiest and the best educated in London, and great fortunes. Two men will be executed, one for having in his possession the implements of coining, the other for robbery with violence. One other was to have been executed, but on further consideration we thought his was not a sufficiently bad case.
16 July 1829. Before the report Madame de Cayla, the Duchess D'Escars, &c., were presented to the King. . . . The Recorder's Report was a very heavy one. All the cases bad, and seven ordered for execution. The King seemed very well. Stratford Canning and Lord Strangford were at the Court, to be presented on their return.[21]

[19] C. F. Greville, *Greville's diary, including passages hitherto withheld from publication*, ed. P. W. Wilson (2 vols., 1927), i. 310 (12 Dec. 1830).
[20] Ibid. 309–10 (1 Jan. 1832).
[21] Ellenborough, *Diary*, 101, 296, 72.

Indeed, Ellenborough's description of the first recorder's report he attended provides us with our most vivid image of the social and political contexts within which Council business was conducted:

> council at 3. Recorder's Report. The first I have attended. The Duke of Buccleuch and Lord Rosslyn were sworn in as Scotch Lord-Lieutenants. The appointment of Lord Rosslyn gives offence to Lord Elgin and Lord Leven, both having larger estates in the country, but it was quite right to give the Lord-Lieutenancy to Lord Rosslyn. I never cease regretting he is not in the Cabinet. The King seemed souffrant from the first; but he was placed in a through air in a very cold room, which chilled even me who was nearest the fire. He fell asleep, and latterly seemed really ill. The Duke fell asleep, too, and looked like death. The Duke of Leeds, Lords Winchester and Conyngham stood behind the King's chair. The Recorder sits on the King's right hand on a stool. We dined with the Chancellor. He gave us a good dinner, and bad plate.[22]

Ellenborough does not mention that at this bizarre meeting five men were sent to their deaths, two for stealing horses and three for house-breaking, one of them aged 17.[23]

And, finally, a nice vignette. In 1826 Wellington told Mrs Arbuthnot that 'he was quite disgusted at the manner in which the business was done', but by this he meant something other than we might expect:

> He said that the Chanchellor *mumbled* thro' the cases in such a way that the King said he could not hear one word. At last the King went to sleep, the President of the Council, Lord Harrowby, went to sleep, & of 46 capital convicts, many of them atrocious cases, *six only were condemned*. The Duke always says that, when he is not in waiting, he will not attend the Recorder's reports, he thinks them such a mockery of justice.[24]

In this view, the Council mocked justice because it sent men to the gallows not too cavalierly but too infrequently: it was the Council's lax *generosity* that upset Wellington most.

2. *Mercy and the king*

Historians have tried to be kind to George IV. Unspeakable though he was in other respects, they agree that he was 'humane'.[25] They all quote Greville's observation that the king was 'always leaning to the side of mercy' against the severer inclinations of Tenterden and the recorder. In

[22] Ellenborough, *Diary*, 47 (3 Mar. 1828).

[23] Cases and outcomes at each Council are listed in PP 1835, xlv.

[24] Bamford and Wellington (eds.), *Journal of Mrs Arbuthnot*, ii. 58–9 (28 Nov. 1826).

[25] C. S. Parker, *Sir Robert Peel . . . from his private correspondence* (1891), 315; Aspinall, 'Grand Cabinet', 339–40; C. Hibbert, *George IV* (1976), 312–14.

the more touching cases, the king insisted on putting the question of life or death to the vote so that it was 'decided by the voices of the majority'—as if this were strange. Now, it is true that the king did sometimes battle to save people. Greville claimed that many owed their lives to his refusal to defer to the lawyers' or the home secretary's advice.[26] The king also wrote to Peel that mercy was 'a word more consoling to the King's mind, than language can express'.[27] But alas for yet another myth, many of the king's postures were less wholesome (or influential) than these commendations convey. George IV was less a humane man than a sentimental or squeamish one. He obeyed, it was rightly said, 'no higher springs of action than emotion'—and then not sympathetic emotion.[28] There was little strength or determination in his gestures on behalf of the condemned, and he saved few necks in the event. He usually found it easier to capitulate to stronger voices—especially to Peel's, as we shall see.

How remote the plights of the condemned were from George's understanding many episodes reveal. As regent, he nursed his gout in the Brighton Pavilion in the winter of 1815–16 while fifty-eight people waited five freezing months in Newgate cells until he returned to London to receive the report. When Brougham denounced the profligacy of a court which, 'when the gaols were filled with wretches, could not suspend for a moment their thoughtless amusements to end the sad suspense between life and death', ministers closed ranks. The attorney-general, Garrow, felt that the condemned had little cause to complain: at least they lived a bit longer. And Castlereagh pretended that the recorder was busy examining the evidence and enquiring into the prisoners' characters. Delays like this were recurrent. Illness often accounted for the king's absence: 'gout in the left knee', for instance, in 1822.[29] That same excuse looked thin, however, when the king came to London in February 1828 for the marriage of the duchess of Kent's daughter, but refused to do so to receive two reports awaiting him: again parliament debated the issue in vain. A solution was found only when the ancient principle that reports had to be received in London was rescinded. From November 1828 most reports were received at Windsor until William IV heard them at St James again.[30]

[26] C. F. Greville, *The Greville memoirs, 1814–1860*, ed. L. Strachey and R. Fulford (8 vols., 1938), i. 304–5 (21 July 1829).

[27] George IV to Peel, 20 July 1823: Peel papers, BL Add. MS 40299, fo. 243.

[28] A. Fonblanque, *England under seven administrations* (2 vols., 1837), ii. 17.

[29] Peel to Eldon, 11 Nov. 1822, Peel papers, BL Add. MS 40315, fo. 70.

[30] Aspinall, 'Grand Cabinet', 340–4.

Royal jealousies and dislikes also held up the merciful business. When in November 1829 the recorder was too ill to attend Council (gout again) and it was proposed that Denman as common serjeant should present the reports in his place, the king exploded. He had not forgiven Denman for acting as Queen Caroline's counsel in his divorce case. 'The king said, if Denman dared to come within his palace, he w^{d} call the Guard himself & order them to turn him out, & raved & stormed like a madman.' He 'threw himself into a terrible tantrum', 'and he was so violent and irritable that they were obliged to let him have his own way for fear he should be ill. . . . So business is at a stand-still, and the unfortunate wretches under sentence of death are suffered to linger on, because he does not choose to do his duty.' Wellington explained to George that 'the making of this report was the privilege of the City of London', so Denman *had* to do it. But Wellington also admitted that 'it did not signify one farthing whether they were kept in prison a little longer or shorter time'. The king thought so too, and the Council met only when the recorder recovered. Of the thirteen men who for over two months had been awaiting their fates in the Newgate cells, two were ticked off for hanging.[31]

This episode made the chancellor, Lyndhurst, think the king was mad. Greville knew better: 'the fact is, that he is a spoiled, selfish, odious beast, and has no idea of doing anything but what is agreeable to himself, or of there being any duties attached to the office he held.'[32] 'No Council yet,' Greville noted a few days earlier: 'the King is employed in altering the uniforms of the Guards, and has pattern coats with various collars submitted to him every day.'[33] And it is Greville who, as usual, leaves us with the most abiding image of the king's pre-occupations in these weeks: 'I was standing close to [the King] at the Council, and he put down his head and whispered, "Which are you for, Cadland or the mare?" (meaning between Cadland and Bess of Bedlam); so I put my head down too and said, "The horse!" and then as we retired he said to the Duke, "A little bit of Newmarket."'[34]

If ministers had to defer to the king's petulance, they fought harder against his squeamishness about hanging. George meant well at the

[31] Eldon to Peel, 10 Feb. 1824: Peel papers, BL Add. MS 40315, fo. 112; Bamford and Wellington (eds.), *Journal of Mrs Arbuthnot*, ii. 314; A. Aspinall (ed.), *The letters of George IV*, 1812–1830 (3 vols., 1938), iii. 448; Greville, *Memoirs*, i. 333 (20 Nov. 1829), 340 (5 Dec. 1829); J. Arnould, *Memoir of Thomas, first Lord Denman* (2 vols., 1873), i. 436; PP 1835, xlv. 161.

[32] Greville, *Diary*, i. 133–5 (16 Jan., 20 Nov., 5 Dec. 1829); cf. *Memoirs*, i. 338–9.

[33] Greville, *Memoirs*, i. 337 (1 Dec. 1829).

[34] Greville, *Diary*, i. 47; C. F. Greville, *The Greville memoirs: A journal of the reigns of King George IV and King William IV*, ed. H. Reeve (3 vols., 1874), i. 148 (21 Dec. 1828).

beginning of his reign and at first he won a point or two. In March 1820 he announced that now he was king he wished to order no prisoner whatever to execution. Sidmouth had to point out that of the seventy-seven names on the next report some were 'of so heinous a nature, that it would be of very ill consequence [if] the offenders should be only transported'. In any case, Sidmouth added, convicts had been executed after assizes who might just as well have merited pardon had the king attended to their cases, and justice surely required consistency: 'the accident of a crime being committed in one County or another ought not to make any difference in apportioning the punishment.' George was persuaded by this implausibly principled argument, although Sidmouth had to repeat it at the Council's meeting. The king was then overruled without apparent difficulty, and 'two utterers of forged notes, two servts. for robbing their masters, one horse-stealer and one cattle-stealer were not included in the reprieve; but it was understood that the former four only should suffer'. Hobhouse, to whom we owe this account, noted that 'the King had recourse to laudanum before he proceeded to business'.[35]

The man who had the shortest way with royal squeamishness was Peel, succeeding Sidmouth as home secretary in 1822. For all his jealous defence of the royal prerogative in the Commons, he more than any preceding home secretary ensured that the prerogative came *de facto* to lie in his, the executive, hands rather than the king's.[36] Peel had an early lesson in the need to stand up to the royal whim. Just before he took office the Council agreed to reprieve a 13-year-old boy, Henry Newbury, condemned to death in February 1822 for stealing a silver teapot valued at £4 from a house in Grosvenor Place. The lad's sentence was commuted to transportation, but Newbury's respectable supporters petitioned for further mitigation. They pointed out that Newbury was the son of a private soldier who had worked for two bricklayers before he was seduced by bad company, and that his masters were willing to take him back into service. Nine very respectable people signed the petition.[37] The king told Peel that he found their pleas 'moving', and that he wanted the boy to be detained in a house of correction 'for such

[35] A. Aspinall (ed.), *The diary of Henry Hobhouse, 1820–1827* (1947), 17 (25 Mar. 1829).

[36] Such was Peel's success in this that in 1828 Lord Colchester, seeking to solve the problem posed by the king's refusals to attend reports in London, suggested that henceforth Old Bailey (as well as assize) appeals might as well be dealt with by the home secretary alone. The Cabinet's agreement to hold Councils at Windsor made the suggestion redundant; but it is doubtful whether such a suggestion could have been made before Peel took office: *Diary and correspondence of Charles Abbot, Lord Colchester* (3 vols., 1861), iii. 548, 551.

[37] HO17.76 (dossier Oh14).

a term as may be deemed expedient'. Newly in office, Peel had to acquiesce.[38] This little victory encouraged George to try harder at the next Council on 18 May 1822, Peel's first as home secretary. This time Peel dug his heels in, entering into a famous fight with the king. How Peel won it merits examination. It brings us closer to the finer textures of the Council's (and Peel's) decision-making than diaries do. We also get closer to the people whose lives were the stakes in the lottery.

3. *The king and Mr Peel*

A history of the pardoning process based on élite diaries is a history with its victims left out. Luckily the petition archive enables us at several points to trace passing references in the gossip of the great to the sad crimes and destinies of the London poor. We could make many choices here; but the fates of eight of the men sentenced at the Old Bailey in April 1822 and dealt with at Peel's first council on 18 May signify more than most. It was around their cases that Peel first outmanœuvred the king. Their story also introduces us to Peel, who deserves the next chapter all to himself. What follows is a narrative study of decision-making at the highest levels. It moved through four stages.

(i) *The Old Bailey death sentences, April 1822*

The fourth Old Bailey sessions of 1822 began on 17 April. In nine days Judges Park and Burrough, Recorder Knowlys, and Knowlys's deputy, Arabin, tried 221 felonies. Park sentenced to death ten of his case-load, Burrough two, the recorder five, and Arabin four. They included two women: one an arsonist, the other a cook who stole £25 worth of goods from her master. Three men were condemned for horse-theft, and two for stealing a lamb worth £1; there was also a forger to the sum of £30; the remaining thirteen were housebreakers and burglars.

Not all of these petitioned. Most knew that their sentences would be commuted because of their youth, character, sex, or the triviality of their offence. Two young burglars, for example, had stolen clothes and sheets amounting to nineteen shillings in value, and one horse-thief was aged only 16. Off the Council let them go to the hulks or New South Wales. A more difficult case was that of a horse-thief Matthew Verney, half-witted, allegedly insane. Insanity had been accepted as a defence long before the McNaghten rules of 1843, but in the 1820s the defence

[38] Parker, *Peel*, 315; Peel papers, BL Add. MS 40299, fo. 50.

was rarely pleaded or accepted at the Old Bailey,[39] and the recorder had bypassed Verney's afflictions at the trial. So the parents petitioned. Verney, aged 45, had fractured his skull when 15 and had twice been committed to St Luke's hospital: 'we have been obliged to have a keeper to attend him and it is well known the whole of his conduct is tinctured with insanity.' The parents were respectable housekeepers, and their petition was formally drawn up if scratchily signed. It was supported by the prosecutor and thirty-six local tradesmen, gentlemen, and clergy, and a hospital certificate. The Council concurred and Verney's name was ticked for reprieve.[40]

It was the remaining thirteen burglars who caused the trouble. Together they had netted only £90 worth of goods, mostly in clothes and tableware.[41] But we know from Hobhouse's diary that the Council considered the crimes of eight of them to be 'serious'. We also know that this was the occasion when Peel first showed his teeth. He wanted all eight hanged.[42]

Those eight petitioned the king. The first two had no chance. Bartholomew (34) and Close (33) had burgled to the value of £7 from a public house.[43] In two petitions[44] Bartholomew admitted 'the justice of his sentence'; but his wife was pregnant and his five helpless children, the youngest 16 months old, depended solely on him. His character hitherto was irreproachable; twenty-seven tradesmen supported this, including the prosecutor and two bakers who had employed Bartholomew—'a[s] good and honest servant as ever I had'. Close's petitions similarly mustered the signature of the prosecutor and two householders to testify to his repentance and his wife's pregnancy too. He added that his previous good character would have been attested by three 'witnesses of respectability' had his case not been unexpectedly heard on the first day of the sessions, and so they could not come. He was a shoemaker, 'ruptured, and of a weakly habit of body'. He begged mercy with all the elaborate tropes of submission and loyalty that the occasion demanded: 'Your Majesty's most humble, dutiful, and Loyal, but unfortunate Petitioner, throws himself upon your Gracious Majesty's Royal consideration and clemency, hoping to receive such relief as the nature of his case in your Majestys paternal kindness may

[39] N. Walker, *Crime and insanity in England*, i. *The historical perspective* (Edinburgh, 1968), 69, 72.

[40] All dossiers considered at this Council are bundled in HO17.76, and this discussion is based on them. Each individual dossier was indexed (e.g. in Verney's case as Qh49): references are to these index numbers.

[41] OBSP 1822: 239–312.

[42] Hobhouse, *Diary*, 87.

[43] OBSP 1822: 241.

[44] Qh48.

think fit to grant unto a most grateful Supplicant and humble Subject.'[45] His appeal was signed by a dozen householders and neighbours. But it was no good. The king and the Council were to agree that Close and Bartholomew should be ticked off for hanging.

(ii) *The appeals of Desmond and Davis; Anson; and Ward*

The remaining petitions caused more trouble because they upset the king somewhat, and also because some attracted the support of disinterestedly benevolent men. In the first case Edward Desmond and James Davis, both aged 17, were sentenced to death by the recorder for burgling a law-stationer's house. They netted a substantial booty of coats, watches, coins, £9 in silver, and two £5 notes. Disturbed by the landlord and caught in the hue and cry, they were tried and condemned with wondrous celerity. Committing their crime four days after the sessions opened, they were tried almost at once: no hope of a prepared defence here.[46] Still, they found support in an unexpected quarter. One of the lodgers whose property was stolen was the chaplain to the high sheriff of Worcestershire and an LL D to boot. Benevolence personified, he visited Desmond in Newgate and found him penitent and his family honest and industrious. He petitioned to say so, and added that neither Davis nor Desmond offered him violence when challenged; the pistol they carried was unloaded. He added that the law should not punish theft without violence as harshly as it did theft with violence. A distinction in punishment would 'have a strong tendency to soften down the barbarity and brutality which persons of their character might otherwise be induced to evidence'. Meanwhile Desmond's own petition stressed that he was 'a fatherless and friendless orphan', who 'being an only son, was in [his parents'] lifetime allowed too much liberty' and so fell into bad company. It was signed by twenty-nine tradesmen and householders convinced of 'the integrity of his repentance'.[47] Davis's petition testified to the respectability of his parents (they had lived in the same parish for twenty years and his father was a glass-blower) and to his own penitence and youth. The prosecutor, his master, the foreman of the jury which tried him, and some dozen tradesmen begged mercy in view of Davis's age. Respectable persons interceded on Davis's behalf too. One Lucius H. Robinson of Covent Garden Chambers even interviewed Peel in person.[48]

The second burglary case had been tried by Park. The Greenford

[45] Qh50. [46] OBSP 1822: 291; *The Times*, 26 Apr. 1822. [47] Qh42. [48] Qh47.

house of a Bond Street linen-draper had been burgled by four men, Henry Naylor (24), George Adams (32), Edward Ward (21), and John Anson (26). They were making off with a £10 clock, £6 worth of silver, and sundry tableware when the prosecutor woke up, fired a pistol at them, and raised a hue and cry as the men ran into the night and hitched lifts on carts to escape. When arrested, scratches on their persons, bullet holes in a coat, and the fact that Adams was carrying a pistol led to easy convictions.[49]

Nothing could save Adams and Naylor, it transpired. In his petition Adams acknowledged his guilt and pleaded that this was a first offence. The two-score friends who scratched their ill-formed signatures on a petition to support him had nothing to add. Naylor's petition likewise pleaded only a former character good enough to attract the supporting signatures of the prosecutor, a German baroness (his employer), and six members of the jury who had tried him, but this availed him nothing.[50] Adams and Naylor were to hang alongside Bartholomew and Close. Four down and four to go.

On Anson's and Ward's behalf, bigger efforts were made; their tales were more elaborate and their social standing more easily affirmed.[51] Anson's petition announced that he 'was for some years employed as a clerk in the offices of several respectable solicitors, but having formed a notion of a seafaring life he engaged in a voyage to the South Sea Fishery, and . . . during his absence the owners became bankrupt and from a failure of the season [he] was upon his return thrown destitute upon the world without any recompense for his services'. The prosecutor supported the petition, along with some dozen others. His mother and stepfather added that Anson 'was always a steady and faithful son' to parents who were 'in an humble sphere of life and of honest and industrious habits'.

Edward Ward, finally, had respectable connections. His father had traded to Oporto until the French invasion compelled him to leave; he had then entered a partnership with the royal wine merchants Delapierre & Co. until the firm failed. His friends procured the boy's admission into Morden College and found respectable situations for his brothers. Edward was 'a handsome lad of gay and easy temper', but with an intellect 'weak almost to imbecility'. As his own petition put it, the death of his mother caused him to fall into 'bad company'. This petition was signed by the solicitor James Harmer, whom we have met

[49] OBSP 1822: 241. [50] Qh45 and 46. [51] Qh44 and 43.

before: he knew the family and added that on his mother's death the youth's father cohabited with 'a female of very bad and immoral habits' whose influence on the boy was pernicious. Additional support came from the City sheriff John Garrett who, not without emotion, wrote to Peel of Ward's 'highly reputable connexions', whom he knew well and who were deeply afflicted by his plight. A City friend and commercial connection of the family wrote that though Ward had appeared before magistrates earlier, he had committed no atrocious act, and that the prosecutor had recommended all the burglars to mercy.

So Desmond, Davis, Anson, and Ward had fair chances. They were young. Their prosecutors or their jurymen and at least one of their victims petitioned against their execution. Their lapses from respectability could be encased within exonerating narratives. One had good connections. And their cases were representative of their times, for they had moved an informed segment of public opinion which thought that execution for crimes without violence was unwarranted.

It is the central point of this discussion that none of this counted for much in Peel's mind: so much (again) for the power of 'opinion'. He came to his first Council meeting as he left his last, determined to hang men even-handedly and as numerously as good order required, and not to make exceptions where distinctions were narrow. So he took a strong line against all eight burglars under report; and initially he won his case for wholesale execution. Thanks to his connections, Ward's case was discussed at some length, and the chief justice was in his favour; but Peel argued that it was 'impossible to spare him without at the same time extending mercy to most of the others, who were less culpable'. Of all eight men condemned for burglary, he convinced the Council that it was 'right for the sake of example to let the law take its course'.[52] At this stage the king apparently agreed. A scaffold for eight men was going to be needed in four days' time.

Then something happened. After the meeting the king's 'mind soon gave way' (as Hobhouse put it). There must be some reason to reprieve some of the eight, he insisted. Vexed, Peel had no option but to embark on a hunt for distinctions between the eight men's culpability. His heart was not in it. He was determined to hang them all if he could. But thanks to the king, this was going to be another cliff-hanger.

[52] Hobhouse, *Diary*, 87.

(iii) *Auctioning lives*

Home Office enquiries to magistrates exposed one useful distinction between Davis and Desmond on the one hand, and Anson on the other. On the very eve of execution, the Bow Street magistrate Birnie reported that Davis and Desmond had appeared before him for stealing a greatcoat and had been acquitted only because the prosecution witnesses were of 'indifferent character', while Desmond was already known to several of his officers 'as keeping bad company and late hours'. Anson was unknown to him, however, and to the Marlborough Street magistrate too.

Peel was not too impressed by this, but he needed one sop for the king's soft-heartedness. He found a better one than Anson, however, when at the last minute Davis betrayed an associate. A Newgate official reported that after 'a good deal of extra conversation' (a euphemism, one suspects), Davis confessed that one Barnes had also been involved in his burglary, and that Barnes had also committed an earlier armed burglary in Westminster. Desmond was presented with Davis's confession and he too admitted Barnes's complicity, 'after which he appeared to be relieved as from a great burthen'.

Peel was prepared to accept Davis's betrayal as a circumstance in his favour. He recommended the king to respite him and to let the other seven hang in the morning. His suggestion did not go down well, however. At seven o'clock that evening the king wrote to thank Peel 'for his active humanity' in reprieving Davis, but 'what', he wrote, 'is to be done concerning his accomplice, Desmond, who is of the same age? Is there any opening for the other poor young man Ward? The king would be truly glad if such could be found.' Another urgent note came from Carlton House at midnight: the king had to say 'after the deepest reflection, that the executions of tomorrow, from their unnatural numbers, weigh most heavily and painfully on his mind'. 'From the entire conviction that he ought to abridge the numbers for execution', he did hope that Desmond and Ward might be saved too. Speaking for himself, he would not hesitate to respite Ward forthwith. He wanted Peel to extend mercy to two others besides Ward, so that, for the same crime, only four would suffer in place of eight. The king trusted in God that this extension of his royal clemency would answer every purpose of justice. He had no hesitation in believing that Mr Peel would readily enter into the king's feelings upon this melancholy occasion.[53]

[53] Peel papers, BL Add. MS 40299, fos. 67ff. With some misquotations, this correspondence with the kings is reprinted in Parker, *Peel*, 315–16.

Peel, doubtless banging his head in frustration if he ever did anything so overt, had no choice but to play for time. Late that night he extended a two-day respite for Ward, Anson, and Desmond. But he let the king know that the Council would need to reconsider the cases and that he had directed Garrett, the sheriff, not to encourage any hope that the capital sentences would be remitted.[54]

At least those excluded from the respite could be killed off without more ado. So at eight o'clock the next morning, on Wednesday 22 May, Bartholomew, Close, Naylor, and Adams were brought out from the Debtor's Door of Newgate, and there they dropped together. Bartholomew bowed to the crowd from the scaffold. Close, the weakly shoemaker with the pregnant wife, was agitated and stared wildly around, exclaiming 'Where's my wife? Let me see my poor dear wife!' 'He was with difficulty kept from disturbing the other prisoners . . . [and] was carried up by two men; he twice fell on his knees, and was obliged to be supported to the scaffold.'[55] As the newspapers always put it, 'an immense concourse of people' attended the spectacle; but it was not a famous execution-day in fact. Outside this present text their only memorial is a broadside and woodcut which showed how the printer was wrong-footed by the Council's tergiversations. Five figures are inserted on the scaffold instead of four; and the text purports to describe the executions of Naylor, Adams, Ward, Anson, Bartholomew, Close, Desmond, and Davis (Plate 40).

Peel then called a Cabinet meeting on the same day. Here a compromise was reached whereby the king's wishes and Peel's might each be half satisfied. It was like an auction. How about six burglars being killed, in lieu of the king's four and Peel's eight? The lucky one to join Davis was to be 17-year-old Desmond, age signifying here. But not Ward and Anson. They were to join Bartholomew, Close, Naylor, and Adams in eternity after all. Peel wrote to the king that this was a unanimous decision, and that he would call on the king personally to explain it.[56] According to Hobhouse, the king acquiesced meekly.[57] It was a bold victory for Peel, and it strengthened his hold over the Council's merciful business from that date on.

[54] Peel papers, BL Add. MS 40299, fo. 70.

[55] *The Times* and *Morning Chronicle*, 23 May 1822, carried identical 20-line reports of the execution, poached also by *Ann. Reg.* 1822: 88.

[56] Peel papers, BL Add. MS 40299, fo. 70.

[57] Parker, *Peel*, 316–17; Hobhouse, *Diary*, 87.

40. *Execution of Henry John Naylor, George Adams, Edward Ward, George Anson, William Bartholomew, John Close, Edward Desmond, and John Davis*, 1822.

The broadside which bore this woodcut commemorated the story told in this chapter. Its depiction of the Newgate gallows was used on many other broadsides, appropriate numbers of bodies being inserted to match those hanged. Here the printer anticipated eight executions, adjusting the inset but not the title when last-minute reprieves cut the number to five. Another man was reprieved subsequently; but the broadside was printed and in circulation by then.

(iv) *A last bid*

The drama was not quite over. With two days to go, private efforts to save Ward and Anson redoubled. They came chiefly from one of those humane sheriffs whose appeals litter the London dossiers in these years. After being obliged by his office to attend the four men's execution on the Wednesday morning, Sheriff Garrett wrote politely to Peel to thank him for the respites for the others. They were 'an act of merciful interposition on the part of Providence and an act of great humanity on the part of His Majesty's advisers', he said. He wrote without apparent irony, though he admitted that 'the apparent fixed determination on the part of His Majesty's council to execute the eight unfortunate criminals had weighed heavily upon my mind'. What he could not credit now, however, was the Cabinet's decision to reprieve Desmond but not Ward and Anson. He proposed to wait on Peel at nine o'clock the next morning to convey his opinions.

We cannot know what was said then, but his letter conveys its drift. Here was one man who in his guts felt the shame of it all:

> Sir!, how soon has the hope of mercy and flattering prospects for two of the criminals vanished, what new light can have burst forth so as to have deemed it necessary to exclude the poor unfortunate (but truly penitent) men Ward and Anson? Can it for a moment be intended to execute so severe a sentence upon them after holding out this gleam of mercy, rare indeed has such a thing taken place as execution after a respite, in the course of nine years that the present Ordinary of Newgate has held his office, *only one solitary instance has occurred, and that in the case of a Rape*. Let me then implore you Sir! as I would supplicate at the throne of grace for mercy on the behalf of these wretched men, not to suffer the vengeance of the law to be executed upon them, under the peculiar circumstances in which they are now placed, that sort of hope having been held out, which in almost every instance has ultimately obtained a mitigation of punishment.

With one day to go before Ward's and Anson's hanging, William Wilberforce was enlisted in the cause. Anson's dossier contains a letter asking Wilberforce to fulfil an earlier promise to see Peel about Ward's plight. Ward's brother also sent Peel a petition: Ward had no criminal record, while the pardoned Davis and Desmond were old offenders recently tried for burglary. The prosecutor likewise petitioned that Ward and Anson had 'already suffered the horrors and pangs of expected death, feelings in themselves almost expiatory of their crime'; both were penitent, and in any case 'Your memorialist humbly represents that the lives of two out of the four burglars may be deemed a sufficient offering to the violated laws of their country.' And the wretched Edward Ward himself wrote that he had 'already suffered all the horrours [*sic*] and pangs of executed death and is now the victim of the bitterest disappointment, being ordered for execution a second time'. Ward played the only card left to him: he would freely and truly answer any questions about the accessories and instigators of the burglary which might be put to him, if that could earn him a commutation.

'*Refused*,' Peel scribbled! Garrett the sheriff was instructed to tell Anson and Ward that they were to hang next morning. This out of pity Garrett could not do. On the fatal day he wrote to Peel that he 'did not personally communicate with the prisoners last evening, but requested the Ordinary of Newgate to do so, it being a task too painful for him to execute'. Anson and Ward were led from their Newgate cells to the scaffold. Throughout the drama so far the press had been reticent. *The Times* and *Morning Chronicle* had reported the proceedings of the April sessions in no more than several lines, and the hangings five weeks later were reported curtly. The proceedings at King in Council and their aftermath were not referred to at all.[58] This was no *cause célèbre*, and 'public opinion' was silent. Now it was all over, *The Times* gave a couple of inches to Ward's 'quivering lip and uplifted eye', and to the fact that although Anson died immediately, 'Ward struggled very hard.'[59] The *Morning Chronicle* did not report the executions at all.

There is much in this story which connects with earlier parts of the book: the support for the condemned from prosecutors and neighbours as well as from relatives; a sheriff's compassion; the grounds for mercy and its opposite. But if the episode signifies in the present context, it is for the light it throws on Peel and on the dominance he established over the Council and the king. This dominance was confirmed when a year later, on 19 July 1823, the king tried it on once more. His sensibilities were moved this time on behalf of one Samuel Miles, sentenced to death by Burrough J. for uttering a forged £5 banknote. Miles was a trickster. Recently acquitted on a charge of theft, he passed the forged note on to a shopkeeper from whom he bought a cabin trunk, volunteering the information that he had just got off a ship from the East Indies and that a fortune was due to follow him. But he was young (only 19, the court had been told, 16 according to evidence in the trial, 17 according to the king), and the jury recommended him to mercy. He also had influential patrons. Hobhouse reports:

> [Miles's] connexions being respectable, made great interest to save his life, and obtained access to Lady C[onyngham] who wrote that she had impressed the king strongly in favour of the culprit and endeavours w[d]. be made to save his life, but she desired that too sanguine expectations might not be entertained, because she had not been very successful in former applications.

Lady Conyngham had 'almost complete control over' the king, however, and the king accordingly desired Peel 'immediately [to] enquire into the case of . . . [this] unfortunate boy, only of seventeen years of

[58] *The Times* only once referred to the meeting of King in Council in 1822, on 24 June, and then in a curt announcement that it had been decided to execute 3 men on the following Wednesday.

[59] *The Times*, 25 May 1822.

age, who the King fears is under sentence of death, and ordered for execution on Monday morning next. . . . The King is anxious to save his life, on account of his *extreme youth*, provided anything can be alleged to commute the sentence.' Peel replied soberly that the case had been considered already and that the law ought to take its course, the case being similar to another for which a man had been executed a few weeks earlier. If Hobhouse is to be believed, this time Peel 'resolved, if the king persisted, to send a respite, and to resign his office'. The threat worked and the king 'ungraciously acquiesced', weakly regretting that there were apparently no gounds for mercy—that 'word more consoling to the King's mind, than language can express'. 'The D. of Wellington undertook to dissuade Lady C. from further interference with the administration of justice', and young Miles duly hanged.[60] Only once more do we hear of royal intervention of this kind, in the case of the Quaker forger Hunton in 1828, whose life the king mildly professed himself to be 'very desirous (if it can be done with any sort of propriety) to save'. But Hunton hanged too.

Throughout the 1820s the strange proceedings of the King in Council attracted little public attention. There was only one noisy indictment of its procedures publicly, and that towards the end of its life. A year after the tories' electoral defeat in 1830, Gibbon Wakefield's *Facts relating to the punishment of death in the metropolis* delivered an excoriating attack *inter alia* on the whole process at Council:

> In London the question of life or death is decided by many and irresponsible judges; who know nothing at all of the case except from report; who conceal the grounds of their decision; who give to rumour the weight of evidence; whose minds are constantly occupied with other and to them far more important matters; and many of whom (including the chief—the First Home Secretary) act as judges here, merely because elsewhere they are skilful politicians.[61]

But in the end it was not public opinion that killed off the Council. Nor is the charming tale true that abolition came with Victoria's accession in 1837, because 'the infliction of the death penalty was a matter which could not with propriety be discussed in the presence of a girl of eigh-

[60] Parker's account of this episode is defective (Parker, *Peel*, 317). This paragraph is based on Peel papers, BL Add. MSS 40299, fos. 241, 243, and 40300, fo. 265; Hobhouse, *Diary*, 104; and OBSP 1823: 336. An appeal dossier on Miles is indexed in HO19, but cannot be traced in HO17.97.

[61] E. Gibbon Wakefield, *Facts relating to the punishment of death in the metropolis* (1831), in M. F. Lloyd Prichard (ed.), *The collected works of Edward Gibbon Wakefield* (1968), 243–5.

teen'.[62] The simpler truth is that in 1837 the recorder's report became redundant because the number of capital sentences had shrunk to near insignificance. The annual average reviewed by the Council dropped from 175 in 1828, of whom 21 were left for execution, to 46 in 1834, all reprieved. By 1837 only 8 people were executed in the *whole* country, all of them murderers, to whom mercy was never extended.[63] It remains only to add that when the whigs abolished the great constitutional instrument of King in Council, no statutory authority could be found for its existence in the first place.[64]

[62] There is no evidence for this charming conceit, repeated, however, in Aspinall, 'Grand Cabinet', 338, and Radzinowicz, i. 112.

[63] PP 1835, xlv. 161, and criminal statistics for 1837, PP 1837–8, xlii.

[64] Aspinall, 'Grand Cabinet', 338.

CHAPTER 21

MERCY AND MR PEEL

1. *The great reformer*

IN 1831 EDWARD GIBBON WAKEFIELD RAISED A KEY QUESTION. Lately released from three years in Newgate for abducting an underage heiress, scales had dropped from his eyes in that dark place. Gentleman though he had been and would become again, he had lost the politeness which inhibited questions as rude as this. This question was in effect: *how could these men bear to send people to the gallows with such equanimity*? He had the home secretary Robert Peel in mind:

> Almost every Old Bailey Sessions subjects the Secretary of State to the pain of denying, absolutely or for a time, the prayer of some heart-broken wretch, who face to face with the judge, half choaked with grief, and perhaps kneeling at his feet, pleads for the life of a father, a husband, or a child. To undergo this pain, not less probably, on the average, than once a month, must, one would think, injure the health, so as to shorten the life, of him who suffers the pain. Yet the late Secretary who had suffered it for a great many years, was, at the end of his career, and in his legislative capacity, adverse to abolishing the punishment of death. . . . Mr Peel's long practice in that one matter had rendered him callous, and had led to his piquing himself on his firmness.[1]

Could 'practised callousness' really be all there was to it? Historians nowadays are unlikely to say anything so unseemly. They would explain Peel's capacity to endure these heart-breaking transactions by talking about his pragmatism, sense of duty, or belief in providential dispensation, high-minded answers which make Wakefield's answer seem childlike or vulgar. But let us see: for clearly there were two quite distinct ways of viewing the world and its retributions at odds here, the humanely empathetic and one quite other than that. For once in this book let us get into the mind of a great hangman. Peel is not only cen-

[1] E. G. Wakefield, *Facts relating to the punishment of death in the metropolis* (1831), in M. F. Lloyd Prichard (ed.), *The collected works of Edward Gibbon Wakefield* (1968), 244.

tral to our subject-matter, looming over so many past chapters as he has; he also offers an acid test of how a powerful segment of the high political culture, holding political power until 1830, thought and felt about the hanging law in its last days.

Peel was famously a man of paradoxes. We are concerned with only one of them. As home secretary from 1822 to 1830 (with a five-month break in 1827–8), he came to dominate the king and his Council as they met to discuss whom to pardon and whom to kill in Newgate. He dominates the provincial petition archive too, across a critical span of years when petitioning probably became more frequent and ambitious as prosecutions, convictions, and hence pardons increased in a climate of heightening expectations about criminal law reform. The final adjudicator upon whom the merciful business of the prerogative converged, it was to him rather than to the king that people came in their elaborate and needful deference, trusting in his clemency as much as in the king's to let them or their friends or protégés escape the noose or transportation. Petitioners found little mercy in Peel, however, and certainly no matching sense of that commonsensical fairness which seems increasingly to have moved lay opinion. Many sad cases hanged; many whose guilt was hardly proved were conditionally pardoned but still transported.

The paradox is that it was the same Peel who did most to raise reformers' expectations. When he took up the Home Office in his thirty-fourth year he was already experienced in office and a formidable debater, essential to the government's unity and recognized as a possible leader: the greatest political fixer of his age. Moreover, this was the man who had no hesitation

> in avowing that he was a strong advocate for the mitigation of capital punishments. He wished to remove, in all cases where it was practicable, the punishment of death; for it was impossible to conceal from ourselves, that capital punishments were more frequent, and the criminal code more severe, on the whole, in this country, than in any country in the world.[2]

As he boasted in 1827: 'Tory as I am, I have the . . . satisfaction of knowing, that there is not a single law connected with my name, which has not had for its object some mitigation of the severity of the criminal law.'[3] Since strictly speaking this was true, most historians have taken him at his word. 'He was humane,' one admirer declares: 'He cared

[2] *The speeches of . . . Sir Robert Peel delivered in the House of Commons* (4 vols., 1853), i. 456, ii. 131 (henceforth *Speeches*).

[3] *Speeches*, i. 509.

intensely about the distress and poverty of the society in which he lived.' His consolidation of the statutes and rationalizations of legal procedure are said to have broken the inertia of centuries. Even if he had not gone on to refurbish conservatism in the 1830s and 1840s, most historians would give him a good press for this alone.[4]

Enough has been said already about equivocal sympathies in this era, the acquiescence in what was and had to be in large reaches of opinion, the strength of inertias and resistances within political and cultural systems, not to mention about Peel's own death-dealing in Council, to prepare us for what follows. We are not going to meet the great criminal law reformer of legend.

Peel's interest in criminal law reform had less to do with repudiating the barbarism of past times than with his interest first in restoring the law's credibility against public attack, and secondly in making it more efficient, even more punitive—more of a terror, not less. Far from dismantling the bloody code, or intending to begin to dismantle it, his statutory consolidations were intended to outflank the wholesale reform advocated by the whig–radicals and to retain the legal values of the *ancien régime* in their essentials. It was not without meaning that his revisions were accepted by the Commons with scarcely a whisper of dissatisfaction and were not even debated in the Lords. When Romilly had earlier mooted much the same measures as Peel implemented, he had been 'overpowered by sarcasms, invectives, and majorities'. But now (Denman wryly noted in 1824) 'suddenly the minister [Peel] proposes the reprobated project as a government measure, and converts, while he laughs at, his former adherents'.[5] The reason is that the main legal structures of the *ancien régime* were safe in Peel's hands.

To give Peel his due, he was candid about his motives. As early as May 1823 Mackintosh's nine resolutions for abolishing the death penalty on several important crimes forced him into the open.[6] He acknow-

[4] R. Blake, *The conservative party from Peel to Churchill* (1970), 664. Cf. N. Gash's view that 'with a reforming Home Secretary the citadel of legislative inertia had been finally outflanked' (*Mr Secretary Peel* (1961), 329). Beattie also claims that Peel's statutes 'swept away' many of the capital provisions in the criminal law and 'brought the criminal law more into harmony with the practice of the courts' (pp. 632, 635). Paul Johnson credits Peel with the most ambitious work of law reform since Henry II, and thinks that Peel regarded the operations of the law 'with horror' and 'was appalled by the frequency with which the rope was employed': *The birth of the modern: World history, 1815–1830* (1991), 865–6.

[5] J. Arnould, *Memoir of Thomas, first lord Denman* (2 vols., 1873), i. 254.

[6] Mackintosh wanted the death penalty removed from forgery, larceny in shops, dwelling-houses and ships, stealing horses, sheep, and cattle; he also wanted the repeal of the Black Acts. It was Mackintosh also in that debate who first proposed that judges be exempted from the necessity of pronouncing the death sentence when it was obvious that they were not going to recommend its execution.

ledged the need for change, but he made it clear that 'the real question . . . was . . . as to degree' and that the degree was going to be very limited. All he was ready to do was to abolish 'every part of the criminal statutes that could not with safety be acted on' or which 'could hardly ever be enforced'. This was in order 'not to lower the effect of the most solemn sentence of the law by [the] indiscriminate application of it', and to 'avoid the mockery of condemning men to death merely because that penalty was attached to the crime which they had committed'. What would mainly be tidied away was the glaring anomaly of sentencing people to death when it was public knowledge that they would not be hanged. Only thus could parliament 'add to the solemnity and *efficiency* of the laws'.[7] Peel kept his word. The capital punishments he did abolish in the following years involved crimes for which the sentence of death was never now executed, either because, like the Black Act or the impersonation of Chelsea pensioners, they were rarely if ever prosecuted, or because conditional pardons on them had become routine. This was a man who was embarking on a holding operation—a tidying-up.

He had no sense of isolation as he did so, and he was right in this. With hindsight it will always be tempting to regard those who defended the old regime as a reactionary rearguard, vainly justifying a penal code whose obsolescence should have been obvious, powerless to contain the tide of opinion against its discretionary severity to which whigs and middle classes were giving mounting voice.[8] Also with hindsight we may discern the structural overloading of the penal code (increasing prosecutions, convictions, and hence too many death sentences) and believe it to be determining.

But the final obliteration of the code was by no means foreseen in the 1820s, any more than parliamentary reform was. Political cultures exhibit strong inertias, and few in that decade thought wholesale change was inevitable. To large sectors of the political nation, including many whigs, the old code continued to make as much sense as it had to Paley in 1785. Peel was justified in taking for granted a strongly conservative sentiment in the country based on 'previous customs and formed habits'. This sentiment had to be catered to, he said in 1827. If parliament overthrew existing penalties too hastily, 'a strong prejudice might arise in the country against measures that were intended for the public

[7] *Speeches*, 243–4, 246, 247, 455–6; my italics.

[8] For just such an implication, see R. McGowen, 'The image of justice and reform of the criminal law in early nineteenth-century England', *Buffalo Law Review*, 32 (1983), 123; and Beattie, p. 631.

good; and thus the great object of justice and humanity might be defeated'. As he put it again in 1830: 'The habits and usages of the country were adapted to and formed on the severity of our code, and he found it necessary to proceed in the mitigation of this severity with great caution. He thought it advantageous to continue the severity of the law in its letter, but gradually to meliorate its practical application.' Out of office in 1832, he still believed that 'too rashly weakening the present protection of property' by the repeal of capital statutes would excite 'the prejudices of society'.[9]

Bathing in the adulation of his party, the judiciary, and much of the press, and taking his reputation as a law reformer very seriously, he felt sure that his prison and legal reforms would secure the basic frameworks of law-enforcement for a foreseeable future. That they left the death penalty on many offences and retained huge areas of judicial and executive discretion in sentencing did not affect his belief that his reforms would 'outlive the fleeting discharge of the mere duties of ordinary routine, and . . . confer some distinction on my name, by connecting it with permanent improvements in the judicial institutions of the country'.[10] And until 1830 this hope was realistic. The whigs in opposition were not a serious threat to government control; many might have been accommodated within a tory government in any case. There was no cause to expect them to become a government imminently. When the tories did fall, it was largely thanks to the revolt of ultra-tories against Catholic emancipation. Peel meanwhile and until then could properly claim that he had the conservative nation behind him.

Even so, hindsight has its say. A decade later Peel's final record was one of failure. The whigs had won the argument after all. Into power they came in 1830 after two generations in the wilderness. Within three years the abolitionist MP Ewart could declare that 'the whole of [Peel's] edifice, erected with so much care and trouble, was tumbling to pieces', soon to 'become its own dilapidated monument'.[11] The annual score of public executions in England shrank to a trickle. Peel's painstaking consolidation of the statutes was unpicked as in 1832 the whigs abolished death for coining, livestock-theft, stealing from dwelling-houses (these comprised one-third of all death sentences passed between 1825 and 1831), and for most forgery too.[12]

Even then Peel did not see the point. Out of office, he protested at

[9] *Speeches*, i. 406, 456, ii. 131, 162–3, 524. [10] Ibid. i. 410. [11] Radzinowicz, i. 606.

[12] The Commons voted to abolish the death sentence on all forgeries, but the Lords insisted on its retention for the forgery of wills, powers of attorney, and transfers of stock.

these 'most dangerous' experiments. He declared that crime was increasing and that he had already established the right deterrent balance between hanging and mitigation.[13] But it was to no avail. Housebreaking was removed from the capital list in 1833—accounting for another third of those capitally sentenced in Peel's last years as home secretary.[14] From 1837, when the whigs repealed more statutes, English people could in theory hang only for a dozen crimes,[15] but in practice only murderers or (up to 1861) attempted murderers were executed in future. In 1841 the capital list was reduced even further.

If Peel opened the way to this great end, he did so unintentionally. Some will feel, then, that there was an odd short-sightedness in him, the reverse of prophetic. Add to this the shortcomings of his law reforms in and of themselves—Bentham thought him a 'weak and feeble' law reformer; Radzinowicz finds his great work of consolidation 'somewhat mechanical'[16]—and the man's reputation in this context begins to look tattered.

Peel had his own personality to contend with. He was a reticent and formal man, remorselessly over-disciplined in boyhood by the upbringing which turned the son of a Lancashire calico-printer into the patrician leader his father intended him to be.[17] He spoke with a Lancashire accent which he fought self-consciously to control. Rigorous parental pressure left its scars. It was always 'easier for him to deal with problems than with people', Gash has written.[18] O'Connell compared his smile to the glint of the silver plate on a coffin-lid. Cold in manner, meticulous in matters of honour and duty, he managed 'his elocution like his temper', Disraeli said, and 'neither was originally good'.[19] Disraeli thought he lacked originality or imagination and made principle serve expediency. This became a familiar charge: Peel was to 'betray' his party twice, by supporting Catholic emancipation in 1829 and by repealing the Corn Laws in 1846. A recent reappraisal by Hilton discerns a rigidity in his thinking, even a dogmatism, which made it hard for him to compromise in a crisis and which could express itself in a defensive arrogance under attack. 'The most faultless of Ministers,'

[13] *Speeches*, ii. 524.

[14] Radzinowicz, i. 602–5.

[15] Notably murder, attempted murder when injuries were caused, rape and carnal abuse of girls under 10, unnatural offences, robbery when attended with cutting or wounding, arson when lives were endangered, piracy when murder was attempted, riot and feloniously destroying buildings, embezzlement by servants of the Bank of England, and high treason. See App. 2 below.

[16] Radzinowicz, i. 574 and n. 24.

[17] C. S. Parker (ed.), *Sir Robert Peel . . . from his private correspondence* (1891), 9 ff.

[18] N. Gash, *Sir Robert Peel: The life of Robert Peel after 1830* (1972), 664.

[19] Blake, *Conservative party*, 18.

Lord Hatherton observed: he gave to politics the same kind of cool attention he had given geometry and mathematics at Oxford.[20]

As home secretary there was much unimaginative dutifulness in him, as his role in Council has revealed. He upheld the rule of law without favour under harrowing pressures, subjected almost daily to impassioned pleas for mercy in the correspondence and affidavits which flooded his desk after each assize circuit and Old Bailey sessions. But we have confirmed Wakefield's verdict that he upheld the law without evident compassion either. He only once wept a passing tear for those he sent to the gallows, but it was a thin tear after all. The applications for mercy were 'very painful' for him to deal with, was what he said.[21] He mainly said, however (and with more conviction), that he would always execute his guardianship of the royal prerogative out of a sense of the highest duty: 'The exercise of mercy ought to be as prompt and as pure as the visitation of justice.'[22]

His detachment was impressive; but as we approach his labours from his supplicants' viewpoints, it is impossible to warm to him as in this spirit his pencils repeatedly scribbled 'Let the law take its course.' In his responses to these documents we meet something more of the workaday politician to whom Professor Beales has referred, something more of that 'passionless observation to events' which was noted in his later years, and rather less of the prophetic and humane statesman depicted by his admirers.[23] And beyond that we often find a knowing disingenuousness as he massaged evidence to support his goals. This was nowhere more evident than in the way in which he put the case for regularizing policing. It is worth dwelling on this, because it tells us how he thought about 'crime', and that will take us to the core of the matter.

2. *Peel and crime*

Peel was the first politician of note to shape and exploit a rich vein of anxiety about the decay of moral values to which increases in 'crime'

[20] B. Hilton, 'Peel: A reappraisal', *Historical Journal*, 22/3 (1979), 589. I am much indebted to Boyd Hilton's unrivalled understanding of Peel, aspects of which he has generously clarified for me in luminous and lengthy personal communications. The last of these, on Peel's attitudes to judicial discretion and the prerogative, elaborating on and correcting my interpretation in the last part of this chapter, arrived as this book went to press. I can only hope that his subtle and ambitious analysis leads to a published study. I have not modified this chapter to take account of it, and am content to be corrected in due time.

[21] *Speeches*, ii. 180.

[22] Ibid. i. 183–4.

[23] D. Beales, 'Peel, Russell and reform,' *Historical Journal*, 17/4 (1974), 878; Hilton, 'Peel', 587–9, quoting Lord Hatherton.

supposedly testify. He was also the first to plead the case to this effect from the criminal statistics, those perilous data which have been harnessed to justify the state's expanding disciplinary powers ever since. Waving statistics which allegedly proved an 'increase in crime', but which actually proved only an increase in prosecutions and committals to trial, he declared in March 1826 that it was

> impossible . . . to contemplate without painful reflections, the state of this country with respect to the number and the increase of criminal offences. It is useless, it is worse than useless, to conceal from ourselves the truth that there is not in this country that security from fraud and depredation which there ought to be in a well-constituted society.

Peel also deployed committal statistics to overcome parliamentary suspicion of despotic police disciplines when he moved for a select committee on the police in February 1828, and yet again in April 1829 in the debate on the Metropolitan Police Improvement Bill. Here he developed the argument with which many have replied to those who fear that too energetic a defence of order must threaten liberty. English liberty was costly, he implied, and moral order was what suffered:

> the freedom of action which is allowed to every man by our law, the absence of any control upon that action through the medium of police establishments, like those which exist in many countries, empowered to act upon vague suspicions, and preventing by unceasing vigilance the commission of offences that would otherwise be completed—such causes no doubt contribute in many instances to favour the early stages of vice in this country.[24]

As he put it again three years later: 'When . . . they talked so much of the liberty of the subject possessed by the people of this country, he was afraid that they gave credit to some parts of the population for the enjoyment of much more liberty than [thanks to the criminal] they actually possessed.'[25] In the police debate of April 1829 Peel attributed the 'frightful' increase in crime to 'the increased mechanical ingenuity of the age; by which the perpetration of crime was aided'. He thus enunciated the doctrine, to be repeated down the generations, that crime was a product not of poverty but of material progress and of its concomitant moral indisciplines: 'All causes . . . of the increased comforts of the people of the country became thus sources of crime; not less from the increased temptation which they necessarily created, than from the increased facilities which they afforded of perpetration and evasion.' Peel completed the Manichaean scenario by defining that

[24] *Speeches*, i. 403. [25] Ibid. ii. 4, 8.

other scapegoat in the evolving English ideology of crime, the 'criminal class'. 'He believed that these criminals were, in almost all instances, trained and hardened profligates—that they had been incited to the commission of crime, by the temptations which the present lax system of police held out to them.'

Peel consistently backed up his vision with 'statistical facts'. But what did the criminal statistics really show? The numbers of people committed to trial at assizes and quarter sessions in England and Wales ran as follows (after low wartime levels before 1815):[26] 1815: 7,818, 1816: 9,091, 1817: 13,932, 1818: 13,567, 1819: 14,254, 1820: 13,710, 1821: 13,115, 1822: 12,241, 1823: 12,263, 1824: 13,698, 1825: 14,437, 1826: 16,164, 1827: 17,924, 1828: 16,564, 1829: 18,671, 1830: 18,107. The average rates per 100,000 of the *male* population for the three five-year periods are as follows: 1815–19: 173.6; 1820–4: 181.9; 1825–9: 214.6. By this measure, adjusted for population growth, prosecuted male offences increased by 4.7 per cent across the first two periods and by another 18 per cent across the second and third periods. These increases were not spectacular in comparison with those which had occurred immediately after the war in 1815, and with those still to come in the 1830s and 1840s. Moreover, Peel himself acknowledged that much of the increase was registered not in crimes of violence or major property crimes but in the misdemeanour of petty larceny;[27] this regularly comprised over two-thirds of all prosecutions at assizes and quarter sessions.

But, whether large or small, the most telling point is that the increases had less to do with changes in collective behaviour or social environment than with Peel's own successes in easing and encouraging prosecutions—an irony which he either was not candid enough or could not afford to acknowledge, even though the easing of prosecution was a consistent goal of his reforms.[28] His Criminal Law Act of 1826 swept away old statutes of 1751, 1754, 1774, and 1818 which had provided for the discretionary payment of some prosecutors' and witnesses' expenses at felony trials, and authorized payment to every prosecutor of 'the actual expenses incurred by him'—at trials for misdemeanour as well as for felony. If there was an ostensibly egalitarian instinct behind

[26] V. A. C. Gatrell and T. Hadden, 'Criminal statistics', in E. A. Wrigley (ed.), *Nineteenth-century society: Essays in the use of quantitative methods for the study of social data* (Cambridge, 1972), table 3: 392. The figures exclude offences dealt with summarily in petty sessions.

[27] *Speeches*, i. 558.

[28] Peel's concern to increase the incidence of prosecution and to inhibit the practice of compromising out of court prompted him to toy with an innovation ahead of his time, that of the use of a public prosecutor: *Speeches*, i. 406.

this,[29] its inflationary effects on crime rates were significant. As Brougham observed a decade later, Peel's Act 'greatly increased the number of commitments and has been the cause of many persons being brought to trial, who ought to have been discharged by the magistrates'. And sure enough larceny prosecutions were boosted from 10,087 in 1825 to 12,014 in 1827 alone.[30] Add to this Peel's success in extending magistrates' jurisdiction over some felonies and misdemeanours in 1827; in establishing safeguards against the squashing of prosecutions on technical quibbles; in abolishing capital punishment on some burglaries and housebreakings and the distinction between grand and petty larceny (this reduced the number of capital felonies and brought all cases of simple larceny within the jurisdiction of lower (and cheaper and speedier) courts)[31]—and a good part of the increase in 'crime' rates in the later 1820s is readily explained. Any remaining part of the increase could be explained in terms of the hardening of public attitudes to lawlessness. This was an era when commercial interest, popular turbulence, and newly apocalyptic meanings attached to crime (which Peel was not the least to inflame) were inducing people to prosecute offences to which blind eyes had anciently been turned. In short, there is no convincing evidence that 'crime' was increasing, even the crimes of the poor to which the law customarily gave its attention.

Peel's easy way with figures was also revealed in his attempts to justify his hanging record. In 1830 he claimed that a comparison of the seven-year periods before and after his assumption of the home secretaryship had shown a marked decline in executions.[32] This was misleading. The seven years before 1822 had been marked by a punitive backlash against post-war disorders. A fairer comparison would have been with the seven years beginning in 1805. An annual average of just under 13 people a year were hanged in London then, and 67 in England and Wales as a whole. This makes Peel's annual average for 1822–8 of 18 hangings per year in London and 63 in the whole country look unremarkable. In 1829, 74 people hanged in England and Wales, and in 1830, the last year of his home secretaryship, 46. Fewer than these had been hanged in years like 1808 (39 executions) or 1811 (45).

[29] 'To withhold public aid from the prosecutor . . . amounts to the frequent denial of all reparation to the poor man, and to the impunity of great offenders': ibid. 405.

[30] Radzinowicz, i. 573; PP 1830–1, xii. 495 ff.

[31] Radzinowicz, i. 582–6; *Speeches*, i. 408.

[32] *Speeches*, ii. 131, 163: Gash, *Mr Secretary Peel*, 486, takes the figures at face value.

3. *An ethical pessimism*

The fact that statistics of these kinds could be massaged by an acknowledged master of detail may seem characteristic of 'a man as uncandid and self-deceiving as Peel'.[33] But it is fruitless to enquire how much he believed in the quantitative evidence he marshalled to support his arguments. The times were innocent in social analysis, and simple-minded criminological positivism of this kind is not unknown today. Moreover, Peel had his own economic principles to defend—against calls for a relaxation of the gold standard, for example. He could not afford to concede that rising crime was attributable to a general distress which might put the credibility of that financial order in jeopardy: far better to blame 'the unruly passions and corrupt natures of human beings'.[34] But he believed in that corruption, too. Peel has not been the only politician to seek confirmation in the criminal statistics of what his ethical pessimism led him to discern in the social order: a Manichaean conflict of good and evil, with evil an ever-present and growing threat. It is in acknowledging the force of this mentality that we come closest to understanding Peel's implacability as he dealt with the appeals which landed daily and weekly on his desk. The appeals were often pitiful to be sure. He found them 'painful'. But what he chiefly saw in them was evidence of man's viciousness and need for redemption: 'It is of no use to deceive ourselves; it is of no use to shut our eyes against what is passing around us, and to praise ourselves upon our superior morality and Christian virtues. It is our duty to look around us, and to provide, so far as in us lies, remedies for the evils with which we are surrounded.'[35]

Peel's religion was, his biographer says, outwardly only 'a simple, rational, pious protestantism'. Hilton extends this by suggesting that along with other liberal tories and moderate evangelicals in the 1820s he shared a Butlerian belief in a morally self-governing universe, shaped, as he himself said, by God's great 'system of social retribution' which rewarded merit and punished vice. Since the divine system was self-regulating, the moral task of the judge and the executive was limited. It was chiefly to facilitate providential retribution by removing man-made impediments to it: but retribution would still take its course without human interference. This might explain why Peel was indiffer-

[33] Hilton's verdict in another context: 'Peel', 605.

[34] B. Hilton, 'The ripening of Robert Peel', in M. Bentley (ed.), *Public and private doctrine: Essays in British history presented to Maurice Cowling* (Cambridge, 1993), 63–84.

[35] *Speeches*, i. 557, 562–3.

ent to the redemption of the vicious in reformative prison regimes. And because he would not have thought that scaffold death was the worst of fates either, he was equally reluctant to extend unnecessary mercies out of mere compassion. His life, *pace* Wakefield, would not shorten as he fulfilled his excruciating duties, because he knew that in the providential universe a sublime *laissez-faire* could usually rule equitably as causes led ineluctably to retributive effects.[36]

Little of this was spelt out. In meditating law and punishment, he preferred to appeal to the 'dictates of common sense'.[37] Since he seldom clarified what the bases of common sense were, and never pinned them to his providential beliefs, he often sounded like a die-hard, especially on punishment. He often lamented the absence of a more flexible system of secondary punishments alternative to death, but his every instinct, given this absence, was to hold on to the old treatments, to make them more efficient—and this could mean harsher. Thus he was not one of those evangelicals who abhorred corporal punishment.[38] Since it was the soul rather than the body that most mattered in his theology, the bodily pain of others could remain remote from his imaginings. Peel trusted in the punitive efficacy of bodily pain. He associated himself with his friend Sydney Smith's dictum in 1822 that the only real test of a good prison system was 'the diminution of offences by the terror of the punishment'. What Smith meant by this was clear:

> nothing but the tread-wheel, or the capstan, or some species of labour where the labourer could not see the results of his toil,—where it was as monotonous, irksome, and dull as possible,—pulling and pushing, instead of reading and writing. . . . There should be no tea and sugar,—no assemblage of female felons round the washing-tub,—nothing but beating hemp, and pulling oakum, and pounding bricks.[39]

Peel similarly told the Commons that the chief benefit of the treadmill lay in its inflicting 'a stigma, and a disgrace, a moral punishment'. The punishment it delivered 'could be graduated'. 'On the whole, he considered the tread-mill an admirable contrivance, and that no system of labour could be devised which was so little liable to abuse.' (Prisoners actually put on weight when subjected to it, he added, so 'the punishment could not have operated injuriously'.)[40] Similarly, when Grey

[36] Hilton, 'Peel', 608–10; B. Hilton, *The age of atonement: The influence of evangelicalism on social and economic thought, 1785–1865* (Oxford, 1988), 229–31.

[37] *Speeches*, i. 397.

[38] Hilton, *Age of atonement*, 215–17.

[39] S. Smith, *Works* (2 vols., 1859), i. 356–7.

[40] *Speeches*, i. 272, 275. The source of this opinion was the set of replies from visiting magistrates to Hobhouse's enquiry of 26 Dec. 1823 on the 'bodily mischief or inconvenience' caused by the

Bennet sought vainly to bring in a Bill to abolish whipping, Peel declared that 'he had always been friendly to the punishment of whipping, when exercised within salutary limits . . . he had not been able to find a single instance of abuse for the last seven years'.[41]

He also agreed (he wrote to Sydney Smith in 1826) that 'salutary terror' should be added to the punishment of transportation, and that more work should be found for those on the prison hulks. He had faith in little else, because the number of convicts which had to be dealt with was 'too overwhelming'. Solitary confinement 'sounds well in theory' but its effects on sensitive intellects could not be foreseen and could be fatal; ignominious public labour on the highways would 'revolt' public opinion; and a long term of imprisonment without hard labour, as in the Millbank penitentiary, was a punishment which, especially in winter, was 'thought by people outside to be rather an enviable one. . . . The present occupants are . . . living too comfortably, I fear, for penitence.'

These were no passing arguments. He repeated them in the debates on punishment for forgery in 1830, and he continued to resist the abolition of death penalties in the abolition debate of 1832 on the grounds that no sufficiently *punitive* and secure secondary punishments could be provided as yet.[42] Peel often implied that his greatest success in the evolution of secondary punishment was the institution of the hulk establishment at Bermuda in 1823, where within three years some three or four hundred convicts were set to exacting public works on the fortifications and harbour, John Noden among them, many to die of yellow fever. And as late as May 1830 he was recommending hulk labour as the most economical method of disposing of convicts, and was planning to make the convict regime in New South Wales even harsher.[43]

Otherwise he did little about the shortcomings in penal provision. His main prison initiative, the Gaols Act of 1823, only consolidated older prison statutes, set up local rate-paid gaols in every county and major town, and regulated prison disciplines and magistrates' inspection. Some classification of prisoners was introduced a year later, though young and first offenders were given no special treatment.[44] The Act's limitations are sometimes excused on the grounds that this was not an

treadwheel. All replies denied any mischief, and many included prison surgeons' attestations to that effect too. One surgeon who gave the opinion that the treadwheel was 'most salutary' in its effects included a report on the death of a prisoner who fell into the wheel and was crushed: HO44.14.

[41] *Speeches*, i. 239–40.

[42] Peel to S. Smith, 24 Mar. 1826, in Parker, *Peel*, 401–2; *Speeches*, i. 166–7, ii. 524.

[43] See the letter to Sydney Smith quoted above; *Speeches*, ii. 156–8.

[44] Gash, *Mr Secretary Peel*, 315–16; Radzinowicz, i. 570–1.

age when government could direct limited administrative and financial resources to prison-building, and that the 1823 Gaols Act at least moved tentatively towards national prison reform. But administrative bureaucracies were no less rudimentary and resistance to central expenditure no less inhibiting in the 1830s, when the whigs got to work more effectively.[45] What was lacking in Peel was the political will or the inspirational drive to do more. God's retributions operated adequately as things were.

In their own time Peel's punitive instincts were unexceptional enough, little though they accorded with those of reformist whigs or of more interventionist evangelicals or Quakers. They were doubtless shared by the fox-hunting and port-swilling backbenchers of his party whom he so deplored but upon whose votes his proposals depended. His instincts were also shared by the judges. The judiciary did not welcome Peel's changes,[46] but they went along with his proposals both because he took them into his confidence (always complimenting them in the Commons for their advice) and because they could trust his instincts as they did their own.[47] For all their evangelical inspiration, his penal policies meshed neatly with the values of a governing class which had itself been cruelly disciplined in hard schools. We have already noted Peel's revealing joke in the whipping debate—that if Grey Bennet's principle were accepted, even 'the punishment of flogging in public schools ought to be abolished by act of Parliament'. This was a laughable proposal, he meant to say; and MPs would have chortled understandingly.

Peel's administrative skills were best witnessed in his work on statutory consolidation. He took most pride in this achievement, and the task he set himself was immense. Hitherto, as he said, criminal legislation had been 'left to the desultory and unconcerted speculations of every man

[45] Cf. Radzinowicz, i. 570–1; Gash, *Mr Secretary Peel*, 315 ff. Admittedly, it had not been forgotten that Millbank penitentiary had cost nearly half a million pounds to build and repair between 1812 and 1822, at a then prodigious cost of £458 per prisoner place. When the model prison of Pentonville was built in 1840–2, costs were kept down to £82,271, or £158 per prisoner place: R. Evans, *The fabrication of virtue: English prison architecture, 1750–1840* (1982), 249, 347–8.

[46] Eldon, the lord chancellor, taught Peel the need for tact when he refused to steer Peel's Prison Bill through the Lords because it was ill-prepared and judges had not been consulted. Eleven months later, he wrote strangely to Peel that he had never even heard of the Prison Bill (Eldon to Hobhouse, Jan. 1823; to Peel, 30 Dec. 1823: Peel papers, BL Add. MS 40315, fos. 83, 108). The chief justice also played hard to get when consulted on the Burglary Bills in 1826: 'his time was too much engrossed to enable him to attend to so important a question' (Eldon to Peel, 21 Jan. 1826, ibid., fo. 239).

[47] Cf. Peel to Lord Chancellor Lyndhurst, consulting him on the Forgery Bill which he circulated also, in confidence, to the chief justice and attorney- and solicitor-generals: 1 Sept. 1829 (ibid. 40316, fo. 56).

who had a fancy to legislate. If an offence were committed in some corner of the land, a law sprang up to prevent the repetition, not of the species of crime to which it belonged, but of the single and specific act.' Hence the theft statutes alone numbered ninety-two. He intended to reduce them to one statute of some thirty pages: 'It appears so conformable to the dictates of common sense, that the law, of which all men are supposed to have cognizance, and which all are bound under heavy penalties to obey, should be as precise and intelligible as it can be made.'[48] His first move was in 1823 to abolish the death penalty for larceny to the value of forty shillings in shops and ships (but not houses), and for the impersonation of a Greenwich pensioner, and other such outmoded capital statutes; the Black Act was repealed too. Four consolidating Acts between 1827 and 1830 dealt with some four-fifths of the old statutes on theft, malicious damage, offences against the person, and forgery. Altogether, Peel claimed, some 278 old statutes were repealed and replaced by eight Acts.[49] This undeniably testified to Peel's way with his party and to his parliamentary skill and energy.

But what was the achievement in real terms? Peel's Acts did somewhat curtail the nominal death penalty, notably for coining and a few property offences.[50] Larcenists in a dwelling-house were allowed to take £5 instead of only £2 worth of goods before they were liable to die (although this only allowed for the inflation of values since Anne's reign, when the old statutes were enacted, and was what Romilly had recommended a decade before).[51] The capital crime of burglary was a little restricted by defining the location of the crime more narrowly. And after much agonizing on Peel's part, the abolition of the old distinction between petty and grand larceny put paid to the provision that a second conviction for a theft above one shilling was capital; such cases could now be tried summarily.[52]

[48] *Speeches*, i. 399–400, 397.

[49] Ibid. ii. 74.

[50] Radzinowicz, i. 582–4, describes these changes fully.

[51] Cf. Peel to Eldon, 14 Apr. 1826: 'It is so desirable to retain the capital punishment upon the offence of stealing in a dwelling house, that I cannot help thinking it would . . . be desirable to raise the amount, the stealing of which makes the offence capital, from forty shillings to five pounds. Forty shillings was the amount fixed in the reign of Queen Anne—equivalent then perhaps to five pounds now. How very rare the instance in which we should execute for stealing in a dwelling house when the sum . . . was less than five pounds. Would it not be a great advantage to take out of the Recorder's Report [to the King in Council] the many slight cases of stealing in a dwelling house on which sentence of death is now passed?' (Peel papers, BL Add. MS 40315, fo. 252).

[52] Peel at first opposed this abolition, proposing to Eldon that the distinction between grand and petty larceny be set at a theft of 5*s*. rather than of 1*s*., for the 1*s*. was set 'probably five hundred years ago'. He thought that this would allow them to justify transporting grand larcenists for 14 years instead of the 7 at present (Peel to Eldon, 14 Apr. 1826: Peel papers, BL 40315, fo. 252.)

But the limits of the overall mitigation are clarified by the fact that the forty-six people hanged in the last year of Peel's home secretaryship died in numbers and for offences little different from those hanged before his Acts came into force. For practical purposes, the key hanging offences remained intact (see Table below).[53]

The number of hangings before and after Peel's Acts

	1824	1830
Arson	1	6
Burglary	13	2
Housebreaking	1	8
Highway robbery	6	5
Forgery	3	0
Coining	1	0
Murder or attempted murder	18	15
Piracy	0	2
Rape, sodomy, etc.	4	7
Sheep- or horse-theft	2	1
TOTAL	49	46

At least nobody was hanged for forgery in 1830, but this was no thanks to Peel. He told parliament in 1823 that the death penalty on forgery did not need mitigation, and right up to 1830 he continued to think that forgers should die. His grounds for this, that forgery was now a rare offence owing to the return to cash payments, begged a question, since rarity was usually thought to be the best reason for mitigation.[54] As it was, seven of the eighteen forgers sentenced to death in 1828–9 were hanged under his signature (one committed suicide); those whose sentences were commuted had mainly committed forgeries of receipts, deeds, or bonds, for which execution was now exceptional.[55] The last man to die for forgery in England was a friendless and obscure customs clerk in financial difficulties named Thomas Maynard, hanged outside Newgate on the last day of 1829, protesting to the end that he had been the dupe of others. 'The law to take its course,' Peel scribbled routinely on the back of the brief mercy petitions which supported him.[56] If forgery hangings were suspended after Maynard's execution, it was only because forgery law was about to come under parliamentary scrutiny.

[53] PP 1830–1, xii. 501. [54] *Speeches*, i. 247. [55] PP 1835, xlv.
[56] OBSP 1829: 847 ff.; petitions in HO17.68.

Peel meant the hangings to be resumed once his task of consolidating the forgery statutes was complete.

There is no reason to believe that when Peel did turn to the forgery statutes it was because he had changed his mind since 1823 or was responding to public pressure. Attacks on the statutes had been well orchestrated ever since the report of the 1819 select committee, and the 1820s had thrown up *causes célèbres* (notably Fauntleroy's and Hunton's) whose executions provoked much debate, as we have seen. But we have also seen that Peel could reasonably claim that opinion on the subject, especially in London, was far from uniform (and Quakers were not that active against them either), and that the statutes acted as a deterrent. Banks which in three days transacted business amounting to £4.7 million had had forgeries committed upon them in 1830 amounting to a paltry £400.[57] So he remained obdurate, his moral view of the question unflinching. He admitted that with the return to cash payments it might no longer be possible to claim that England's commerce depended on the awful protection of death, but

> when he recollected . . . the magnitude of the gain, the great temptation, the difficulty of the detection, that there were no confederates necessary, and no violence to alarm people . . . he thought [these circumstances] invested this crime with a peculiar and exclusive character—a character which belonged to no other species of crime against which the legislature had to guard.

Add to that the fact that forgery was usually committed by persons 'of ability and information', and the difficulties of punishing the culprit otherwise than with death would be apparent. Punish such men with solitary confinement and low diet, and they became sick or insane. Put men of education on the hulks, and public sympathy would be aroused. Transport them, and their natural talents would ensure that in New South Wales they would prosper. 'The men accustomed, as forgers generally were, to all the comforts and many of the luxuries of life, were not likely to be influenced so much by the fear of the punishment of transportation and imprisonment, as of death.'[58] Although the abolition of death on forgery 'would unquestionably free him from many very painful applications' for mercy, his main interest was in rationalization, not abolition. The only reason he had delayed addressing the task of forgery consolidation was that it was 'very difficult'. Each public department of state would have to consolidate its own particular statutes; and then how did you distinguish between those forgeries on which the

[57] *Speeches*, ii. 180. [58] Ibid. 130–3, 163, 166–7.

death sentence should be retained and those on which it might be mitigated? There was much behind-the-scenes debate on all this.[59] In the event his 1830 Forgery Consolidation Bill was consistent with his overall strategies. It confined itself to removing the death penalty from the capital forgeries which seldom resulted in execution, but retained it for the forging of 'everything which represented money'.[60] In other words, men like Maynard were meant to continue to die in future.

It was not to last, however. In the third reading of the Forgery Bill in June 1830 Peel suffered the sharpest defeat in his law reform career, for Mackintosh carried his amendment that forgery be punished only by transportation. Relinquishing the measure to Mackintosh, Peel was ungracious: 'his sentiments remained entirely unchanged, and he believed they would soon have reason to repent the decision to which they had just come'.[61] But Mackintosh's triumph in 1830 portended Peel's greater defeat as, within a few years, his new statutory edifice was dismantled wholesale.

Peel's effectiveness as a penal and law reformer was compromised both by his ethical pessimism and by his belief that society could survive in turbulent times only if secular authority was resolutely defended. Both these things determined his reverence for the law. His deference to the judiciary, from Lord Chancellor Eldon downwards, is the most obvious manifestation of this. What is less often recognized is his passionate defence of the central principles of the old regime: the royal prerogative of mercy and the discretionary element in judicial and executive power. These were precisely the points at which the bloody code was at its most vulnerable, its credibility undermined by the fact that by the late 1820s 95 per cent of the condemned had to be pardoned to prevent an excess of executions. Even members of the king's Council began to worry about this, as we have seen.

Yet almost as soon as he took office Peel had cause to rebuff whig and radical attacks on the prerogative, and he did so fiercely. When Sir

[59] Attorney-General Scarlett recommended a distinction between forgeries in which 'the mischief of a successful utterance is generally without remedy' because 'they were immediately convertible into money', and those, like the forgery of wills, deeds on real estates, receipts for money, orders for goods, etc., in which 'the discovery of the forgery in general defeats its object'. Peel broadly followed this advice: Scarlett to Peel, 31 Aug. 1829: Peel papers, BL Add. MS 40399, fos. 312–13. For correspondence on these difficulties between Peel, Hobhouse, Gregson, and the lord chief justice, see ibid., fos. 314, 375, 410, 419.

[60] Peel kept the death penalty for the forging of wills, 'an offence of easy commission and great enormity': Peel to lord chief justice, 29 Dec. 1829, ibid., fo. 410.

[61] *Speeches*, ii. 181.

Francis Burdett moved to address the king in April 1822 for a remission of sentence on the imprisoned radical Henry Hunt, Peel was on his feet at once. The prerogative of mercy was 'exclusive', he proclaimed with rare passion. It was never to be interfered with by parliament: that 'would have the effect of enlarging the functions of the democratic part of [the] constitution far beyond its useful and natural boundary'.[62] He pushed this view further in the great debate on Mackintosh's abolitionist motion a year later. After reviewing the ratios of death sentences to conditional pardons over past years, he asked: 'Would it be fair . . . to take away the discretion by which these punishments had been thus apportioned; or could they hope to make a law so precise in all its provisions as to substitute it with effect?' It would, he said, 'be impossible to establish any code of laws which would prevent the necessity of a discretionary power on the part of the executive'. He advised the House 'to do away [with] any suspicion of the deficiency of the laws, inferred from the disproportion between the number of convictions and of executions'. Then he paraphrased Paley's famous distinction between the certainty of divine justice and the uncertainty of human justice (see Chapter 19) in these terms:

> It might be hard to say to a man, that his life should be valued at a particular rate, depending upon local or temporary expediency. But this was the very reasoning upon which law was founded. On what other ground could they pretend to inflict capital punishments? It was not that they, in the deficiencies of human nature, were able to determine that which could only be effected by a tribunal above—the exact degree of moral turpitude attached to each particular offence. But while mankind were constituted as they were, having to struggle with all the imperfections of their senses, this was the last mode which legislation could devise for the preservation of civil order.[63]

While the tories remained in office, this deference to ancestral ideology continued to underpin the criminal law. Only subsequently would it succumb to that quite different image of justice to which so many petitioners were giving voice in the 1820s, as we have seen, and writers too, who had come to dismiss the Peelite cosmology as cant. Journalistic attacks on the old legal system were coming to be loaded with newly confident sarcasms. 'There is as much fashion in what is termed justice as in bonnets or sleeves,' Fonblanque wrote in 1831: 'The judge's cap is as capricious as the ladies'. Sometimes the trimmings are

[62] *Speeches*, 183–4.

[63] Ibid. 244–6. Cf. Paley on the distinction between human and divine justice, *Principles of moral and political philosophy* (1785), 521, and Ch. 19, above.

blood red, sometimes the sky blue of mercy is in vogue. . . . The severity of the law opens a field for arbitrary power under the plausible names of justice or of mercy.'[64] As McGowen has written, the debates on the criminal law in the 1820s expressed two quite contradictory legal philosophies. Peel, hailed as a great penal reformer, was the most committed protagonist of the old order. He tried to conserve as much of it as he could.

Was Peel's callousness merely 'practised', then, as Wakefield said? It hardly seems so simple now. It is more polite to speak of his pessimism about human nature, his belief in providential dispensation, his respect for constitutional continuities, his concern for hierarchical order. Yet to a man like Wakefield—and not only him—this would seem all of a piece really, endorsing the central charge. More than one petitioner in the 1820s told Peel bluntly that 'a familiarity with scenes of blood has disqualified your august presence [from] rightly appreciating the life of man.'[65] However high-mindedly, Peel still let more people hang in the 1820s than any predecessor in office, and he meant to go on killing them; for many the ethical sense in this was opaque and fading. Within a couple of decades Peel's brand of providentialism was to give way to a warmer incarnational theology, concerned as much with bodily resurrection as with the soul's salvation, as Hilton shows. Meanwhile, to many, Peel's was already looking like a cold and outmoded faith, befitting the man but not the changing times.

[64] A. Fonblanque, 'Capital punishments' (1831), in his *England under seven administrations* (2 vols., 1837), ii. 159–60.

[65] HO17.88/I (Hunton's case).

EPILOGUE

1868: ENDING THE SPECTACLE

1. *A civilizing moment*

THE LAST MAN TO BE HANGED PUBLICLY IN ENGLAND WAS THE Irish Fenian Michael Barrett. He was executed outside Newgate on 27 May 1868 for blowing up a wall of Clerkenwell prison in an attempt to rescue other Fenians imprisoned there; he killed fifteen people instead. Two days later, after months of debate, parliament passed the Act to Provide for the Carrying out of Capital Punishment in Prisons. The first person hanged behind prison walls—at Maidstone on 13 August—was Thomas Wells, an 18-year-old Dover railway porter who had murdered his stationmaster after a reproof. The first hanging inside Newgate followed on 8 September, when an 18-year-old coffee-house servant, Alexander Mackay, was executed for murdering his master's wife. Thereafter executions were concealed from public view until the last two in Manchester and Liverpool in 1964.

We know what MPs thought they were doing in dismantling the age-old way of executing capital justice. Most took it for granted that public hanging was 'a disgrace not only to civilization but to our common humanity', as one put it. Its abolition would be consistent 'with the humane legislation of the past thirty years', declared another. To yet another, 'barbarous and demoralising' public executions were 'not in accordance with the spirit of the age'. The *Daily Telegraph* also said that hangings were 'a fragment of mediaeval barbarism': 'wisdom and humanity' were about to end the 'ugly practice', and 'public decency' would be achieved at last. Historians have agreed. Radzinowicz linked the abolition of public hanging to 'the growth of humanity'; another historian has hailed 1868 as a 'landmark in the more humane treatment of criminals'.[1]

[1] PD3 cxc (1868), c. 1135 ff., cxci (1868), c. 1045; Radzinowicz, iv. 352; Cooper, p. 178.

We cannot deny that 1868 was a civilizing moment in British history (the Act extended to Scotland too). It ended an age-old plebeian festival, always rowdy, often cruel, and increasingly shocking. Belatedly and tacitly it endorsed the realization that the hanged were individuals, not mere emblems of a corruption in the body politic whose deaths it made political sense to display. It allowed that their terrors were only too easy to imagine, and that it might disturb the peace if people had too many chances to imagine them. The moment also advanced the great civilizing enterprise of removing others' miseries from view. It sanitized capital punishment by having it executed henceforth in private, as if it were an operation which required a clinician's concentration. Urban decorum and the nation's civility were advanced by these processes.

None of this, however, means that 1868 marked a *humane* moment in British history. A civilizing process may redeploy, sanitize, and camouflage disciplinary and other violence without necessarily diminishing it. Respect for the criminal's humanity was not intrinsic to the shift, and it is only because people have always needed to believe otherwise that the words 'humanity' and 'civilization' still cohabit within progressive narratives. Inside prisons for a century yet, murderers continued to be strangled on ropes too short or ropes too long, dying more dreadfully in private than in public, in chilly proceedings with crowd support withdrawn. News of disasters still leaked out from behind those high walls: the decapitation achieved by the hangman Berry when he used too long a rope; the demise of one Burton in Durham gaol in 1883, when on his first drop the rope caught under his elbow so that he had to be hauled up and hanged all over again; the hanging of Conway in 1892 on a long drop which nearly severed his neck.[2] Hiding horrors did not end them, although authorities got better at hushing them up.

Upon this huge change in policy three of the book's themes converge. They define what this epilogue has to say. First, there was the purposeful negation of 'sympathy' in the deepening Victorian consensus that hanging for murder should continue. Secondly, there was none the less the anxiety, shame, and squeamishness about hanging which many if not all polite Victorians still embarrassingly felt. And, thirdly and consequently, there was their intensifying dislike of that sardonic scaffold crowd which often and intolerably mirrored both what the law did to people and what it was that the good people were sanctioning. To

[2] H. Potter, *Hanging in judgment: Religion and the death penalty in England* (1993), 102–4.

exclude the crowd from the ceremony was, we argue, the most urgent reason why public executions were abolished; despite the rhetoric, 'humanity' was neither here nor there.

2. *Hanging for murder*

There would not be much anxiety, shame, and squeamishness to discuss in this chapter if you took most Victorian pronouncements about hanging at face value. Much of this book has lent its ear to 'new voices' on the subject of punishment, but it has not been denied that the old punitive voices retained their vigour. Those who defended retributive killing sounded very sure of themselves by the mid-nineteenth century. They were helped by the fact that the excesses of the bloody code were past and after Victoria's accession only murderers hanged. The likes of Wakefield and Thackeray might protest as they pleased, but appeals to the sympathetic imagination lost energy in these conditions; indignant voices of their kinds were fading. In parliament a phalanx of abolitionists led by Ewart made brave stirs at first, scoring ninety votes for total abolition of capital punishment in 1840. But outside support was fickle, not least among the literary sages. Wordsworth supported the death penalty in his 1839–40 sonnets on the subject; Macaulay called abolitionists 'effeminate'; Dickens, Thackeray, Carlyle, and possibly Mill, supported total abolition in the early 1840s, but they soon changed their minds, with Dickens attacking scaffold crowds rather than hanging, and Thackeray falling silent after his fine flourish in 'Going to see a man hanged'. By 1850 Carlyle's manic rhetoric caught the dominant mood by inveighing against 'Beneficence, Benevolence, and the people that come together to talk on platforms and subscribe five pounds' to stop murderers paying for their sins. Most polite people would have assumed that a deep wisdom spoke in this windy tosh:

> the one answer to [the felon] is: 'Caitiff, we hate thee . . . not with a diabolic but with a divine hatred. God himself, we have always understood, "hates sin", with a most authentic, celestial, and eternal hatred. A hatred, a hostility inexorable, unappeasable, which blasts the scoundrel, and all scoundrels ultimately, into black annihilation and disappearance from the sum of things.' . . . 'Revenge,' my friends! revenge, and the natural hatred of scoundrels, and the ineradicable tendency to *revancher* oneself upon them, and pay them for what they have merited: this is forevermore intrinsically a correct, and even a divine feeling in the mind of every man.[3]

[3] *Latter-day pamphlets*, 2, 'Model prisons' (1850), in *Collected works* (n.d.), 51, 53, 65–7.

If this was not enough for abolitionists to stomach, mid-century debate also filled with Old Testament advice on the wisdom of eye-for-eye retributions. Not for nothing has Potter noted that without evangelical and episcopal support capital punishment would have ceased far sooner than it did. 'He that sheddeth man's blood by man shall his blood be shed,' Lord Shaftesbury told the nation, to explain why he voted against abolition in 1849. There was a bit of worrying about contrary advice to turn the other cheek, but Jehovan principles generally won the day. 'The very foundation of civil society rested upon the declaration that [the death penalty] was God's institution', the bishop of Oxford announced confidently in 1856: 'the civil government bore the sword on God's behalf.'[4]

If this reference took its audiences back to the previous century, so did arguments that executions should be made more horrible, not less. Fielding's aphorism was quoted: 'a murder behind the scenes, if the poet knows how to manage it, will affect the audience with greater terror than if it was acted before their eyes.'[5] The Lords' select committee of 1856 and the royal commission a decade later both advised this principle, as did righteous bodies like the evangelical Revd John Clay, among others. Clay recalled how at an execution he attended years before, four doomed men had been inconsiderate enough to upset 'intelligent persons' by protesting innocence and singing hymns as they awaited their fates: 'If those men had been executed within the walls of the gaol, such a scene would not have taken place. The bravado or determination to take the very last chance of life, I think, would not have been exhibited. I think their firmness, so to call it, would have given way.'[6] It was in this same warm spirit that Home Secretary Gathorne Hardy introduced the 1868 Bill.[7] The 'mystery and indefiniteness' of private execution would help deterrent ends, he said, since 'the criminal class have a greater dread of death on a private than on a public scaffold'. This at least was honest, avoiding the conceit that strangling people privately was kinder than strangling them in public, a projection from domesticated bourgeois culture with little meaning for plebeians. The new policy was *meant* to strike at the crowd support which had been the most humane thing about *ancien régime* executions for those who lived their lives on the streets.[8] Hidden hangings were also recommended because their

[4] Cooper, pp. 50, 112; Potter, *Hanging in judgment*, p. vi.

[5] *An enquiry into the late increase of robbers*, ed. M. R. Zirker (Oxford, 1988), 169.

[6] SC 1856: 31.

[7] PD3 cxc (1868), c. 1127–30.

[8] Admittedly, the abolitionist Charles Gilpin thought in 1868 that this was a subject impossible to

enhanced terror might better propel the condemned to penitence: they would be 'beneficial to the criminals themselves', as Gathorne Hardy added, wondrously blind to the meanings of words.

These punitive dispositions were finally reinforced by anxieties that the presumed growth of 'crime' ensued from past mitigations of prison punishment and the abolition of transportation. Focused by a handful of London garrottings in 1862, these anxieties swelled to a panic which toughened penal policies after that date. Clever people as well as gullible capitulated to this climate. When J. S. Mill voted for the death penalty in parliament in April 1868, it was because he too deplored the 'mania' for mitigating punishments while hardened criminals multiplied. He discerned 'an enervation, an effeminacy, in the general mind of the country' which must be resisted, he said.

Mill was no reflexive hanger. He seems to have opposed capital punishment in 1830, and is thought to have done so in 1842;[9] and as an MP he campaigned against the execution of the Fenians Barrett in London and Burke in Ireland. None the less, when he spoke in the Commons to oppose the abolitionists' last-ditch motion to have done with capital punishment altogether, his arguments took on the predominant colourings. In comparison with braver and more manly times, he declared, people nowadays were too shocked by death. 'It is not human life only, not human life as such, that ought to be sacred to us, but human feelings,' he murkily announced: 'the human capacity of suffering is what we should cause to be respected, not the mere capacity of existing.' And the suffering caused by life imprisonment must far outweigh that entailed in 'the short pang of a rapid death'. Execution was not only the most appropriate and impressive punishment for murder; it was also (wait for it) the 'least cruel'. Taking a murderer's life was not to show disregard for life, but respect for it. And if by mischance innocent lives were sometimes taken—well, in this country, where rules of evidence were if anything too favourable to the prisoner, and judges and juries were uncommonly scrupulous and merciful, such cases were 'extremely rare'.[10]

Uttered in the comfortable tropes of the sage who had never seen either noose or prison, Mill's weighty pontifications helped to defeat the

have an opinion on: while some of the condemned showed bravado in public, 'others, especially women, regarded the publicity as the worst part of the punishment'. PD3 cxc (1868), c. 1136.

[9] See 'Attempts to save the ex-ministers' (1830), in A. P. and J. M. Robson (eds.), *Collected works of John Stuart Mill*, xxii. *Newspaper writings* (Toronto, 1986), 163–8; M. St J. Packe, *The life of John Stuart Mill* (1954), 465.

[10] PD3 cxci (1868), c. 1047–55.

amendment for total abolition by 127 votes to 23—a sad vote for abolitionists compared to that of 1840. This enabled parliament to move on to pass the half-measure which pre-empted abolition for a century yet—just as abolitionists feared it would. Abolitionism could make no headway in this climate. The Society for the Abolition of Capital Punishment was suspended in the 1860s for lack of support, and it disbanded after the 1866 royal commission recommended private executions. It resurfaced as the Howard League for Penal Reform. How efficiently the 1868 Act silenced abolitionism thereafter is proved by small signs as well as big. When the first hidden execution took place in Maidstone, it was in a town 'so notorious for its anti-capital punishment feeling,' the *Star* reported at the time, 'that lawyers will tell you it is most difficult to get a jury to return an adverse verdict against a prisoner on trial for his life.' Who could ever have said such a thing about any English town in the ensuing century?

So much, it would seem, for anxiety, squeamishness, and shame, not to mention for that sympathetic ground the past century had laid—apparently all negated in the deepened conviction that deterrence and the earthly replication of divine order depended on rope. Doubtless most of the Victorian middle classes went along with this, either hating murderers or settling for what had to be in an imperfect world; many would have given the question no thought at all. So what more can be said? Where are those anxious subtexts we look for? They are there, though we must change tack to find them.

If it is true that polite people were uneasy about hanging before 1830, it would be odd if Victorians felt otherwise. After all, 347 murderers were hanged publicly in England and Wales between 1837 and 1868.[11] This number was enough to ensure that executions remained a familiar urban spectacle, but not so familiar that they could ever be dismissed as commonplace, as might once have been true. Their relative infrequency and their limitation to murder made them if anything more awesome than hitherto: thus it was more likely, not less, that anxieties would converge upon them. Moreover, from the royal commissions on the criminal law in the 1830s on to the parliamentary debates of 1868, it was officially known and said (though suppressed) that some people had hanged in innocence. The knowledge was also available (if repressed) that the spectacle tapped into disturbing fantasies and visceral curiosity

[11] Plus a trickle in Scotland, though at one-twentieth the English rate: see App. 2, below.

about how others died. The politer the person, the less likely such knowledge would break through his or her self-control. Instead, repudiated feelings would as like as not be projected upon and associated with the scaffold crowd. In the form of disgust, anxiety could usually be discharged that way. Still, repressed feelings do tend to erupt somehow or other, and as it happens we need not dig very deeply into Victorian execution discourses to find revealing excesses of pleading—and odder things besides. This was still a 'feeling' generation all right—which is why it fought so hard to regulate dangerous emotions. In some, the tension between a so-called ethical agreement about hanging and the sympathetic imagination caused unbearable fractures. In others, the intuition nagged that hanging was obscene, negating the civility they flaunted. In others again, fastidiousness did battle with old curiosity, which *would* raise its head. Here was space enough for disturbances to develop.

3. *Squeamishness and released sensation*

If Mill said pompous things about capital punishment, he said a finely prescient thing about the civilizing process. With the spread of civilization, he wrote in 1836, 'the spectacle, and even the very idea, of pain, is kept more and more out of sight of those classes who enjoy in their fulness the benefits of civilization'. Thanks to 'a perfection of mechanical arrangements impracticable in any but a high state of civilization', the infliction of pain can be delegated 'to the judge, the soldier, the surgeon, the butcher, and the executioner'. 'It is in avoiding the presence not only of actual pain, but of *whatever suggests offensive or disagreeable ideas*, that a great part of refinement consists.'[12] All this makes our point with wonderful precision.

A century-long marriage of sensibility and scientific rationalism meant that more imaginative beings could not put the hanged man's or woman's gurglings, kickings, and urinations into the theorized distances in which they had once been helpfully located. Ever less protected by the ascribed symbolic meanings which had anciently veiled the condemneds' tremulous and suffering bodies, imaginative witnesses, if they came too close, would now partake of the body's experiences vicariously, and be shaken by them. Thackeray did in 1840, we saw; and London sheriffs did too. We met Sheriff Garrett in 1822 unable to tell

[12] J. S. Mill, 'Civilization' (1836), in Robson (eds.), *Collected works*, xviii. *Essays on politics and society* (Toronto, 1977), 130–1 [my emphasis].

Ward and Anson that their appeals had failed; and Sheriff Booth in 1828 unable to attend Abbott's execution after he had fought for Abbott's life.[13] While Courvoisier awaited his hanging in the Newgate press-yard in 1840, one sheriff asked for his autograph, but the other simply wept. At many executions thereafter sheriffs announced that they were too sickened to attend. And not only sheriffs: at Barrett's execution a policeman fainted. In 1856, moreover, Samuel Wilberforce, bishop of Oxford, deplored the squeamish contagion which induced home secretaries to use the royal prerogative a little too freely to let too many murderesses off the noose, however heinous their crimes. The good bishop wanted them all killed forthwith, but clearly even politicians were past tolerating that by now.[14]

That hidden executions would appease this squeamish culture sooner or later was inevitable; but the business of obliterating scaffold realities long predated 1868. No more chopping off of traitors' heads after 1820, as we have seen, while the condemned were stopped from bearing their ropes to the scaffold in 1824. Gibbeting and anatomizing went in the early 1830s, and other aggravated punishments followed, along with exposure of the condemned at last sermons or to aristocratic voyeurs a decade later. So much did decorum become the order of the day that black curtains began to be draped around scaffold bases so that 'little of the convicts could be seen after the withdrawal of the bolt': the soiling clothes, kicking legs, and hangman pulling from below must be hidden.[15] So necessary did these concealments become that in 1868 they were carried into prisons, although they had no public purpose there. When Mackay was hanged inside Newgate in September drapery hid the scaffold lest the officials 'see the executioner turn the screw which lowers the drop'.[16] Their eighteenth-century counterparts would have snorted at these elaborations—and rightly so, because they consulted the feelings not of victims, but of watchers. The drops remained as short as ever. 'The sufferings of the murderer were dreadful, *but on this we will not dwell*,' as the *Annual Register* wrote revealingly on Mackay's death.

Was this squeamishness a necessary price to pay for civility? Hardly so. Some nations retained public executions without being uncivilized. In France the guillotine worked publicly until 1939; a provincial decapitation could draw 10,000 spectators in 1912—'camping overnight in tents in the public square, renting window space in hotels overlooking the

[13] See Chs. 20 and 15. [14] RC 1866: 149, 222; Cooper, p. 112. [15] *Ann. Reg.* 1867: 156–61.
[16] *Daily Telegraph*, 9 Sept. 1868.

guillotine'; and nobody seems to have been disgusted.[17] Anglo-Saxons needed these evasive defences rather peculiarly, it seems. Some saw their absurdity and mocked them bitterly. Defeated in the parliamentary vote of April 1868, the abolitionist MP Charles Neate parodied the parliamentary victors by announcing brightly that in hanging 'he wished to limit manual interference by one fellow-creature with another'. So what about gassing murderers with carbonic acid, or offering them the means to commit suicide in the decent privacy of their cells? Anything to 'prevent the painful necessity of having the hands of the hangman about the culprit's neck'.[18] The Commons cried Neate down for this indecency; but it was not long before the truth beat his fiction hollow. A few weeks later the *Daily Telegraph*, deploring the old gallows 'with its clattering drop and its occasional necessity for tugging at the culprit's legs', thought it 'high time that we were rid of Mr Calcraft as a recognised member of the body politic'. A guillotine would do the killing better, if only that machine were not so French. Perhaps the device recently tested in America would do the trick. With this, 'the mere touch of a spring . . . is enough to heave the criminal up to a great height, and break his neck instantaneously. . . . The pinioning is done by the prison warders. The act is mutual, and no individual odium is incurred. One touch of the sheriff's foot does the rest.'[19] We know what was foreshadowed here. Thanks indeed to Mill's 'perfection of mechanical arrangements', in US penitentiaries today poison is injected into veins from machines hidden in adjacent rooms. With remote levers pulled, hands staying clean, eyes not watching, executioners dissociate themselves from what they do, and in their own minds escape contamination.

Primal curiosity and excitement will usually find displaced ways of venting themselves. And sure enough they erupted in mid-century scripts like this one, the *Leader*'s report of the poisoner Palmer's hanging in Stafford, 1856:

> And now the hangman grasps the rope—Palmer bends his head—the noose is slipped over—his face grows yet more ghastly—his throat throbs spasmodically—he moves his neck round, as a man with a tight collar—the hangman is hurrying off the drop—he suddenly bethinks himself of the cap—turns back—clutches at the criminal's right hand, as if asking for pardon—'God bless you, goodbye,' says the prisoner, in a low, distinct voice—the cap or white bag is

[17] G. Wright, *Between the guillotine and liberty: Two centuries of the crime problem in France* (Oxford, 1983), 167–73.

[18] PD3 cxci (1868), c. 1063.

[19] *Daily Telegraph*, 27 May 1868.

pulled over his head—the peak blows out from his chin by the violent and rapid respiration—another second, the bolt is drawn, down falls the drop with a slight crash—the arms are thrown up from the elbow, with the hands clenched—the body whirls round—the hangman from below seizes the legs—one escapes from his grasp, and by a mighty spasm is once drawn up—the chest thrice heaves convulsively—the hangman looses his hold—the body again whirls round, then becomes steady, and hangs a dull, grey, shapeless mass, facing the newly risen sun.[20]

It might be said of such accounts that the conventions of the empathetic narrative were released here; and this text certainly exemplifies the practised poignancy of execution reporting by this date. As detail piles upon detail, all closely observed, pathetic effects are skilfully achieved. But, alas, this is formulaic and artful writing really, aiming at no goal beyond the gratuitously sensational. Boswell's generation attended such events in person and watched closely, out of a curiosity candidly confessed; and when they described what they saw, it was never in lip-smacking terms, but to monitor their own sympathetic responses. By contrast, this writing was voyeuristic while it pretended to be something else. It matched Newgate novels, gothic melodramas, and penny dreadfuls in its sensational pursuit. It was like the crime literature whose purpose (as Mrs Oliphant wrote of *The woman in white*) was to thrill the reader's 'nerves' and chill him or her by 'a confused and unexplained alarm'. This, though, was no fiction. It partook of what Fonblanque in 1837 termed 'the diseased appetite for horrors', in a society in which newspapers were making 'murder so much their staple of interest, and such large profits of it, that we have for some time apprehended that, in the event of a scarcity of subjects, the proprietors would find their account in . . . committing their own murders'. In this writing money spoke.[21]

Earlier press reporting, in the 1820s say, had been restrained by comparison with this. Reporters had dwelt in increasing detail on the crime, on the prisoner's preparation for death, on the demeanour of the crowd, and on the final drop. But they still paid lip-service to execution's 'solemnity' and kept the experience distanced in a tide of clichés. Prison scenes were 'melancholy', shackling rooms 'gloomy', the approaching

[20] Cited in T. Boyle, *Black swine in the sewers of Hampstead: Beneath the surface of Victorian sensationalism* (1990), 89. I am indebted to Boyle's analysis of reports on the Palmer hanging, although his works to different effect from mine.

[21] Ibid. 90; A. Fonblanque, 'The diseased appetite for horrors', in his *England under seven administrations* (2 vols., 1837), i. 194; Mayhew, i. 229; R. D. Altick, *Evil encounters: Two Victorian sensations* (1986), ch. 5.

hour of eight 'awful', victims 'unhappy' as they waited to be 'launched into eternity', and they 'ceased to exist' after 'a few convulsive struggles'. Perhaps through this protective fog some readers like Maria Edgeworth experienced the event's implied sensation, but not many, one suspects. By mid-century, however, the *Observer* or *Weekly Chronicle* would boast twenty columns and sometimes a supplement on major murders, the details, as Mayhew observed, 'being written in a most honest deprecation of the morbid and savage tastes to which the writer is pandering'. The monstrous other now dominated crime reporting, fiction, and melodrama. Hence those Greenacre or Good chap-books which we noted in an earlier chapter, with their half-naked female corpses being sawn up in cosy parlours in lascivious and shocking display.

The dishonesty in these genres lay in the fact that you would not catch polite people *attending* such scenes as Palmer's hanging, read about them as they might. In Stafford that day 'scarcely any women disgraced themselves by being present', and although the Potteries and Black Country 'poured forth their thousands, for at most of the neighbouring works the operatives had claimed a holiday', it was noted that the thousands who swarmed to the execution by train and cart wore 'heavy shoes'. The biggest audience was still a polite one, tucked away in distant clubs and drawing-rooms, consuming images in safety, as pornography is always consumed.

That said, these ghastly literary effects were probably not always easily achieved. Reporters were familiar with arguments against capital punishment. They knew the innocent sometimes hanged, and were not unanxious about their tasks. Sometimes there is a tautness in their writing, as if they were unsure about its reception. Their straining justifications for hangings and their emphasis on the atrocity of crimes gave some games away. When Barrett faced the gallows, such doubts attached to his guilt that the Home Office ordered respites so that his alibis could be investigated. Doubts stuck, especially in plebeian minds. Hence most reports on his execution gave as much space to the Clerkenwell explosion as to his strangulation. Likewise describing London's first hidden execution, *The Times* did 'not wish to dwell upon . . . the painful scene' when Mackay took minutes to choke to death. His crime was the 'substance of the short narrative of his death on which the public *should really dwell*'.[22]

[22] *Daily Telegraph*, 27 May 1868; *The Times*, 9 Sept. 1868.

Tension is also discerned in a report of the first private hanging, of the youthful railway porter Wells at Maidstone.[23] So little of the horror escapes this reporter's gaze that he must have felt its implications at one or another level. But he opts for the easy way out by the end. They dug a hole inside the prison yard for Wells to drop into, erecting the scaffold over it, he tells us. On this Wells was to hang watched by the governor, chaplain, under-sheriff, surgeon, and sixteen press reporters, of whom ours was one. Early on Wells's last morning he dressed in his best—his porter's uniform, no less; and he pinned a brave though wilting flower to his coat in the faintest echo of a long-forgotten Tyburn gaiety. Here there was no crowd to play to, and the 'fear and dread of death were strong' in him instead; and so the poor man fainted and could 'take nothing in refreshment but a cup of tea'. Everyone was almost as upset as Wells himself. The chaplain prayed in accents 'broken by his deep emotion', and the hangman squeezed Wells's pinioned hands 'by way of farewell'. To the scaffold Wells was led at last: and there 'the dreadful stillness was broken, and the quivering lips babbled forth, very drearily and piteously, two lines of a simple hymn he had learnt in prison'. When the drop fell, he struggled on the noose for two minutes:

> That he was dressed in the everyday uniform of a railway porter, with the initials of the company worked on his coat collar, a faded bit of flower in his buttonhole, only served by contrast to make more repulsive the linen-clothed face, already swelling under its covering; the clutched hands, already crawled over by flies; the distorted neck, already showing abrasion by the rope.

That there was shocking and shocked knowledge in this witnessing is apparent, and at first the reader is held by it. The faded flower, the quivering lips, that piteous hymn! But the reporter cannot move on to the indictment which this knowledge threatens; his own or his readers' confusion forbids that. Hence gratuitous images multiply nervously, and false notes begin to be struck: the swelling face and the flies. Hanging was 'an odious, sickening, and repulsive spectacle', he continues, and now he cannot stop: 'Strong men turn pale and avert their eyes when the drop falls; and peculiarly hateful is the dull, dead sound—unlike any other sound we ever heard—of the falling body suddenly stopped in its descent. Again there is no more degrading sight upon earth than that of a man swinging to and fro on the gallows like a scarecrow.' At the end conventional pieties save him. The law must be validated when all is said and done. So hanging, he reminds us, is at least now 'robbed of

[23] *Police Service Advertiser*, 22 Aug. 1868.

many of its most repulsive accompaniments': he means the scaffold crowd. Only twenty or so people waited outside the prison for the black flag to be hoisted, so Wells's death was 'no longer sport for the scum of the populace'. His last prayers were 'not drowned by the curses and laughter of prostitutes and thieves', nor was he able 'to simulate the bravado of those distinguished ruffians of old times who thought they died like men when, in fact, they only died like brute beasts'. And so to his conclusion. Altogether, the reporter proclaims in relief, the new system was '*wholesome, sound, and humane*'!

4. *Scapegoating the crowd*

How this reporter escaped from his knowledge about the law's brutality can hardly be misread. He discharged it upon the scaffold crowd, and at its expense. He was not thinking hard as he did this, but he had good reason not to. For decades past the crowd had been made to bear the odium of the law's brutality, just as the hangman had: so in 1868 it was easy to assume that with the crowd's abolition an obscene ceremony was in some magical sense cleansed. This reflex carried unconscious meanings which are our final theme.

To put what follows in perspective we first retrace some steps: for the attack on all manner of crowds—and not only this one—had been under way for well over a century before 1868. In its broadest sense it sought to establish those real and symbolic distances between respectable and vulgar purlieus which marked the increasing status differentiations of an economically dynamic, polarized, and urbanizing society. The attack on scaffold gatherings was sustained by this as well.[24]

The repudiation of the scaffold crowd had difficulties to deal with, however. It could not be as easily corralled by magistrates or moral reformers as festive, political, or sportive crowds were. It retained unique functions not easily disavowed. It assembled to learn moral lessons, to affirm the law's sovereign power, and to bear witness to the law's open processes. Belief in this last and quasi-libertarian function was enduring. It explains why executions continued to be public when the need for penal display in symbolic excess had otherwise faded: the belief was a resistant one, in short. When a hanging was transferred from Tyburn to Bethnal Green in 1769, the sheriffs objected lest it 'make

[24] R. W. Malcolmson, *Popular recreations in English society, 1700–1850* (Cambridge, 1973), ch. 6; P. Borsay, *The English urban renaissance*; R. D. Storch (ed.), *Popular culture and custom in nineteenth-century England* (1982), 5–8; B. Bushaway, *By rite: Custom, ceremony and community in England, 1700–1880* (1982), ch. 7.

way for private executions, and for all those dreadful consequences with which private executions are attended, in every country where they have been introduced'.[25] A full century later this precedent was approvingly cited in parliament as opponents of private executions insisted (however vainly) that 'the principle of the law was, that the execution of a person was the act of the whole nation', and that the people must 'see their own punishment carried out', since it was 'the wise practice for centuries to make the people feel that the law was the expression of their own judgement and will'. Also, they added, the poor man 'had a right to be hanged in public' that he might declare innocence or admit guilt, while private executions would fan the recurrent popular suspicion that 'any man with £1,000 might escape from capital punishment, even after sentence had been pronounced'.[26]

In the eighteenth century these principles were at odds with mounting anxiety about the crowd's threats to street order and social segregation. The outcome was usually compromise—in the ending of the Tyburn procession most notably. As fashionable estates developed north of Oxford Street and close to the Edgware Road (at the foot of which Tyburn stood), the trustees of Edgware Road and of the New (Marylebone) Road were already petitioning in the 1750s that the movement of 'merchandise and manufactures [was] very much incommoded by the great concourse of idle and disorderly persons who usually attend all executions'. 'The same disorderly persons', the petition continued, 'are likewise greatly complain'd of by persons of quality and distinction who inhabit the great squares and streets adjoining and in the way to that place, up to which or very near the spot buildings are now erected and inhabited by persons of rank and fortune.'[27] In 1768–9, similarly, grandees like Grosvenor, Chesterfield, Kerry, Shrewsbury, Rochford, Sussex, Portland, Barrington, Rockingham, Perceval, and Hillsborough were among the 125 proprietors of estates near Tyburn who petitioned that they were

> greatly annoyed and disturbed by the vast concourse of people that always assemble there upon days of execution whereby great tumults disturbances

[25] *Gentleman's Magazine* (1769), p. 613, cited by P. Langford, *A polite and commercial people: England 1727–1783* (Oxford, 1989), 298.

[26] In 1856 Lansdowne remembered how, as a boy, he heard the repeated rumour that Dr Dodd walked the streets of London years after his hanging. In the 1860s Fauntleroy was said to be living in France after secret resuscitation; and lately a stuffed figure of Tawell was thought to have been hanged in that wealthy murderer's stead: Cooper, p. 113; PD3 cxc (1868), c. 1133–4.

[27] Corporation of London RO, *Historical papers*, i. 43 (113A). Petition undated, but the reference to the New Road, constructed in 1756 to develop the broad front from Edgware Road to Tottenham Court Road, suggests the late 1750s.

riots and nusances [*sic*] happen, and not only your petitioners are prevented from going in and out of their houses and prejudiced in their property but also the public roads leading from London to Oxford and Paddington are for several hours so thronged that neither horseman nor carriages can pass or repass without the greatest difficulty and danger, as the widest part of the street at the said place of execution doth not exceed seventy feet.

They wanted the gallows relocated 'at a convenient distance from the town so as not to annoy the inhabitants thereof'. The spot suggested was 'where the two roads from Hampstead and Kentish Town met the road from Tottenham Court' (the site of the present Camden Town tube station): this was closer to Newgate, the road was wide, and only two public houses were nearby.[28]

In the event two evangelical sheriffs, Barnard Turner and Thomas Skinner, settled Tyburn's fate. They brought the gallows to Newgate in 1783. And in these men the moral indictment of the crowd found its first effective voice. Galvanized by the shock of the Gordon riots and inspired by Howard's *State of the Prisons*, they set about reforming Newgate and the Compters of the Poultry and of Wood Street.[29] Their understanding of the psychology of moral contamination put Tyburn in their sight-lines too:

How far has it happened, that an unthinking passenger, of the lower rank of life, is attracted, by an innocent curiosity, to make one of the crowd, and insensibly drawn on to add to the confusion, without knowing it, till he catches the prevailing sentiments of his company, and becomes indifferent to the spectacle. . . . His want of feeling becomes habitual, the heavy stroke of justice no longer makes an impression on his mind, and he proceeds to treat morality with levity and the law with contempt.[30]

Then they wheeled out the motifs we saw Mandeville rehearsing in 1725 and which were to be restated for another eighty years yet. The crowd, they wrote, exhibited 'indecent levity' and 'wantonness of speech . . . profane jokes, swearing, and blasphemy'. As the 'sordid assemblage of the lowest among the vulgar' mocked 'the awful sentence of the law',

[28] Corporation of London RO, Misc. MS 21.31.

[29] They compiled new rules on hygiene and chapel attendance, forbidding gaming, drinking, prostitution, blasphemy, and garnish, arguing that the solitude and separation of prisoners 'would increase the punishment of a jail' and help 'moral amendment'. 'When the mind was alienated from every resource of external amusement, and driven by necessity into self, the most rational expectations might be formed of improvement. The natural abhorrence, which exists in every individual, to Vice, when he has an opportunity of viewing it in its true aspect and deformity, would compel these men to seek for consolation, when none of their more usual comforts were practicable, from Religion and new Principles': Turner and Skinner, *An account of some alterations and amendments attempted in the duty and office of sheriff* (1784), 38–9.

[30] Ibid. 25–6.

'the ends of public justice [are] defeated; all the effects of example, the terrors of death, the shame of punishment, are all lost.' Executions should be conducted 'with becoming form and solemnity, if order were [to be] preserved'.[31] A solution was offered when the demolition of houses between the Old and Little Bailey outside Newgate opened an area in which the sheriffs calculated that 5,000 people might assemble in controlled conditions. They put the idea to Lord North and Lord Chief Justice Mansfield, who in turn put it to the other judges. Everyone thought it was a good idea, so it was here that the first ten felons were strangled on 9 December 1783.[32]

The problems that Tyburn had posed were far from settled by its abolition, however. Bigger, more intrusive, and more opinionated crowds assembled at Newgate than had ever assembled at Tyburn. As they swilled into the City's heartland, tightening police and military supervision was required all the way down to 1868. Not surprisingly, hostilities were quickly resumed. They widened as middling people joined the attack, measuring their own advance to respectability against the crowd's barbarism. When Francis Place heard in the 1820s that Dr Johnson and Boswell had actually lamented the abolition of the Tyburn procession, he could hardly believe his ears: they 'ought to have known better, gabbling with the vulgar garrulity of two old apple women'. To show how times had improved Place combed the *Newgate calendars* for 'savage and inhuman' gallows and pillory stories from the 'bad olden times'. Tyburn festivals had been nothing more than 'a low, black-guard merry-making', he decided: 'the manners and feelings of the populace who permitted such conduct' were 'as disgraceful to a nation pretending to civilization, as any thing can well be conceived to be'.[33] We have registered elsewhere what was under way in constructions like these. Place read the history of his times as a progress from barbarism to civility, because that reading replicated what he discerned in his own biographical maturing. This same projection helps to explain the repudiation of scaffold spectacles by all manner of people lucky enough to swim with a rising economic or social tide.

Many sensitive souls in the eighteenth century revealed anxiety about scaffold crowds, then; but it was among Victorians that this bore the fullest fruit. Among them a newly energetic, plausible, and fruitful displacement came into play. As brutal passion was attributed to the crowd

[31] Ibid. 22–4. [32] HO42.3, fos. 205–7, 524; *Gentleman's Magazine*, 2 (1783), 991, 1060.
[33] Place papers, BL Add. MS 27826, fos. 106–7, 88–9, 33.

which watched scaffold killings (rather than to those who ordered, inflicted, or justified them), self-controlling people in effect began to project disowned aspects of their own feelings on to a scapegoat crowd which they could then self-gratifyingly castigate.

Two vignettes show how reflexively and easily these projections were achieved. In November 1849 Charles Dickens and the cartoonist John Leech made up a party to watch Mr and Mrs George Manning hanged on the roof of Horsemonger Lane gaol (south of the Thames) for murdering their lodger.[34] They hired part of a house from which to watch, getting 'the whole of the roof (and the back kitchen) for the extremely moderate sum of Ten Guineas, or two guineas each', Dickens reported. It was a bargain, because this was to be a famous execution. The Mannings were the first married murderers to hang together since 1700. Moreover, she was a Belgian; and to this injury she had added insult by shouting during her trial: 'There is no law nor justice to be got here! Base and degraded England!' When she chose black satin for her hanging dress that material at once went out of fashion for the season. Some 40,000 people attended her and her husband's demise.

By 1849 Dickens had reneged on his abolitionist principles, but there is no doubting his shock at the deaths he watched. Three years later he recalled being haunted for weeks by images of 'those two forms dangling on the top of the entrance gateway—the man's, a limp, loose suit of clothes as if the man had gone out of them; the woman's, a fine shape, so elaborately corseted and artfully dressed, that it was quite unchanged in its trim appearance as it slowly swung from side to side.' He even had to make a detour past Horsemonger Lane gaol in the hope of eradicating the image from his mind. What is most striking, however, is that this is all he *did* write about the hangings themselves. Significant for our argument was his instinct to put his distress and disgust elsewhere: 'the horrors of the gibbet and of the crime which brought the wretched murderers to it *faded in my mind before the atrocious bearing, looks, and language of the assembled spectators*.' A few days after the execution he wrote one of his famous denunciations of those spectators for *The Times* (using much the same sentence structure as that used in his denunciation of the crowd at Courvoisier's execution in 1840, cited on p. 60 above):

> The shrillness of the cries and howls that were raised from time to time . . . made my blood run cold. . . . When the two miserable creatures who attracted

[34] This discussion follows P. Collins, *Dickens and crime* (1964), 235–40.

all this ghastly sight about them were turned quivering into the air, there was no more emotion, no more pity, no more thought that two immortal souls had gone to judgment, no more restraint in any of the previous obscenities, than if the name of Christ had never been heard in the world.

The Mannings' death is also remarkable for the way in which Leech's commemoration of it in *Punch* replicated Dickens's evasion knowingly.[35] *The great moral lesson at Horsemonger Lane gaol* (Plate 41) might be the first (and only?) drawing of a public execution in which the gallows was not so much reduced to marginal significance (a traditional emphatic device) as utterly removed from view. Horsemonger Lane gaol is visible in the background, but the picture is cut off above the doorway to omit the scaffold and the hanging bodies on the roof. Since the frame is alive with its own energy, we yield to the point thus implied. A rough-featured and jolly crowd fills the picture, just as Dickens described. Waiting for the execution, they smoke, laugh, scuffle, grimace, gossip; children play pranks; and harassed constables keep order. This close focusing induces us to accept what would be unacceptable on a moment's thought, that the watching crowd matters more than what they watched, and that it is the crowd that is unlovely—forget the hanging. Beneath the picture nine parodic verses in canting speech—'The lesson of the scaffold, or The ruffian's holiday'—drive the message home. The verses come to their climax as the drop falls and the bodies dangle:

> Hurrah! you dogs, for hangin', the feelings to excite;
> I could ha' throttled Bill almost, that moment, with delight.

The second vignette is of a similar evasion, this time on the part of the Lords' select committee on the mode of carrying out capital punishment in 1856. It happened that the committee met in May, only a few weeks after one of the more gruesomely botched executions of the century. The murderer Robert Bousfield had thrown himself into the fire in his condemned cell on the night before his hanging and had burnt himself badly. Next morning, with his face bandaged like some ghastly mummy, he had to be assisted up the scaffold ladder and seated on a chair while the noose was put round his neck and the ordinary prayed. When the drop fell, Bousfield mustered strength enough three times to hook his toes on to the edge of the platform, scrabbling desperately until Calcraft the hangman yanked on his legs from below. At last he died to the hooting and jeers of the angry crowd. Who was the

[35] *Punch*, 17 (1849), 210.

210 PUNCH, OR THE LONDON CHARIVARI.

The Great Moral Lesson at Horsemonger Lane Gaol, Nov. 13.

THE LESSON OF THE SCAFFOLD; OR, THE RUFFIAN'S HOLIDAY.

We was havin a kevarten wen Bill he says, says he,
"To-morrow is the hanging-match; let us go and see."
I was game for anything: off we set that night;
Ha! the jolly time we spent until the morning light.

'Neath the timbers whereupon the conwicts wos to die,—
(And ugly black the gallows looked atween us and the sky)—
More than thirty thousand on us shouted, yelled, and sung,
Chaffin about murder, and going to be hung.

Each public-house was all alight, the place just like a fair;
Ranting, roaring, rollicking, larking everywhere,
Boosing and carousing we passed the night away,
And ho! to hear us curse and swear, waiting for the day.

At last the morning sunbeams slowly did appear,
And then, ha, ha! how rum we looked, with bloodshot eyes and blear:
But there was two good hours at least afore the hanging yet,
So still we drained the early purl, and swigged the heavy wet.

Thicker flocked the crowd apace, louder grew the glee,
There was little kids a dancin, and fightin for a spree;
But the rarest fun for me and Bill, and all our jolly pals,
Was the squeakin and squallin and faintin of the gals.

"Time's up!" at last cries Bill. "Why, sure, it ain't to be a sell!
Never. It can't be, I should think. All right! There goes the knell!
See, here they come, and no mistake, Jack Ketch and all his crew:
The Sheriffs, Parson, and—that's them! Hats off in front, there, you!

"Quick, Jack's about it. There he's got the fust beneath the beam;
And now, the other! Not a start, a tremble, or a scream!
All's ready. There they stand alone. The rest have gone below.
Look at him—look—he's at the bolt! Now for it! Down they go!"

'Twas over. Well, a sight like that afore these eyes of mine
I never had—no sort of mill, cockfightin, or canine.
Hurrah! you dogs, for hangin, the feelins to excite;
I could ha throttled Bill almost, that moment, with delight.

But, arter all, what is it? A tumble and a kick!
And, anyhow, 'tis seemingly all over precious quick,
And shows that some, no matter for what they've done, dies game!
Ho, ho! if ever my time comes, I hope to do the same!

41. John Leech, *The great moral lesson at Horsemonger Lane gaol, November 13* (1849).

Leech's cartoon was drawn after he and Dickens had watched the Mannings hang. It is remarkable for its omission of the scaffold on the prison roof and its concentration on the raucous crowd instead—as if to stress that it was the crowd that was barbarous, not the law which did the hanging. The canting verses endorsed the point.

murderer here? It was the crowd's question, and the crowd had an answer. It shouted that word at the hangman and sheriffs in the accusation which rang out so often at these events, and which 1868 was meant to silence for ever.

By their lordships, the question the crowd posed was sedately sidestepped, of course. That murderers must hang was obvious by now, and beyond debate. Better therefore look to how the crowd behaved than to what the law inflicted: the crowd must constitute the problem, not the hanging. What followed was predictable. The crowd was characterized by 'all kinds of levity, jeering, laughing, hooting, whistling, whilst the man is coming up—while he is still suffering—whilst he is struggling, and his body is writhing'. 'There was no evidence of any moral impression'; 'low jesting and indecent ribaldry' accompanied the hanging, and 'the hubbub and uproar' were 'most disgusting'. People went to executions 'as they would to a theatre, or the exhibition of a bull-fight'. Their tendency was merely 'to harden those who see' them.[36] In these terms the committee recommended the abolition not of hangings, but of opportunities for crowds to watch hangings. Once again, in effect, executioners (or their masters) contrived to dissociate themselves from what they did.

The evasions evidenced here were not wholly efficient, however. They left two problems unresolved, and only the abolition of the public scaffold would settle them. First, although at many executions the crowd's misbehaviours were real enough and satisfyingly deplorable, there was always some knowledge—Thackeray evinced it at Courvoisier's hanging, for example—that what the polite classes saw in the crowd was not its real demeanour, but a projection of their own aggression and excitement. Innumerable crowd reports, less sharply noted than Thackeray's, expressed what witnesses wanted to see rather than what they did see. The second problem was the urgent one, and more difficult to deal with too. It arose at the many hangings where vocal elements in the crowd were so far from being 'brutalized' that they refused to endorse the proceedings' legitimacy at all. What was intolerable in such cases was that some of the crowd saw through the law's pretensions more clearly than the polite people did, commenting sardonically on a tableau which they refused to accept as their own. We saw this in Chapter 3, in those eloquent scaffold demonstrations which élite remembering so neatly oblit-

[36] SC 1856, p. i.

erated from the record: that protest against Samuel Wright's execution in 1864, for example—with its plebeian petitions, public meetings, and deputation to the Home Office, its black-bordered handbills denouncing 'one law for the rich and another for the poor', its boycotting of the hanging, the drawing of neighbourhood blinds on the fatal morning, and the cries of 'Judicial murder' from the few who did attend. We also recall Sir George Grey's sense then that these representations sought to 'terrorize' the government. No wonder Wright had to hang.

Protests like these were not rare events, however thinly the crowd's enemies have commemorated them. To the contrary, they seem to have become more frequent as time passed. Many working people were highly politicized by mid-century, better able than their masters, sometimes, to discern the quasi-political considerations which sent some to their doom and saved others. At the last public execution of all, the *Daily News* might as usual see in the crowd only 'an incarnation of evil passions and perverted sympathies' which showed 'neither real pity for the criminal nor horror of his crime'. But this tired reading could not hide the fact that the crowd's behaviour had another point to it entirely, as the paper inadvertently conceded: 'The bastard pride in [the condemned man's] animal courage and the brutal delight that he died game made the law and its ministers seem to them the real murderers, and Barrett to be a martyred man.' And indeed this crowd did cry out at Barrett's executioner as its predecessors had at countless earlier hangings: 'Shame!—Down with him!—Bah, bah, murderer, bah!' And, when the hangman came to cut down the body an hour later, they called: '"Come on body snatcher! Take away the man you've killed!" &c. . . . amid such a storm of yells and execrations as has seldom been heard even from such a crowd.'[37]

Clearly, the public scaffold's abolition had to come about sooner or later; and we can now sum up the reasons why. In the short term, the timing was determined by the election of a government ready to implement the recommendations of the 1866 royal commission; perhaps the 1867 Reform Act also helped. The Introduction has advised caution on this latter claim: it is ill-supported by public or private statements. But the coincidence is suggestive; awesome displays of state power might well have been thought less necessary now that enfranchised urban householders could be presumed to endorse the devices and agencies of state. Of longer-term causes, the most obvious would seem to be squeamishness—

[37] *Daily News, The Times*, 27 May 1868.

that avoidance of 'whatever suggests offensive or disagreeable ideas' which Mill applauded, in which 'a great part of refinement consists'. The abolition of public execution certainly advanced that long process of social sanitization from which we all benefit (or believe we benefit): by catering to a deepening collective embarrassment and shame, 1868 was a key moment in the civilizing process. Yet it has been a premiss of this book that shifts in sensibility of this order usually ensue from rather than induce redeployments of power, so it is to two other causes that we should give priority. One was narrowly political. As MPs knew, abolishing public executions ensured that capital punishment for the worst of crimes would continue, its deterrent horror enhanced through its invisibility. Total abolitionists were neatly outflanked by this manœuvre. The crowd was outflanked too. This brings us to the heart of the matter. As they grew wiser, humble people must be denied opportunities for deriding great people's pretensions; abolition defen-ded polite Victorians' representation of their own civility by silencing the plebeian mockery of that representation. Abolition may be said to have been achieved *chiefly* by reference to that adverse image of the scaffold crowd which had taken shape over the previous century. It was not for nothing that when the home secretary introduced the abolition Bill in 1868, he pointed away from the noose's violence and towards that of the crowd, achieving a reflexive displacement so often reiterated that it won automatic assent.[38]

Nowadays it seems easy to see through the elaborate self-delusions of Victorian élites. Certainly as this book ends we may agree that abolishing the public scaffold had little to do with the visceral horrors of hanging a man or woman, let alone with the alleviation of those horrors, whatever Victorians claimed. Appeals to humanity encased their policy, but the state's retributive power continued to override imaginative compassion, and the horror continued behind prison walls for a century yet. This, however, is not to charge Victorians with hypocrisy in their evasive fictions. Since their illusions, myths, and practices were no more self-serving than ours are we can hardly expect them to have discerned truths which we have learned to see only recently, painfully, and dimly. When all is said and done, these final verdicts must bear as strongly on our sense of ourselves as they do on past times: that Victorians' civility

[38] The public spectacle was 'objectionable and horrible' because it was 'nothing put a public amusement', Gathorne Hardy told the Commons. It attracted 'the very worst classes', many 'on the very road to the gallows', who attended 'only to make a jest of what is to every worthy person a subject of the gravest regret'. Far from assimilating the scene's moral messages, they 'keep one another's courage up by singing low songs and laughing at low jests'. PD3 cxc (1868), c. 1127–41, cxci (1868), c. 1033–63.

only veneered the state's violence over; that in hiding penal violence they consulted their own feelings and not those of the punished; and that within the secret prison power was to be—and is—wielded more efficiently than ever it had been at Tyburn.

APPENDIX I

THE PETITION ARCHIVE

(I) A NOTE ON THE SAMPLE

I have used some hundred or so of the more complex appeals for mercy for the years 1800–30, concentrating on the 1820s because only after 1819 were dossiers comprehensively retained in the Home Office papers. The cases come from the judges' reports (HO47) for 1784, 1822, 1823, 1827, 1828, and 1829, and from the bundled dossiers in HO17 and PC1 for 1819–35. The sample is random but not systematic, so this is not the exhaustive study the archive deserves. The geographical distribution of appeals, their success or failure, and the identities, ages, and genders of appellants and their supporters need deeper research.

None of the early nineteenth-century appeals has been consulted before. But a very few late eighteenth-century appeals have been used by historians debating the class interests embedded in the criminal law. Douglas Hay used a handful from the 1780s to argue *inter alia* that 'the claims of class saved far more men who had been left to hang . . . than did the claims of humanity'. King contested this by quantifying the grounds on which mitigations of sentence were recommended in 136 judges' reports bound in the two volumes for 1787 and 1790. Tabulating the 'factors mentioned', he showed that judges justified mercy recommendations by referring more often to prisoners' previous records, youth, or numbers of dependants than to the respectability of their connections.[1]

I have not sought to comment on this debate directly, partly because it is now more than a little tired, and partly because the early nineteenth-century archive is too vast and the cases too rich to be reduced usefully to the bald 'factors mentioned' which the thinner eighteenth-century dossiers invited. Long attention is often needed before a single case reveals itself fully, and issues are seldom inscribed in black and white. A quantified study might still reflect what the historian wants to find. Whether or not the claims of class are disproven in the appeal process depends on how literally you read class,[2] or how literally

[1] D. Hay, 'Property, authority and the criminal law', in D. Hay *et al.* (eds.), *Albion's fatal tree: Crime and society in eighteenth-century England* (1975); P. King, 'Decision-makers and decision-making in the English criminal law, 1750–1850', *Historical Journal*, 27 (1984).

[2] Neither Hay nor those associated with him were unaware of the multiple faces of the law, benign as well as malign: see E. P. Thompson, *Whigs and hunters: The origin of the Black Act* (1975), ch. 10. But this has not cooled the heat with which the consensual elements in the relationship between law and people have been insisted upon by their critics. See (among many others) J. Langbein, 'Albion's fatal flaws', *Past and Present*, 98 (1983), and 'The criminal trial before the lawyers', *University of Chicago Law Review*, 45 (1978); J. Brewer and J. Styles (eds.), *An ungovernable people: The English and their law in the seventeenth and eighteenth centuries* (1980); J. Innes and J. Styles, '"The crime wave": Recent writings on crime and criminal justice in eighteenth-century England', *Journal of British Studies*, 25 (1986).

you read 'factors mentioned' without cognizance of the full contexts of cases. The eighteenth-century dossiers omit much material. The more comprehensive textures of decision-making which are revealed in the letters, affidavits, memos, etc. available for the 1820s suggest that the apparent infrequency of references to the respectability of prisoners' connections or patrons in early reports meant that that question had either already been considered or otherwise conveyed to the judge, or was palpable in character evidence or the literacy of petitions or the designations of those who signed them. Petitions signed clumsily or with crosses hardly ever worked. When it was not palpable, decision-making encompassed the question of felons' or their supporters' characters very openly, as nearly every case discussed in this book shows. It would be surprising if it were otherwise in so hierarchical a society.

My main reason for eschewing a 'systematic' survey will be clear to those who have read the microhistories in the book. It is that the most interesting revelations in appeals lie between the lines of single cases rather than within aggregations of many—in phrasings and images which flow around a story uncensored, revealing feelings, assumptions, and attitudes which are not always conscious. I have explained, justified, and applied this in Chapter 17. I have discussed the numbers of appeals, their purposes, and their authors in Chapters 6 and 16.

(II) A NOTE ON THE ARCHIVE

Before 1782 petitions for mercy were kept in State Papers Domestic: SP36 (1727–60) and SP37 (1760–82). Some others (mainly eighteenth-century) survive in Home Office papers HO42 and HO44. Others (for the years 1819–44, but unindexed) survive in Privy Council papers: PC1. A series covering the years 1784–1830 was initiated after the separation of the Home and the Foreign Departments of State in 1782; this was bound in 75 annual volumes and titled 'Judges' Reports': HO47. The vast majority of appeals, however, including all those against Old Bailey sentences, were stored separately in bundles.

The early ones are apparently lost, though a few are randomly kept in HO42 and HO44. The bundles survive in full for the years 1819–39: HO17.1 (131 bundles) and 1839–54: HO18.2 (381 bundles). The individuals they concern are indexed alphabetically in separate register books, HO19. By the 1820s these bundles included many dossiers which contained judges' reports. The bound series of judges' reports in HO47 was discontinued in 1830, presumably because the distinction had become meaningless. Some petitions were retained in the Privy Council papers (PC1, as above) because the Privy Council retained a nominal control of the transportation colonies and acted as a court of appeal for sentences delivered in other colonies (most of the appeals retained there concern prisoners who were about to be or who had been transported and colonial murders, though through some accident of sorting a few of its dossiers are of the same domestic kind as those preserved in the Home Office papers). HO6 contains judges' formal recommendations for mercy for 1816–40, along with lists of London capital convicts discussed at the King in Council and summaries of petitions, if any, and their outcome. The numbers and contents of these documents, and the conditions of their production, are discussed in Chapter 6.

APPENDIX 2

EXECUTION AND MERCY STATISTICS

1. Annual prosecution, sentencing, and execution statistics, and occasional synoptic tables for felonies tried at assizes and quarter sessions in England and Wales and at the Old Bailey, were published in Parliamentary Papers from 1805 onwards. There are no national figures before 1805, but the 1819 committee on criminal law published data on death sentences, pardons, and executions in London and Middlesex from the Old Bailey in the eighteenth century and up to 1804 (figures for the home circuit were published too).

Tables 1 and 2 illustrate the relatively low Old Bailey execution rates in the earlier eighteenth century; their peak in the 1780s and contraction thereafter; the next major peak in national as well as London executions after the peace of 1815; the high levels of conviction and execution sustained throughout the 1820s, matching those of the first decades of the century; the increasing percentages of capital sentences that were commuted (chiefly to transportation) by the prerogative of mercy; and the rapid collapse of executions in the 1830s.

TABLE 1. *Capital convictions and executions, London and Middlesex (Old Bailey), 1701–1834*

Years	Capital convictions	Executions	% pardoned
1701–25	471	156	67
1726–50	300	125	58
1751–60	168	98	42
1761–70	509	246	52
1771–5	429	151	65
1776–80	386	174	55
1781–5	687	293	57
1786–90	501	208	58
1791–5	364	98	73
1796–1800	416	95	77
1801–05	398	43	89
1806–10	428	53	88
1811–15	673	85	87
1816–20*	1058	140	87
1821–5	730	96	87
1826–30	857	89	90
1831–4	428	12	97

TABLE 2. *Capital convictions and executions, England and Wales, 1805–40**

Years	Capital convictions	Executions	% pardoned
1805	350	68	81
1806–10	1874	286	85
1811–15	2760	374	86
1816–20**	5853	518	91
1821–5	5220	364	93
1826–30	6679	307	95
1831–5	4984	207	96
1836–40	1181	51	96

* Including London and Middlesex.

** Data for 1819 not published. These figures are the average of the 4 known years multiplied by 5.

Sources: See Table 1.

2. Extrapolation from the Old Bailey figures permits a rough estimate of national executions (and death sentences) for the period 1770–1830 by the following calculations. Across quinquennia 1806–30 the proportion of Old Bailey executions to the total in England and Wales increased as shown in Table 3.

TABLE 3. *Old Bailey executions as proportion of England and Wales, 1806–30*

	Old Bailey executions	England and Wales executions	*Old Bailey* %
1806–10	53	286	19
1811–15	85	374	23
1816–20	140	518	27
1821–5	96	364	26
1826–30	89	307	29
TOTAL	463	1849	25

The appendices of the 1819 committee give an Old Bailey total of 1,101 *executions* for 1770–1804 inclusive. If, following either the 1806–30 or the 1806–10 ratios above, we assume that this was either 25 or 19 per cent of the unknown

Footnotes to Table 1 (*opposite*)

*Data for 1819 not published. These figures are the average of the 4 known years multiplied by 5.

Sources: PP 1819, appendices (for Old Bailey data); PP 1826–7, xix. 186 ff.; PP 1830–1, xii. 495 ff.; PP 1835, xlv. 21 ff.; annual judicial statistics.

national total, we can put national executions for 1770–1804 at between 4,404 and 5,795. Add to these figures the 1,918 national hangings recorded for 1805–30, and the estimated England and Wales total for the whole period 1770–1830 ranges between 6,322 and 7,713 executions—probably nearer the higher total than the lower. Given that Old Bailey *death sentences* comprised 23 per cent of the England and Wales total in 1806–10, similar calculation results in an estimated England and Wales total of 36,566 death sentences for the period 1771–1830.

3. The distribution of executions in England and Wales for 1805–32 by type of felony is contained in Table 4.

TABLE 4. *Executions by felony, England and Wales, 1805–32*

	1805–18		1819–32	
	No.	%	No.	%
Murder	202	19	193	19
Attempted murder	49	5	58	6
Rape	47	4	55	6
Sodomy	28	3	22	2
Arson	29	3	52	5
Burglary, housebreaking	216	21	203	21
Larceny in dwelling-house	24	2	41	4
Robbery	116	11	160	16
Horse-, sheep-, cattle-theft	82	8	96	10
Coining	10	1	12	1
Forgery, uttering	204	20	78	8
Other	28	3	23	2
TOTAL	1035	100	993	100

4. In 1832 capital punishment was abolished for stock-theft, larceny up to £5 from a dwelling-house, coining, and forgery (except of wills and powers of attorney to transfer stock). In 1833 it was abolished for housebreaking; in 1834 for returning from transportation; in 1835 for sacrilege and letter-stealing by post-office servants. In 1837 the death penalty was restricted to: murder; attempted murder; rape and carnally abusing girls under 10; sodomy; burglary attended by violence to the person; robbery with cutting or wounding; arson of dwelling-houses or ships when lives were endangered; piracy when murder was attempted; showing false signals to cause shipwreck; setting fire to HM ships of war; riot and feloniously destroying buildings; embezzlement by servants of the Bank of England; high treason. In 1841 the death penalty was

removed from rape and felonious riot. In 1861 treason and murder became the only statutory capital crimes. In practice, after 1837 only murderers were hanged (along with three who attempted it, before 1841): between 1837 and 1868 inclusive, 347 murderers were executed in England and Wales.[1]

[1] Execution data up to 1854 come from annual criminal statistics; for 1855–63 from RC 1866: 661; for 1864–8 from PP 1871, lviii.

INDEX OF PERSONS

Numbers in italics refer to plates

Abbott, Charles (Lord Tenterden, chief justice) 434, 502, 505, 506, 509, 511–12, 541, 546, 548, 550
Abbott, James (attempted murderer) 413–15, 435
Adams, George (burglar) 557–60, *561*
Adolphus, John (barrister) 309, *310–11*
Ainsworth, William Harrison (novelist) 126, 131, 145–8, 184
Aldini, J. (anatomist) 256
Alexander, Sir William (judge) 504
Allen, William (Quaker) 333, 394, 398, 404
Anderson, James (robber) 441–2
Angelo, Henry (fencing master) 61–2, 93, 158, 253, 276–7, 292–3, 309
Anson, George (burglar) 557–63, *561*
Arbuthnot, Mrs (diarist) 311–12, 314, 550
Ariès, Philippe (historian) 77
Ashford, Mary (murdered) 336
Astbury, William (coiner) 525
Austen, Jane (novelist) 242, 299
Avershaw, Lewis (highwayman) 33, 268

Bach, Johann Christian (musician) 64
Bagehot, Walter (writer) 11
Ballantine, William (serjeant) 113, 142, 247, 384, 387, 430, 504
Bamford, Samuel (radical) 309, 435, 523
Barclay family (Quaker bankers) 400, 405
Barrett, Michael (Fenian) 24, 46, 54, 589, 596, 599, 609
Barry, John (Quaker) 502
Bartholomew, William (burglar) 555–60, *561*
Beattie, John (historian) vii, 523, 524, 536
Beccaria, Cesare (philosopher) 331, 332, 339, 383, 516, 520
Bell, John (boy murderer) 1–4, *4*, 296
Bellingham, John (murderer) 57, 98, 103, 249
Bennet, Grey (MP) 435, 507, 578, 579
Bentham, Jeremy (philosopher) 18, 22, 73, 233, 237, 327, 383, 516, 520
  influence of 329, 331, 332, 383, 571
Berkeley, George (philosopher) 234
Best, William Draper (Lord Wynford; judge) 103, 268, 420, 424–9, 473, 501, 502, 503, 541–2
Bewick, Thomas (engraver) 176, 184–5, *186*, 283, *284*
Bickersteth, Revd Edward (evangelical) 58
Birmingham, Thomas (highwayman) 34, 48
Blackstone, William (jurist) 203, 314, 327, 332, 346, 383, 434, 500, 518, 520
Blake, William (poet) 13
Blandy, Mary (murderess) 51
Bonner (publisher) 158–9, 165
Borrow, George (novelist) 114
Boswell, James (diarist) 45, 62, 124 n., 242, 260, 261–2, 281, 344, 383, 430, 598, 604
  on curiosity and sympathy 251, 285–92
  and John Reid 91, 171, 254, 289–91
  and Paul Lewis 287–91, 293
  wants terror in executions 286–7
Bousfield, William (murderer) 38, 51, 99–100, 114, 606–8
Bowman, Christian (coiner) 36, 166, 319 n.
Brandreth, Jeremiah (Pentrich leader) 158, 312–13, 317, 320, 537
Braudel, Fernand (historian) 238
Bright, John (MP) 22
Bronfen, Elisabeth (critic) 243
Brougham, Lord (whig) 329, 520, 548, 551, 575
Brunt, John Thomas (conspirator) 305–6, 307, *310–11*
Buller, Sir Francis (judge) 507
Burdett, Sir Francis (radical) 186, 584
Burke, Edmund (statesman) 13, 20, 73, 231, 344, 419, 518
Burke, Peter (historian) 91, 94
Burrough, Sir James (judge) 419–20, 503, 506, 511, 526, 540, 554
Butler, Joseph (theologian) 234, 273, 576
Buxton, Thomas Fowell (reformer) 362, 371, 395, 401
Byron, Lord (poet) 230, 249–50, 253, 261

Calcraft, William (hangman) 50–1, 68, 99–100, 103, 597, 606
Campbell, Lord (judge) 506, 512
Carlyle, Thomas (sage) 5, 233, 297, 380, 591
Castlereagh, Lord (statesman) 551
Catnach, Jemmy (publisher) 142, 144–5, 160–1, 168, 176
Charles I 300, 314
Charlotte, Princess 313–14
Chesterfield, Lord 262
Christian, Edward (justice) 508
Clare, John (poet) 175
Clay, Revd John (evangelical) 60, 592

Close, John (burglar) 555–60, *561*
Cobb, Richard (historian) 448
Coigly, James (Irish rebel) 318
Coke, Sir Edward (jurist) 2, 315–16
Coleridge, Sir John Taylor (judge) 501–2, 511
Coleridge, Samuel Taylor 95, 252, 335, 338
Collier, John Payne (critic) 120–1
Colquhoun, Patrick (magistrate) 18, 124, 237
Conyngham, Lady (king's favourite) 563–4
Cook, James (murderer) 248, 257, 268–9, 381–2
Corder, William (murderer) 39, 58, 69, 159, 173, 253–4, 256–8
Cotton, Revd Henry (ordinary) 43–5, 66, 305, 308, 355, 361, 438–9
  Newgate ministry of 384–9
Courvoisier, François (murderer) 24, 57, 60, 65–7, 101, 114, 158, 167, 263, 277, 281, 605
  effigy of 115–17, *118*
  Thackeray on execution of 294–7
Cranworth, Lord (lord chancellor) 60, 96
Cruikshank, George (artist) 255, 263, 361, 364, 365, 384
  and gallows emblems 184–9, *188*, *513*
Cruikshank, Robert 66, 255, 356, *357*, 384
Cureton, Elizabeth (prosecutrix) v, 447–93
Curtis, James (preacher) 253, 380

Dagoe, Hannah (burglar) 291–2, 293
Damiens (regicide) 14, 253, 262, 276
Darby family (Quakers) 400, 405, 450, 457, 483, 485
d'Archenholz, Johann Wilhelm (writer) 36
Darwin, Charles (scientist) 272–4
Davidson, W. (conspirator) 308, *310–11*, 380
Davis, John (burglar) 556–60, *561*
Defoe, Daniel (novelist) 147, 329
Denman, Thomas (chief justice) 329, 505–6, 509, 510, 512, 524, 536, 537, 552, 568
Desmond, Edward (burglar) 556–62, *561*
Despard, Edward Marcus (traitor) 50–1, 85, 88, 98, 104, 298 n., 316–17, 319, 320, 541
Dickens, Charles (novelist) 117, 131, 146, 156, 195, 233, 297, 332, 368, 435, 509, 514
  on fascination with gallows 239, 242
  and scaffold crowds 59–60, 62, 74–5, 265–6, 591, 605–6
Diver, Jenny (alias of Mary Young) 117
Dixon, Hepworth (barrister) 109–10, 156
Dodd, Revd William (forger) 62, 64, 83, 206, 275, 281, 283, 290 n., 292–4, 408, 602 n.
Drummond, Revd Hay 342, 343, 346, 347

Eaton, Daniel Isaac (radical) 89 n., 383
Eden, William (statesman) 327
Edgeworth, Maria (novelist) 259, 599
Edwards (spy) 304, 308
Edwards, Edward (surgeon) 480–3
Egan, Pierce (humorous writer) *66*, 131
Eldon, Lord (lord chancellor) 320, 332, 360, 502, 512, 543, 546, 547, 583
Elias, Norbert (social theorist) 14, 17
Ellenborough, Lord (chief justice) 320, 332, 363, 364, 369, 401, 500–1, 506, 508–9, 531, 541
Ellenborough, Lord (younger, lord privy seal) 510, 547–8, 549–50
Ellis, Benjamin (robber) 424–7, 440
Erskine, Henry (whig) 334
Evelyn, John (diarist) 245
Ewart, William (MP) 22–3, 269, 297, 329, 443, 507, 570, 591

Fairburn, John (publisher) *85*, 365–6
Fane, Robert Cecil 311, 314
Farmer, J. S. (ballad collector) 126 n., 130
Fauntleroy, Henry (forger) 35, 45, 57, 74, 101, 117, 253, 387–8, 435, 547, 582, 602 n.
  petitions for 209, 212, 409, 413
Fenning, Eliza:
  crime and trial of 358–9
  execution of 36, 38, 65, 83, 102, 263, 339, 353–6
  funeral of 88–9, 356
  and Hone 363–7, 436
  memory and opinion on 113, 363, 367–70, 440
  petition for 210
  piety of 354–5, 361, 380
  portrait of 255, 356, *357*
Ferrers, Earl (murderer):
  execution of 62, 117, 245–6, 253
  funeral of 85–6, 89
  Tyburn procession of 33–4, 38, 52, 305
Fielding, Henry (novelist) 23, 36, 281, 292, 326, 592
Fielding, Sir John (magistrate) 134–5, 145
Fonblanque, Albany (journalist) 584, 598
Forde, Revd Brownlow (ordinary) 383
Forster, William (Quaker) 362, 398
Fortey, W. S. (publisher) 142–3
Foster, Revd Henry (preacher) 376, 377
Foucault, Michel (social theorist) 14–17
Fox, Charles James (statesman) 62, 352, 518
Frankland, Colonel William (MP) 319
Freud, Anna (psychoanalyst) 266
Freud, Sigmund (psychoanalyst) 17, 238–40, 261, 264, 275
Fry family (Quakers) 400–1, 404, 405

Fry, Elizabeth 8, 226, 328, 338, 371
and death penalty 373–4, 377, 391, 404–5
in Newgate 380, 389, 400
sensibility of 372, 391–4

Garrett, John (sheriff) 558, 561–3, 595
Garrow, Sir William (judge) 319–20, 434, 504, 525, 527, 539–40, 551
Gaselee, Sir Stephen (judge) 3, 509
Gash, Norman (historian) 571
Gay, Peter (historian) 270
George IV, King 95, 269, 546
and mercy 550–4, 558–60, 563–4
and Peel 553–64
see also Prince of Wales
Géricault, Théodore (artist) 178, 182, 250 n.
Gillray, James (caricaturist) 185–8, 514
Gilpin, Charles (MP) 64, 103, 443
Gilpin, William (painter) 190
Godwin, William (philosopher) 125, 339, 344, 345, 353, 364
Goldsmith, Oliver (novelist) 329, 364
Good, Daniel (murderer) 70, *71*, 114, 159, 255, 272, 599
Graham, Sir Robert (judge) 503–4, 535
Greenacre, James (murderer) 39, 68–70, *72*, 114, 159, 191, *193*, 599
Greville, Charles (clerk to Council) 11, 548–9, 550–1, 552
Grey, Sir George (home secretary) 102, 609
Grose, Francis (antiquary) 59, 129, 131
Grose, Mary (farmer's wife) 245–6
Grose, Sir Nash (judge) 342, 508
Gurney family (Quakers) 400, 401, 404, 405
Gurney, Hudson (Quaker) 411
Gurney, Sir John (judge) 512
Gurney, Joseph John (Quaker) 73, 374, 404

Hackman, Revd James (murderer) 171, 251, 262, 275, 282, 283–4, 292
Haggerty, Owen (murderer) 34, 57, 63, 355, 369, 415, 436
Hale, Sir Matthew (jurist) 471–2
Haley (free-thinker) 42, 45, 384, 386, 388
Hamilton, Elizabeth (author) 393
Hammett, Sir Benjamin (MP) 317, 319 n.
Hanway, Jonas (reformer) 327, 330, 397
Hardy, A. E. Gathorne (home secretary) 248–9, 592–3, 610 n.
Hardy, Mary (farmer's wife) 246
Hardy, Thomas (novelist) 266
Harmer, James (attorney) 209, 330, 362, 369, 409, 414, 435–9, 443, 557–8
Harris, Edward (robber) 209, 436–9
Harris, Phoebe (coiner) 38, 73, 337
Hartley, David (philosopher) 229, 263, 328
Hartley, Robert (highwayman) 35, 38, 47, 167
Harwood, Joseph (robber) 40–5
Haskell, Thomas (historian) 17, 232 n., 398
Hatfield, John ('Keswick Imposter') 35, 47–8, 252–3, 335
Hay, Douglas (historian) vii, 32, 612
Haydon, Benjamin (painter) 272
Hayes, Catherine (murderess) 317
Hazlitt, William (essayist) 339
Heath, John (judge) 501
Heath, William (caricaturist):
Merry England 193, *194*
Hibner, Esther (murderess) 68, 74, 89, 336
Hilton, Boyd 372, 571, 576, 585
Hindley, Charles (collector) 368
Hobhouse, Henry (under-secretary) 99, 506, 535, 538, 553, 555, 558, 564
Hogarth, William 140–1, 195, 254–5
appropriation of motifs in *Idle 'prentice* 30, 172, 190–2, *191*, *192*, 193
and gallows emblem 177–8, 183–5, 188–9, *184*, *185*
Idle 'prentice executed at Tyburn 64, 183, 190, 246
The four stages of cruelty 183, *185*, 256
Holloway, John (murderer) 34, 38, 57, 63, 355, 369, 415, 436
Holloway, John (Sussex murderer) 69, 81
Holroyd, Sir George Sowley (judge) 472, 502
Holyoake, George Jacob (radical) 60, 214
Hone, William 168, 187, 263, 330, 340, 358, 364–5, 435, 436, 499, 531
and Eliza Fenning 356, 360, 363–7
'little books' of 189, 365, 513
and opinion 369–70
Hood, Thomas (poet) 160–1
Hotham, Beaumont (judge) 512
Howard, John (reformer) 226, 237, 327, 330, 331, 339, 371, 380, 397, 603
Hullock, Sir John (judge) 502
Hume, David (philosopher) 232, 234–5, 285, 286, 290
Hunt, Henry (radical) 584
Hunt, Leigh (writer) 247, 363
Hunton, Joseph (forger) 35, 64, 101, 117, 220, 404, 406, 409–13, 415–16, 564, 582
Hurle, Anne (forger) 38, 48, 49, 97
Hutcheson, Frances (philosopher) 234

Ings, James (conspirator) 305–6, *310–11*

Jebb, Dr John (writer) 333, 343

Jenkins, James (Quaker) 403
Jerdan, William (journalist) 430
Jerrold, Douglas (editor) 297
Jessop, John (barrister) 431–3
Jobling, William (murderer) 268–9
Johnson, Samuel (essayist) 13, 55, 230, 251, 289, 292, 326, 344, 375, 407, 408
 and Dr Dodd 292–4
 on justice 516
 on sympathy 286
 on Tyburn procession 37, 286, 604
Johnston, Robert (robber) 50, 97
Jones, Mary (shoplifter) 165, 332

Kelly, Sir Fitzroy (attorney-general) 22, 443
Ketch, Jack (hangman) 114, *116*, 121–2, *122*, *188*
King, Peter (historian) vii, 206, 524, 612
Knatchbull, Sir C. (JP) 528
Knight, Charles (publisher) 248
Knowlys, Newman (recorder) 441, 509–10, 511, 547, 554

La Motte (spy) 317
Labrousse, C. E. (historian) 238
Lachlan, Mrs (evangelical) 382
Lamb, Charles (essayist) 274–5, 352
Lansdowne, Lord (home secretary) 432, 433
Laqueur, Thomas (historian) 91, 95–7
Laurie, Sir Peter (sheriff) 54, 277
Lawrence, Sir Thomas (painter) 255
Leech, John (artist) 195, 605–6, *607*
Lewis, Paul (highwayman) 36, 40, 287–91, 293
Lichfield, C. H. (benefactor) 425–6
Lieven, Princess 309
Liverpool, Lord (prime minister) 299, 309, 311
Lloyd, Corbyn (Quaker) 359, 362
Lloyd, Edward (publisher) 159
Lloyd, Sarah (thief) 339, 340–53, *351*, 357
 execution and funeral of 341, 348–9
Locke, John (philosopher) 229, 328
Lofft, Capel (reformer) 326, 339, 343–4, 436
 and Sarah Lloyd 340–1, 345–53
 sensibility of 339, 345, 353
 victimization of 349–53
Loughborough (lord chancellor) 21 n., 319 n., 418 n.
Ludlam, Isaac (Pentrich leader) 158, 312–13
Lyndhurst, Lord (lord chancellor) 512, 552

McGowen, Randall (historian) vii, 227, 329, 585
Macaulay, Thomas Babbington (as barrister) 260, 430–1, 591
Mackay, Alexander (murderer) 589, 596, 599
Mackenzie, Henry (novelist) 252
Mackintosh, Sir James (MP) (reformer) 5, 22, 233, 226, 332–4, 401, 403, 419, 430, 510, 516–17, 568, 583, 584
McLean, James (highwayman) 38, 83, 158, 163, 254, 260
Madan, Martin (writer) 532–3
Mahoney, Joseph (housebreaker) 413
Maistre, Joseph de 95
Maitland, Frederick (lieutenant-general) 98
Malcolm, Sarah (murderess) 117 n., 254–5
Mandeville, Bernard (author) 36, 38, 59, 61, 95, 252, 603
Manning, George and Maria (murderers) 57, 74, 114, 159, 173–4, 195, 255, 605–6, *607*
Mansfield, Lord (chief justice) 604
Marks, Lewis (caricaturist) 193
Martin, Sarah (evangelical) 381
Matthews, Henry (traveller) 253
Maugham, Robert (legal writer) 434
Mayhew, Henry (journalist) 65, 122–3, 159, 169, 171–4, 599
Maynard, Thomas (forger) 413, 581
Melbourne, Lord (statesman) 507, 549
Meredith, Sir William (MP) 165, 293, 326, 332, 334
Miles, Samuel (forger) 563–4
Mill, James (philosopher) 233
Mill, John Stuart (philosopher) 234, 297, 371, 375
 on civilization 270, 595, 610
 on death penalty 591, 593
Milnes, Monckton (MP) 277, 295
Monro, Alexander (sr.: anatomist) 45–6, 289–90
Monro, Alexander (jun.: anatomist) 272–3
Montagu, Basil (reformer) 68, 84, 340, 362, 364, 365, 395, 398, 399, 401, 438 n.
Moore, Tom (author) 131
More, Hannah (evangelical) 125, 161–3, *162*, 167, 228, 373, 380–1
More, Sir Thomas 498 n.
Morris, George (robber) 441–2

Naylor, Henry (burglar) 557–60, *561*
Neate, Charles (MP) 597
Neild, James (evangelical) 383
Newbury, Henry (housebreaker) 553–4
Newdegate, Charles (MP) 75
Noden, John (rapist) v, 204, 206, 447–93, 524–5, 578
North, Lord (premier) 604

Oliphant, Mrs (novelist) 598
Oliver (spy) 313, 349, 531

Opie, Amelia (novelist) 270–1

Paine, Tom (radical) 125, 231, 343, 344
Paley, William (archdeacon) 327, 330, 360, 363, 369, 396, 419, 443, 519, 523–4, 569
  defensiveness of 263, 518
  on divine and human justice 517–19, 584
  justifies capital code 202–3, 498
Palmer, Dr William (poisoner) 58, 114, 597–9
Park, Sir James (judge) 220, 416, 431, 473, 504, 511, 525, 528–9, 535, 540, 554, 556
Parker, George (actor) 129, 132
Parr, Dr Samuel (whig writer) 231, 333, 367, 499
Patmore, Coventry (poet) 60, 246
Pattern, John (rapist) 206
Peel, Robert (home secretary) 22, 42, 200, 205, 207, 221, 396, 411, 415, 428, 437, 493, 506, 521, 525, 540, 541, 544
  and forgery 406, 581–3
  and gentry 525–9
  and King in Council 546–7, 553–65, 567
  and law reform 237, 402, 405–6, 440, 497, 507, 511, 567–71, 574–5, 579–83
  mercilessness of 438, 441–2, 558, 562, 566–7, 571–2, 575, 576–8, 585
  on police and crime 237, 572–5
  and prerogative 543, 553, 567, 572, 583–4
  and providence 566, 576–7, 585
Pepys, Samuel (diarist) 62, 157, 239, 244–5, 251, 253, 262, 268, 292
Phillips, Charles (writer) 11, 113, 368
Phillips, Sir Richard (reformer) 114, 364, 398, 401, 524
Pitts, Johnny (publisher) 133, 141, 143, 147, 151, 160, 168, 191, 192, 353–4
Place, Francis (radical) 64, 94, 126, 158, 231, 285, 436
  and ballads 123–9, 131, 141, 149–55, 161
  and criminals 117, 133–4, 171, 604
Plumb, J. H. (historian) 77
Pollock, Sir J. Frederick (attorney-general) 441, 535
Portland, duke of (home secretary) 342, 343, 346–7, 348, 350, 352
Potter, Harry (historian) 592
Poulter, John (robber) 129
Price, Hanbury (coiner) 34
Priestley, Joseph (scientist) 229, 339
Prince of Wales 193, 332, 363, 502, 503, 507, 551
  see also George IV

Queen Caroline 160, 552
Queen Victoria 95, 564–5

Radzinowicz, Sir Leon (historian) vii, 11, 53, 293, 571, 589
Rann, John (highwayman) 33, 36, 40, 47, 133–4, 144
Reid, John (sheep-stealer) 86, 91, 171, 254, 289–91
Reynolds, Sir Joshua (artist) 261–2, 285
Richardson, Samuel (novelist) 36–7, 61, 67, 229, 230, 242
Robinson, Mary (maid of Buttermere) 335–6
Rogers, Samuel (poet) 10
Romilly, Sir Samuel (whig reformer) 19, 21, 22, 47, 84, 226, 247–8, 327, 328, 330–2, 359, 367, 400, 403, 505
  on capital statutes 330–3, 340, 398, 401, 430, 500–1, 568, 580
  on treason 318–20
Ross, W. G. (singer) 142–3
Rousseau, Jean-Jacques 230, 332
Rowlandson, Thomas (artist) 96, 281–2
  and gallows images 178, 179, *180*, *181*, 190
  Last dying speech *172*
  The dissection 264–5, *265*
Roxburgh, duke of (bibliophile) 254, 263
Russell, Lord John (statesman) 23, 101, 296, 297, 511
Russell, Whitworth (prison inspector) 83
Ryder, Sir Dudley (judge) 36, 508

Sale (murderer) 109–10, 123, 133, 138, 144
Savage, Robert (housebreaker) 35
Selwyn, George (wit) 253, 262, 276
Sewell family (Quaker) 411, 412
Shaftesbury, Lord (3rd: philosopher) 234
Shaftesbury, Lord (7th: philanthropist) 592
Sharp, Cecil (ballad scholar) 143
Shelley, Percy Bysshe (poet) 313–14, 514
Shepherd (publisher) 158, 165
Sheppard, Jack (housebreaker) 40, 87, 126, 130–3, 138, 140–1, 145, 147, 254
Sherwood, Mrs Mary (novelist) 247
Sidmouth, Lord (home secretary) 205, 212, 308, 354, 362, 373, 404, 422, 553
Silvester, Sir John (recorder) 359–60, 364, 367, 509
Sinclair, George (reformer) 395, 403
Skinner, Thomas (sheriff) 603
Smith, Adam (philosopher) 87, 231, 233, 235, 286, 498
Smith, Charles (writer) 189
Smith, John Thomas (artist) 263
Smith, Madeleine (murderess) 368
Smith, Sydney (wit) 22, 215, 417, 440, 441, 499, 506, 534, 535, 539, 542, 577, 578

Southey, Robert (author) 14, 75, 95, 111–12, 233, 268
Stewart, Dugald (philosopher) 13, 71, 235
Strutt, William (Quaker) 405
Stuart, Jack (beggar) 79–80
Suffield, Lord (reformer) 269, 400
Sullivan, Margaret (coiner) 337–8
Swift, Jonathan(satirist) 32

Thackeray, William Makepeace (novelist) 24, 113, 180, 189, 263, 281, 591
  and Courvoisier 294–7
  on gallows wit 274, 279
  on scaffold crowd 62, 66–7
Thistlewood, Arthur (conspirator) 299, 302–3, 304, 307, 309, *310–11*, 320
Thomas, Keith (historian) 231
Thompson, Edward (historian) 220
Thornhill, Sir James (painter) 254
Thornton, Abraham (tried for murder) 336
Thrale, Mrs 261, 285
Thurtell, John (murderer) 53 n., 58, 114, 117, 159, 160, 255, 435
Tidd, Richard (conspirator) 306, 309, *310–11*
Tindal, Sir Nicholas (chief justice) 166
Tocqueville, Alexis de 503, 506
Townley, George (murderer) 103
Turner, Barnard (sheriff) 603
Turner, William (Pentrich leader) 158, 312–13
Turpin, Dick (highwayman) 34, 86, 126, 133, 146, 147

Vaughan, Sir John (judge) 452, 467–8, 474, 479–80, 483, 492, 504, 511, 535
Vaux, J. H. (ex-convict) 127, 146
Villette, Revd John (ordinary) 383

Wakefield, Edward Gibbon (writer) 6, 21, 209, 219, 328, 379, 380, 386, 389, 591
  on death 75–6
  on executions 102
  on John Bell 2, 296
  on King in Council 564–5
  on Newgate 42–3
  on opinion 397, 441
  on Peel 566, 572, 585
Wall, Joseph (Governor) 48, 75, 101, 186, 263
Walpole, Horace (diarist) 33, 38, 62, 95, 117 n., 158, 253, 254–5, 260, 262, 293 n.
Ward, Edward (burglar) 557–63, 561
Warner, Dr John (scholar) 284
Warton, Thomas (writer) 253
Watkins, John (writer) 361, 362, 363–7
Wedgwood, Josiah (potter) 405
Wellington, duke of (statesman) 193, 550, 552, 564
Wells, Thomas (murderer) 589, 600–1
Wesley, John (preacher) 378–9, 532
Whitbread, Samuel (MP) 84
Wiener, Martin (historian) 18
Wilberforce, Samuel (bishop) 592, 595
Wilberforce, William (evangelical) 402, 518, 562
Wild, Jonathan (thief-taker) 133, 147, 384
Wilde, Edward Archer (attorney) 441
Wilkes, John (radical) 237, 326
William IV, King 551
Williams, John (Ratcliffe Highway murderer) 84–5, *85*, 255
Wilson, James (traitor) 48, 83–4, 98
Wind, Edgar (critic) 315
Witts, Revd Francis (diarist) 58, 271–2
Wollstonecraft, Mary (author) 260
Wood, Matthew (alderman) 247, 308, 401, 435
Wood, William Page (later Hatherley, lord chancellor) 247, 384, 394–5, 512
Woodforde, Revd James (diarist) 282–3
Woodward, Revd George (diarist) 283
Wordsworth, Dorothy 252, 352
Wordsworth, William (poet) 5, 95, 248–9, 252, 335, 352, 591
Wright, Samuel (murderer) 102–3, 210, 609
Wrightson, Keith (historian) 78

York, duke of 356, 507, 509

SUBJECT INDEX

Numbers in italics refer to plates; asterisks refer to the Index of Persons

Abergavenny 538
abolitionists 22–4, 164, 269, 297, 444, 591, 593–4, 610
Ampthill 102
anachronism and judgement 55, 499, 514
anatomization 23, 38, 69, 79, 84, 87, 244, 255–8, *265*, 273, 284, 596
anger 219–21, 438–9
animals 231, 334
  hanging of 283, *284*
Annual Register 67, 355, 596
anonymous letters 87, 220–1, 412, 415, 416, 420–1
anti-slavery 230, 334, 343, 371, 372, 399, 400, 405
anxiety 259, 262–3, 594, 599–601
appeal, court of 207 n., 208, 442
appeal system, criticisms of 432, 442
aristocrats 232, 278–9, 418, 602
  at executions 65–6, *66*, 253–4, 308, 596
  and sentencing 526–8
arson 170, 341, 420
assizes 57–8, 90, 201, 270–3, 430, 466–8, 505, 506–7, 532–3, 537
Association for Preserving Liberty and Property 125
attorneys 209–10, 433–5
  see also barristers; Harmer*; lawyers
Aylesbury 59

bail 470, 534
ballads:
  audiences for 127–33, 282
  bawdy 136–8
  contents of 133–8
  'flash' 123–55
  pastiche 145–8, 277
  suppression of 124–6, 161–3
  Tyburn 138–44, 276
Bank of England 187–9, *188*, 406, 409, 435
Bank restriction note 187–9, *188*
Bardell v. *Pickwick* 509, 532
Barnaby Rudge 332
barristers 209, 332, 423, 430–3, 434, 502, 505, 506, 520 n., 537
  see also attorneys; defence, constraints on; lawyers
Bartholomew Fair 92–3, 120, 130, 132
Beggar's Opera 38, 74, 120, 147, 195, 254, 288
belief, *see* piety
benevolence 226, 230, 233, 234, 258, 326, 333, 339–40, 390–1, 499, 514, 526
bestiality 419–20
Bethlam Hospital 239, 260
Birmingham 34, 168
Black Act 500, 569, 580
Blackwood's Magazine 295, 368
'bloody code', *see* capital statutes
body metaphors:
  and punishment 83–5, 260, 329, 516, 534, 590, 595
  and treason 300, 314, 315–18, 321
branding 16
Bristol 39, 158–9, 165, 166, 169
broadsides:
  morality of 156–7, 163–9
  'patterers' of 168, 171–4, *172*, 282
  publishers of 141–5, 147, 158–61, 168, 170
  sales and purchasers of 159–62, 168–75
  see also gallows: emblem; woodcuts
Buckingham 525
buggery 100, 171, 420–1, 441
burkers 57, 68, 114, 247
burning, *see* women: burning of
Bury St Edmunds 50, 58, 409
  see also Corder*; Lloyd, Sarah*
Bushel's case (1670) 523

Caleb Williams 353
callousness 76–9, 251, 284
Cambridge 58
'cant' 127–33, 145–7, 277, 606
capital statutes 7, 201–2, 500
  as 'bloody code' 20–1, 201–2, 237, 376, 403, 583
  campaigns and opinion against 326–30, 516, 522
  justifications of 202–3, 263, 279, 281
  limits of campaigns against 396–416, 419
  material pressures on 18–21, 227, 522, 569, 610
  petitions against 401–2, 403, 405, 408
  repeal chronologies 21–5, 237–8, 570, 591
  repeal or consolidation of 9, 295, 332, 439, 570–1, 579–83, 617–18
  see also Peel*; Quakers
Captain Swing 193, *194*
Carlisle 95
Carmarthen 50
Carnarvon 422

carnival and execution 29, 92–9, 119–20
Cato Street conspirators 47–8, 57, 113, 253, 435, 503
  authorities and executions of 307–9
  burial of 88, 308
  crowd at executions of 307
  curiosity about 305, 309, *310–11*, 318;
    depictions of executions of 300–4, *301*, *303*
  last hours of 298, 305–6
  military at executions of 98, 302, 307–8
  and opinion 299–300, 303–5, 307, 312
  polite classes at executions of 277, 309–12
ceremony 90, 94–5, 532–3
  see also assizes
chap-books 70, 119, 300
character, *see* sentencing
Chartism 168, 298
Cheap Repository of Tracts 125, 161–3, *162*
Chelmsford 512
Chester 209, 501
children:
  at executions 246–50
  executions of 4, 512
  in Newgate 385
Church of England and hanging 373, 518, 592
civility, *see* squeamishness; sympathy
civilization 11–12, 24–5, 227, 270, 294–7, 590
Civilization and its discontents 17
civilizing process 11–12, 17, 240–1, 589–90, 595, 610
Clarissa 229
class, *see* criminal law: and property interests
Classical dictionary of the vulgar tongue 129
Coalbrookdale 454–9
  men and women in 449, 450, 459–65, 490–1
coining 34, 36 n., 166, 319 n., 525
commercialization 67, 94
  see also broadsides
community and opinion 447–50
condemned, the:
  behaviour of 355, 383, 592 n.
  dress of 35–6, 83, 290, 353
  effigies and portraits of 115–17, *118*, 254–5, 295, 356, *351*, *357*
  E. Fry on behaviour of 35, 38, 110, 377–9
  letters of 354, 362
  ministry to 308, 355, 361, 377–83
  terror, anger, drunkenness of 37–40, 43, 383, 386–7, 395, 438–9, 453, 560, 600
  visiting of 83, 254, 260, 287–8, 373, 380–2, 405
  see also dying 'game'; Harwood*; Wesley*
consent in criminal law, *see* criminal law
conservativism 5, 231, 319, 502, 531, 583
  see also Peel*; tories
counsel, *see* defence, constraints on
counter-revolution 238, 320, 327, 345
Court of Aldermen 383, 388, 509
Court of Common Council 362, 403, 409, 437, 440, 509, 510
crime, causes of 573–4
crime, increase in 18, 572–5, 593
  see also prosecutions: rise of
'criminal classes' 41, 131–2, 284, 574, 592
criminal law:
  access to 213–14, 529
  as arena of negotiation 449–50, 515–16, 522
  consent in 515, 520, 570
  consent in, in broadsides 156–7, 163–8, 175
  consent in, of crowd, *see* scaffold crowds
  consent in, of middling classes 285, 369
  and property interests vi, 19, 498, 529–30, 612–13
culture and material conditions 14, 227, 231–2, 281, 294, 296–7, 610
  see also capital statutes: material pressures on
curiosity about execution 83, 239–40, 242–6, 250–8, 305, 309, 594–5, 597
  suppression of 258, 259–61, 365
  see also Boswell*; Cato Street; emotions: 'primal'; shame
custom 450, 462

death:
  denial of 78–80, 109–11, 274
  élite attitudes to 230, 237, 239, 243, 251, 328
  popular attitudes to 74–89, 110–11, 119, 377–8
  see also Boswell*; callousness; curiosity; Darwin*; 'defences'; emotions: 'primal'
death sentences 428, 452, 479
  numbers of 7, 20, 501, 544–5, 565, 615–18
decapitation, *see* executions: Roman; treason
decorum 240, 243, 252, 258, 260–1, 280, 596
defence, constraints on 359, 431–3, 443, 534–5, 537, 539
  see also barristers; trials
'defences' (psychological) viii, 76, 110, 216–17, 240, 258, 266–7, 279
  see also death: denial of; deference; denial; gallows: wit; 'learnt ignorance'; piety; squeamishness
deference:
  as 'defence' 215–7
  in mercy petitions 199–200, 203, 208, 213–21
  'real' *v.* instrumental 111, 196, 198, 214–21
deists 305
  see also rationalists
denial as 'defence' 261, 267–72, 314
Derby, *see* Pentrich

diaries 291
Dictionary of national biography 369
discretion, *see* sentencing, discretion and character in
disembowelling, *see* treason
dissenters 230, 343, 391
divine justice, *see* God; justice: ideas of
doctors 480–3
Dover 528, 589
drawing on hurdles, *see* treason
dreams of scaffold 81, 272–4
Durham 590
dying 'game' 33–7, 109–11, 119, 138, 144, 276, 305–6, 609

Ecclesiological Society 76–7, 79
Edinburgh 45, 50, 68, 91, 97, 289
effigies and portraits, *see* condemned
embezzlement 528
emotional learning 13–14, 233–6, 416
  see also sympathy
emotions:
  displacement of 243, 270, 273, 595, 597–8, 601, 604–8, 610
  history of vi, ix–x, 6
  and neurosis 240–1, 258, 261
  'primal' emotions 71–3, 228, 238–9, 261, 279, 597
  repression of 24, 226, 229, 240–1, 259–61, 264, 267 n., 270–1, 595
  see also 'defences'; public opinion
empathy 2, 13, 226, 267, 285, 291, 353, 372
Enlightenment 226, 227, 328, 516
Essex 214, 431
euphemism 29, 207, 276
evangelicals 23, 37, 226, 228, 230, 330, 362, 371–2, 437, 579, 603
  and death penalty 371–5, 380, 389, 394–5, 396–8, 412–13, 576–7
  earnestness of 240, 252, 258, 261, 278
evidence 429, 473, 519, 523
Ewart's Act (1836) 443
Execution Dock 30, 268
executions:
  by crime 7, 100–1, 617
  effects of when hidden 23–4, 37, 54, 270, 327, 592–3
  familiarity with 282–4, 284
  and ignominy 83–4
  illustrations of 96–7
  of the innocent 360, 440, 441–3, 593, 594
  methods of 51–4, 593–4
  mismanagement of 50–1, 54, 166, 590
  numbers in 1820s 7, 9, 32, 497, 545, 575, 581, 585
  numbers in Europe 7–9, 16 n., 327
  numbers and frequency of 7, 9–10, 20, 30–2, 202, 239, 298 n., 565, 570, 615–17
  numbers in Scotland and Ireland ix, 8, 50, 590
  by occupation 8, 280
  rituals in 80–9
  Roman 249–40, 286–7
  speed of 442
  suffering in 45–52, 245, 249, 257, 306, 600
  in USA 12, 594
  see also carnival; condemned; humanity; miscarriages of justice; murder; public execution, abolition of
Exeter 219

facetiousness 114, 276–9
'fairness', *see* justice
Fenians 48, 99, 593
  see also Barrett*
flogging, *see* women: whipping of
Forgery Bill (1830) 402, 583
forgery, death for 19, 187, 398, 401–2, 405–6, 408, 409, 413, 435, 580–3
  see also Bank of England; Fauntleroy*; Hunton*
Frankenstein 281
French Revolution 320, 332, 344
funerals 78–80, 85–6, 88–9, 349

gallows:
  emblem of 112–13, 157, 175–96
  wit 261, 272–9, 305
  see also scaffold; woodcuts
'game' death, *see* dying 'game'
Gaols Act (1823) 578–9
garotting 593
gentility, *see* decorum
Gentleman's Magazine 20, 52, 124, 262, 507, 544
gentry 225, 231–2, 343, 418, 519
  and sentencing 525–9
gibbeting 23, 87, 179, 180, 246, 248, 267–9, 381–2, 596
Glasgow 60, 83, 298
Gloucester 58, 100, 168, 214, 511
God 396, 517–19, 543, 584, 592
God's indignation against sin 378
Gordon riots and rioters 603
  ballad on 133
  executions of 10, 20, 31–2, 294
  mercy petitions of 208, 209–10, 218
gossip 113, 355, 362, 458, 463–5, 487
  see also oral tradition
grand jury 467, 525, 526

Great expectations 435
guillotine 9, 18 n., 46–7, 52, 596–7

hanging, *see* execution
Hanging not punishment enough 263, 518
hangman *116*, 303, 317, 590, 597
  execration of 99–100, 307, 609
  fees of 356
  see also Calcraft*; Ketch*
Hertford 58, 99
highwaymen in ballad 134–5, 144
History of the Fairchild family 247
Home Office 197, 308, 437, 425, 439, 510, 512
  investigations by 206, 425, 599
  judicial independence of 520, 523, 525–9
home secretary 269
  see also mercy prerogative; Peel*
Horsemonger Lane gaol 30, *31*, 85, 88, 102, 605, 606, *607*
Hounslow Heath 40
Howard League for Penal Reform 594
hulks 200, 207, 216–17, 219, 221, 427, 484, 493, 578
humane principles 6, 11–12, 18, 234, 290 n., 371, 377
humanitarian narratives, *see* narratives
humanitarianism 6, 11, 53, 241, 261, 295, 377
  as fashion 293
  language of 227, 326, 418, 590, 591, 600–1, 610
  and personality 17, 232 n., 398
  and private execution 54, 589–90, 592, 600–1
humanity-mongers viii, 5, 233, 380
Huntingdon 395, 511

identification:
  with hanged 280–1, 285, 288–92, 293–4, 294–7, 389–91, 394–5
  processes of 235–6, 242, 258, 294, 331, 391, 595
  see also sympathy
indignation 418, 424, 428, 431–3
inequality 213–14, 280 n., 612–13
infanticide 165, 167, 169–70, 336, 389, 508
Ingoldsby legends 278
insanity 230, 365, 371, 554–5
Ipswich 158, 170

Jack Hall, chimneysweep 34, 140–4, 185 n.
Jacobites 15, 36, 314, 317
judges:
  arrogance of 431–3, 508–9
  bad tempers of 532–3
  incompetence of 509–10, 535–6
  independence of 504–5, 520, 525–9
  mediocrity of 502–6
  mentalities of 498–9, 506–8, 510–11, 512–14, 518–19, 530–1, 579
  and mercy prerogative 201, 206, 208, 484, 500
  'reports' of 612–13
  salaries of 505
  satirized 193, 369, 499–500, *500*, 504, 514
  tears of 167, 292–3, 508
  see also sentencing, discretion and character in
jury:
  independence of 519, 520, 523, 523–5
  reluctance to hang 285, 332, 400, 406–7, 416, 508, 522, 524
  weaknesses of 359, 523, 531, 537
justice:
  'fairness' as basis of 18, 203, 326, 330, 369, 417, 423–4, 436, 516, 517, 520, 542
  ideas of 5, 196, 449–50, 514, 515–19, 539, 566–7, 584
  rhetoric of 520
  rough 198, 359, 360–1, 530–42
  rough justice exposed 366–7, 375–6, 417–18, 431–3, 441–2, 539
  sense of 234–5, 516
  see also Paley*; sentencing; trials

Kennington Common 30, 33, 56, 81
Kent 34, 528
Keswick, *see* Hatfield*
King in Council 31, 42–3, 360, 388, 421, 430, 438, 510, 520, 542
  anxieties in 543–4, 547–9
  and Cato Street 307–8
  duties and members of 200, 201, 204, 205, 207, 544–6
  preoccupations of 549–50, 552
  procedures of 546–7, 554–65
  see also George IV*; Peel*; Wakefield*

Ladies' Society for Visiting Prisons 373, 380, 405
Lancashire 198
lawyers 19, 329, 429–30, 443, 468, 504 n., 522, 523, 536–7
  see also attorneys; barristers; trials
'learnt ignorance' 240, 261, 270–2
Leicester 381
Libel Act (1792) 523
liberty 498, 500, 519, 573
literacy 117–19, 138, 209–12
Liverpool 589
love 402, 499, 508
Luddites 98–9
  see also Pentrich

Madame Tussaud's 114–17, *118*, 255, 295
magistrates 88 n., 470, *513*, 520–2, 525, 528–9, 559, 575
Maidstone 1, 35, 47, 167, 594, 600
Man of feeling, The 252
Manchester 99, 168, 589
marxists 530
material conditions, *see* capital statutes; culture
medical profession 480–3
see also anatomization
Memoirs of modern philosophers 393
memory:
collective 236, 314, 316, 321
of criminals 111–19, 246–9
mentalities 440, 441, 542
'resistant' 236–8, 267, 279, 319–21, 497–8, 499, 518–19, 530–1, 534–5, 569, 601
and treason 300, 314, 316, 319
see also community; judges; microhistories
mercy:
campaigns for 340–53, 353–70, 422–9, 431–3, 436–9, 447–93, 557–63
gentry and 525–9
recommendations to 342, 415, 421, 453, 522, 523, 524, 541, 556, 563
mercy petitions:
costs and difficulties of 208–12
deference in 199–200, 203, 208, 213–21
described and exemplified vi, 197–200, 204–5, 210–12, 415, 419–23, 483–5, 612–14
failure of 207–8, 438–9
from middle classes 206, 292–4, 403, 409, 411, 417–29
from relatives 41, 555, 556
numbers and ambition of 203–7, 369–70, 491, 567
perjury in 218–19
rhetoric in 212–13, 412, 555–6
mercy prerogative:
administration of 200–7, 544
and home secretary 201, 204–7
and judges 201, 206, 208, 484, 500
and justice 520–2
and sovereignty 200, 203, 317, 347, 519, 543
see also King in Council; Peel*
Methodists 355, 361, 378–9, 383
see also condemned: ministry to; Wesley*
microhistories vi, 424, 447–9, 613
middle classes 225–6, 229, 231–2
see also mercy petitions
middling classes 62–4, 126, 225–6, 364, 436, 604
and consent in criminal law 285, 369
and sympathy 285, 300, 364, 595
see also jury: reluctance to hang
military at executions 97–9, 302, 307–8, 313, 604
see also state, power of
miscarriages of justice 539
see also executions: of innocent; justice: rough; trials
Moll Flanders 147
Monthly Magazine 340, 347
morbidity 243, 258, 264–6, *284*
Morpeth 39
mortality 14, 77
MPs, independent 9, 22, 297, 443
murder:
execration of murderers, *see* scaffold crowds: discriminations and ghoulishness of
executions for 10, 100–1, 297, 591–4, 610
Ratcliffe Highway murders 84–5, *85*, 158, 255
music-hall 142–3, 146
Murder Act (1752) 255, 267

narratives:
'humanitarian' ix, 1–3, 13, 229, 230, 236, 257, 305, 314, 318, 332–3, 339, 370, 371, 598
'impeded' 236–8, 569
of progress 11–12, 226–7, 236–8, 297, 318, 325, 330, 371, 377, 569, 590
see also mentalities: 'resistant'
necrophilia 262, 264–6
nerve theory, *see* sensibility
New Grub Street 242
Newgate:
condemned cells in 42
condemned sermons in 43–5, 65, 83, 253, 305, 378, 388
disorder in 385
keeper of 200, 209, 439
mock trials in 93
new drop at *30*, 52–4, *53*, 57, 145, 249, 277, 596
reform of 383–4, 603
see also aristocrats; condemned; Ladies' Society for Visiting Prisons; ordinary of Newgate; Tyburn: women, prison-visiting by
Newgate calendars 113–14, *115*, 132, 147, 157–8, 247, 300, 336, 604
Newgate Monthly Magazine 386
newspapers:
and opinion 299, 300, 312, 337–8, 346, 361–3, 397, 403, 416, 441, 521, 534, 540, 563
reporting by 55, 100, 191, *191*, 259–60, 282, 598–601
Bury Post 348, 353
Daily News 609
Daily Telegraph 589, 597
Day 356, 361

newspapers (*cont.*):
  Examiner 355, 361, 363, 396
  Huntingdon and Cambridge Gazette 521
  Illustrated London News 255
  Leader 597–8
  Morning Chronicle 312, 352, 361, 563
  Morning Herald 67, 312
  Morning Post 68, 312
  Observer 360, 361, 599
  Public Ledger 360
  Salopian Journal 468
  Shrewsbury Chronicle 468
  Star 594
  Times, The 3–4, 42, 48, 67, 109, 327, 360, 415, 441, 468, 510, 563, 599, 605
  Universal Daily Register 124, 327, 337
  Weekly Chronicle 191, 193, 599
  Weekly Despatch 436, 437
Norwich 58, 173, 282, 379, 410, 418
Nottingham 39, 58, 538, 540

Observations on man 328
Old Bailey 30, 40, 201, 206, 604, 615–18
  justice at 30, 201, 441, 509–10, 533, 536–7, 544–5
oral tradition 112–13, 117–19, 138, 140–8
  see also gossip; literacy; memory
order 519
ordinary of Newgate 355, 382–4, 563
  Accounts of 114, 157 n., 336, 383
  see also Cotton*; Forde*; Villette*
'others':
  criminals as 100–1, 228, 242, 263, 280–1, 284–97, 336, 599
  traitors as 281, 300, 305, 314
Oundle 520
Oxford 51, 282

pain 14, 77
Pamela 242
Panglossian histories 530
pardons:
  conditional 202, 207, 492–3
  free 207, 442, 521, 522, 538, 542
  increase in 21, 440, 497, 543, 544, 547, 583–4, 615–16
  see also King in Council; mercy petitions; mercy prerogative
partial verdict, *see* jury: reluctance to hang
penal reform narrative 326–8
penal reformers 18, 22, 226–7, 232, 325–7, 331–4, 343–5, 377, 383–4
  groupings of 328–30
  see also Quakers
penitence 37, 226, 374–5, 378–9, 386–7, 593
Pennington Heath 318
Pentez riots 97
Pentrich disturbances, executions for 158, 312–14, 317–18, 503
Peterloo 56, 168, 192, 299, 435, 513, 523
petitions, *see* capital statutes; mercy petitions
petty treason, *see* women: burning of
Philadelphia 9, 399
Philanthropist, The 333, 399
phrenology 254–5, 257
piety:
  and belief 226, 373–4, 382, 391
  as 'defence' 374–6, 382, 387–9
  and sensibility 372, 389–95
pillory 69–70, 77, 89 n.
pious perjury, *see* jury: reluctance to hang
Pitt's 'terror' 531
place de Grève 14, 47, 253, 276
poaching 145, 207 n., 521, 526, 527, 538
police 237, 279, 297, 437, 498, 573
polite classes, *see* middle classes
pornography 70, 71, 72, 598–9
Prelude, The 248–9, 335
prerogative, *see* mercy prerogative
printing press 158
prints, audiences for 188–90
  see also broadsides
prisons 10, 112–13, 237, 327, 372, 377, 380, 383, 397, 497, 578–9, 603
Privy Council 545, 613
progress, *see* narratives: of progress
prosecutions:
  cost of 20, 520, 574
  malicious 214, 437–8, 441, 459–60
  reluctant 19, 285, 328, 400, 406–8, 416
  rise of 18–21, 213–14, 297, 407–8, 497, 573–5
  social biases in 214
psychohistory 238 n., 266
public execution, abolition of (1868) 23–4, 249, 263, 269, 297, 589–91
  debates on 592–4, 597, 602, 610
public opinion 293, 332, 338, 396–7, 408, 441, 443
  as a construct 325–6, 396–7, 416
  and Eliza Fenning 367–70, 440
  emotion in 326, 331–4, 336, 339–40, 370, 408, 525
  and policy 18–21, 227, 419, 439–42, 497, 558, 564, 582
  see also community; Quakers
Punch 195, 297, 606, 607
Punch and Judy 94, 112, 119–23, *122*, 140, 143

punishment:
aggravated 15, 596
ethical values in 5, 18 n., 156–7, 161–8, 175, 237, 261, 296, 518–19, 595
as political tactic 15–16, 299, 590, 601–12, 610
proportion in, *see* justice
purposes of 10, 589; Paley on 202, 517–8, Boswell on 287, Forde on 383, Peel on 584 Portland on 347
secondary 238, 577
and social control 18 n., 518–19
and state formation 16–17, 24, 55, 238
see also anatomization; gibbeting; military at executions; sensibility; state, power of; treason

Quakers 18–19, 226, 237, 270, 328, 330, 362, 390–5, 443, 450, 579
character of 390–1, 398, 400
and death penalty 372–5, 397–8, 401–2
and forgery 408, 413, 582
as propagandists 372, 397, 400–5
and selective interest in death penalty 403–5, 408, 413–16, 441
see also Fry*; Hunton*; piety; public opinion
Quarterly Review 502

radicals 22–3, 168, 186–8, 192–4, 226, 231, 329–30, 364–5, 369, 409, 443, 523
see also Hone*
rape 174, 198, 449, 451–2, 464–6, 469, 486, 501, 527
law on 470–4
Ratcliffe Highway, *see* murders
rationalists 226, 328, 331–2, 343–4
rationalization 320–1, 498
recorder of London 40–3, 201, 362, 396, 403, 509, 540, 545–6
see also Knowlys*; Silvester*
recorder's report 43, 386, 545, 548, 550, 551, 565
Reform Acts (1832, 1867) 9, 22, 23, 610
regression 264, 267 n.
repression, *see* emotions
respectability 612–13
see also sentencing
resuscitation of hanged 91, 273, 289–90
Riot Act 307
rioters 39, 103, 166, 204
Romanticism 230, 335–6
royal commissions:
on capital punishment (1866) 62, 592, 610
on criminal law (1836–41) 23, 436, 441–2, 533, 534–5, 594
rule of law 515–16, 519–20, 522, 530–1
see also trials

sadism 264
Sam Hall, see *Jack Hall*
scaffold:
beliefs about 80–1, 82
dreams about 81, 272–4
scaffold crowds:
consent withheld by vii, 90–1, 99–105, 111, 608
depictions of *30*, 190–2, *190*, *191*, *192*, *193*, 195, 301–3, *301*, *303*, 605–6, *607*
discriminations of 99–105, 156–7, 168, 307, 313, 349, 606–9
élite hostility to 23–4, 59–61, 269, 312, 590, 595, 601–10
excitement of 71–3, 239, 240
functions of 601–2
ghoulishness of 67–74
military control of 97–9, 302, 307–8, 313, 604
scapegoating of 240, 601, 610
size of 7, 56–8, 337, 355, 605
social composition of 61–7, 239, *239*
Scottish radicals 298–9
see also Wilson*
select committee:
on capital punishment (1856) 62, 75, 368, 592, 606–8
on criminal law (1819) 328, 401–3, 406, 435, 582
sensibility 6, 17, 238, 240–1, 278
cultures of 226, 228, 293, 331, 364–5, 391–5
and nerve theory 229, 231, 235, 251–2, 255–6, 328–9
origins and expressions 13, 228–32, 267, 287
and punishment 226–7, 251, 331, 522, 610
see also Lofft*; piety; sadism; sympathy
sentencing, discretion and character in 511–12, 515, 519, 522, 527, 539–42, 548, 570, 584, 613
see also judges; justice: rough; trials
servants and crime 169–71, 342, 350, *351*, 360, 367, 428–9
shame 17, 230, 240, 258, 261–3, 272, 274, 365, 590, 610
sheep-stealing 395, 528–9
sheriffs:
and abolition of Tyburn 96, 603–4
as champions of condemned 41, 200, 209, 414–5, 422, 435, 441–2, 558, 561–3
duties of 346–7
sensibility of 65, 269–70, 388, 561–3, 595–6
Shoplifting Act 500
Shrewsbury 58, 94, 103, 204, 420, 447, 521
sin 373–4, 376, 378, 381, 515, 591
skimmington rides 92, 112, 453, 462
slavery, *see* anti-slavery

social differentiation 126, 225–6, 601, 604
Society for:
  Abolition of Capital Punishment 594
  Constitutional Information 343
  Diffusion of Knowledge upon Punishment of Death 352, 398–401, 405, 427, 437, 450
soldiers and sailors 198, 419, 422, 542
Somerset 419
spies 304, 308, 349, 531
squeamishness 64–5, 240, 244, 273, 297, 551, 590, 596
  and civility 17, 225, 595–7, 610
  as defence 267–70
Stafford 58, 597
state, power of 91, 94–9
  see also military at executions; punishment: and state formation
statistics 573–5, 581, 615–18
  see also death sentences; executions; mercy petitions
Stirling 104, 298
suicide 84–5
superstition, *see* scaffold: beliefs about
Surgeon's Hall 33, 284
Surrey 214
Sussex 69, 81
sympathy:
  and civility 24, 226, 228, 231–2, 240, 267, 299–300, 312, 595
  limits on 228, 238, 280–1, 563
  'natural' 234–5
  'pre-modern' 2, 234, 285, 407, 422–3, 512
  as social bond 226–7
  and social distance 225, 228, 237, 240, 260, 280–1, 285–97, 412–16
  teachers of, 13–14, 231, 233–6
  see also Boswell*; empathy; evangelicals; identification; middling classes; 'others'; Quakers; sensibility

taboo vii, 29–30, 240, 260
Tait's Magazine 145
tears 13, 167, 230, 285, 292, 407, 596
thefts 423, 500–1, 574–5, 580
Thetford 282
tories 22, 238, 332, 518–19, 547, 584
  see also conservatism
torture 15–16
transportation 7, 10, 200, 201, 205, 207, 326, 427, 429, 484, 578, 593
  judges on 501, 503
treadmill 577
treason:
  crowds at executions for 104, 313, 318, 321
  forms of execution for 83, 85, 281, *301*, *303*, 306, 308, 312–3, 315–21
  trials and traitors 228, 298, 531
  see also body metaphors; Cato Street; mentalities; women: burning of
trials:
  defendants' limited rights in 515, 531–2, 534–6, 537
  disorder in 531–3
  lawyers' absence from 536–7
  speed of 536–9, 544
  witnesses' absence from 212 n., 538
Tyburn 8, 10, 54, 55, 291
  hangings abolished at 96, 263, 602–4
  procession to and crowd at 32–40, 61–2, *190*, 293

victims of crime 5, 281–2
Visit to the Philadelphia prison 399
Vocabulary of the flash language 127, 146
visiting condemned, *see* condemned

Waltham Forest 424
Warwick 525
Westminster Review 329
Whickham 78–9
whigs 22, 332, 344, 440, 441, 442 n., 443, 526, 548–9, 565, 569, 570–1, 579
whipping 338, 507, 578, 579
Wilkesites 92, 231, 504
Wimbledon Common 33
Winchester 541
Wolverhampton 217
women:
  attending executions 65, 68, 74, 80–1, *82*, 101, 260, 309–12, 599
  attitudes to 164–6, 230–1, 243, 260, 334–8, 414, 507–8
  burning of 7, 16, 36, 248, 264, 316, 317, 319 n., 337, 338
  curiosity about executions 245–6, 254, 259–60, 270–1
  executions of 8, 65, 264, 270, 337, 596
  in Newgate 385, 389–90
  prison-visiting by 254, 260, 373, 380–2, 405
  whipping of 338
  see also Coalbrookdale: men and women in; Fenning*; Lloyd*; rape; servants and crime
woodcuts 157–8, 160, 183–4, 302, 354, 560
  and 'image-magic' 175–8, *177*
  see also gallows: emblem of
Worcester 527

Yarmouth 381, 409
York 34, 166, 178